TOP 1000 SCIENTISTS FROM THE BEGINNING OF TIME TO 2000 AD

TOP 1000 SCIENTISTS FROM THE BEGINNING OF TIME TO 2000 AD

A Reader's Guide

Philip Barker

The Book Guild Ltd
Sussex, England

The Book Guild Ltd.
25 High Street,
Lewes, Sussex

First published 1999

Set in Times
Typesetting by
Acorn Bookwork, Salisbury, Wiltshire
Printed in Great Britain by
Antony Rowe Ltd, Chippenham, Wiltshire

A catalogue record for this book is
available from the British Library

ISBN 1 85776 405 6

Dedicated to my mentor and teacher, Sir Derek Barton, one of the towering figures in twentieth century science, who continues to inspire budding scientists

Philip Barker

PREFACE

Science has entered the public domain as never before. It is therefore natural that the desire to learn about practitioners of science has shown a corresponding development. More than any other group of professionals, it is the scientists who have shaped the twentieth century and, in all probability, will continue to determine our immediate future. Curiously, despite this, they are probably the least studied group.

Although it is easy to name the top political figures or artists, identification of the top scientists is fraught with difficulties because of the magnitude of the task involved. As indicated in Professor Jordanova's foreword, efforts have been made in this direction most notably the *Dictionary of Scientific Biography*, but also, *Larousse Dictionary of Scientists* and *Cambridge Biographical Dictionary*. These, however, are primarily for scholars or historians of science. An average lay person with a desire to learn about the top figures who have shaped science would find these volumes extremely tedious.

It is this audience that the author has tried to address. While relying heavily on these volumes, the author has attempted to extract the top 1000 names in science that *should be known to everyone* with a serious interest in the role of science in the society. Each figure is profiled with salient personal and professional details. In this task, the author was assisted enormously by a survey he conducted in eighty university departments around the world. Although it cannot be said with authority that these names represent the top 1000 of science, it can be asserted that any effort to define the leaders could not succeed without them.

The author would like to express his gratitude to his colleagues in the US, particularly Columbia University for their assistance and encouragement. Inestimable gratitude is also due to Professor Jordanova, President, British Society for History of Science and Fellow, Royal Historical Society for writing the foreword and for his helpful comments.

Philip Barker

FOREWORD

Practitioners of the sciences, medicine and technology have long been interested in the lives of their predecessors and colleagues, often as exemplary demonstrations of how a good life, dedicated to the acquisition of natural knowledge, should be lived. Inevitably sheer curiosity has also played a part in the enthusiasm for biography. Yet, what we mean by 'biography' remains unclear. Is it the basic facts of the person's life, information that we can put to a variety of uses? Is it an understanding of how someone comes to study the natural world, and of their thought patterns? Is it a tale of heroism, designed to inspire others? It can be all of these, and which has been given emphasis depends on the time and the place. In the eighteenth and nineteenth centuries, scientific biographies were largely written by other practitioners, family members or by those with a special interest in the field. Only more recently have professional historians of science become more involved with the genre. Over the last thirty years the nature and status of scientific biography has changed markedly. The first volume of the *Dictionary of Scientific Biography* appeared in 1970; edited by the distinguished historian of French science, Charles Gillispie, it included most of the main figures in various fields of science, although it was quite selective in so far as medicine was concerned. In the 1970s many historians were suspicious of biography because they feared it uncritically celebrated scientific achievement and took individuals out of context. Since then, however, it has become clear that critical biographies are possible, which, precisely because they do focus on a single individual, can locate them in a rich historical setting. Furthermore, the emotional mode of biographies does not have to be sycophantic, but can be carefully judged, offering historically informed judgements about an individual's significance. Recent work on Marie Curie, Charles Darwin and Michael Faraday demonstrates the point.

Any collection of scientific biographies will also face two challenges. The first is deciding whom to include. Given the huge numbers of people who have been or are scientists, technologists and medical practitioners, this will always have an element of the arbitrary to it, with the final list carrying selectors' values. Gender is a particular issue here. The second is defining the boundaries of science, medicine, and technology, which are inevitable fluid, with fields such as psychology still having an uneasy status. The important issue here is to recognise that contests about what counts as 'science', like those about who counts as a 'great scientist', are inevitable, and far from being unnecessary and tiresome, they are the very stuff of which reputations are made, fields defined, and change fostered. So long as we are as explicit as possible about criteria of judgement, these challenges should not pose undue difficulties.

There is a huge general interest in everything to do with science at the moment. It is not possible here to explore the origins and consequences of this trend, but it does have clear implications for this volume. The lives of those who create authoritative knowledge of the natural world form one of the main bridges between communities of highly specialised experts and the wider public, and for this reason, if for no other, we should foster the study of scientific biography. The human complexities of science, medicine and technology may not have simple lessons to teach, but they can be extremely informative, occasions for generating a deeper understanding of what is, after all, one of the most significant phenomena of human history.

Ludmilla Jordanova
University of East Anglia
President, British Society for the History of Science, 1998–2000

A

ABRAHAM, Sir Edward Penley
(1913–   )

English biochemist, born in Southampton. Educated at Queen's College, Oxford, he received a Rockefeller Foundation travelling fellowship (1938–1939) and worked in Stockholm before joining the Sir William Dunn School of Pathology at Oxford to study wound shock. There he joined Chain and Florey to purify penicillin. Chain had isolated penicillin as a powder, and while trying to prepare a sample suitable for X-ray analysis by Dorothy Hodgkin, Abraham found that it readily crystallized as the sodium salt, thereby providing the crucially sought confirmation of its purity. During 1950–1960 he isolated the antibiotic cephalosporin C from the mould *Cephalosporium acremonium*; like penicillin it acts on the bacterial cell wall and is inhibited by penicillinase. Cephalosporin antibiotics are used against bacteria that cause throat infections, boils, abscesses, diphtheria and typhoid fever (particularly for patients allergic to penicillin), but unfortunately possess undesirable side effects and may cause allergies and kidney damage. Abraham was elected FRS in 1958, and served as Professor of Chemical Pathology at Oxford from 1964 to 1980. He was knighted in 1980.

ADAMS, John Couch
(1819–1892)

English astronomer, born in Lidcot, Cornwall. He entered St John's College, Cambridge in 1839, graduated as Senior Wrangler and first Smith's Prizeman (1843), and was elected to a fellowship at his college. He took up the problem of the unexplained irregularities in the motion of the planet Uranus, assuming that these were due to the existence of an unknown perturbing body in the space beyond Uranus, and by September 1845 he had derived elements for the orbit of such a body. Meanwhile, in France, Le Verrier had tackled the same problem, and making the same assumption as Adams, derived both the mass and the orbit of the hypothetical new body (1846). Le Verrier transmitted his results to the Berlin Observatory where Galle searched for and discovered (23rd September 1846) an object whose motion defined it as a planet. A major controversy arose between France and England on the question of priority in this great discovery, but in the course of time it was recognized that Adams and Le Verrier had each solved the problem independently. The new planet received the name Neptune. In 1847 Adams was offered, but declined, a knighthood when Queen Victoria visited Cambridge, and in 1848 he was awarded the Copley Medal of the Royal Society. In 1858 he was appointed

Lowndean Professor of Astronomy and Geometry at Cambridge, where he spent the rest of his life pursuing problems in mathematical astronomy: his work on the motion of the Moon earned him the Gold Medal of the Royal Astronomical Society (1866). As president of that society he had the satisfaction of presenting its Gold Medal to Le Verrier for his theories on the motions of the four great planets (1876). Three years after his death a portrait medallion of Adams was placed in Westminster Abbey near the grave of Isaac Newton.

ADANSON, Michel
(1727–1806)

French botanist, born in Aix-en-Provence. He grew up in Paris, where he studied theology, classics and philosophy, and attended classes in the Jardin du Roi, taught by Bernard Jussieu, who influenced his approach to plant classification. He travelled to Senegal in 1748, and during six years there he collected thousands of botanical and zoological specimens. On his return, he published *Histoire naturelle du Senegal* (1757) and *Les familles naturelles des plantes* (1763–1764). In the latter work he analysed the theoretical basis for the classification of plants in natural order, reflecting their real similarities and differences. He was critical of Linnaeus's reliance on a single character – fructification – as the basis for natural classification, and suggested instead that trial and experience should determine the most significant characters by which particular groupings of plants should be made. By giving detailed critical descriptions of each family, he anticipated the modern multi-character approach in taxonomy. The baobab genus of African savannah trees, *Adansonia*, is named after him.

ADDISON, Thomas
(1793–1860)

English physician, born near Newcastle and educated at Edinburgh University (MD, 1815) and London. He held positions at several hospitals and dispensaries, including Guy's Hospital, where for many years he was the leading medical teacher and diagnostician. He collaborated with Bright on an unfinished textbook of medicine (1839), and left outstanding descriptions of many diseases and their pathological signs, including pneumonia, tuberculosis, 'Addison's disease' (insufficiency of the suprarenal capsules), and 'Addison's anaemia' (now known as pernicious anaemia). His *On the Constitutional and Local Effects of Disease of the Suprarenal Capsules* (1855) was a model of clinical investigation; he noted the peculiar bronzing of the skin of patients suffering from the effects of adrenal insufficiency, and through post-mortem examinations, showed that symptoms were associated with destruction of the gland by tuberculosis, cancer or another cause. He hid a nervous, retiring personality behind a haughty exterior, and preferred the wards and post-mortem room of the hospital to the more lucrative possibilities of private practice.

ADRIAN, Edgar Douglas, 1st Baron
(1889–1977)

English physiologist, born in London, and one of the founders of modern neurophysiology. He trained in the Physiological Laboratory, Cambridge where, after medical studies at St Bartholomew's Hospital London and war service, he became a lecturer and also a Fellow of Trinity College, Cambridge. Much influenced by his tutor Lucas, he devoted his career to the study of the nervous system and carried out important work in designing and building equipment to amplify and record the minute electrical impulses of nerves. In collaboration with Zotterman and then Brok, he recorded the electrical activity of single, isolated nerve fibres, from sensory and subsequently motor nerves. He showed that there is only one kind of nervous impulse (neural information being conveyed by variations in the frequency at which those impulses are transmitted), the 'frequency code' which is a fundamental characteristic of nerves. In an extensive examination of different sensory systems in many different animals, he investigated the mechanisms of peripheral functioning of receptors and sense organs in response to a stimulus, and then followed the processes of information transmission into the central nervous system, where he studied the recording and analysing of sensory information. He also developed techniques to study and understand the gross electrical activity of the brain, electroencephalography (EEG), used clinically for the study of epilepsy and other brain disorders. For his work on the function of neurons he shared the 1932 Nobel Prize for Physiology or Medicine with Sherrington. He was appointed Professor of Physiology at Cambridge (1937–1951), Master of Trinity College, Cambridge (1951–1965), Vice-Chancellor (1957–1967) and subsequently Chancellor of the university (1968–1975). He also served in many other capacities, including those of Foreign Secretary (1945–1950) and President (1950–1955) of the Royal Society.

AGNESI, Maria Gaetana
(1718–1799)

Italian mathematician and scholar, born in Milan, the daughter of a professor of mathematics at Bologna and the eldest of 21 children. Educated privately, she was a child prodigy, speaking six languages by the age of 11. She published books on philosophy and mathematics, and her mathematical textbook, *Istituzioni analitiche* (1784), became famous throughout Italy. One of the few women mathematicians to gain a reputation before the 20th century, she assimilated the work of many different authors and developed new mathematical techniques. She is best known for her description of a versed sine curve which, following an early mistranslation of Italian, became known as the 'witch of Agnesi'.

AIRY, Sir George Biddell
(1801–1892)

English astronomer, born in Alnwick, Northumberland. He entered Trinity College, Cambridge in 1819, graduated as Senior Wrangler and Smith's Prizeman (1823) and was elected a Fellow of Trinity College (1824). He became Lucasian Professor of Mathematics in 1826, and Plumian Professor of Astronomy and Director of the Cambridge Observatory in 1828. His researches on optics, in which he applied the wave theory of light to analysis of the structure of images of point sources formed by lenses, earned him the Copley medal of the Royal Society (1831), and his laborious investigations in planetary theory were recognized by the Royal Astronomical Society's award of its Gold Medal (1833). Two years later he was appointed Astronomer Royal and Director of the Greenwich Observatory. Finding this establishment in a poor and ineffective state, Airy saw to its complete reorganization, transforming it into the finest meridian observatory in the world. He established magnetic and meteorological departments in Greenwich and used submarine telegraphy to determine the longitudes of various observatories internationally, in this way achieving worldwide acceptance of the Greenwich zero meridian. He pioneered the transmission of telegraphic time signals for the railways. Also important was his determination of the mean density of the Earth through pendulum experiments at the top and bottom of deep mines. He was President of the Royal Society in 1871, four times President of the Royal Astronomical Society, and received honours and decorations from all over the world, including the Prussian *Pour le Merite* and membership of the French Legion of Honour. He was knighted in 1872.

AKI, Keliti
(1930–)

Japanese-American seismologist, born in Yokohama. He was educated at the University of Tokyo where he received his BSc in 1952 and PhD in geophysics in 1958. He was a Research Fellow at Caltech (1958–1960), then returned to Tokyo (1960–1962, 1964–1966) before moving permanently to the USA in 1966, becoming a US citizen in 1976. He was a Professor of Geophysics and then R. R. Schrock Professor at MIT (1966–1984). Since 1984 he has been W. M. Keck Professor of Geological Sciences and Science Director at the Southern California Earthquake Center. He introduced the concept of seismic moment, a measure of the extent of an earthquake (1966), then in 1967 a scaling law for seismic spectra. He pioneered strong motion seismology which has been of great importance to civil engineering (1968) and in 1969 discovered coda waves, which describe the ground vibration after primary waves have spread out from the earthquake source. He also developed seismic tomography (three-dimensional seismic modelling) techniques (1974), and produced the first quantitative analysis of the earth tremors accompanying a volcanic eruption. His Aki-Larner method allowed theoretical calculation of seismic motion in a layered medium with an irregular interface. He published the popular textbook *Quantitative Seismology: Theory and Methods* (2 vols) in 1980.

ALBERTUS MAGNUS, St, Count of Bollstädt
(c.1200–1280)

Philosopher and theologian, known as the 'Doctor Universalis', born in Lauingen. He studied in Padua, entered the newly founded Dominican order, and taught in the schools of Hildesheim, Ratisbon and Cologne. From 1245 to 1254 he lectured in Paris, and in 1254 he became provincial of the Dominicans in Germany. In 1260 he was named bishop of Ratisbon. Albertus excelled all his contemporaries in the breadth of his learning. In his scientific works he was a faithful follower of Aristotle as presented by Jewish, Arabian and western commentators; in voluminous commentaries on the writings of Aristotle, he comprehensively documented thirteenth-century European knowledge of the natural sciences, mathematics and philosophy. He also engaged in alchemy, although his works express doubts about the possibility of transmutation of the elements, and gave a detailed description of the element arsenic.

ALCMAEON
(fl.520 BC)

Greek philosopher, born in Crotona, Italy. Although writing on medical subjects, Alcmaeon was a philosopher and almost certainly not a physician. Like other pre-Socratics, he displayed wider interests that included astrology and meteorology and standard philosophical questions like the immortality of the soul. A pupil of Pythagoras, he made an early statement of the Pythagorean doctrine that health was a positive stage that depended on maintenance of humoral opposites (dry/wet, hot/cold, bitter/sweet) in their proper balance. Alcmaeon was concerned with the internal causes of diseases. Rather like the followers of Hippocrates, he divided these causes into disorders of environment (climate conditions), of nutrition and of lifestyle (excessive or insufficient exercise, etc). It appears that Alcmaeon was an early pioneer of dissection.

ALFVÉN, Hannes Olof Gösta
(1908–)

Swedish theoretical physicist, a pioneer of plasma physics, born in Norrköping. Educated at Uppsala University, he joined the Royal Institute of Technology, Stockholm (1940), becoming Professor of Electronics in 1945 and Professor of Plasma Physics in 1964. He moved to the University of California in 1967. Alfvén carried out pioneering research on ionized gases, or plasmas, and their behaviour in electric and magnetic fields. He made important contributions to the theory, describing how magnetic fields can sometimes become 'frozen' into plasmas, effectively binding the ions to the magnetic field lines passing through. He also predicted the existence of 'Alfvén waves' in plasmas (1942), which were later observed. His theories have led to important advances in the analysis of the motion of particles in the Earth's magnetic field and stellar structure, and in the attempts to develop nuclear fusion reactors. He shared the Nobel Prize for Physics with Nel in 1970.

ALHAZEN (Ibn al Haythem)
(c.965–1040)

Arab mathematician, born in Basra, Iraq. He wrote a work on optics (known in Europe in Latin translation from the thirteenth century) giving the first account of atmospheric refraction and reflection from curved surfaces, and the construction of the eye. He constructed spherical and parabolic mirrors, and it was said that he spent a period of his life feigning madness to escape a boast he had made that he could prevent the flooding of the Nile. In later life he turned to mathematics and wrote on Euclid's treatment of parallels and on Apollonius's theory of conics.

AL-KHWARIZMI, Abu J'far Muhammad ibn Musa
(c.800–850)

Arab mathematician who wrote in Baghdad on astronomy, geography and mathematics. He wrote an early Arabic treatise on the solution of quadratic equations, synthesizing Babylonian solution methods with Greek-style proofs of their correctness for the first time. His writings in Latin translation were so influential in transmitting Indian and Arab mathematics to medieval Europe that the methods of arithmetic based on the Hindu (or so-called Arabic) system of numeration became known in medieval Latin, by corruption of his name, as 'algorismus', from which comes the English 'algorithm'; the word 'algebra' is derived from the word *al-jabr* in the title of his book on the subject.

ALLEN, Sir Geoffrey
(1928–)

English chemical physicist and industrial chemist. He received his PhD at the

University of Leeds, and following two years at the National Research Council in Ottawa, Canada, he joined the University of Manchester where he eventually became Professor of Chemical Physics. He then moved to professorships at Imperial College of Science and Technology. Allen's principal scientific interests lie in the chemistry and physics of polymers. In 1981 he joined Unilever after 25 years of academic life, and since then he has held various industrial appointments including Director of Unilever and Courtaulds, and Executive Advisor to Kobe Steel Ltd. Through the Society of Chemical Industry he is known as a spokesman for the subject of chemistry to the general public. He was Chairman of the Science Research Council in 1977, and in 1980 he was responsible for the inclusion of engineering in the title to form the Science and Engineering Research Council. He was elected FRS in 1976, knighted in 1979, and is a Vice-President of the Royal Society and a Senior Vice-President of the Institute of Materials.

ALPHER, Ralph Asher
(1921–)

American physicist, born in Washington DC, known for his theoretical work concerning the origin and evolution of the universe. After studying at George Washington University, he spent World War II as a civilian physicist, and later worked at Johns Hopkins University and in industry. Together with Bethe and Gamow, he proposed in 1948 the 'alpha, beta, gamma' theory which suggests the possibility of explaining the abundances of chemical elements as the result of thermo-nuclear processes in the early stages of a hot, evolving universe. These ideas were developed to become part of the 'Big Bang' model of the universe. Also in 1948, he predicted that a hot 'Big Bang' must have produced intense electromagnetic radiation which would have 'cooled' (or redshifted), and this background radiation was in fact observed in 1964 by Penzias and Robert Wilson.

ALTER, David
(1807–1881)

American physicist and inventor, born in Westmoreland, Pennsylvania. He graduated from the Reformed Medical College in New York City in 1831. A prodigious inventor, he worked on electric clocks, invented an electric telegraph (1836) and discovered methods of purifying bromine and producing coal oil from coal. One of the earliest investigators of spectra, he suggested that each element has a characteristic spectrum, and pioneered the use of the spectroscope in determining the chemical constitutions of gases and vaporized solids. His theories were later confirmed by Bunsen and Kirchhoff.

ALTMAN, Sidney
(1939–)

Canadian biologist, born in Montreal. He was educated at MIT and the University of Colorado, Boulder, and started his career as a teaching assistant at Columbia University (1960–1962). After holding several fellowships, he became assistant then associate professor (1971–1980) at Yale University, where since 1980 he has been Professor of Biology. In the early 1970s, Altman studied the process by which an RNA transcript of DNA 'matures' to form transfer RNA, the amino acid carrier in protein synthesis. He found that the precursor RNA was specifically cleaved at both ends by ribonuclease P, an enzyme which contains both RNA and protein components and is bound to the ribosome.

Although he initially thought both components were essential for enzyme activity, parallel studies by Cech (1982) led him to discover in 1983 that the RNA component of ribonuclease P alone catalyses the maturation of transfer RNA. Other examples of RNA catalysis have since been described, giving further confirmation of one of the many predictions of Crick and suggesting a possible route for the emergence of a stable, self-replicating living system prior to evolution. For his pioneering research, Altman shared the 1989 Nobel Prize for Chemistry with Cech. He contributed to and edited *Transfer RNA* (1978).

ALTOUNYAN, Roger Edward Collingwood
(1922–1987)

British physician and medical pioneer, inventor of the anti-asthma drug Intal and the 'spinhaler'. Born in Syria of Armenian–English extraction, he spent his summer holidays with his four sisters in the Lake District, where they met the author Arthur Mitchell Ransom (1884–1967) and became the real-life models for the children in his *Swallows and Amazons* series of adventure books. After qualifying as a doctor at the Middlesex Hospital in London, he practised for a while in Aleppo, Syria; he returned to England in 1956 to join a pharmaceutical company, where he worked in his own time to develop the drug Intal to combat asthma, from which he suffered. He was thus an early experimental subject. A pilot and flying instructor during World War II, he developed the spinhaler device to inhale the drug, based on the aerodynamic principles of aircraft propellors.

ALVAREZ, Luis Walter
(1911–1988)

American experimental physicist and Nobel Prize winner of exceptionally wide-ranging talents. Born in San Francisco, he studied physics at the University of Chicago where he built one of the first Geiger counters in the USA and together with Compton used it to study cosmic rays. He joined Lawrence at Berkeley University in 1936 where he worked on nuclear physics; here he discovered electron capture by nuclei, tritium radioactivity and the magnetic moment of the neutron. During World War II he worked at MIT, where he invented a radar guidance system for landing aircraft in conditions of poor visibility. After the war he returned to Berkeley where he became Professor of Physics in 1945. There he worked on developing the bubble chamber technique invented by Glaser in 1956 for studying particle interactions. The first bubble chambers were small, a few inches in diameter, but Alvarez was convinced that this was a useful technique and built a series of larger bubble chambers culminating in a 'large' bubble chamber 72 inches long. He used these to carry out a range of particle physics experiments in which a large number of 'resonances', particles that live only for a very short time, were identified. These results helped to show a pattern in the properties of the resonances which ultimately led to the quark model invented by Gell-Mann and Zweig. In 1968 Alvarez was awarded the Nobel Prize for Physics for this work. He applied physics and ingenuity to a variety of problems: he used cosmic X-rays to show that Chephren's Egyptian pyramid had no undiscovered chambers; he showed that only one killer was involved in the assassination of President Kennedy; and he founded two companies to make optical devices, including the variable-focus spectacle lens which he invented for his own use. With his geologist son Walter (b. 1940) he studied the catastrophe of 65,000,000 years ago which killed the dinosaurs. Based on radiotracer analysis, they proposed that its cause was the impact on Earth of an asteroid or comet.

AMONTONS, Guillaume
(1663–1705)

French instrument-maker and experimental physicist, born in Paris. Amontons was incurably deaf from adolescence. The son of a lawyer, he studied geometry, mechanics and physical science, then drawing, surveying and architecture in order to qualify himself for employment in public works. He designed many instruments including a hygrometer (1687), a 'folded' barometer (1688), a conical nautical barometer (1695) and various air thermometers. To determine longitude at sea he suggested in 1695 a new clepsydra (or water clock). In 1699, recently elected to the Academy of Sciences, he published a study of mechanical friction: and experiments relating to his 'fire-mill' (a hot-air engine) showed that a constant volume of air heated from room temperature to that of boiling water increased its pressure by one third. In 1702 Amontons found that for any constant mass of air, the increase in pressure was proportional to temperature rise, irrespective of initial pressure. Gay-Lussac later demonstrated explicitly the general relationship between temperature and volume for all gases. In 1703 Amontons defined the 'extreme cold' of his thermometric apparatus as that point at which the air would exert no pressure, hinting at an unattainable absolute zero of temperature. Fahrenheit was particularly indebted to his investigation of the thermal expansion of mercury (1704). Amontons died from gangrene in 1705.

AMPÈRE, André Marie
(1775–1836)

French mathematician, chemist and physicist, whose name is given to the basic SI unit of electric current (ampere, amp). He was born in Lyons, where his father was a wealthy merchant. Soon after his birth, the family moved to the nearby village of Poleymieux; today the house is a national museum. Following the educational theories of Rousseau, he was allowed to educate himself by means of his father's considerable library. He had a phenomenal memory, and his mathematical abilities became apparent at an early age. He was appointed to a lectureships in mathematics at the École Polytechnique in Paris in 1803, appointed Inspector-General of the Imperial University in 1898, and elected to the Chair of Experimental Physics at the Collège de France in 1824. Although he contributed to a number of fields (notably what today we would call physical chemistry), he became best known for laying the foundations of the science of electrodynamics through his theoretical and experimental work, following Oersted's discovery in 1820 of the magnetic effects of electric currents, in *Observations électro-dynamiques* (1822) and *Théories des phénomnes électro-dynamiques* (1830). He derived an expression for the force between small elements of two current-carrying conductors, showing that the force is inversely proportional to the square of the distance between them. Ampère was elected FRS in 1827. Of sensitive disposition, the execution of his father in the civil disturbances of 1793 resulted in a prolonged nervous breakdown. The death of his first wife in 1804, and a catastrophic second marriage, were blows from which he never fully recovered.

ANAXAGORAS
(c.500–428 BC)

Greek philosopher, born in Clazomenae. For 30 years he taught in Athens, where he had many illustrious pupils. His scientific speculations may have provided the pretext for his prosecution for impiety (he held that the Sun, Moon and stars were huge incandescent rocks), and he was banished from Athens for life. He withdrew to Lampsacus, on the Hellespont, and died there. His most celebrated, but obscure, cosmological doctrine was that matter is infinitely divisible into particles, which contain a mixture of all qualities, and that mind (*nous*) is a pervasive formative agency in the creation of material objects.

ANAXIMANDER
(c.611–546 BC)

Greek philosopher, born in Miletus, successor and perhaps pupil of Thales. He posited that the first principle was not a particular substance like water or air but the *apeiron*, the infinite or indefinite. He is reputed to have used the gnomon (a sundial with a vertical rod) to measure the lengths of the seasons, by fixing the times of the equinox and solstices. He is credited with drawing the first map of the inhabited world, and he recognized that the Earth's surface must be curved, although he visualized it as a cylinder rather than a sphere. He was aware that the heavens appeared to revolve about the Pole Star, and therefore pictured the sky as a complete sphere. No trace of his scientific writings has been found, but he is credited with many imaginative scientific speculations, for example that the Earth is unsupported and at the centre of the universe, that living creatures first existed in the waters of the earth, and that human beings must have developed from some lower species that more quickly matured into self-sufficiency.

ANDERSON, Philip Warren
(1923–   )

American physicist, born in Indianapolis. He studied antenna engineering at the Naval Research Laboratories during World War II and then at Harvard, where he received his PhD in 1949. As a student under van Vleck he worked on pressure broadening in microwave and infrared spectra. He joined Bell Telephone Laboratories in 1949, carrying out theoretical work on the electrical properties of disordered systems. From considerations of wave propagation in disordered materials he demonstrated that it is possible for an electron in these materials to be trapped in a small region (1958). This process, later known as 'Anderson localization', became a cornerstone to the understanding of the electronic behaviour of disordered materials and contributed to their extensive exploitation in applications that today include amorphous silicon solar cells, thin film transistors and xerography. Anderson investigated minimagnets and revealed the microscopic origin of magnets in bulk materials. His work also included theoretical studies of superfluidity found in helium-4 below 2 kelvin, and he clarified the meaning of the Josephson effect. He was appointed Assistant Director of Bell Telephone Laboratories in 1974, moving the following year to Princeton's University to become Professor of Physics. For his theoretical work on the electronic structure of magnetic and disordered systems he received the 1977 Nobel Prize for Physics jointly with Mott and van Vleck.

ANDERSON, Tempest
(1846–1913)

English ophthalmic surgeon, traveller and volcanologist, born in York. He was educated at St Peter's, York, and University College London. After developing a particular interest in the eye, he became an ophthalmic surgeon at York County Hospital. He made his own cameras and lenses, and enthusiastically promoted the use of photography in geology. He travelled extensively, including visits to Martinique, St Vincent, other islands of the West Indies, and to the Philippines, compiling an early photographic record of volcanic phenomena. With Flett, he was commissioned by the Royal Society to investigate the eruptions of La Soufrire and Mt Pele. His volcanological works included *Volcanic Studies in Many Lands* (1903), and he published many articles in medical journals.

ANDREWS, Thomas
(1813–1885)

Irish physical chemist, born in Belfast. From 1828 onwards he studied, in succession, chemistry in Glasgow and Paris, and medicine in Dublin, Belfast and Edinburgh. Andrews then practised as a physician in Belfast and taught chemistry at the Royal Belfast Academical Institution. In 1845 he was appointed Vice-President of the recently established Queen's College, Belfast, and from 1849 to 1879 he occupied the Chair of Chemistry there. He was awarded the Royal Medal of the Royal Society in 1844 and elected FRS in 1849. Throughout his career he was constantly engaged in research. His most important work (around 1860) was on the pressure, volume and temperature relationships of carbon dioxide, which established the existence of the critical temperature of a gas, above which it cannot be liquefied. He also made extensive and precise studies of heats of chemical reaction, and demonstrated that ozone (discovered by Schönbein in 1840) is an allotrope of oxygen.

ANFINSEN, Christian Boehmer
(1916–)

American biochemist, born in Monessen, Pennsylvania. He was educated at Harvard, and from 1939 to 1944 was American Scandinavian Foundation Visiting Investigator at the Carlsberg Laboratory, a leading institute of protein chemistry, where he was influenced by Linderstrom-Lang. During 1947–1948 Anfinsen worked with Theorell at the Medical Nobel Institute before moving to the National Institutes of Health in Bethesda, Maryland, where he served as head of laboratories from 1950 to 1982. At the time when Sanger was elucidating the primary structure of the protein insulin, Anfinsen developed the combined application of enzymic and chemical hydrolysis for the preliminary fragmentation of the protein under study and the chromatographic identification of peptides that were joined by disulphide (cystinyl) bonds. He concluded that the 128 amino acids of ribonuclease formed a single peptide chain with four disulphide bonds and a single N-terminus. The final primary sequence of ribonuclease reflected the combined research of several workers including Moore and Stein, with whom he shared the Nobel Prize for Chemistry in 1972. From 1955 he studied the secondary and tertiary structures of ribonuclease, as defined by Linderstrom-Lang, by observing the effects of partial proteolytic digestion and urea denaturation. Impressed by the recent discovery that part of a protein hormone could be removed without loss of biological activity (1948), Anfinsen extensively surveyed the importance of various regions of ribonuclease in refolding and reoxidizing the fully reduced denatured enzyme to restore activity. This led him to discover an enzyme that assists reoxidation of the cysteine residues to give the correct disulphide interactions. Anfinsen related his observations to molecular biology and evolution in *The Molecular Basis of Evolution* (1959). Since 1958 he has been a co-editor of the important review series *Advances in Protein Chemistry*.

APOLLONIUS OF PERGA
(fl.250–220 BC)

Greek mathematician, known as the 'Great Geometer'. He was the author of the definitive ancient work on conic sections which laid the foundations of later teaching on the subject. The *Conica* is essentially a unified treatment of conics, with no special treatment of ellipses, parabolas and hyperbolas; some later writers claim that the work is written in a Greek version of coordinate geometry. Four of its books survive in Greek, three in Arabic translation only (translated by Halley), and the eighth and last is lost. Apollonius also wrote on various geometrical problems, including that of finding a circle touching three given circles, but most of these works are lost and we know about them only through later commentators. He put forward two descriptions of planetary motion, one in terms of epicycles and the other in terms of eccentric motion.

APPLETON, Sir Edward Victor
(1892–1965)

English physicist, born in Bradford. He showed exceptional promise from an early age, and matriculated at the University of London at the age of 16. Trained at St John's College, Cambridge, he was appointed assistant demonstrator in experimental physics at the Cavendish Laboratory in 1920. His researches on the propagation of wireless waves led to his appointment as Wheatstone Professor of physics at London University (1924–1936). In 1936 he returned to Cambridge as Jacksonian Professor of Natural Philosophy. In 1939 he became Secretary of the Department of Scientific and Industrial Research, and in 1949 he was appointed Principal and Vice-Chancellor of Edinburgh University. His work revealed the existence of a layer of electrically charged particles in the upper atmosphere (the Appleton layer) which plays an essential part in making wireless communication possible between distant stations, and was also fundamental to the development of radar. He was elected FRS in 1927, appointed President of the British Association for the Advancement of Science in 1953, knighted in 1941, and awarded the Nobel Prize for Physics in 1947 for his contributions 'in exploring the ionosphere'.

ARBER, Werner
(1929–)

Swiss microbiologist, born in Gänichen. In 1949 he commenced his studies at the Swiss Federal Institute of Technology. In 1953 he moved to the University of Geneva as a research assistant in the department of biophysics, where he helped develop techniques for the preparation of bacteriophages (the viruses that attack bacteria), obtaining his PhD in 1958. In 1962 he discovered the mechanism of host-cell induced variation, involving two types of enzymes, endonucleases and methylases. After a period at the University of Southern California, he returned to Geneva and then moved to Basle as

Professor of Molecular Biology from 1970. Arber proposed that when bacteria defend themselves against attack by phages, they use selective enzymes, which he called 'restriction enzymes', which cut the phage DNA and do so at specific points in the DNA chain. Such enzymes clearly gave the option of securing short lengths of DNA; if these could then be joined in specific ways, then securing an 'unnatural DNA' should be possible, and with it entry to the new field of so-called genetic engineering. Through the efforts of many groups, especially in the USA, this was brought to full fruition in the 1970s, with valuable results such as the preparation of monoclonal antibodies for use in clinical diagnosis and treatment of disease. Arber shared the Nobel Peace Prize for Physiology or Medicine in 1978 with Hamilton Smith and Nathans.

ARCHIMEDES
(c.287–212 BC)

Greek mathematician, born in Syracuse, the most celebrated of the ancient mathematicians and one of the greatest of all times. He almost certainly studied at Alexandria. In popular tradition he is remembered for the construction of siege-engines against the Romans and the Archimedean screw still used for raising water; he demonstrated the powers of levers in moving large weights, and in this context made his famous declaration 'Give me a firm place to stand, and I will move the Earth'. His importance in mathematics, however, lies in his discovery of formulae for the areas and volumes of spheres, cylinders, parabolas, and other plane and solid figures, in which the methods he used anticipated the theories of integration to be developed 1800 years later. He also used mechanical arguments involving infinitesimals as an heuristic tool for obtaining the results prior to rigorous proof. He founded the science of hydrostatics, studying the equilibrium positions of floating bodies of various shapes. In legend he discovered 'Archimedes principle' while in the bath and ran into the street with a cry of 'Eureka!' ('I have found it'); the principle states that a wholly or partly immersed body in a fluid displaces a weight of fluid equal to its own weight. His astronomical work is lost. His work, some of which only survives in Arabic translation, combines an amazing freedom of approach with enormous technical skill in the details of his proofs. He was killed at the siege of Syracuse by a Roman soldier whose challenge he supposedly ignored while immersed in a mathematical problem.

ARGAND, Aimé
(1755–1803)

Swiss physicist and chemist, born in Geneva, who lived for a time in England. In 1784 he invented the Argand burner, the first scientifically designed oil-burner for the purpose of illumination. It consisted of a cylindrical wick supported between two concentric tubes, the circular flame being supplied with air on both sides, within a glass chimney which protected it from extraneous draughts. Argand lamps were soon brought into use for lighthouse illumination, and were later improved by Fresnel who introduced burners with as many as four concentric wicks.

ARGAND, Jean-Robert
(1768–1822)

Swiss mathematician, born in Geneva. He gave his name to the Argand diagram, in which complex numbers are represented by points in the plane. By profession a book-keeper, his work was largely independent of that of the more famous mathematicians of his time.

ARISTARCHOS OF SAMOS
(fl.c.270 BC)

Greek astronomer who worked in Alexandria. He is famous for his theory of the motion of the Earth, maintaining not only that the Earth revolves on its axis but that it travels in a circle around the Sun, anticipating the theory of Copernicus. He was also a practical astronomer; he used a method for determining the relative distances of the Sun and Moon which involved observing the angular separation of the two bodies in the sky at the Moon's first quarter. His result, that the Sun is about 20 times more distant than the Moon, although greatly in error due to the crudeness of his method, was the first attempt to make this important observation. He inferred correctly that as the Sun and Moon are almost of the same apparent size, their dimensions are in proportion to their distances.

ARISTOTLE
(384–322 BC)

Greek philosopher and scientist, one of the most important and influential figures in the history of Western thought. He was born in Stagira, a Greek colony on the peninsula of Chalcidice, the son of the court physician to the King of Macedon (who was the father of Philip II and the grandfather of Alexander the Great). In 367 BC he went to Athens and was first a pupil then a teacher at Plato's Academy, where he remained for 20 years until Plato's death in 347 BC. Speusippus succeeded Plato as head of the Academy and Aristotle left Athens for 12 years. He spent time at Atarneus in Asia Minor (where he married), at Mytilene, and in about 342 BC he was appointed by Philip of Macedon to act as tutor to his son Alexander. He finally returned to Athens in 335 BC to found his own school (called the Lyceum due to its proximity to the temple of Apollo Lyceius), where he taught for the following 12 years. His followers became known as 'peripatetics', supposedly because of his restless habit of walking up and down while lecturing. Alexander the Great died in 323 BC and there was a strong anti-Macedonian reaction in Athens; Aristotle was accused of impiety, and perhaps with the fate of Socrates in mind, he took refuge at Chalcis in Auboea, where he died the following year. Aristotle's writings represented an enormous, encyclopaedic output over virtually every field of knowledge: logic, metaphysics, ethics, politics, poetry, biology, zoology, physics and psychology. The bulk of the work that survives actually consists of unpublished material in the form of lecture notes or students' textbooks, which were edited and published by Andronicus of Rhodes in the middle of the first century BC. Even this incomplete corpus is extraordinary for his range, originality, systematization and sophistication. In Greek biology, Aristotle represented the highest point; the successors of his era (such as Pliny the Elder) were little more than uncritical compilers. He began the history of zoological taxonomy, incorporating considerable factual detail in his writings, much of it from the Hippocratic School but a considerable amount based apparently on personal observation. His main work of descriptive zoology is *Historia animalium*, but there are also many references in *De partibus*, *De generatione* and other works. Although interested in biodiversity, he had no interest in classification *per se*, and nowhere lists the main taxonomic groups he recognized. He developed a *scala naturae*, although he was clearly opposed to the concept of evolution. His descriptions of a developing chick marked the founding of embryology. Major Aristotelian themes of natural philosophy included the theory that the Earth is the centre of the eternal universe. He taught that everything beneath the orbit of the Moon is composed of earth, air, fire and water, and is subject to generation, destruction, qualitative change and rectilinear motion; everything above the orbit of the Moon was composed of ether, and subject to no change but circular motion. Aristotle believed that all material things can be analysed in terms of their matter and their form, the latter factor constituting their essence, and that many scientific explanations are properly teleological. His work exerted an immense influence on medieval philosophy (especially through St Thomas Aquinas), Islamic philosophy (especially through Averros), and indeed on the whole Western intellectual and scientific tradition. In the Middle Ages, he was referred to simply as 'the Philosopher', and the uncritical and religious acceptance of his doctrines was to hamper the progress of science until the Scientific Revolution of the sixteenth and seventeenth centuries. The works most widely read today include the *Metaphysics* (the book written 'after the *Physics*'), *Nicomachean Ethics, Politics, Poetics*, the *De anima* and the *Organon* (treatises on logic).

ARMSTRONG, Henry Edward
(1848–1937)

English chemist, born in London. He studied chemistry at the Royal College of Chemistry and began research with Sir Edward Frankland, but received his PhD from Leipzig in 1870 for work with Kolbe. He returned to the UK to become professor at the London institutions, which later became Imperial College, and remained there until his retirement in 1911. He did pioneering work in a number of areas of organic chemistry, particularly the structure and reactions of benzene and naphthalene compounds. He also speculated on the state of ions in aqueous solution. However, he is most frequently remembered for the inspirational nature of his teaching. He was elected FRS in 1926.

ARNOLD, Joseph
(1782–1818)

English botanist, born in Beccles, Suffolk. He studied medicine at Edinburgh, graduating in 1807 with a thesis on dropsy. In his youth he showed an interest in botany and contributed articles to the *Gentlemen's Magazine*. In 1808 he entered service with the British navy as a ship's surgeon and was surgeon on the *Hindostan* to Sydney. William Bligh (1754–1817) the captain on the return voyage, promised Arnold an introduction to Banks; this offer seems never to have been taken up. In January 1815 he sailed on the *Northampton* as surgeon to female convicts bound for Botany Bay. During this voyage he collected insects in South America and Australia for his friend Alexander McLeay (1767–1848). Unfortunately, his collections and most of his journals were lost on the return voyage when the *Indefatigable* caught fire at Baravia. He visited Java with Sir Thomas Stamford Raffles, who later invited Arnold to accompany him to Sumatra as naturalist. On 19 May 1818 at Pulau Lebar (Sumatra), Arnold discovered the largest flower known, measuring a yard across and weighing 15 pounds. This was named *Rafflesia arnoldii* by Robert Brown. Arnold died shortly afterwards, of fever, at Padang in Sumatra.

ARP, Halton Christian
(1927–)

American astronomer, born in New York City. After receiving his first degree at Harvard and his doctorate at Caltech, he spent the majority of scientific life at the Mount Wilson and Palomar Observatory. In 1956 he was the first to establish the relationship between the maximum luminosity of novae and the rate at which this luminosity declines. In 1966 he produced an *Atlas of Peculiar Galaxies*, and in the same year he suggested that the redshifts associated with distant galaxies are not related to velocity, with the implication that Hubble's law cannot be used to measure distances; this view never became widely accepted. Arp has done much to relate radio galaxies to their optical counterparts. He is also famous for his work on the high-luminosity variable starts (RV Tauri stars) that are found in globular clusters. In addition hc studies pulsating novae, particularly those in the Andromeda galaxy.

ARRHENIUS, Svante August
(1859–1927)

Swedish physical chemist, born in Wijk, near Uppsala. In 1876 he entered the University of Uppsala, where he studied both chemistry and physics, but found the conditions there for pursuing such studies were unsatisfactory and in 1881 he migrated to Stockholm. His doctoral thesis on the experimental determination and theoretical interpretation of the electrical conductivities of dilute solutions of electrolytes was barely accepted by the University of Uppsala, whose scientific establishment was apparently outraged by its novel ideas. Arrhenius was fortunately already highly regarded by Ostwald, van't Hoff and other prominent physical chemists, and he was awarded a travelling scholarship by the Academy of Sciences. He spent five years visiting the laboratories of Europe in which physicochemical studies were pursued. In 1891 Arrhenius became a lecturer in the Stockholm Hgskola; he was promoted to professor in 1895. In 1902 he received the Davy Medal of the Royal Society, and he was elected a Foreign Member of the society in 1911. He was awarded the Nobel Prize for Chemistry in 1903. He became Director of the Nobel Institute of Physical Chemistry in Stockholm in 1905, a post he held until a few months before his death in 1927. The theory of electrolytic dissociation, as expounded by Arrhenius in his thesis, was clarified, extended and consolidated in a paper in 1887. In addition to explaining electrolytic conductivity, the theory rationalized many other properties of electrolyte solution, notably osmotic pressure (studied by van't Hoff) and catalysis by acids (studied by Ostwald). The other main contribution of Arrhenius to physical chemistry was his formulation in 1889 of the dependence of the rate coefficient of a chemical reaction upon temperature (the Arrhenius equation). In later life he applied the methods of physical chemistry to the chemistry of living matter, and he was also interested in astrophysics, particularly the origins and destinies of stars and planets.

ARTIN, Emil
(1898–1962)

Austrian mathematician, born in Vienna. He studied in Leipzig, and taught at Göttingen and Hamburg before emigrating to the USA in 1937, where he held posts at Indiana and Princeton before returning to Hamburg in 1958. His work was mainly in algebraic number theory and class field theory, and has had great influence on modern algebra. After Hilbert had written his decisive summary of nineteenth-century number theory in 1894, attention was focused on what is known a Abelian class field theory. Takagi, the first modern Japanese mathematician, brought this line of research to a conclusion, and on the basis of his work Artin then solved the most important problem raised by Hilbert: the existence of a general reciprocity law. This discovery by Artin completed a line of inquiry begun by Gauss which was central to the theory of numbers. Artin also wrote a classic description of modern Galois theory. Recently his work on braids (a part of knot theory) caught the attention of particle physicists.

ASHBURNER, Michael
(1942–)

English geneticist. He was educated at Churchill College, Cambridge, where he received a BA in 1964, a PhD in 1968 and an ScD in 1978. He remained at Cambridge as a research assistant from 1966 to 1968. He then became a university demonstrator (1968–1973), lecturer (1973–1980), Reader in Development Genetics (1980–1991) and Senior Research Fellow at Churchill College (1980–1990). Elected FRS in 1990, he has been Professor of Biology at the University of Cambridge since 1991, and throughout his career has held many research fellowships and visiting professorships abroad. Ashburner is best known for his work on the definition and regulation of the heat shock genes of *Drosophila melanogaster*, the fruit fly. Heat shock genes are activated by a short burst of increased temperature, and are therefore a useful model system for studying gene control.

ASTBURY, William Thomas
(1889–1961)

English X-ray crystallogapher, born in Longton, Stoke-on-Trent. He studied at Cambridge and with Sir William Bragg's team at University College London (1920–1921) before becoming a lecturer in textile physics at the University of Leeds (1928), where he held the new Chair of Biomolecular Structure from 1945. In 1926 he began taking X-ray diffraction photographs using, not crystals, but natural protein fibres such as those of hair, wool and horn. Using photographic techniques he had helped to develop, he showed that diffraction patterns could be obtained which changed when the fibre was stretched, or wet. On this basis he classified fibrous proteins into the keratin-myosin-elastin group and the collagen group comprising the proteins of connective tissue, such as tendons, cartilage, fish scales and gelatin. The latter group do not stretch, though they contract considerably upon soaking in hot water. Astbury's interpretation of their molecular structures proved incorrect, but his pioneer work laid the basis for important work which followed, most notably Pauling and R.B. Corey's discovery that the contracted structure of hair corresponded to the protein alpha helix while the beta-keratin stretched structure, found naturally in silk fibroin, corresponded to the anti-parallel pleated sheet. The alpha helix was also found to occur in haemoglobin by Perutz, and together with the parallel pleated sheets, constitutes the protein secondary structure in the classification of Linderstrom-Lang. Astbury probably coined the phrase 'molecular biology' and with Florence Bell he attempted the first hypothetical structure for the key genetic material DNA (1938).

ASTON, Francis William
(1877–1945)

English physicist, born in Birmingham. Educated at Birmingham and Cambridge universities, he was noted for his work on isotopes. Following work done by J. J. Thomson, Aston developed the mass spectrograph (1919), with which he identified 212 naturally occurring isotopes. He stated the 'whole number rule' which observes that all isotopes have very nearly whole number masses relative to the defined mass of the O isotope. He won the Nobel Prize for Chemistry in 1922 for this work. The Aston dark space, in electronic discharges, is named after him.

ATANASOFF, John Vincent
(1903–)

American physicist and computer pioneer, born in Hamilton, New York the son of an electrical engineer who had emigrated from Bulgaria. He was educated at the University of Florida, Iowa State College and the University of Wisconsin, where he received a PhD in physics in 1930. Aged 27 he became a theoretical physicist at Iowa State College. In 1942, with the help of Clifford Berry, a talented graduate student, he built an electronic calculating machine – the ABC (Atanasoff-Berry-Computer) – one of the first calculating devices utilizing vacuum tubes. In 1941, on a visit that was to have important repercussions. Mauchly visited Atanasoff and discussed his work. Neither Atanasoff nor Iowa State applied for patents: Atanasoff failed to appreciate the importance of his work and, caught up in World War II, he let the project drop and dismantled the ABC in 1948. He took no further interest in building computers, although his ideas did enter the mainstream of computer development through Mauchly, who was influenced by Atanasoff in constructing the ENIAC. In a controversial sequel in 1972, a landmark court case involving the ENIAC patents ruled that Atanasoff – not Eckert and Mauchly – was the true originator of the electronic digital computer.

ATIYAH, Sir Michael Francis
(1929–)

English mathematician, born in London. Educated in Egypt and Manchester Grammar School, he graduated from Trinity College, Cambridge, in 1952. After lecturing at Cambridge and Oxford he became Savilian Professor of Geometry at Oxford (1963–1969), and in 1966 was awarded the Fields Medal (the mathematical equivalent of the Nobel Prize). After three years at Princeton, he returned to Oxford as Royal Society Research Professor in 1973. He became Master of Trinity College and President of the Royal Society in 1990. One of the most distinguished British mathematicians of his time, he has worked on algebraic geometry and algebraic topology. He worked on K-theory, a cohomology theory for vector bundles, and then on differential operators between bundles, which led him to the index theory of differential operators, a rich blend of analysis and topology. His work has been taken up by many collaborators and students in the UK and abroad. With his students, Russian collaborators, and with Raoul Bott at Harvard, he worked on the theory of 'instantons' (which give a mathematical description of objects that behave like both particles and waves) and on Yang-Mills theory. Most recently he has taken up the mathematics of conformal field theory, where he has been particularly concerned with bridging the gap between mathematicians and physicists. He was elected FRS in 1962 and knighted in 1983.

AUER, Karl, Baron von Welsbach
(1858–1929)

Austrian chemist, born in Vienna, who invented the gas mantle which bears his name and carried out important work on the rare metals. Auer isolated the elements neodymium and praesodymium. Continuing his researches, he found that a mixture of thorium nitrate and cerium nitrate becomes incandescent when heated. He used this mixture to impregnate gas mantles, thereby making gas lighting cheaper and more efficient. His mantle is still used in kerosene lamps. Auer also made osmium filaments for use in electric light bulbs. These lasted longer than their carbon counterparts and were the forerunners of the cheaper tungsten filaments used today. In addition he developed the cerium-iron alloy known as 'Auer metal' or 'mischmetal'. The first improvement over the flint and steel in use since medieval times, it is still used to strike sparks in cigarette lighters and gas appliances.

AUERBACH, Charlotte
(1899–)

German-British geneticist, born in Krefeld. She was educated in Berlin, before attending university courses in Berlin, Würzburg and finally Freiburg (under Spemann), graduating in 1925. She then took up school teaching before starting her PhD course at the Kaiser Wilhelm Institute under Otto Mangold. In 1933 all Jewish students were forbidden to enter the university and she moved to Edinburgh, completing her thesis at the Institute of Animal Genetics. In the late 1930s she studied mutation with Hermann Müller, and was the first to discover chemical mutagenesis, arising from her work on the affects of nitrogen mustard and mustard gas on *Drosophila*. Chemical mutagenesis thereafter became her main research, with particular emphasis on the biological side and the kinds of mutations induced. She was appointed as a lecturer in genetics (1947) and reader (1957) at Edinburgh, and served as professor from 1967 to 1969. She was one of the very first to work out how chemical compounds cause mutations and to compare the differences between the actions of chemical mutagens and X-rays, contributing many papers and books on the subject. These include *Genetics in the Atomic Age* (1956), *The Science of Genetics* (1961) and *Mutation Research* (1976). She was elected FRS in 1957 and awarded the Royal Society's Darwin medal in 1976.

AUGER, Pierre Victor
(1899–)

French physicist, born in Paris. He was educated at the École Normale Suprérieure and later became professor at the University of Paris. Auger discovered that an atom can de-excite from a state of high energy to a lower energy state non-radiatively, by losing one of its own electrons rather than emitting a photon. Electrons emitted in this way are known as Auger electrons. Sources of Auger electrons are useful for calibration of nuclear detectors as the electrons are monoenergetic, whereas beta electrons are emitted with a range of energies. He went on to investigate the properties of the neutron following its discovery by Chadwick, and later worked on cosmic-ray physics. He discovered extended air showers, also known as Auger showers, where the interaction of cosmic rays with Earth's upper atmosphere produces cascades of large numbers of secondary particles, he showed that the primary particles impinging on the upper atmosphere have energies in the region of 10 – 10 giga-electronvols. The origin of these high-energy cosmic rays is still not understood today.

AVERROËS, Ibn Rushd
(1126–1198)

The most famous of the medieval Islamic philosophers, born in Cordova, son of a distinguished family of jurists. He was himself Kadi (judge) successively at Cordova, Seville and in Morocco, and wrote on jurisprudence and medicine in this period as well as beginning his huge philosophical output. In 1182 he became court physician to Caliph Abu Yusuf, but in 1185 was banished in disgrace (for reasons now unknown) by the caliph's son and successor. Many of his works were burned, but after a brief period of exile he was restored to grace and lived in retirement in Marrakesh until his death. The most numerous and the most important of his works were the *Commentaries on Aristotle*, many of them known only through their Latin (or Hebrew) translations, which greatly influenced later Jewish and Christian writers and offered a partial synthesis of Greek and Arabic philosophical traditions.

AVERY, Oswald Theodore
(1877–1955)

Canadian-born American bacteriologist, responsible for a key step in the genesis of molecular biology. Born in Halifax, Nova Scotia, he studied medicine at Colgate University and spent his career at the Rockefeller Institute Hospital, New York (1913–1948). He soon became an expert on pneumococci, and in 1928 was intrigued by a claim that a non-virulent, rough-coated strain could be transformed into the virulent smooth strain in mouse serum, by the mere presence of some of the dead (heat-killed) smooth bacteria. Avery confirmed this result, and went on in 1944 to show that the transformation is actually caused by a deoxyribonucleic acid (DNA), a chemical present in the dead bacteria. Cautiously, he did not go on to suggest that the informational molecules which carry the whole reproductive pattern of any living species (the genes) are simply DNA; this idea emerged slowly around 1950 and forms the central concept of molecular biology.

AVOGADRO, (Lorenzo-Romano) Amedeo (Carlo)
(1776–1856)

Italian (Piedmontese) physicist and chemist, born in Turin. The son of a lawyer, civil servant, and one-time senator of Piedmont in the Sardinian Kingdom. Amedeo succeeded his father as Count of Quaregna in 1787. He trained in his father's profession, graduating as doctor of ecclesiastical law in 1796. While busy as a lawyer and public administrator he acquired a scientific education. From 1806 he abandoned his official positions completely to concentrate on science, and soon became Professor of Mathematics and Physics at the College of Vercelli (1809). In 1819 he was elected to the Turin Academy of Sciences. When the first Italian Chair of Mathematical Physics was established at Turin one year later, Avogadro was appointed to the post, which he held until 1822 and again from 1834 until his retirement in 1850. He published widely on physics and chemistry, founding his work upon an increasingly unpopular caloric theory of heat elaborated most extensively in his *magnum opus*, *Fisica de'corpi ponderabili* (1837–1841). In 1811, seeking to explain Gay-Lussac's law of combining gaseous volumes (1809), he formulated the famous hypothesis that equal volumes of all gases contain equal numbers of molecules when at the same temperature and pressure (Avogadro's law). He also introduced the idea of a polyatomic molecule. The hypothesis was practically ignored for around 50 years, a neglect partially explained by Avogadro's social, intellectual and geographical isolation. Initially close to French scientific traditions, he was increasingly out of touch with the main currents of European thought. But in 1860 at the Karlsruhe Chemical Congress, Cannizzaro circulated a work which attempted to systematize inorganic chemistry using a restatement of Avogadro's hypothesis in contemporary terms. Universal acceptance did not come until the 1880s.

AXELROD, Julius
(1912–)

American pharmacologist, born in New York City. He qualified in biology and chemistry, and worked for some years as a laboratory analyst before entering research. He obtained his PhD degree at the age of 45. In 1949 he joined the section of heart chemistry of the National Institutes of Health (Associate Chemist 1949–1950; Chemist 1950–1952; Senior Chemist 1953–1955) and then moved as chief of the pharmacological section of the Clinical Sciences Laboratory at the National Institutes for Mental Health (1955–1984). His inquiries focused on the chemistry and pharmacology of the nervous system, especially the role of the catecholamines, adrenaline and noradrenaline. Originally intrigued by a report that abnormalities of their metabolism might be implicated in schizophrenia, he examined the chemical transformations that they underwent in normal neural tissue. He determined in great detail the mechanisms of action of noradrenaline in nerve cells and discovered a new enzyme, catechol-o-methyl transferase (COMT), an essential regulator of noradrenaline in the body. His work accelerated investigations into the links between brain chemistry and psychiatric disease, and in the search for psychoactive drugs. He was joint winner of the 1970 Nobel Prize for Physiology or Medicine with Ulf von Euler and Katz.

AYRTON, Hertha, originally Phoebe Sarah Marks
(1854–1923)

English physicist, born in Portsea, near Portsmouth, and educated in mathematics at Girton College, Cambridge. She is best known for her work on the motion of waves and formation of sand ripples, and the behaviour of the electric arc. Her collected papers describing her work on the latter were published as *The Electric Arc* (1902). After extensive research on arc lamps (with her husband William Ayrton), including cinema projector lamps and search lights, she took out several patents; her improvements to searchlight technology were put into practice in aircraft detection during both world wars. During World War I she invented the Ayrton fan for dispersing poison gases – this invention was later adapted to various other applications, including improvement of ventilation in mines. Ayrton was nominated for fellowship of the Royal Society in 1902, but refused on the grounds that she was a married woman. She received the society's Hughes Medal in 1906.

AYRTON, William Edward
(1847–1908)

English engineer and inventor, born in London. He studied mathematics at University College London, and electricity at Glasgow under Kelvin. He joined the Indian telegraph service in 1868, and soon devised a method of detecting faults which was of great benefit in the maintenance of overland telegraphic communication. In 1873 he was appointed to the Chair of Natural Philosophy and Telegraphy at the new Imperial Engineering College in Tokyo, where he established laboratories for the teaching of applied electricity, the first of their kind. Returning to London in 1879 he became Professor of Physics and Electrical Engineering successively at the City and Guilds of London Institute, Finsbury Technical College (1881), and Central Technical College, South Kensington (1884). With his colleague John Perry (1850–1920) he invented the first absolute block system for electric railways (1881), the first electric tricycle (1992), and many electrical measuring instruments. They published jointly some 70 scientific and technical papers between 1876 and 1891. His first wife, Matilda Chaplin (1846–1883) was a pioneer woman doctor; his second, Hertha Ayrton, continued his work on the electric arc and other inventions.

B

BAADE, (Wilhelm Heinrich) Walter
(1893–1960)

German-born American astronomer, born in Schröttinghausen. He studied at the universities of Münster and Göttingen, and from 1919 to 1931 worked at the Hamburg observatory at Bergedorf. In 1931 he moved to the Mount Wilson Observatory in California. Baade's major interest concerned the stellar content of various systems of stars. He discovered the existence of two discrete stellar types or 'populations', characterized by blue stars in spiral galaxies and fainter red stars in elliptical galaxies. In 1944, helped by the wartime blackout in the Pasedena area, he succeeded in resolving into stars with the 100-inch telescope the centre of the Andromeda galaxy M31 and of its two companions M32 and NGC205. In the 1950s, using the 200-inch telescope on Mount Palomar which had come into operation in 1948, he began a systematic survey of the positions of the recently discovered radio sources which led to a number of optical identifications. His work on stellar systems earned him the Gold Medal of the Royal Astronomical Society in 1954. After retirement from California he became Gauss Professor at the University of Göttingen (1959).

BABBAGE, Charles
(1792–1871)

English mathematician, born in London. Educated at Trinity and Peterhouse colleges, Cambridge, he spent most of his life attempting to build two calculating machines. The first, the 'difference engine', was intended for the calculation of tables of logarithms and similar functions by repeated addition performed by trains of gear wheels. A small prototype model was described to the Royal Astronomical Society in 1822, and earned the award of the society's first Gold medal. Babbage was then granted money by the government to build a full-sized machine, but in 1842, after some £17,000 of public money and £6,000 of his own had been invested without any substantial result, government support was withdrawn (an unfinished portion of the machine is now in the Science Museum, London). Meanwhile Babbage had conceived the plan for a much more ambitious machine, the 'analytical engine', which was designed not just to compute a single mathematical function but could be programmed by punched cards, like those in the Jacquard loom, to perform many different computations. The cards were to store not only the numbers, but also the sequence of operations to be performed. The idea was too ambitious to be realized by the mechanical devices available at the time, but can now be seen to be the essential germ of the electronic computer. Babbage held the Lucasian Chair of Mathematics at Cambridge (1828–1839), although he delivered no lectures.

BACON, Francis, Baron Verulam of Verulam, Viscount St Albans
(1561–1626)

English philosopher and statesman, born in London, the younger son of Sir Nicholas Bacon. He entered Trinity College, Cambridge, and in 1576 Gray's Inn, being called to the bar in 1582. He became an MP in 1584. On the accession of James VI and I (1603) he sought royal favour by extravagant professions of loyalty by planning schemes for the union of England and Scotland, and making speeches in parliament to prove that the claims of the king and parliament could be reconciled. For these services he was knighted (1603) and made a commissioner for the union of Scotland and England. In 1612 he offered to manage parliament for the king and, in this capacity, was promoted in 1613 to the attorney-generalship. In 1616 he became a privy councillor, in 1617, lord keeper, and in 1618 Lord Chancellor, being raised to the peerage. In 1621, however, he was accused of taking bribes in his capacity as a judge; the evidence was incontrovertible, and it was the end of his political career. He died five years later deep in debt. Bacon's attempt to review and classify all branches of human knowledge he called 'the grand instauration of the sciences', and as a first step towards this objective he published in 1605 the *Advancement of Learning*; he never succeeded in bringing this scheme to completion. His philosophy was also described in *De Augmentis Scientiarum* (1623), a latin expansion of the *Advancement*, and *Novum Organum* (1620). In this he repudiated the deductive logic of Aristotle and his followers, and stressed the importance of experiment and inductive reasoning in interpreting nature, as well as the necessity for proper regard for any possible evidence which might run counter to any held thesis. He described heat as a mode of motion, and light as requiring time for transmission, but he was generally behind the scientific knowledge of his time. His greatness consists in his insistence on the facts, that man is the servant and interpreter of nature, that truth is not derived from authority, and that knowledge is the fruit of experience; and in spite of the defects of his method, the impetus he gave to future scientific investigation is indisputable. He was the practical creator of scientific induction. As a writer of English prose and a student of human nature, he is seen to best advantage in his essays. In his fanciful *New Atlantis* he suggested the formation of scientific academies, and by the year 1645 weekly meetings were being held which led to the foundation in 1660 of the Royal Society of London for Improving Natural Knowledge.

BACON, Roger
(c.1214–1292)

English philosopher and scholar, probably born near Ilchester, Somerset. He studied at Oxford and Paris, and began to gain a reputation for diverse and unconventional learning in philosophy, magic and alchemy which led to the soubriquet *Doctor Mirabilis* ('Wonderful Teacher'). Around 1247 he joined the Franciscan Order, and soon afterwards returned to Oxford to develop his interests in experimental science. He attempted the compilation of an encyclopaedia of universal knowledge, but even in his day the task was beyond the capabilities of any one man. He seems to have been uncomfortably outspoken, quarrelsome and uncompromising in his views, and he suffered rejection, censorship and eventually imprisonment at the hands of the Order for the heresy of his 'suspected novelties'; he died in Oxford soon after his release from 14 years of incarceration by the Order in Paris. He has been (mistakenly) credited with scientific inventions like the magnifying glass and gunpowder, but he certainly published some remarkable speculations about lighter-than-air flying machines, mechanical transport on land and sea, the circumnavigation of the globe, and the construction of microscopes and telescopes. His views on the primacy of mathematical proof and on experimentalism have often seemed strikingly modern, and despite surveillance and censorship from the Franciscans he published many works on mathematics, philosophy and logic whose importance was only recognized in later centuries.

BAEKELAND, Leo Hendrik
(1863–1944)

Belgian-born American chemist, one of the founders of the plastics industry. He was born in Ghent and studied there and at other Belgian universities, and became Professor of Physics and Chemistry in Brugge in 1885. He emigrated to the USA in 1889 and founded a chemical company to manufacture one of his inventions, photographic printing paper which could be used with artificial light. Subsequently he made the first synthetic phenolic resin from the condensation products of phenol and formaldehyde. The condensation of organic molecules had been described by Baeyer in 1872, but it was Baekeland who first put this phenomenon to industrial use. The resin, known as Bakelite, replaced hard rubber and amber as an insulator and its success led to the founding of the Bakelite Corporation in 1910. Baekeland wrote on many topics in organic chemistry and electrochemistry, and also on the reform of patent law. He was elected president of the American Chemical Society in 1924, and received many academic honours at home and abroad. He died in Beacon, New York.

BAER, Karl Ernst Ritter von
(1792–1876)

Estonian-born German naturalist and pioneer in embryology. Born in Piep of a wealthy family of Prussian origins, Baer graduated in medicine at Dorpat in 1816 and later studied at Berlin, Vienna and Würzburg. He was appointed professor at Königsberg (1817–1834), where he undertook his most important research, and from 1834 he investigated the mammalian ovary, especially the small follicles discovered therein by Graaf in 1673. Through dissecting a colleague's pet dog, Baer showed that the Graafian follicle contained a microscopic yellow structure which was the egg (ovum) rather than merely, as thought since the Greeks, the mingling of male and female seminal fluids. Investigating embryo development, Baer was the first to differentiate the notochord, the gelatinous cord that in vertebrates becomes the backbone and skull; he also drew attention to the neural folds which develop into the central nervous system. Investigating morphogenesis, Baer stressed that the embryos of various species initially share highly analogous and even indistinguishable forms. Differentiation does not occur till later. This was to be known as the 'biogenetic law': in embryonic development, general characters appear before special ones. In the development process, Baer showed that the embryo of a higher creature passes through stages resembling phases in the unfolding of lower animals.

This idea (formulaically stated as ontogeny recapitulating phylogeny) proved important to comparative anatomy, embryology and evolutionary theory. Baer himself was no evolutionist, but his teleologist notion of *Entwicklungsgeschichte* (development history) was influential in paving the way for evolutionary styles of thinking.

BAILEY, Liberty Hyde
(1858–1954)

American horticulturalist and botanist, born in South Haven, Michigan. He became assistant to Asa Gray at Harvard University. He was appointed Professor of Horticulture and Landscape Gardening at Michigan State (1885) and Cornell (1888), and founded the Bailey Hortorium of New York State College in 1920. This was a remarkable private herbarium of 125,000 plants and a library of 3,000 books. Much of his work was concerned with identification of cultivated plants. He published many papers on *Carex* (sedges), *Rubius*, *Vittis* (vines), brassicas and palms. He also experimented on crosses and varieties within the genus *Curcurbita*, the important genus which includes melons, squashes and cucumbers. He edited various works such as the *Standard Cyclopedia of Horticulture* (1914–1917), later revised into the single-volume *Hortus* and coined the term 'cultivar'. His *Manual of Cultivated Plants* was published in 1923. He was President of the American Association for the Advancement of Science, of the Botanical Society of America, the International Congress of Plant Sciences, and the American Society of Horticultural Science.

BAILY, Francis
(1774–1844)

English astronomer, born in Newbury, Berkshire. He started his working life as a banker and went on the stock exchange in 1799. Astronomy was a hobby encouraged by his acquaintance with Priestley. In 1820 he was a founder member and the Vice-President of the Royal Astronomical Society. Elected FRS in 1821, he retired from business in 1825 and fitted his house in Tavistock Place, London, with an observatory. Baily's accurate catalogue of nearly 3,000 star positions led to the award of the Gold Medal of the Royal Astronomical Society in 1827. On 15 May 1836 he observed a total eclipse of the Sun from Jedburgh, Scotland. Just before the sun disappears behind the Moon the sunlight appears as a brilliant cluster of spots. Baily's description of this phenomenon was so graphic that these appearances of residual segments of the solar disc between the mountains of the Moon have become known as Baily's beads. His description of the solar corona after the 1842 eclipse was the first realistic description of this region of the solar outer atmosphere. Baily also studied ancient eclipses, and knowing the rate at which the orbital parameters of the Moon changed, he dated the eclipse that occurred during the battle between the Medes and Lydians as 30 September 610 BC. He improved Cavendish's apparatus, and after a series of delicate measurements between 1838 and 1842, he concluded that the Earth has a mean density of 5.66 times that of water.

BAIRD, John Logie
(1888–1946)

Scottish electrical engineer and television pioneer, born in Helensburgh. He studied electrical engineering in Glasgow at the Royal Technical College (now the University of Strathclyde) and the University of Glasgow. Poor health compelled him to give up the post of engineer to the Clyde Valley electric power company, and after a brief career as a sales representative and an unsuccessful business venture in the West Indies, he settled in Hastings (1922) and began research into the possibilities of television. Hampered by continued ill-health and lack of financial support, he nevertheless built his first television apparatus almost entirely from scrap materials, and in 1926 gave the first demonstration in London of a television image. The following year he transmitted primitive television pictures over the telephone lines from London to Glasgow, and 1928 the signals he broadcast from a station in Kent were picked up in the USA. His 30-line mechanically scanned system was adopted by the BBC in 1929, being superseded in 1936 by his 240-line system. In the following year the BBC chose a rival 405-line system with electronic scanning made by Marconi-EMI. Other lines of research initiated by Baird in the 1920s included radar and infrared television ('Noctovision'); he continued his research up to the time of his death and succeeded in producing three-dimensional and coloured images (1944) as well as projection onto a screen and stereophonic sound.

BAKER, Herbert Brereton
(1862–1935)

English physical chemist, born in Livesey, near Blackburn. He was educated at Balliol College, Oxford, where his tutor was Dixon, to whom he acted as assistant (1883–1885). From 1886 to 1904 he was a schoolmaster, mainly at Dulwich College, where he contrived to engage in both teaching and research, and then returned to Oxford as Dr Lee's Reader in Chemistry at Christ Church. He was Professor of Chemistry at Imperial College, London from 1912 to 1932. Baker was elected FRS in 1902, appointed CBE in 1917, received the Davy Medal of the Royal Society in 1923, and was President of the Chemical Society (1926–1928). His researches were almost entirely on the effect of intensive drying on chemical systems and he made many remarkable observations, e.g. that in the complete absence of moisture, phosphorus or sulphur may be distilled in oxygen without reaction. The interpretation of much of his work was controversial.

BALFOUR, Francis Maitland
(1851–1882)

Scottish embryologist, Grandson of the Marquis of Salisbury and younger brother of Arthur Balfour, the future prime minister. Francis Balfour came from an illustrious family. Edinburgh-born, he showed precocious promise in natural history while still at Harrow School. A student at Trinity College, Cambridge, he received a first in natural science in 1871, and embarked upon physiological research under Foster, producing in 1878 notable work on elasmobranch development that displayed the qualities of precision and close observation for which he later became renowned. In the same year he was elected FRS. After completing a two-volume *Treatise on Comparative Embryology* (1880), he was awarded a medal by the Royal Society in 1881, and in the following year Cambridge created for him a chair in animal morphology. A keen climber, he died in a fall on an unconquered peak in the Chamonix district before the year was out. One of numerous embryologists examining morphogenesis in the light of Baer's research programme and of Charles Darwin's evolutionary theory, Balfour was distinguished by painstaking microscopic accounts of the development process and by his aversion to the grander phylogenetic philosophizings of contemporaries such as Haeckel.

BALL, John
(1818–1889)

Irish botanist and alpinist, born in Dublin. Taken to Switzerland when seven years old, he was profoundly affected by the view of the Alps from the Jura. The following year, at Ems, he spent much time trying to measure the height of the hills with a mountain barometer. After a period at Cambridge University, Ball visited Sicily and published a valuable paper on its botany. In 1852 he became Liberal MP for Carlow and advocated many measures which later became law, including the disestablishment of the Church of Ireland. As Colonial Under-Secretary from 1855 to 1857, he influenced the government in its decision to prepare a series of Floras of the British colonies and possessions. He was the first President of the Alpine Club (1857) and author of the *Alpine Guide* (1863–1868). In 1871 he accompanied Sir Joseph Dalton Hooker and George Frederick Maw to Morocco and investigated the flora of the Great Atlas. His *Spicilegium Florae Maroccanae* (1877–1878) was the earliest work on the Moroccan flora. In 1882 he visited South America. Ball proposed a theory on the antiquity of the alpine flora, and another that most endemic South American plants originated on a hypothetical ancient mountain range in Brazil.

BALL, Sir Robert Staywell
(1840–1913)

Irish mathematician, born in Dublin, son of the secretary of Dublin Zoo. He obtained a scholarship to Trinity College, Dublin, where he studied mathematics and experimental physics. In 1865 he became tutor to Lord Rosse's son at Parsonstown (now Burr), the home of the then largest telescope in the world. 1867 saw Ball appointed to the Chair of Applied Mathematics and Mechanics at the Royal College of Science, Dublin, and in 1874 he became the Astronomer Royal for Ireland and Andrews Professor at Trinity College. He was elected FRS in 1873 and knighted in 1886. In 1890 he moved to Cambridge as Lowndean Professor and Director of the Observatory, and remained there until he died. Ball's main scientific interest was in the movement of rigid bodies about fixed points, i.e. the mathematical theory of screws. He was also interested in the connection between this theory and the quaterion theory of the linear vector function. Astronomically, Ball is remembered as a great public lecturer – 'a lecturer whose science and wit and playfulness combined can absolutely rivet any audience from a savant to a little child' (H. Montagu Butler, 1913), and his *Popular Guide to the Heavens* was published in many editions between 1892 and 1955.

BALTIMORE, David
(1938–)

American molecular biologist, born in New York City, joint winner of the 1975 Nobel Prize for Physiology or Medicine with Temin and Dulbecco. As a high school student, Baltimore met Temin at the Jackson Laboratory, Bar Harbor, Maine. He received a BA in chemistry from Swarthmore College, Pennsylvania, and a PhD at Rockefeller University, New York. In 1972 he became professor at MIT. In 1990 he was appointed President of Rockefeller University, but he resigned in 1991 following a controversial legal battle over an allegedly fraudulent paper in the journal *Cell*; Baltimore's name had appeared on the paper, although he himself was not charged with misconduct. Over the period 1965–1968 he worked on viral genetics with Dulbecco at the Salk Institute, California. Temin had suggested that certain viruses, like the Rous sarcoma virus which stores its genetic code in RNA, could insert viral genes into the DNA of the cells that the viruses infected. This suggestion was contrary to the then-accepted idea that DNA produced RNA which then formed protein. Baltimore discovered an enzyme which could make DNA from RNA, called 'reverse transcriptase', in 1970. Viruses with RNA as the genetic code, such as the virus associated with AIDS, are now known as 'retroviruses'. With reverse transcriptase, scientists can manipulate the genetic code. Concerned about the misuse of genetic engineering, Baltimore supported a moratorium on research in the mid-1970s. In the 1980s he chaired a US National Academy of Sciences committee on AIDS.

BANACH, Stefan
(1892–1945)

Polish mathematician, born in Krakow. He studied at Lvov, where he became a lecturer in 1919 and professor in 1927. During World War II he was forced to work in a German institute for research on infectious diseases: although he returned to work at Lvov University after the war, his health was ruined and he died less than a year later. Banach is regarded as one of the founders of functional analysis, and his book *Théorie des opérations linaires* (1932) remains a classic. He founded an important school of Polish mathematicians which emphasized the importance of topology and real analysis. His name is attached to a class of infinite dimensional linear spaces, whose elements are usually functions, in which a concept of length is defined. These spaces occupy an important place in the study of analysis, and are increasingly applied to problems in physics, especially in particle physics, where they are natural generalizations of the formalism of Hilbert and von Neumann.

BANKS, Sir Joseph
(1744–1820)

English botanist, born in London, and educated at Harrow, Eton, and Christ Church, Oxford. In 1766 he made a voyage to Newfoundland collecting plants, and between 1768 and 1771 accompanied James Cooks' expedition round the world in a vessel, the *Endeavour*, equipped at his own expense. He later took part with Solander in an expedition to Iceland. In 1778 he was elected President of the Royal Society, an office which he held for 41 years. His friendship with King George III led him to persuade the King to turn Kew Gardens into a botanical research centre. He founded the African Association, and the colony of New South Wales owed its origin mainly to him. His desire to cultivate the Pacific bread-fruit in the West Indies led to the *Bounty* expedition under Captain Bligh. He was also responsible for the introduction of the Mango from Bengal, and many fruits of Ceylon and Persia, and he suggested the growing of the Chinese tea plant in India. His name is commemorated in the genus *Banksia*. His significance lies in his far-reaching influence, rather than through any single personal contribution to science. He facilitated the work of others, partly through his wealth, which he dispensed generously in scientific causes, and also through his official positions. He was a strong advocate of the need for close international contact between scientists, even in time of war, such as the French and American revolutions. Banks was made a baronet in 1781.

BANTING, Sir Frederick Grant
(1891–1941)

Canadian physiologist, born in Alliston Ontario. He grew up on a small farm, and studied medicine at Toronto University. During World War I he served as a surgeon in the Canadian army Medical Corps. Soon after he returned, he established a surgical practice, but as this was not very successful, he became a demonstrator at the Western University Medical School where he conducted his first medical research. Several attempts had already been made by other workers around the world to prepare a pancreatic extract for the treatment of diabetes, but they had all met with little success. Banting came up with a possible new method and was advised to approach Macleod, Professor of Physiology at Toronto and an expert on carbohydrate metabolism, who allowed him to use his laboratory. With Best, a recent graduate in physiology and biochemistry, and later with the biochemist Collip, a practical method was devised and extracts of pancreas of sufficient purity were produced to allow the first successful clinical trials in January 1922. Soon afterwards a means of commercial preparation of insulin was devised and it became the principal means of treating diabetes. Banting was made Professor of Medical Research at Toronto in 1923 and was awarded the 1923 Nobel Prize for Physiology or Medicine jointly with Macleod. Banting shared his part of the prize with Best. He established the Banting Research Foundation in 1924, and the Banting Institute at Toronto in 1930. He was knighted in 1934. A pioneer in aviation medicine, he was killed in a wartime air crash.

BARANY, Robert
(1876–1936)

Austrian-born physician and otologist, born in Vienna. He graduated from the University of Vienna in 1900 and undertook further studies in internal medicine, psychiatric neurology, neurology and surgery in Frankfurt, Heidelberg, Freiburg and Vienna. In 1903 he joined the staff of the University of Vienna ear clinic. He extended earlier work on the inner ear which had been carried out on animals, and pioneered the study in humans of the inner ear's balancing apparatus. He was able to correlate dizziness with objective bodily factors such as eye movements, and the reactions of muscles. He proved the connection between the balancing apparatus and the brain, making it possible for equilibrium disturbances and vertigo to be investigated systematically. Barany volunteered for service during World War I, hoping to be able to put his ideas about the treatment of brain wounds to the test. In 1915, while in a Russian prisoner-of-war camp in Siberia, news reached him that he had been awarded the 1914 Nobel Prize for Physiology or Medicine. He was released and in 1917 became director of the oto-rhinolaryngology clinic at the University of Uppsala in Sweden.

BARCROFT, Sir Joseph
(1872–1947)

Irish physiologist, born in Newry, County Down. He read natural sciences at King's College, Cambridge, and graduated in 1897. He stayed in the Physiological Laboratory at Cambridge for his entire career, becoming professor in succession to John Langley in 1926. He retired in 1937, although he continued to work in the laboratory until appointed head of the newly created Animal Physiology Unit of the Agricultural Research Council in 1942. His early academic career was disrupted during World War I, when he worked in chemical warfare at Porton Down. His research work concentrated on understanding respiratory function, studies that included whole animal physiological experiments: observations on humans at extremes of altitude on expeditions in the Andes and in simulated laboratory conditions: and important biochemical work on the respiratory pigment haemoglobin and its binding and dissociation with oxygen. Many of these contributions are summarized in *The Respiratory Function of the Blood* (1st edition 1914; 2nd edition in two volumes, 1925–1928) and in 1934 he published *Features in the Architecture of Physiological Functions*, which summarized his views of integrative physiology. In the final years of his life, apart from wartime service again at Porton, he focused on the physiology of the developing foetus. Using a Caesarean section technique, Barcroft measured foetal blood volume and placental blood flow, the transfer of gases across the placental membrane, the control of respiration and movement and growth parameters of the foetus, findings that he encapsulated in his final book, *Researches on Pre-Natal Life* (1946). He was elected FRS in 1910 and knighted in 1935.

BARDEEN, John
(1908–)

American physicist, born in Madison, Wisconsin. He studied electrical engineering at Wisconsin and worked as a geophysicist at the Gulf Research Laboratories for three years, before obtaining his PhD in mathematical physics at Harvard (1936). He joined a new solid-state physics group at Bell Telephone Laboratories in 1945. Together with Brattain and Shockley he developed the point-contact transistor (1947), for which they shared the Nobel Prize for Physics in 1956. Their early device, made by placing two metal contacts only 0.005 centimetres apart on a germanium surface, paved the way for the electronics revolution that continues today. Bardeen was professor at Illinois University (1951–1975), and with Cooper and Schrieffer he was awarded the Nobel Prize for Physics again in 1972 for developing the Bardeen-Cooper-Schrieffer, or CS, theory. The observation that the temperature at which a metal becomes superconducting is inversely proportional to its atomic mass suggested to Bardeen that the effect of the vibrations of the atoms in the metal lattice on the electrons (electron-phonon interactions) must be involved. Cooper meanwhile had shown that in a metal two electrons could form a resonant state, known as a 'Cooper pair', and these ideas provided the first satisfactory theory to explain why certain metals lose their resistance to electric current at low temperatures. Bardeen was the first recipient of two Nobel physics prizes.

BARNARD, Christian Neethling
(1922–)

South African surgeon, born in Beaufort West. He graduated from Cape Town

Medical School. After a period of research in the USA, he returned to Cape Town in 1958 to work on open-heart surgery and organ transplantation. In December 1967 at Groote Schuur Hospital he performed the first successful human heart transplant. The recipient, Louis Washkansky, died of pneumonia 18 days later, drugs given to prevent tissue rejection having heightened the risk of infection. A second patient, Philip Blaiberg, operated on in January 1968, survived for 594 days.

BARNARD, Edward Emerson
(1857–1923)

American astronomer, born in Nashville, Tennessee. Born into a poor family, he had little education in his early years, but became experienced in photographic techniques during work in a portrait studio and developed a strong amateur interest in astronomy. After discovering a number of comets and becoming skilled in astronomical work, he became both a teacher and a student at the observatory of Vanderbilt University. He moved to Lick Observatory in 1887 and was appointed Professor at Yerkes Observatory of the University of Chicago in 1895. Following a systematic photographic survey of the sky, he correctly concluded with Maximilian Wolf that those areas of the Milky Way which appear to be devoid of starts, or 'black nebulae', are in fact clouds of obscuring matter. His wide-ranging research included studies of novae, binary stars and variable stars. He discovered the fifth satellite of Jupiter (1892), later named Amalthea, and identified the star with the greatest known apparent motion across the sky, now known as Barnard's star (1916).

BARROW, Isaac
(1630–1677)

English mathematician and divine, born in London. He was educated at Charterhouse and Trinity College, Cambridge, where he was elected a Fellow in 1649. His royalist sympathies and leaning towards Arminianism prevented him from obtaining the professorship of Greek until 1660. He travelled abroad (1655–1659), became Professor of Geometry at Gresham College, London (1662), and the first Lucasian Professor of Mathematics at Cambridge (1663), but he resigned in 1669 to become Royal Chaplain and was succeeded by Isaac Newton. He founded the library of Trinity College, Cambridge, when he became Master in 1673. Barrow published Latin versions of Euclid and Archimedes, and lectures on optics (which were soon superseded by Newton's), as well as extensive theological works and sermons. In his original work in mathematics he anticipated aspects of the theories of different calculus, which began to develop at the end of the seventeenth century.

BARTHOLIN, Caspar, the Younger
(1655–1738)

Danish anatomist and politician, son of Thomas Bartholin the Elder. Born in Copenhagen, he studied medicine in Holland and France. He became an expert anatomist. He was the first to describe the greater vestibular glands in the female reproductive system ('Bartholin's glands') and the larger salivatory duct of the sublingual gland ('Bartholin's duct'). Caspar Bartholin was the last and possibly the most eminent of a line of Bartholins. His grandfather, the first Caspar Bartholin (1585–1629) studied medicine at Copenhagen, Basle and at

Padua with Fabrizio, returning in 1611 to a chair at Copenhagen and increasingly specializing in theological studies. Caspar Bartholin the Elder was the father of two eminent scholar-scientists, Thomas Bartholin the Elder and Erasmus Bartholin.

BARTHOLIN, Erasmus
(1625–1698)

Danish physician, physicist and mathematician, son of Caspar Bartholin the Elder (1585–1629) and brother of Thomas Bartholin the Elder. He studied medicine at Leiden and Padua, and was appointed Professor of Medicine and Mathematics at Copenhagen in 1656. In 1669 he discovered that when an objective is viewed through Iceland feldspar (calcite), a double image is produced, but he was unable to explain this. Huygens, Isaac Newton and Fresnel all contributed to the explanation of the phenomenon of double refraction, which is basic to the understanding of polarization.

BARTON, Sir Derek Harold Richard
(1918–1998)

English chemist, born in Gravesend. He received his undergraduate and postgraduate training at Imperial College, London, and obtained his PhD in 1942 for work under the direction of Sir Ewart Jones. After two years in military intelligence and a year with Albright and Wilson, he returned to Imperial College as assistant lecturer and then ICI Fellow. In 1949 he spent a year at Harvard while Robert Woodward was on sabbatical, and produced a seminal paper on the relationship between conformation and chemical reactivity for which he was awarded the Nobel Prize for Chemistry in 1969 (shared with Hassel). In 1950 he moved to Birkbeck College, and worked on a number of structural problems, in triterpenois and steroid chemistry. This continued when he moved to the Chair of Organic Chemistry in Glasgow in 1955 and back to Imperial College in 1957, but by, 1960, X-ray crystallography had largely replaced degradative studies in the determination of structure, and Barton turned his attention to synthetic and biosynthetic work. Important results were obtained in a number of areas, particularly in the biosynthesis of steroids. He also pioneered the use of photochemical reactions in synthesis. In 1977 he was appointed Director of the French National Centre for Scientific Research (CNRS) Institute for the Chemistry of Natural Substances in Gif-sur-Yvette and he became active in designing new reactions, some involving free radicals and others ligand-coupling reactions based on bismuth. After 11 very productive years in France, in spite of much administrative work and time given to the appreciation of *haute cuisine* he was appointed Distinguished Professor at Texas A & M University where his research has continued unabated. During the course of his career he has received nearly every honour possible for an organic chemist, including election to the Royal Society in 1954. He was knighted in 1972 and made an Officer of the French Legion of Honour in 1985.

BARTRAM, John
(1699–1777)

American botanist, born in Marple, near Darby, Pennsylvania, and educated locally. Considered the 'father of American botany'. Linnaeus called Bartram 'the greatest natural botanist in the world'. Bartram inherited an uncle's farm, but sold this and purchased another on the banks of the Schuylkill river at Kingessing, near Philadelphia.

There, he became a successful small farmer and also built up an unrivalled collection of North American plants, which he began selling to European botanists and horticulturists through the English woollen draper and botanist Peter Collinson (1694–1768). This unsuccessful business allowed Bartram to visit Virginia, the Allegheny Mountains, the Carolinas and elsewhere in search of plants. In 1743 the British Crown commissioned him to visit the Indian tribes of the 'League of Six Nations'; the results were published as *Observations on the Inhabitants, Climate ... from Pensilvania to Onondago, Oswego and the Lake Ontario, in Canada* (1751). In 1765 he was named King's Botanist, and explored Florida, where he discovered *Franklinia altamaha* (named after his friend Benjamin Franklin). His son William (1759–1823) was also a botanist; his best-selling *Travels* (1791) strongly influenced English Romanticism. He also published *Observations on the Creek and Cherokee Indians, 1789* (1853).

BASOV, Nikolai Gennadiyevich
(1922–)

Russian physicist, who invented masers and lasers, born in Voronezh. He served in the Red Army during World War II, and studied in Moscow. He joined the Lebedev Physics Institute in Moscow as a laboratory assistant in 1948, became deputy director (1958–1973), and was appointed director in 1973. His work in quantum electronics – more specifically, the interaction between matter and incident electromagnetic waves – provided the theoretical basis for the development of the master (Microwave Amplification by Stimulated Emission of Radiation), subsequently successfully producing numerous types of these devices (1960–1965) using a variety of methods to excite the materials to a state in which they produced the laser effect, including exposure to electron beams and light sources (optical pumping). In 1968 he used powerful lasers to produce thermonuclear reactions. For his work on amplifiers and oscillators used to produce laser beams he was awarded the 1964 Nobel Prize for Physics jointly with his colleague Prokhorov and the American physicist Townes.

BASSI, Agnostino Maria
(1773–1856)

Italian botanist and pioneer bacteriologist, born in Lodi. Educated at Pavia, his work on animal diseases partly anticipated that of Pasteur and Koch. As early as 1835 he showed after many years' work that a disease of silkworms (muscardine) is fungal in origin, that it is contagious, and that it can be controlled. He proposed that some other diseases are transmitted by micro-organisms.

BATES, Henry Walter
(1825–1892)

English naturalist, born in Leicester. With his friend Wallace he left to explore the Amazon in 1848, and after Wallace's return to England in 1852, he continued alone until 1859 when he returned with 14,700 specimens, including almost 8,000 species of insect new to science. In 1861 he published *Contributions to an Insect Fauna of the Amazon Valley*, in which he described the phenomenon now known as Batesian mimicry. In this, harmless, edible species of animal are found resembling others which may be distantly related and are distasteful or poisonous, and thus gain protection from predators. This discovery came during the Darwinian controversy and provided strong evidence in favour of natural selection.

His travels are described in *The Naturalist on the River Amazon* (2 vols, 1863). In 1864 he became assistant secretary of the Royal Geographical Society.

BATESON, William
(1861–1926)

English geneticist, born in Whitby, Yorkshire, and educated in natural sciences at Cambridge. He introduced the term 'genetics' in 1909, and became the UK's first Professor of Genetics at Cambridge (1908–1910). He left to become Director of the new John Innes Horticultural Institution (1910–1926). His interest in heredity was stimulated by a trip across the Asian Steppes (1886–1887) to investigate the relationship between variation in environment, salinity in lakes, and variation in populations of shellfish. He went on to produce the first English translation of Mendel's work. Bateson believed in discontinuous variation and was in sharp and acrimonious disagreement with Pearson and the biometricians, who advocated a statistical continuous approach to hereditary variation. Bateson showed that some genes are inherited together, a process now known as 'linkage'. He played a dominant part in establishing Mendelian ideas, but was a major opponent of chromosome theory. In Bateson's view, there was no material link between the chromosome and any physical characteristic of the organism. Although an ardent evolutionist, he was opposed to Charles Darwin's theory of natural selection, as the small changes demanded by the theory seemed insufficient to account for the evolutionary process. He wrote *Mendel's Principles of Heredity: A Defence* (1902) and *Problems of Genetics* (1913).

BATTERSBY, Sir Alan Rushton
(1925–)

English chemist, born in Leigh. Having obtained his BSc from the University of Manchester (1946), then he moved to the University of St Andrews where he received his PhD in 1949. From 1948 to 1953 he was on the staff at St Andrews and during this period he spent time at the Rockefeller Institute, New York, and at the University of Illinois, before moving to the University of Bristol in 1954. In 1962 he was invited to the newly created second Chair of Organic Chemistry at the University of Liverpool. From 1969 to 1992 he was professor at the University of Cambridge. During the period 1954–1968 he worked on the chemistry of alkaloids and terpenes, but he is most famous for research on the haem pigments, cytochromes, chlorophylls and vitamin B12. He has shown that the building block for all these naturally occurring substances, porphobilinogen, is polymerized by the enzyme deaminase to give a linear tetrapyrrole followed by an enzyme-catalysed ring closure. He has also identified several of the precorrins on the biosynthetic pathway leading to vitamin B12. Battersby was elected FRS in 1966 and knighted in 1992, but remains Professorial Fellow of St Catherine's College.

BAYLISS, Sir William Maddock
(1860–1924)

English physiologist, born in Wednesbury. Bayliss was apprenticed to a local general practitioner before beginning his studies at University College London (UCL). He received a BSc in 1882 but failed his second MB in anatomy, and after that he began to concentrate on physiological studies. He went to Oxford in 1885 and received a degree in physiology in 1888, before returning to UCL to serve under Sharpey-Schafer. Bayliss remained at UCL for the rest of his life, from 1912 as Professor of General Physiology. Much of his research was conducted in collaboration with Starling. They worked on electrophysiology, the vascular system and intestinal motility. In studies of pancreatic secretion they showed that the discharge is induced by a chemical substance which they called secretin, produced in the duodenum and carried to the pancreas in the blood. During World War I Bayliss worked on wound shock. His celebrated book *Principles of General Physiology* first appeared in 1914 and became known as the 'textbook for professors'. In 1903 he took an action for libel against Stephen Coleridge, secretary of the National Antivivisection Society, who had accused Bayliss of carrying out experiments on unanaesthetized animals. He won £2,000 damages and presented the sum for the furtherance of research in physiology. He was elected FRS in 1903, and knighted in 1922.

BEADLE, George Wells
(1903–1989)

American biochemical geneticist. Born on a farm in Wahoo, Nebraska, he became interested in agricultural genetics as a student. He studied the genetics of maize and the fruit fly (*Drosophila*), and popularized the use of the bread mould *Neurospora* in research. At Stanford University (1937–1946) he collaborated closely with Tatum. They irradiated spores of Neurospora, isolated mutants and investigated their nutritional requirements. They developed the idea that specific genes control the production of specific enzymes; for example, spores unable to grow unless provided with vitamin B6 were found to have a mutation in the gene which metabolized vitamin B6. Beadle and Tatum shared the 1958 Nobel Prize for Physiology or Medicine with Lederberg. In the 1950s, as President of the American Association for the Advancement of Science (AAAS), Beadle worked for more openness in scientific research and the AAAS agreed not to hold meetings in cities which practised racial segregation. He was professor at Caltech (1945–1961), and President of Chicago University from 1961 to 1968.

BEAUMONT, William
(1785–1853)

American army surgeon, born in Lebanon, Connecticut. His pioneering study on *Digestion* (1833) was based on experiments with a young Canadian patient, Alexis St Martin, who was suffering from a gunshot wound that had left a permanent opening in his stomach, and which Beaumont treated. Between 1824 and 1833 Beaumont used St Martin as a willing experimental subject. He confirmed the chemical nature of digestion, established that the presence of food stimulated the secretion of gastric juice and directly observed the role of alcohol in the causation of gastric inflammation (gastritis). His *Experiments and Observations on the Gastric Juice and the Physiology of Digestion* (1833) was the first major American contribution to physiology. It was reprinted in Edinburgh and translated into German. Beaumont retired from the army in 1839 and went into general practice in St Louis, Missouri.

BECQUEREL, Alexandre-Edmond
(1820–1891)

French physicist, born in Paris, son and assistant of Antoine César Becquerel. At

18 he was admitted to both the École Polytechnique and the École Normale Supérieure, but he decided instead to join his father at the Natural History Museum of Paris. In 1852 he was appointed to the Chair of Physics at the Conservatoire des Arts et Métiers, taught chemistry at the Société Chimique de Paris (1860–1863), and eventually succeeded his father at the Museum in 1878. He did research into electricity, magnetism and optics. He measured the properties of electric currents, and demonstrated in 1843 that Joule's law governing the production of heat generated by an electric current applied not only to solids, but also to liquids. He measured the electromotive force of the voltaic pile by means of an electrostatic balance designed by his father. He also investigated diamagnetism, the magnetic properties of oxygen and solar radiation. In the course of his experiments on light, he constructed the 'actinometer' (an instrument that determined the intensity of light by measuring the amount of electric current produced in photochemical reactions), and a phosphoroscope while investigating luminescence.

BECQEUREL, Antoine Henri
(1852–1908)

French physicist, born in Paris, he studied at the École Polytechnique and the school of bridges and highways. In 1876 he started teaching at the Polytechnique and later he succeeded his father Alexandre-Edmond Becquerel in the Chair of Physics at the Natural History Museum. He was an expert in fluorescence and phosphorescence, continuing the work of his father and grandfather. Following the discovery of X-rays by Röntgen, Becquerel investigated fluorescent materials to see if they also emitted X-rays. He exposed a fluorescent uranium salt, pitchblende, to light and then placed it on a wrapped photographic plate. He found that a faint image was left on the plate, which Becquerel believed was due to the pitchblende emitting the light it had absorbed as a more penetrating radiation. However, by chance, he left a sample that had not been exposed to light on top of a photographic plate in a drawer. He noticed that the photographic plate also had a faint image of the pitchblende. After several chemical tests he concluded that these 'Becquerel rays' were a property of atoms. He had, by chance, discovered radioactivity and prompted the beginning of the nuclear age. His work led to the discovery of radium by Marie and Pierre Curie and he subsequently shared with them the 1903 Nobel Prize for Physics.

BEDNORZ, (Johannes) Beorg
(1950–)

German physicist, born in West Germany. He graduated at Münster (1976) and was awarded his PhD by the Swiss Federal Institute of Technology in Zürich (1982). He subsequently joined the IBM Zürich research laboratory and with Alex Müller, who had considerable experience in the field of superconductors, investigated a new range of material based on oxides. This inspired departure from the more conventional materials based on mixtures of different metals (intermetallic compounds) yielded results in 1986. They announced that they had observed superconductivity at a temperature 12 kelvin higher than the current record of 23 kelvin, which had been held by an alloy of germanium and niobium since 1973. At a stroke a whole new family of superconducting compounds had been revealed. Laboratories around the world rushed to reproduce and improve upon these results, and within two years observations of superconductivity at 90 kelvin in materials made of mixtures of yttrium, barium and copper oxide were confirmed. With superconductors now operating above the boiling point of liquid nitrogen, an inexpensive and plentiful coolant, the practical applications of superconducting devices were set to multiply enormously. Bednorz was awarded the 1987 Nobel Prize for Physics jointly with Müller only one year after the announcement of their discovery.

BEHRING, Emil von
(1854–1917)

German bacteriologist and pioneer in immunology, born in Hansdorf, West Prussia. He entered the gymnasium at Hohenstein in East Prussia in 1865. Enrolling in the Army Medical College in Berlin in 1864, he obtained his medical degree in 1878, and in 1881 moved to Pozna as an assistant surgeon. During this period he became interested in disinfectants, particularly iodoform. In 1888 he went to join Koch's Institute of Hygiene in Berlin. His major contribution was the development of serum therapy against tetanus and diphtheria (1890) in the first application of the science of serology, in collaboration with Kitasato. Diphtheria was one of the most common fatal diseases affecting children at the time, and Behring's work was instrumental in counteracting the disease. In later years, he tried to develop a tuberculosis antitoxin, but had to admit defeat. Much of his work in this area concerned the relationship between tuberculosis occurring in humans and that in cattle. Unlike Koch, he believed the forms to be identical. Although today they are not considered the same, the cattle form may be transmitted to humans, and Behring's recommendations for reducing the occurrence of the disease in animals and for disinfecting milk were important public health measures. He later became Professor of Hygiene at Halle (1894–1895) and at Marburg (from 1895), and was awarded the first Nobel Prize for Physiology or Medicine in 1901. During World War I, the tetanus vaccine developed by him helped to save so many German lives that he received the Iron Cross, very rarely awarded to a civilian.

BELL, Alexander Graham
(1847–1922)

Scots-born American Inventor, born in Edinburgh, son of Alexander Melville Bell. Educated at Edinburgh and London, he worked as assistant to his father in teaching elocution (1868–1870). In 1871 he moved to the USA and became Professor of Vocal Physiology at Boston (1873), devoting himself to the teaching of deaf-mutes and to spreading his father's system of 'visible speech'. After experimenting with various acoustic devices he produced the first intelligible telephone transmission with the famous words to his assistant 'Mr Watson, come here – I want you' on 5 June 1875. He was granted three patents relating to his invention between 1875 and 1877, successfully defended them against Elisha Gray and others, and formed the Bell Telephone Company in 1877. In 1880, with the proceeds of the Volta Prize awarded to him by the French government, he established the Volta Laboratory for research into deafness, and his work there resulted in the photophone (1880) and the graphophone (1887). He also founded the journal *Science* (1883). After 1897 his principal interest was in aeronautics: he encouraged Samuel Langley, and invented the tetrahedral kite.

BELL, Sir Charles
(1774–1842)

Scottish anatomist, surgeon and neurophysiological pioneer, born in Edinburgh, the son of a Church of England clergyman. Bell developed his interest in medicine from his elder brother, John, who became a celebrated surgeon and anatomist. Charles studied at Edinburgh University, moving to London in 1804, where he became proprietor of an anatomy school and rose to prominence as a surgeon. In 1812 he was appointed surgeon to the Middlesex Hospital, in 1828 becoming one of the co-founders of the Middlesex Hospital Medical School. An interest in gunshot wounds led him to treat the wounded from the battle of Corunna; after Waterloo, he organized a hospital in Brussels. A fine drafsman, he was energetic in teaching anatomy to artists. His *Essays on the Anatomy of Expression in Painting* (1806) illustrated the science of physiognomy. Knighted in 1831, in 1836 he was appointed Professor of Surgery at Edinburgh. Today he is remembered for his pioneering neurophysiological researches, first set out in his *Idea of a New Anatomy of the Brain* (1811). Bell demonstrated that, far from being single units, nerves consist of separate fibres sheathed together. Above all he showed that fibres convey either sensory or motor stimuli but never both – fibres transmit impulses only in one direction. His explorations of the sensorimotor functions of the spinal nerves triggered a bitter and prolonged priority dispute with the French Physiologist Magendie. Bell's experimental work led to the discovery of the long thoracic nerve (Bell's nerve). We speak of 'Bell's palsy' as a result of his demonstration that lesions of the seventh cranial nerve could create facial paralysis.

BELL, Thomas
(1792–1880)

English naturalist, born in Poole, Dorset. A dental surgeon at Guy's Hospital (1817–1861), he lectured in zoology and became Professor of Zoology at King's College, London, in 1836. He was Secretary of the Royal Society, President of the Linnaean Society, and first President of the Ray Society (1844). His *British Stalk-eyed Crustacea* (1853) remains a standard work on British crabs and lobsters. He edited the *Natural History and Antiquities of Selborne* (1877), by White whose house he purchased.

BELL BURNELL, (Susan) Jocelyn, neé Bell
(1943–   )

English radio astronomer, born in York, co-discoverer of the first pulsar. She was educated at the universities of Glasgow and Cambridge, where she received her PhD in 1968. She later joined the staff of the Royal Observatory, Edinburgh, and became the manager of their James Clerk Maxwell Telescope on Hawaii. She was awarded the Herschel Medal of the Royal Astronomical Society in 1989, and since 1991 has been Professor of Physics at the Open University. In 1967 she was a research student at Cambridge working with Hewish when they noticed an unusually regular radio signal on the 3.7 metre wavelength radio telescope. This turned out to be the first discovery of a pulsar (PSR 1919 + 21, with a period of 1,337 seconds). Within the first few weeks of discovery the correct conclusions were drawn; the object is stellar as opposed to a member of the solar system, and is a condensed neutron star. The papers by Franco Pacini (1967) and Gold (1968) laid the theoretical foundation.

BENACERRAF, Baruj
(1920–)

Venezuelan-American immunologist, born in Caracas. In 1940 his family moved to the USA and two years later he entered the Medical College of Virginia. As an asthma sufferer in childhood, Benacerraf became interested in immune hypersensitivity. In 1949 he took a research job at the Broussais Hospital, Paris, and in 1956 he returned to the USA to a post at the New York University School of Medicine. Here he became Professor of Pathology in 1960. In New York his research concentrated upon the cells involved in the body's defence against antigens – substances which are foreign to the body. He found that some animals responded to simple synthetic antigens by producing antibodies while others did not, and went on to show that this was genetically determined. He named the genes involved the immune-response genes. Later work clarified the role of the genetically determined structures on the surfaces of cells that regulate immunological responses to diseased cells and in organ transplants. For this work he shared the 1980 Nobel Prize for Physiology or Medicine with Dausset and George Snell. Benacerraf moved to the National Institutes of Health in 1968 and to Harvard Medical School in 1970.

BENIOFF, Victor Hugo
(1899–1968)

American geophysicist, born in Los Angeles of a Russian father and Swedish mother. He was educated at Pomona College, California (AB 1921), in physics and astronomy. He first worked at Lick Observatory (1923–1924) and in 1924 moved to the Carnegie Institution in Washington where he developed a system of seismic recording drums. These gave an unprecedented accuracy of 0.1 second and formed the basis of the now famous Caltech seismic network. His bent for instrument design continued with the variable reluctance seismograph and linear strain seismograph (in service from 1931). These made possible the precise determination of seismic travel-time, the discovery of new seismic phases, extended the recordable magnitudes down to teleseismic (distant) events and enabled first motion determinations, which give the directions of the fault breaks caused by earthquakes. Later these instruments were recommended for monitoring nuclear tests. He received his PhD at Caltech in 1935 for work on these inventions. When the Seismological Laboratory was transferred from Carnegie to Caltech in 1937, Benioff became assistant professor in seismology and professor from 1950, and collaborated with Gutenberg and Charles Richter. From about 1950 (15 years prior to other workers) he became interested in general problems in earthquake mechanisms and global tectonics, developed several new analytical techniques and amassed evidence of earthquakes around the Pacific. In 1954 he published evidence of epicentres deepening away from the trench and extending down to 700 kilometres below Japan (similar to those identified by Wadati). These seismically active down-going crustal slabs beneath trenches are now known as Wadati-Benioff zones.

BENTLEY, Charles Raymond
(1929–)

American geophysicist and glaciologist, born in Rochester, New York. He studied physics at Yale (BS 1950) and geophysics at Columbia (PhD 1959), where he participated in research cruises in the Atlantic with Ewing. He was research geophysicist in Columbia (1952–1956), spent two years in Antarctica and participated in the particular scientific effort for the International Geophysical Year (1957–1958). Since then he has worked at the University of Wisconsin-Madison, becoming A. P. Crary Professor there in 1968. In west Antarctica he discovered that the ice rests on a floor far below sea-level; thus the region would be open ocean if the ice melted. His interest in Antarctica has continued with pioneering work on applying geophysical techniques to determine the physical characteristics of the ice. He is one of the founders of radioglaciology, pioneering methods of using electromagnetic waves to measure the thickness and properties of the ice. Using seismic and gravity techniques, he has contributed to the understanding of Antarctic isostasy and crustal structure. In the 1980s his research group discovered and investigated the soft, deformable sediment beneath a west Antarctic ice stream that is now believed to be a critical factor in local glacial dynamics.

BENZER, Seymour
(1929–   )

American geneticist, born in New York City. He studied physics at Purdue University, Indiana, and taught biophysics there until 1965, when he moved to Caltech. He was a member of the Phage Group, set up by Delbruck, Hershey and Luria to encourage the use of phage as an experimental tool, and isolated more than 1,000 mutants of the rII phage virus of the bacterium *Escherichia coli*. He showed that genes and proteins were co-linear, i.e. that a change in a gene led to a change in the protein it coded for. He also introduced the term 'cistron' for the smallest section of a DNA strand comprising a functional gene. In 1961 he showed that some sections of the DNA strand are more susceptible to mutations than would be expected by chance; these sites became known as 'hot spots', and were later found to be associated with modified nucleic acids in the DNA strand. Since the 1960s, Benzer has moved away from bacterial genetics to research the genetics underlying the behaviour of the fruit fly, *Drosophila*.

BERG, Paul
(1926–   )

American molecular biologist, born in Brooklyn, New York City. He was educated at Pennsylvania State and Western Reserve universities, although his studies were interrupted by World War II, which he spent in the US navy. He later became Professor of Biochemistry at Stanford (from 1959) and at Washington (from 1970). Since 1985 he has been Director of the Beckman Center of Molecular and Genetic Medicine at Stanford. In the 1960s, Berg purified several transfer RNA molecules (tRNAs – each tRNA carries a specific amino acid through the cell to the ribosome for assembly into a protein). In the following decade, he worked with simian virus 40 and the bacterial virus lambda phage, and developed techniques to cut and splice genes from one organism into another. Concerned about the effects of mixing genes from different organisms, Berg organized a year-long moratorium on genetic engineering experiments, and in 1975 chaired an international committee to draft guidelines for such studies. These techniques are now widely used in biological research. In 1978 Berg enabled gene transfer between cells from different mammalian species for the first time. He shared the 1980 Nobel Prize for Chemistry with Sanger and Walter Gilbert.

BERGERON, Tor Harold Percival
(1891–1977)

Swedish meteorologist, born in Godstone, near London, of Swedish parents. After graduating from Stockholm University in 1916, he joined the Bergen School under Vilhelm Bjerknes (1922) and obtained his doctorate at Oslo University (1928). Bergeron researched the occlusion process whereby air in the warm sector of a depression is forced upwards. He also described the 'Arctic front' and explained different types of fronts, including the formation of waves on a stationary front, by studying surface and upper air observations in detail. An inspired teacher, his visits to the USA, Russia, Yugoslavia and elsewhere did much to promote the Bergen School ideas internationally. He returned to the Swedish Meteorological and Hydrological Institute in 1936 and later became professor at Uppsala (1947–1961). With a collaborator, he was responsible for the currently accepted theory that the usual process for the initiation of rain is for ice crystals and water vapour to coexist in clouds. Later he studied the effects of topography on precipitation, both on the mesoscale and on the scale of mountain ranges and synoptic systems, as well as the precipitation mechanism in hurricanes. He received the International Meteorological Organisation Prize (1866) and was awarded the Symons Gold Medal of the Royal Meteorological Society.

BERGSTRÖM, Sune Karl
(1916–)

Swedish biochemist, born in Stockholm. He studied medicine and chemistry at the Karolinska Institute in Stockholm, and taught chemistry at Lund (1948–1958) before returning to the Karolinska Institute as professor (1948–1981). In 1952, following the development of techniques of introducing radioactive atoms into cholesterol by Konrad Block, Bergström demonstrated the formation of the bile acid ester, taurocholic acid, from cholesterol and made major contributions towards solving the metabolic pathway involved by observing the consequences of injecting hypothetical intermediates into a rat. He studied the metabolism of pythocholic acid from snake bile (1960), and found it to be derived from cholesterol by the same general metabolic pathway. He also isolated and purified the prostaglandins, complex fatty acids produced by nearly all mammalian cells. Their physiological effects include lowering blood pressure and the promotion of contraction by intestinal and uterine muscle, for which purpose they are employed to promote foetal abortion. Together with his former student Samuelsson and Vane, Bergström was awarded the 1982 Nobel Prize for Physiology or Medicine for his work on prostaglandins.

BERNAL, John Desmond
(1901–1971)

Irish crystallographer, born in Nenagh, County Tipperary. Educated by the Jesuits at Stonyhurst College, Lancashire, he won a scholarship to Emmanuel College, Cambridge, and from the first showed himself a polymath (with the lifelong nickname 'Sage'). He developed modern crystallography and was a founder of molecular biology, pioneering work on the structure of water. He progressed from a lectureship at Cambridge to become Professor of Physics and then Professor of Crystallography at Birkbeck College, London (1937–1968). It was here that he did his major pioneering work, taking X-ray photographs of biologically important molecules, amino acids, proteins, sterols and nucleoproteins. He was convinced that from an understanding of their physical molecular structure would come a clearer insight into the way the living processes worked. His researches were dominated by one purpose, namely the search for the origin of life and he included among his major works *The Origin of Life* (1967). His wartime service, in close association with Lord Louis Mountbatten in Combined Operations, involved abortive attempts at creating artificial icebergs to act as aircraft carriers, as well as working on bomb tests and contributing substantially to the scientific underpinning of the invasion of the European continent. A communist from his student days, he was active in international peace activity during the Cold War and supported Lysenko in the USSR when his destruction of Soviet genetics drove J. B. S. Haldane out of the British Communist party. Bernal's hopes for communism's possibilities for science were first shown in his *The Social Function of Science* (1939) and *Marx and Science* (1952).

BERNARD, Claude
(1813–1878)

French physiologist, born near Villefranche, Beaujolais. The child of poor vineyard workers, Bernard was educated in church schools and at 19 apprenticed to a Lyons pharmacist. After an unsuccessful attempt to shine as a dramatist, he abandoned the theatre and opted for a professional qualification, choosing medicine. Trained in Paris, he became assistant at the Collège de France to the physiologist Magendie, and in this work developed a passion for experimental medicine. Marriage to a rich paris physician's daughter who brought with her an ample dowry obviated the need for Bernard to practise medicine and allowed him to pursue his researches (the marriage was miserable). In 1855 he succeeded to Magendie's Paris post. Thereafter his career was a stream of promotions and honours, including the Legion of Honour (1867); at heart, Bernard remained a country lover, and he returned as often as possible to his beloved Beaujolais. His greatest contribution to physiological theory was the notion that life requires a constant internal environment (*milieu intérieur*); cells function best within a narrow range of osmotic pressure and temperature when bathed in a fairly constant concentration of chemical constituents. A fearless vivisector, Bernard made wide-ranging experimental discoveries in many areas of physiology. He studied the neurophysiological action of paralysing poisons like curare and demonstrated their use in experimental medicine. Exploring digestion, he researched the operation of enzymes in gastric juice and demonstrated the role of nerves in controlling gastric secretion. He showed how carbohydrates needed to be transformed into sugars before they could be absorbed. He probed the role of bile and pancreatic juice in the digestion of proteins. On the basis of such researches he concluded the nutrition is an intricate process, involving various stages and chemical rather than physical transformations. He discovered glycogen, probed sugar production by the liver and the problem of diabetes. He also discovered the vasomotor and vasoconstrictor nerves, and examined the inhibitory action of the vagus nerve upon the heart. Bernard further developed Lavoisier's ideas on animal heat. He showed that the oxidation producing animal heat is indirect, occurring in all tissues and not simply in the lungs. Bernard stands as one of the founders of modern physiological research. Combining experimental skill with a partiality for theory, he was notably innovative and one of the great masters of productive research. His *Introduction to the Study of Experimental Medicine* (1865) is a classic account of the new biology, concerned to lay bare the elementary conditions of the phenomena of life, and to demonstrate the absence of qualitative differences between the normal and the pathological. Diseases, Bernard claimed, have no ontological reality; they are merely distortions and extreme forms of regular physiological processes and states.

BERNOULLI, David
(1700–1782)

Swiss mathematician, born in Gröningen, son of Jean Bernoulli. He studied medicine and mathematics at the universities of Basle, Strasbourg and Heidelberg, and became Professor of Mathematics at St Petersburg (1725). There he took up the Petersburg paradox in probability, a coin-tossing game which intuitively should command a small entry price but in theory demands an infinite one. In 1732 he returned to Basle to become Professor of Anatomy, then was appointed to a professorship in botany and finally physics. He worked on trigonometric series, mechanics and vibrating systems. In his best-known work he pioneered the modern field of hydrodynamics. His *Hydrodynamica* (1738) explored the relationships between pressure, density and velocity in flowing fluids, and anticipated the kinetic theory of gases, pointing out that pressure could result from the bombardment of a surface by particles of matter and that pressure would increase with increasing temperature. He solved a differential equation proposed by Jacopo Riccati, now known as Bernoulli's equation.

BERNOULLI, Jacques or Jakob
(1654–1705)

Swiss mathematician, born in Basle, brother of Jean Bernoulli. He studied theology and travelled in Europe before returning to Basle (1682), where he became professor in 1687. He investigated infinite series, the cycloid, transcendental curves and the logarithmic spiral. His analysis of the catenary, the curve formed by a non-elastic flexible string when suspended at each end, was applied to the design of bridges. In 1690 he applied Leibniz's newly discovered differential calculus to a problem in geometry, and introduced the term 'integral'. His posthumously published *Ars conjectandi* (1713) was an important contribution to probability theory, and included his discovery of the 'law of large numbers', his permutation theory and the 'Bernoulli numbers', coefficients found in exponential series'. A logarithmic spiral was at his request carved on his tombstone in Basle cathedral, with the motto in Latin 'Though changed I rise the same'.

BERNOULLI, Jean or Johann
(1667–1748)

Swiss mathematician, born in Basle, younger brother of Jacques Bernoulli. He studied medicine and graduated in Basle (1694), but turned to mathematics, and became professor at Gröningen (1695) and Basle (1705). He wrote on differential equations, both in general and with respect to particular curves, such as the clamped beam and the hanging chain, finding the length and area of curves, isochronous curves and curves of quickest descent. He was a quarrelsome man and a bitter rivalry developed between him and his brother. After his brother's death he took the leading role on Leibniz's side in the priority dispute with Isaac Newton regarding the invention of the calculus. He founded a dynasty of mathematicians which continued for two generations, and was employed by the Marquis de l'Hospital to help him write the first text book on the differential calculus.

BERNSTEIN, Richard Barry
(1923–1990)

American physical chemist, born in Long Island, New York. He was educated at Columbia University, New York, and from 1944 to 1946 worked on the Manhattan Project to separate uranium-235. From 1948 to 1953 he was Assistant Professor of Chemistry at Illinois Institute of Technology, Chicago, and from 1953 to 1963 he advanced from Assistant Professor to Professor of Chemistry at the University of Michigan. After 1963 he was professor in succession at the universities of Wisconsin (Madison) and Texas (Austin), Columbia University, and finally the University of California (Los Angeles) from 1983 until his death. Bernstein received many awards, including the Hishelwood lectureship at Oxford (1980), the Debye Award of the American Chemical Society (1981), the Willard Gibbs medal of the same society (1989) and a National Medal of Science (1989). His researches were in chemical kinetics and reaction dynamics by molecular beam scattering and laser techniques. By these techniques he elucidated much detail of what occurs when molecules collide and react. He co-authored (with Raphael Levine) *Molecular Reaction Dynamics* (1974), and also wrote *Chemical Dynamics via Molecular Beam and Laser Techniques* (1982). In 1955 he was the pioneer of 'femtochemistry', the study of reactions which take place on timescales of the order of 10 to the power of –12 seconds.

BERTHELOT, (Pierre Eugene) Marcellin
(1827–1907)

French chemist and politician, born in Paris. His scientific education was acquired largely in the private laboratory of Pelouze, and he became demonstrator at the Collège de France in 1851. He defended his thesis on the chemistry of glycerin in 1854, and in 1858 graduated as a pharmacist. In 1859 he was appointed to the Chair of Organic Chemistry at the École de Pharmacie but continued to work in a laboratory at the Collège de France. His research covered many topics, but within organic chemistry, he is best known for his work on the synthesis of alcohols, aromatic compounds and turpentine derivatives. In a pioneering study, Berthelot and Pan de Saint-Gilles examined the kinetics of esterification, the results of which were significant in the formulation by Guldberg and Waage of the law of mass action. His later experimental work laid the foundations for the science of thermochemistry. In the last stage of his professional life he wrote extensively on the history of early chemistry. He also did much to initiate the chemical analyses of archaeological objects. During the siege of Paris in the Franco-Prussian War (1870–1871), Berthelot played a leading role in the defence of the city and was elected to the Senate in 1871. After a number of minor posts he became Minister of Education for two years (1886–1887), and very briefly, Foreign Minister (1895). However, he never achieved in politics the same eminence as in his scientific work. He received the Legion of Honour in 1861 and succeeded Pasteur as Secretary of the French Academy of Sciences in 1889.

BERTHOLLET, Claude Louis, Comte de
(1749–1822)

French chemist, born in Talloires, Savoy, who made major contributions to the revolutionary advances in chemistry which took place in the second half of the eighteenth century. He studied medicine at Turin and Paris, later becoming private physician to Phillip, Duke of Orleans. Berthollet helped Lavoisier with his researches on gunpowder and also with the creation of a new system of chemical nomenclature, which is still in use today. He accepted Lavoisier's antiphlogistic doctrines, but disproved his theory that all acids contain oxygen by analyzing hydrocyanic acid. Bertholet was the first chemist to realize that there is a connection between the manner in which a chemical reaction proceeds and the mass of the reagents; this insight led others to formulate the law of definite proportions later stated by Proust. He also demonstrated that chemical affinities are affected by the temperature and concentration of the reagents. In the course of his studies of chlorine, he failed to find a way to use potassium chlorate in gunpowder, but made a successful bleach by dissolving chlorine in a solution containing sulphuric acid and various salts. He made the results public so that the mild and agricultural land reserved for traditional bleachfields could be put to better use, and the use of chlorine in bleaching spread rapidly in France and abroad. Following Priestley's discovery that ammonia is composed of hydrogen and nitrogen, Berthollet made the first accurate analysis of their proportions. He was elected to the Academy of Sciences in 1781. During the French Revolution, Berthollet helped to reorganize the coinage, agricultural policy and higher education. When the Academy was reconstituted as the Institut de France in 1793, Berthollet was one of the first members to be elected. In 1796 he led a delegation to Italy to choose works of art for the galleries in Paris. After accompanying Napoleon to Egypt he remained there for two years, helping to reorganize the educational system. He voted for Napoleon's deposition in 1814, and on the restoration of the Bourbons, was created a count. He died in Arcueil.

BERZELIUS, Jons Jacob
(1779–1848)

Swedish chemist who made fundamental advances in atomic theory and electrochemistry. He was born in Vafversunda in East Götland. He studied medicine at Uppsala, and worked as an unpaid assistant in the College of Medicine at Stockholm before succeeding to the Chair of Medicine and Pharmacy in 1807. Subsequently he travelled abroad to meet Humphry Davy, Oersted, Berthollet and other leading chemists. In 1815 he was appointed Professor of Chemistry at the Royal Caroline Medico-Chirurgical Institute in Stockholm, returning in 1832. Soon after Volta's invention of the electric battery, Berzelius began in 1802 to experiment with the voltaic pile, which was to revolutionize physics and chemistry by providing scientists with a reliable, continuous source of current for the first time. Working with Wilhelm Hisinger he discovered, by 1803, that all salts are decomposed by electricity. Berzelius went on to suggest that all compounds, organic as well as inorganic, are made up of positive and negative components, a theory which laid the foundation for our understanding of radicals. In 1803 he and Hisinger discovered cerium. Later Berzelius also discovered selenium and thorium; he was the first person to isolate silicon, zirconium and titanium, and made detailed studies of other rare metals. His greatest achievement, however, was his contribution to atomic theory. His work with salts led him to consider whether elements were always present in compounds in fixed proportions, a question much debated at the time, e.g. by Berthollet and Proust. To this end he analysed some 2,000 compounds between 1807 and 1817. Matching the idea of constant proportions with Dalton's atomic theory, and persuaded of the central importance of oxygen from his studies of Lavoisier's work, he drew up a table of atomic weights using oxygen as a base. In 1818 he published the weights of 45 of the 49 elements then known, revising this table in 1828. In the course of this research he devised the modern system of chemical symbols to replace the chaotic system which had grown piecemeal over a long period of time. Berzelius also made significant contributions to organic chemistry, for example by his analysis of organic acids. He studied isomerism and catalysis, phenomena which both owe their names to him, as does protein with Berzelius recognised as fundamental to all life and named from the Greek *proteios*, meaning 'primary'. A pioneer of gravimetric analysis, Berzelius has had few rivals as an experimenter. As a result of the poverty of his early years, he had to improvise much of his apparatus, and some of his innovations are still standard laboratory equipment, e.g. wash bottles, filter paper and rubber tubing. He also greatly improved blowpipe techniques. Despite the fact that he suffered ill-health for most of his life, he also managed to write extensively and his works were translated into many languages. He wielded enormous influence until the last two decades of his life, which were marred by controversies with younger chemists. Berzelius's achievements brought him many honours. He was elected to the Stockholm Royal Academy of Sciences in 1808 and became its secretary in 1818. He was awarded the Gold Medal of the Royal Society of London, and in 1835 he was created a baron by Charles XIV. He died in Stockholm.

BEST, Charles Herbert
(1899–1978)

Canadian physiologist, born in West Pembroke, Maine, the son of a physician. Best graduated in physiology and biochemistry from the University of Toronto in 1921. As a research student at Toronto in 1921 he helped Banting to isolate the hormone insulin, used in the treatment of diabetes. Banting divided his share of the 1923 Nobel Prize for Physiology or Medicine with Best. In 1925 Best obtained a medical qualification and in 1929 succeeded Macleod as Professor of Physiology at Toronto. When Banting was killed in 1941, Best replaced him as Director of the Banting and Best department of medical research. Besides this involvement with the isolation of insulin, he also enjoyed considerable success during his later career he discovered choline (a vitamin that prevents liver damage) and histaminase (the enzyme that breaks down histamine), and introduced the use of the anticoagulant, heparin. He continued to work on insulin, showing in 1936 that the administration of zinc with insulin can prolong its activity.

BHABHA, Homi Jahangir
(1909–1966)

Indian physicist, born in Bombay and educated there before entering Cambridge (1927) where his tutor was Dirac. In India during World War II, a readership was created for him at the

Indian Institute of Science in Bangalore, where he became professor in 1941, the same year as he was elected FRS. In 1945 he became Director of the Tata Institute for Fundamental Research in Bombay and then Director of the Indian Atomic Energy Commission. When this was reorganized as the Department of Atomic Energy (1948) he was its first secretary. In the same year he was appointed Director of the Atomic Energy Research Centre, renamed the Bhabha Atomic Research Centre after his death in a plane crash. In his early career he derived a correct expression for the cross-section (probability) of scattering positrons by electrons, a process now known as Bhabha scattering. He co-authored a classic paper on the theory of cosmic-ray showers (1937), which described how primary cosmic rays (mainly high-energy protons from outer space) interact with the upper atmosphere to produce the particles observed at ground level. This paper also demonstrated that a very penetrating component of cosmic rays observed underground could not be electrons. These particles were identified as muons (1946). In 1938, in a letter to *Nature*, Bhabha was the first to conclude that the lifetimes of fast, unstable cosmic-ray particles, such as muons, would be increased due to the time dilation effect that follows from Einstein's special theory of relativity. The experimental verification of this is often cited as one of the most straightforward pieces of evidence supporting special relativity.

BISHOP, (John) Michael
(1936–)

American molecular biologist and virologist, born in Yoek, Pennsylvania. He was educated at Gettysburg College where he obtained a BA in 1957, and at Harvard Medical School where he was awarded his MD in 1962. He began his career as an intern then resident in general medicine at Massachusetts General Hospital (1962–1964) and subsequently became a research associate in virology at the National Institutes of Health in Washington DC from 1964 to 1966. He remained in Washington as assistant then associate professor until he moved to San Francisco in 1972. He has been Professor of Microbiology and Immunology at the university of California Medical Centre since 1972, and Director of the G. W. Cooper Research Foundation since 1981. Bishop has received many major awards, including the Nobel Prize for Physiology or Medicine in 1989 (jointly with Varmus) for his discovery of oncogenes. Oncogenes are normal cellular genes whose product has some roles in the normal growth and development of all mammalian cells. However, certain faults in oncogene regulation, for example, over production as a result of the influence of a virus, or faulty expression due to a mutation within the gene, can severely damage the growth and differentiation of the affected cell type and thus cause cancer. An understanding of the function of oncogenes is therefore a crucial step in combating all types of cancer.

BLACK, Sir James Whyte
(1924–)

Scottish pharmacologist, born in Uddington. After graduating in medicine at St Andrews University he became an assistant lecturer in physiology there (1946–1947), a lecturer at the University of Malaya and then senior lecturer at Glasgow Veterinary School (1950–1958). Following appointments at ICI Pharmaceuticals (1958–1964) and the Wellcome Research Laboratories (1978–1984) as Director of Therapeutic Research, he was appointed Professor of Pharmacology and departmental head at University College London (1973–1977); since 1984 he has held similar posts at King's College Medical School, London. In 1962 while at ICI, he discovered the drug netherlide (Alderlin R) which, by binding β adrenergic receptors, blocks the action of catecholamines in changing cardiac tension without inducing the sympathomimetic activity (which causes brachycardia) associated with other drugs. This opened the way to new treatments for certain types of heart disease (angina, tachycardia) and led to the development of the safer, more effective drug propranolol. At the Wellcome Laboratories Black synthesized burimamide and cimetidine, new drugs with which he distinguished two classes of histamine receptor, enabling specific suppression of stomach acid secretion for the treatment of ulcers. He was elected FRS in 1976 and knighted in 1981, and shared with Elion and Hitchings the 1988 Nobel prize for Physiology or Medicine.

BLACK, Joseph
(1728–1799)

Scottish chemist who discovered carbon dioxide and the phenomena of latent and specific heat, one of the most influential chemists of his generation. He was born in Bordeaux, where his father was a wine merchant, and after attending school in Belfast studied medicine at Glasgow and Edinburgh universities. In 1756 Black succeeded Cullen as Professor of Medicine at Glasgow, and 10 years later he moved to the Chair of Medicine and Chemistry at Edinburgh. In his MD thesis of 1754, Black showed that the causticity of lime and the alkalis is due to the loss of 'fixed air' (carbon dioxide), which is present in limestone and the carbonates of these alkalis, but which is driven off by heating. Black found that this new air supported neither life nor combustion, and with this discovery was the first person to realize that there are gases other than air, a fundamental advance which was to dramatically affect the future development of chemistry. At the same time, his experimental method, in which he weighed the amounts of carbonates and alkalis accurately at every stage, laid the foundations of quantitative analysis. Black then turned his attention to changes of state. Having observed that ice takes a long time to melt, even on a sunny day, he showed experimentally that solids need heat to turn them into a liquid at the same temperature and also that the heat required for any given mass is unique to every substance. In the same way, liquids need characteristic amounts of heat to turn them into gases. Black called the heat necessary for changes of state 'latent heat'. He also discovered that the same weights of different substances require different but constant amounts of heat to raise them through the same temperature difference, a phenomena he christened 'specific heat'. Black was a successful physician and industrial consultant and, in his Edinburgh days, was widely consulted on problems such as bleaching and dyeing, iron making, ore analysis, fertilizers and water supplies. Famed as a teacher, his chemistry classes drew students from all over the UK and from Europe and America, contributing largely to the lustre of the university in the greatest days of the Scottish Enlightenment. He died in Edinburgh.

BLALOCK, Alfred
(1899–1964)

American surgeon, born in Culloden, Georgia. He received his medical education at Johns Hopkins University, and his postgraduate training there and at

Vanderbilt University Hospital. He joined the staff at Vanderbilt (1925–1941) and Johns Hopkins (1941–1964), where he pioneered the surgical treatment of various congenital defects of the heart and its associated blood vessels, many of which could be recognized by the presence of cyanosis in infants, and performed the first 'blue baby' operation, cooperating with the paediatrician Taussig. He also did important experimental work on the pathophysiology of surgical shock and its treatment by transfusion of whole blood or blood plasma, and was the first to treat *myassthenia gravis* by removal of the thymus gland. His collected scientific papers were published in 1966, edited by one of his pupils.

BLOEMBERGEN, Nicolaas
(1925–)

Dutch-born American physicist, born in Dordrecht. Educated at the universities of Utrecht and Leiden, he received his PhD in 1946 then joined the staff of Harvard, where he has been Gordon McKay Professor of Applied Physics since 1957. His early interest was in nuclear magnetic resonance which he studied under Nobel laureate Purcell. He later pioneered methods of three-level and multi-level pumping to energize masers, introducing a modification to Townes's early design enabling the maser to work continuously rather than intermittently. His work on non-linear interactions of radiation with matter yielded a theoretical framework which still remains in place. He also investigated methods of using laser light to selectively excite and break a single bond in a molecule. For his many years of research that contributed to the development of the laser, an invaluable tool for probing the structure of matter, he shared the 1981 Nobel Prize for Physics with Schawlow and Kai Siegbahn. His books included *Nonlinear Optics* (1965).

BLUMBERG, Baruch Samuel
(1925–)

American biochemist, born in New York City, joint winner of the 1976 Nobel Prize for Physiology or Medicine with Gajdusek. He studied at Columbia University and at Oxford, and became Professor of Biochemistry at the University of Pennsylvania in 1964. Blumberg discovered the 'Australia antigen' in 1964 and reported its association with hepatitis B. The finding was very rapidly applied to screening blood donors. His study of the distribution of the HBV virus, as it is known, in the population revealed that apparently healthy people could carry and transmit the live virus, and led to ethical and employment problems associated with screening nurses, physicians and other welfare employees. HBV is widespread in Vietnam, Thailand and elsewhere in south-east Asia, and further problems emerged over the adoption of Vietnamese children in America. In 1969 Blumberg introduced a protective vaccine, now widely used. It was also discovered that persistent infection with HBV is associated with nearly all primary hepatocellular carcinomas; effective treatment has now been achieved.

BODMER, Sir Walter Fred
(1936–)

English geneticist, born in Frankfurt, Germany, and educated at Clare College, Cambridge, where he received his BA in 1956, and an MA and PhD in 1959. He remained at Cambridge as a demonstrator in the genetics at Stanford University, and later became Professor of Genetics at the University of Oxford

(1970–1979). He was appointed Director of Research at the Imperial Cancer Research Fund in 1979 and since 1991 has held the post of Director-General there. During his career he has also served as Chairman of the BBC Science Consultative Group (1981–1987). President of the Royal Statistical Society (1984–1985) and Vice-President of the Royal Institute (1981–1982). He was elected FRS in 1974 and knighted in 1986. Bodmer has published extensively on the genetics of the HLA histocompatibility system. This system is responsible for distinguishing self from non-self in the animal body and an understanding of its complexities has been vital in the remarkable progress in transplant surgery. He also published extensively on somatic cell genetics, cancer genetics and human population genetics.

BOERHAAVE, Hermann
(1668–1738)

Dutch physician and botanist, born in Voorhout, near Leiden. He studied theology and oriental languages, and took his degree in philosophy in 1689. But in 1690 he began the study of medicine, and in 1709 was appointed Professor of Medicine and Botany at Leiden. This post included supervision of the botanic garden, to which he added more than 2,000 species in the first 10 years, often utilizing men in the service of the Dutch East and West Indies companies as seed and plant collectors. He distributed many of these introductions to fellow botanists in many other countries. He published catalogues of the garden, describing the floral structure of the plants. The two works on which his medical fame chiefly rests, *Institutiones Medicas* (1708) and *Aphorismi de Cognoscendis et Curandis Morbis* (1709), were translated into Arabic. He pointed out that both plants and animals show the same law of generation, and by 1718 he was teaching sex in plants, his international stature ensuring widespread acceptance of these ideas. In 1724 he also became Professor of Chemistry, and his *Elemmenta Chemiae* (1724) is a classic. Meanwhile patients came from all parts of Europe to consult him, earning him a fortune.

BOHR, Niels Henrik David
(1885–1962)

Danish physicist, born in Copenhagen. Educated at Copenhagen University, he moved to England to work with J. J. Thomson at Cambridge and Rutherford at Manchester, and later returned to Copenhagen as professor (1916). While at Manchester he developed a theory of the hydrogen atom that could explain the observed spectral lines. This was based on two recent discoveries; Rutherford's evidence for the nuclear atom and the idea that energy is quantized, as suggested by Planck and Einstein. Using the nuclear atom model, Bohr applied the restriction that the electrons orbiting the nucleus could follow only certain allowed orbits. This corresponds to a set of discrete allowed electron energy levels and the observed spectral lines could be interpreted in terms of transitions between these levels. Initially the theory, which seemed to discard classical physics, was not well received but in addition to explaining the observed lines of the hydrogen spectrum it also predicted some new lines. When they were subsequently observed the model was accepted. Bohr's model was later shown to be a solution of the Schrödinger equation. During World War II he escaped from German-occupied Denmark and assisted atom bomb research in the USA, returning to Copenhagen in 1945. He later worked on nuclear physics and developed the liquid drop model of the nucleus used by Bethe and Weizscker to explain the stability of some nuclei and nuclear fission. Like Rutherford, Bohr influenced a great many young physicists including Fermi and Landau. He was founder and Director of the Institute of Theoretical Physics at Copenhagen (1920–1922), and was awarded the Nobel Prize for Physics in 1922. His son, Aage Bohr, won the 1975 Nobel Prize for Physics.

BOK, Bart Jan
(1906–1983)

American astronomer, born in Hoorn, the Netherlands, and educated at the universities of Leiden and Croningen. In 1929 he was awarded a fellowship which allowed him to go to Harvard, USA where he spent 25 years, becoming Associate Director of the Harvard College Observatory to its Director Shapley and in 1947 Robert Wheeler Wilson Professor of Astronomy. In 1957 he succeeded Woolley as Director of the Mount Stromlo Observatory in Australia. He was responsible for the choice of the Siding Springs Mountain in New South Wales as the location of the major southern hemisphere observatory for future large instruments. He returned to the USA in 1966 to the post of Director of the Stewart Observatory in Tucson, Arizona, which he held until his retirement in 1970. Bok's lifelong interest, pursued largely in collaboration with his wife Priscilla Fairfield Bok (1896–1975) was in the structure of our galaxy, in particular in the distribution of stars and interstellar matter in regions of potential star formation. He made a special study of small dark clouds, the Bok globules, which he showed to contain enough material for future condensations into star clusters, a result confirmed by later millimetre-wave observations.

BOLTZMANN, Ludwig
(1844–1906)

Austrian physicist, born in Vienna, Boltzmann studied at the University of Vienna where he obtained his doctorate in 1867. From 1869 he held professorships in mathematics and physics at Graz, Vienna, Munich and Leipzig. He was a popular lecturer, numbering Nernst amongst his many students. Although his interests were diverse, encompassing physics, chemistry, mathematics, philosophy and even aeronautics, he is most celebrated for the application of statistical methods to physics and the relation of kinetic theory to thermodynamics. In 1866 he had tried to find a general proof of the second law of thermodynamics using only mechanical principles. Two years later his study of thermal equilibrium incorporated the statistical methods recently introduced by Maxwell in deriving a velocity distribution for colliding gas molecules. Boltzmann extended Maxwell's theory, treating the case when external forces were present, to derive the 'Maxwell-Boltzmann distribution'. He assumed that all possible ways of apportioning energy amongst molecules were equally probable, this was his first paper on the equipartition of energy. The 'H-theorem' (1872) indicated that a gas would tend towards the equilibrium state. In 1877 he presented the famous 'Boltzmann equation' (later carved on his tombstone) relating thermodynamic 'entropy' (S) and the statistical distribution of molecular configurations (W): $S = k \log W$, where k is Boltzmann's constant. This equation showed how increasing entropy corresponded to increasing molecular randomness. Other work dealt with electromagnetism; in

1872 Boltzmann's experiments on dielectrics confirmed predictions of Maxwell's theory, which he went on to promote on the Continent. Boltzmann wrote on viscosity and diffusion (1880–1882) and in 1884 derived the law for black-body radiation found experimentally by Stefan, his teacher in Vienna. Boltzmann's work was unified by a commitment to an atomic theory of matter but the atomists came under attack from positivists in Vienna led by Mach. Partly because of the unpopularity of his views, Boltzmann suffered severe depression from 1900 and tragically killed himself while on holiday in 1906.

BONDI, Sir Hermann
(1919–)

Austrian mathematical physicist, born in Vienna. He moved to England in 1939 to become an undergraduate at Trinity College, Cambridge. In 1942 he started research into radar at the Admiralty Signals Establishment, and after World War II returned to Cambridge, becoming a Fellow of Trinity College. He became a British citizen in 1947, moved to King's College, London in 1954 as Professor of Applied Mathematics, and was knighted in 1983 to become Master of Churchill College. He served with great distinction as Director General of the European Space Research Organisation (1967–1971), as Chief Scientific Advisor to the Ministry of Defence and as Chairman of the National Environmental Research Council (1980–1984). Scientifically, Bondi is best known for his seminal book on cosmology published in 1952 and for his proposal, with Hoyle and Gold, that the universe is in a steady state, matter being continuously created to fill the gaps left by the expansion. In the 1950s this theory stimulated a great deal of research in cosmology. It fell out of favour when it was discovered that the universe was more dense in the past, and encountered serious difficulties in 1965 following the discovery of the microwave background radiation by Penzias and Robert Wilson. In 1962 Bondi wrote a keynote paper showing how the emission of gravitational waves is a necessary consequence of Einstein's general theory of relativity.

BORDET, Jules Jean Baptiste Vincent
(1870–1961)

Belgian physiologist and an authority on serology, born in Soignies. He received his MD from the University of Brussels in 1892, and in 1894 went to Paris to work in Metchnikoff's laboratory at the Pasteur Institute. In 1901 he became Director of the Pasteur Institute in Brussels, where he remained until he was succeeded by his son, Paul, in 1940. Bordet studied the mechanics of bacteriolysis, which he concluded was due to the action of two substances: a specific antibody, heat-resistant to 55 degrees Celsius and present only in immunized animals, and a non-specific, heat-labile substance found in both unvaccinated and vaccinated animals, which he identified as Hans Buchner's 'alexin'. Bordet's work in this field made possible new techniques for the diagnosis and control of infectious diseases. With his brother-in-law, Octave Gengou, Bordet discovered the whooping cough bacillus, extracted an endotoxin, and prepared a vaccine (1906). In 1909 he isolated the germs for bovine peripneumonia and avian diphtheria. He was awarded the 1919 Nobel prize for Physiology or Medicine.

BORN, Max
(1882–1970)

German physicist, born in Breslau (now Wroclaw, Poland). He was educated at the universities of Breslau, Heidelberg, Zurich and Göttingen, and was appointed Professor of Theoretical Physics at Göttingen (1921–1933), lecturer at Cambridge (1933–1936) and Professor of Natural Philosophy at Edinburgh (1936–1953). In 1925, with his assistant Pascual Jordan, he built upon the earlier work of Heisenberg to produce a systematic quantum theory based upon matrix mechanics. He showed that the waves in Schrödinger's wave equation were probability waves, so that the state of a particle (e.g. its energy or position) could only be predicted in terms of probabilities. From this he deduced the existence of quantum jumps between discrete states, such as when an excited electron in high energy level of an atom 'falls' back to a lower level with the emission of a photon. This led to a statistical approach to quantum mechanics. He shared the 1954 Nobel prize for Physics with Bothe for their work in the field of quantum physics.

BOSE, Sir Jagadis Chandra
(1858–1937)

Indian physicist and botanist, born in Mymensingh, Bengal (now in east Pakistan), the son of a deputy magistrate. After his schooling at St Xavier's, a Jesuit College in Calcutta, he went to London to study medicine, but transferred to Cambridge University when he was awarded a scholarship and graduated in natural science in 1884. He returned to Calcutta where he was appointed Professor of Physics at Presidency College. Bose became known for his study of electric waves, their polarisation and reflection, and for his experiments demonstrating the sensitivity and growth of plants, for which he designed an extremely sensitive automatic recorder. In some of his ideas he foreshadowed Wiener's cybernetics. He founded the Bose Research Institute in Calcutta for physical and biological sciences in 1917, was knighted in the same year and became the first Indian physicist to be elected FRS (1920).

BOSE, Satyendra Nath
(1894–1974)

Indian physicist, born in Calcutta and educated at Presidency College there. He became professor at Dacca University before being appointed to another chair at Calcutta in 1952. Later he was appointed National Professor by the Indian government (1959) and President of the National Institute of Sciences of India (1949–1950). In 1924 he succeeded in deriving the Planck black-body radiation law, without reference to classical electrodynamics. Einstein generalised his method to develop a system of statistical quantum mechanics, now called Bose-Einstein statistics, for integral spin particles for which Dirac coined the term 'bosons'. Bose also contributed to the studies of X-ray diffraction and the interaction of electromagnetic waves with the ionosphere.

BOWDITCH, Henry Pickering
(1840–1911)

American physiologist, born in Boston into an old, genteel New England family. Bowditch entered Harvard in 1857, interrupting his studies to serve in the American Civil War (1961–1965), and being wounded in battle in 1863. He obtained his MD from Harvard in 1868 and spent three years in Europe, studying experimental physiology and microscopy first in France (where he decried the physiology teaching) and then in Germany with such teachers as Ludwig. Returning to America, he obtained a teaching post in Harvard in physiology, and sought to introduce German laboratory and research methods. Bowditch went on to produce important experimental work on cardiac contraction, on the innervation of the heart, on the vascular system, and on the reflexes, including the knee-jerk. He later became a pioneer of anthropometry, emphasising the importance of nutrition and environment (rather than heredity) in child growth. He helped build up the Harvard physiology department, instituting important reforms in medical education while Dean of the Harvard Medical School (1883–1893). He was a founder of the American Physiological Society (1887).

BOYLE, Robert
(1627–1691)

Anglo-Irish experimental philosopher and chemist, born in Lismore Castle, Munster. The 14th child of the Earl of Cook, Boyle was raised amongst the Anglo-Irish aristocracy. He studied at Eton from 1635 and three years later commenced a Grand Tour of Paris, Geneva and Florence. He arrived home in the summer of 1644 to find his father dead. Settling on his inherited estate of Stalbridge, Dorset, he combined studies in natural philosophy with the composition of moral and religious essays. From 1645 Boyle took an active part in the meetings of the anti-scholastic Invisible College, precursor of the Royal Society of London. He returned to Ireland briefly, dabbled in anatomical dissection (1652–1654), and then moved to Oxford where he created an experimental laboratory. Ably assisted by Hooke, later Royal Society curator of experiments, Boyle enlisted a versatile new air-pump (the 'machina Boyleana', c.1659) to investigate the characteristics of air and of the vacuum: these investigations appeared in *New Experiments Physico-Mechanicall, Touching the Spring of the Air* (1660). A second edition (1662), responding to criticisms of Thomas Hobbes and others concerning the existence and nature of the vacuum, included 'Boyle's law' which states that the pressure and volume of gas are inversely proportional. The air-pump became a powerful symbol of the new 'experimental philosophy' promoted by the Royal Society from its foundation (1660). Boyle's voluminous writings exemplified the society's experimental programme but were, to some, intolerably prolix and invited caricature, notably from Jonathan Swift. *The Sceptical Chemist* (1661) attacked both alchemical principles (salt, sulphur and mercury) and Aristotelian elements (earth, water, air and fire). Boyle proposed elements which were primitive and simple, perfectly unmingled bodies: the practical limit of analysis and constituents of all things. The *Origin of Forms and Qualities* (1666) elaborated this corpuscular 'mechanical philosophy', wherein natural phenomena were explained in terms of matter in motion; shape, number and motion of particles alone determined the diverse properties of substances. Boyle's researches in hydrostatics, the chemistry of colours, calcination of metals, properties of acids and alkalis, crystallography and refraction all served to demonstrate this approach. In 1668 he took up residence with his sister, Lady Ranelagh, in London, remaining there for the rest of his life. As a director of the East India Company and Governor of the Society for the Propagation of the Gospel in New England since 1661, he worked for the diffusion of Christianity, circulating at his own expense translations of the Scriptures. He endowed the Boyle Lectures, 'for proving the Christian Religion against notorious Infidels'.

BRAGG, Sir (William) Lawrence
(1890–1971)

Australian-born British physicist, born in Adelaide, son of Sir William Bragg. He was educated at Adelaide University from the age of 15, and Trinity College, Cambridge, where he discovered the Bragg law (1912), which describes the conditions for X-ray diffraction by crystals: $n\lambda = 2d \sin \theta$, where λ is the X-ray wavelength, d is the separation of the layers of atoms forming the crystal planes, θ is the angle of incidence of the X-rays and n is an integer. This was a significant advance in theoretical chemistry, leading to the first method of studying the exact positions of atoms in crystal interiors. He later collaborated with his father in the study of crystals by X-ray diffraction and continued it as a professor at Manchester and then at Cambridge (from 1938). Like his father, he became Director of the Royal Institution (1954–1965) and did much to popularize science. They shared the 1915 Nobel Prize for Physics. Lawrence Bragg became Professor of Physics at Victoria University, Manchester (1919–1937), and succeeded Rutherford as head of the Cavendish laboratory in Cambridge (1938–1953). There he supported Crick and James Watson in their work using X-ray crystal studies to deduce the helical structure of DNA, so creating molecular biology and revolutionizing biological science. He was knighted in 1941.

BRAHE, Tycho
(1546–1601)

Danish astronomer, the greatest astronomical observer of the pre-telescope era. Born into a noble family at Knudstrup, south Sweden (then under the Danish crown), he studied mathematics and astronomy at the University of Copenhagen and then at famous centres of learning in Germany (1562–1569). Since 1563 when he observed a conjunction of Jupiter and Saturn, he had realised that there were serious errors in the astronomical tables then in use, and that there was a grave need for reliable star positions. He returned to Denmark with a celestial globe acquired in Germany, and constructed his own improved positional instruments. In 1572 a brilliant nova or new star appeared in the constellation of Cassiopeia (now known to have been a super-nova) which Tycho observed for many months, fixing its position with such accuracy that he could declare it to be more distant than the Moon. His publication of this finding in *De Nova Stella* (1573) made his name. In 1576 King Frederick II granted him generous funds to construct and maintain an observatory on the island of Hven in the Sound. This magnificent establishment which Tycho named Uraniborg ('Castle of the Sky') was equipped with exquisite sextants, quadrants and armillary spheres, as well as accurate clocks, described in his *Astronomiae instauratae Mechanica* (1598). Tycho's observations of the comet of 1577 proved that it was more distant than Venus and moved around the sun, contrary to the accepted notion that comets were local atmospheric phenomena. Work at Uraniborg over more than 20 years resulted in Tycho's catalogue of 777 stars and observations of the Sun and planets, with positions given to an accuracy of one or two arcminutes (posthumously published). In 1596 Tycho's royal patron died. Bereft of support, Tycho left Denmark and at the invitation of the emperor Rudolf II settled at the castle of Benatky near Prague (1599), where he set up some of his smaller instruments, though he made little use of them: he died two years later. During that last phase of his life he was joined by a young assistant Kepler who was to make fruitful use of his master's observations. Unlike his younger contemporary Galileo, Tycho Brahe did not subscribe to the Copernican doctrine of a Sun-centred planetary system, arguing with good scientific reasoning that his observations revealed no annual shift in the positions of stars, as would be expected by a moving observer. The explanation, of course, was that such a shift (parallax) was incomparably smaller than the smallest angle he was able to measure. The Tychonic system envisaged the planets – other than the Earth – revolving around the Sun, with that entire assembly revolving around a stationary Earth. Tycho Brahe's great talents were accompanied by a hot-tempered and arrogant manner; he lost most of his nose in a duel at the age of 19 following an argument over a mathematical detail, and wore a false silver nose for the rest of his life. His tomb in Prague is marked by his effigy clad in armour with his right hand resting on his celestial globe.

BRIDGMAN, Percy Williams
(1882–1961)

American physicist and philosopher of science, born in Cambridge, Massachusetts. The son of a journalist, Bridgman entered Harvard in 1900 and remained there, becoming Hollis Professor of Mathematics and Natural Philosophy (1926), Higgins Professor (1950) and, on his retirement, Professor Emeritus (1954–1961). Soon after completing his PhD in 1908 he initiated experiments on the properties of solids and liquids under high pressure, research for which he was awarded the Nobel Prize for Physics in 1946. These difficult practical investigations eventually elevated the pressures attainable from 6,500 to approximately 400,000 atmospheres. Bridgman designed much of his own equipment: a seal which actually improved with higher pressures left the limits of experiment essentially dictated only by container strength. Studying thermal and electrical conductivity, compressibility, phase changes, and the often strange physical properties of liquids and solids under extreme conditions, he obtained a new form of phosphorus and demonstrated that at high pressures, viscosity increases with pressure for most liquids. Much of this research was relevant to geophysics. In a manner closely allied to his dominant experimentalist viewpoint, Bridgman became deeply concerned with the foundations of his subject and the nature of theorising in physics. The 'operationalist' approach, laid out in *The Logic of Modern Physics* (1927), asserted the identity of any concept with the set of operations (physical and mental) involved in its experimental measurement. Bridgman hoped to bring this contentious but fruitful mode of understanding to bear upon relativity theory and quantum theory; and later to the social, political, psychological and religious domains. When in 1961 Bridgman found himself increasingly debilitated and incurably ill with cancer he took his own life.

BRIGHT, Richard
(1789–1858)

English physician, born in Bristol. He studied medicine in Edinburgh, London, Berlin and Vienna. His *Travels from Vienna through Lower Hungary* (1818) recorded interesting observations on gypsies and diplomacy in Vienna at the close of the Napoleonic period. From 1820 he was on the staff at Guy's Hospital, London, where he helped found the *Guy's Hospital Reports*, was a successful teacher, and made many careful clinical and pathological observations. His classic *Reports of Medical Cases* (2 vols, 1827–1831) contain, *inter alia*, his description of kidney disease ('Bright's disease'), with its associated oedema and protein in the urine. He also left important observations on diseases of the nervous system, lungs and abdomen, and was much sought after as a consultant physician, becoming Physician Extraordinary to Queen Victoria. His projected textbook of medicine, in collaboration with Addison, was never completed.

BRITTEN, Roy John
(1919–)

American molecular biologist, born in Washington DC. He graduated with a BS from the University of Virginia in 1941 and received his PhD from the University of Princeton in 1951. Since then he has been a staff member in the Department of Terrestrial Magnetism at the Carnegie Institution of Washington. From 1973 to 1981 he was also Senior Research Associate of Caltech, and he was appointed Distinguished Carnegie Senior Research Associate in Biology in 1981. He is a Fellow of the American Academy of Arts and Sciences, and a member of the US National Academy of Sciences. Britten discovered, with Davidson, repeated DNA sequences in the genomes of higher organisms. Using recently invented nucleic acid hybridization technology, they showed that the genomes of higher organisms contain much more DNA than is required to code for specific genes giving rise to protein. This DNA is organized into unique, single-copy DNA sequences (coding for single genes), moderately repetitive DNA (coding for gene families), and highly repetitive sequences which are repeated hundreds of thousands of times in the genome. The role of this highly repetitive DNA remains unknown, though the positioning of these sequences flanking most genes has led to speculation that they may play a part in the co-ordinate regulation of genes.

BROCA, (Pierre) Paul
(1824–1880)

French surgeon and anthropologist, born in Sainte-Foy-la-Grande, Gironde. Educated at the University of Paris, where he received his MD in 1849. Broca became assistant professor at the Faculty of Medicine in 1953. He first located the motor speech centre in the brain (1861), since known as the convolution of Broca, and did research on prehistoric surgical operations. His anthropological investigations gave strong support to Charles Darwin's theory of the evolutionary descent of man.

BROWN, Michael Stuart
(1941–)

American molecular geneticist, born in New York City. He was educated at the University of Pennsylvania, and since 1977 has been Paul J. Thomas Professor of Generics at the University of Texas and Director of the Center for Genetic Diseases. While a medical intern at Boston General Hospital early in his career (1966–1968), he met Joseph Goldstein, and with Goldstein moved to Texas, where they began to work on cholesterol metabolism. Cholesterol is produced by mammalian cells as well as being taken up into cells from food, and is carried in the bloodstream by proteins called LDLs (low-density lipoproteins). Brown worked on the genetic disease Hypocholesterolemia, which results in abnormally high levels of cholesterol in the bloodstream due to the failure of cells to regulate the production rate. He found that sufferers from the disease lack a receptor on their cell surfaces to which the LDLs bind, thereby stopping the production of cholesterol. In 1984 Brown and Goldstein elucidated the gene sequence which codes for the LDL receptor, and opened up the possibility of synthesizing drugs to control cholesterol metabolism. They were jointly awarded the 1985 Nobel Prize for Physiology or Medicine.

BROWN-SQUARD, Edouard
(1817–1894)

French physiologist. Born in Port Louis, Mauritius, the son of a Philadelphia sea captain and a French mother, Brown-Squard studied at Paris, receiving his MD in 1848. He practised medicine in the USA, and was briefly professor at Virginia Medical College in Virginia. Having for many years divided his time between the USA, London and Paris, and making his living through private practice, he was appointed Professor of Physiology at Harvard (1864), at the School of Medicine in Paris (1869–1873), and the Collège de France from 1878, succeeding Bernard, upon whose research programme Brown-Squard based many of his own investigations. He proved an ingenious physiological researcher, experimenting in particular on blood, muscular irritability, animal heat, the spinal cord, and the nervous system, where his work on the vasoconstrictive nerves proved especially significant. He also demonstrated the artificial production of epileptic states through lesions of the nervous system. A pioneer of endocrinology, he proved that removal of the adrenal glands would always produce death in animals. He claimed to have rejuvenated himself with extracts from freshly-killed dog testicles, thereby paving the way for the later experiments of Voronoff.

BRUNEL, Isambard Kingdom
(1806–1859)

English engineer and inventor, born in Portsmouth, son of Sir Marc Isambard Brunel (1769–1849). In 1823, after two years spent at the Collège Henri Quatre in Paris, he entered his father's office. He helped to plan the Thames Tunnel, and in 1829–1831 designed the Clifton Suspension Bridge, which was completed only in 1864 with the chains from his own Hungerford Suspension Bridge (1841–1845) over the Thames at Charing Cross. He designed the *Great Western* (1838), the first steamship built to cross the Atlantic, and the *Great Britain* (1845), the first ocean screw-steamer (now preserved at Bristol). The *Great Eastern*, until 1899 the largest vessel ever built, was constructed to his design in collaboration (strained at times) with John Scott Russell from whose yard in Millwall the 'Great Ship' was launched at the second attempt in January 1858, three months late, and 40 years ahead of the technology of the time. In 1833 he was appointed engineer to the Great Western Railway, and constructed all the track, tunnels, bridges and viaducts on that line. He chose the 7-foot 'broad gauge' for the track, when almost every other railway at the time had adopted George Stephenson's 4 feet 8½ inches or less, but eventually the Great Western had to conform, and the last broad gauge train ran in May 1892. Among docks and harbours constructed or improved by him were those of Bristol, Monkwearmouth, Cardiff and Milford Haven.

BUNSEN, Robert Wilhelm
(1811–1899)

German chemist and physicist, born in Göttingen. He studied at the University of Göttingen and for his PhD produced a Latin dissertation on hygrometers. He succeeded Wöhler as professor at Kassel, went to Marburg in 1838, and after a short period at Breslau, followed Leopold Gmelin as professor at Heidelberg (1852), where he remained until his retirement. He was a talented experimentalist, although the eponymous burner, for which he is best known, is a modification of something developed (in England) by Faraday. He did invent the grease-spot photometer, a galvanic battery, an ice calorimeter and actinometer. One of his most significant contributions to modern science was his use of spectroscopy to detect new elements. By this technique, and in collaboration with Kirchhoff, caesium and rubidium were discovered in the mineral waters of Durkheim. His most important work was his study of organo-arsenic compounds such as cacodyl oxide. These compounds were seen at the time to give support to the radical theory of chemical combination proposed by Lavoisier and Berzelius. Following the partial loss of the sight of one eye during an experiment, he forbade the study of organic chemistry in his laboratory.

BURKITT, Denis Parsons
(1911–1993)

British surgeon and nutritionist, born in Enniskillen, Northern Ireland, and educated at Dublin University. Service with the Royal Army Medical Corps took him to Uganda, where he worked as a general surgeon after World War II, and began a series of clinical and epidemiological observations on a common childhood cancer found there (from 1957). It behaved as if it were infectious and subsequent research showed that the cancer – now known as Burkitt's lymphoma – was caused by a virus. Burkitt's other major contribution related the low African incidence of coronary heart disease, bowel cancer and other diseases to the high unrefined fibre in the native diet. He became one of the leading apostles of fibre in Western diets, which earned him the nickname 'the bran man'. He was elected FRS in 1976, a rare honour for a surgeon.

BURNET, Sir (Frank) Macfarlane
(1899–1985)

Australian immunologist and virologist, born in Traralgon in eastern Victoria. Trained in medicine at Melbourne University, he graduated in 1922, and proceeded to the higher degree of MD in 1924. After postgraduate work in Melbourne, Burnet moved to London, earning his passage as a ship's surgeon, to work in bacteriological research in the Lister Institute, before returning to Melbourne in 1928 to the Walter and Eliza Hall Institute for Medical Research. He remained there until 1965, becoming Assistant Director in 1934 and Director in 1944. However, he returned to London for three years in 1931, to work at the National Institute for Medical Research under the direction of Dale. Here Burnet began working on viruses, and perfected the technique of cultivating viruses in living chick embryos. For the next 20 years he made important contributions to understanding the chemistry and biology of many animal viruses, especially the influenza virus. His work on viruses and immunization stimulated his interest in immunology, and from the end of the 1950s he turned his attention to immunological problems, especially the phenomenon of graft rejection. His work transformed the understanding of how the entry of foreign substances (antigens) into the body results in the production of specific antibodies which bind and neutralize the invader. Since there is an enormous number of potential antigens, a satisfactory theory was required to account for the manufacture of these highly specific antibodies to deal with each of them. Convinced by the theories of Jerne that all possible antibodies already existed, Burnet postulated his 'clonal selection theory'. This suggested that antibodies were present on specialized white blood cells, the lymphocytes, each of which carried a unique antibody. When the white blood cell bound to an invading antigen, it would be stimulated to reproduce, thus producing large quantities of its unique antibody. He predicted that if an embryo was injected with an antigen, tolerance to it would be induced and no antibodies would subsequently be produced against it. Medawar's work on immunological intolerance in relation to skin grafting and organ transplants provided the experimental evidence to support Burnet's theory, and in 1960 the two men shared the Nobel Prize for Physiology or Medicine. Burnet was elected FRS in 1942, knighted in 1951, and awarded the Order of Merit in 1958.

C

CALVIN, Melvin
(1911–)

American chemist, born in Minnesota of Russian immigrant parents. He became Professor of Chemistry at the University of California (1947–1971) and head of the Lawrence Radiation Laboratory there (1962–1980). In 1948 he helped elucidate the Thunberg-Wieland cycle by which some bacteria, unlike animals, synthesize four-carbon sugars, and hence glucose, from acetate as shown by Konrad Bloch. The idea that the cycle, operating in reverse, might fix carbon dioxide gas led him to investigate this process in photosynthesis (1950). When *Chlorella* a green alga, was exposed to radioactive carbon dioxide in the dark, radioactivity was transferred to succinate, fumarate, malate and other compounds before being found in glucose. A brief spell of illumination caused the radioactivity to appear in the triose phosphates and sugar phosphates that are now associated with the pentose phosphate pathway. The outcome was the Calvin cycle whereby carbon dioxide interacts with ribulose diphosphate giving, via several reactions, various sugar phosphates and regenerating ribulose diphosphate ready for a repeat of the cycle. The pathway occurs in all photosynthesizing organisms. For this work he was awarded the Nobel Prize for Chemistry in 1961.

CAMPBELL, William Wallace
(1862–1938)

American astronomer, born in Hancock County, Ohio. He joined the Lick Observatory in California in 1891, became its director (1901–1930), and was also President of the University of California (1923–1930). He is best known for his work on the radial velocities of stars. This project started in 1896, and in 1928 he published a catalogue of nearly 3,000 radial velocities. These data were subsequently used for the study of galactic rotation. In 1922 he produced confirming evidence for Einstein's general theory of relativity, and the work done by Eddington, when he measured the bending of the beam of starlight that just skims the Sun's surface during a solar eclipse. He led seven expeditions to study solar eclipses, and elucidated the Sun's motion through our galaxy.

CANTOR, Charles Robert
(1942–)

American molecular geneticist, born in Brooklyn, New York City. He was educated at Columbia University and the University of California at Berkeley, where he received his PhD in 1966. He has taught at Columbia University since 1966, and since 1981 has been Chairman of the Department of Genetics and Development at the university's College of Physicians and Surgeons. In 1984 Cantor developed pulse field gel electrophoresis for the separation of very large DNA molecules; this technique has been an essential tool in examining DNA structures at the chromosome level. He is currently Director of the Human Genome Project at the University of California at Berkeley. This worldwide project is designed to completely map the human genome, first by creating a series of manageable chunks with restriction enzymes and then by DNA sequencing. It will therefore be possible to learn the amino acid sequence of genes as yet unknown (possibly important disease-causing entities) and to identify DNA sequences which may be important in regulation of genetic processes.

CARLSON, Chester Floyd
(1906–1968)

American inventor, born in Seattle. He graduated in physics from Caltech in 1930, then took a law degree and worked as a patent lawyer in an electronics firm. He began to experiment with copying processes using photoconductivity and by 1938 had discovered the basic principles of the electrostatic 'xerography' (from the Greek *xeros* meaning 'dry' and *graphein*, 'to write') process. Between 1939 and 1944, 20 companies refused his patent. In 1947 an agreement was reached with a small photographic company, Haloid, which later became Xerox. Around 12 years later (1959) the first copier was brought onto the market, the Xerox 914. The copying process involved using high voltages to electrically charge an insulating sheet of material. This sensitised it to small amounts of light and by focusing an image of an original document onto the sheet a secondary electrostatic image of the page was created. By selectively picking up black pigment, the insulating sheet became the plate from which the copy was printed. Carlson's copiers were marketed worldwide by the Xerox Corporation, who for many years had little competition, making profits of $15 billion in 1987. His invention made him a multi-millionaire.

CARSON, Hampton Lawrence
(1914–)

American evolutionist, born in Philadelphia, Pennsylvania. Educated at the University of Pennsylvania, he began his career at Washington University, where he was later appointed professor (1956–1971). From 1971 to 1985 he was Professor of Genetics of the University of Hawaii. Carson is best known for his work on the evolution of new species in Hawaiian *Drosophilidae* (the fruit fly). He developed the idea of the 'founder effect' (originally proposed by Mayr), using chromosomal inversions as markers of relationships, devising probable phylogenetic lineages. These ideas were summarized in a *festschrift, Genetics, Speciation and the Founder Principle* (1989).

CAVENDISH, Henry
(1731–1810)

English natural philosopher and chemist, born in Nice. Born into an aristocratic family, Cavendish was educated at Peterhouse College, Cambridge (1749–1753). Leaving without a degree, he lived in Paris for a year and then, returning to London, equipped a laboratory and library for his own and others' use. He was an active and well-known Fellow of the Royal Society from 1760. Family wealth, augmented through a substantial inheritance at the age of 40, enabled him to devote an increasingly reclusive life entirely to scientific pursuits. In 1765 he investigated specific and latent heats, obtaining results which were only published much later. Soon afterwards he demonstrated chemical and physical methods for analysing the distinct 'factitious airs' of which normal atmospheric air was composed (1766). Amongst these were 'fixed air' (carbon dioxide), and 'inflammable air' (hydrogen) which Cavendish isolated. Interpreting his findings within the traditional phlogiston theory of combustion first propounded by Georg Stahl, he believed this 'inflammable air' to be phlogiston itself. In 1784 he ascertained that hydrogen and oxygen, when caused to explode by an electric spark, combined to produce water which could not therefore be an element. Similarly, in 1795 he showed nitric acid to be a combination of atmospheric gases. Furthermore, a small proportion of the air remained after prolonged sparking, an inert fraction much later identified as the element argon. The famous 'Cavendish experiment' (1798) employed a torsion balance apparatus devised by Michell to estimate with great accuracy the mean density of the Earth and the universal gravitational constant. Only a small part of Cavendish's researches were made known during his lifetime. In 1771 a theoretical study of electricity had appeared and later Cavendish ingeniously confirmed the inverse square law of attraction, but it was not until 1879 that his electrical manuscripts (covering statics and dynamics) were edited and published by Maxwell. The Cavendish laboratory (established 1871) in Cambridge was named in his honour.

CELCIUS, Anders
(1701–1744)

Swedish astronomer, born in Uppsala. Celsius taught mathematics and became Professor of Astronomy at the University of Uppsala (1730) where his grandfather, father and uncle had all held academic positions. Between 1732 and 1736 he travelled widely in Europe, visiting many centres of observational astronomy. While in Nuremberg he published an aurora borealis compendium (1733). In 1736 he took part in an expedition organized by the Paris Academy of Sciences to measure an arc of meridian at a northern latitude, showing the flattening of the Earth's poles. He also published speculations on the distance of the Earth from the Sun, and measured the relative brightness of stars. Celsius was responsible for the construction (1740) and subsequent direction of the Uppsala observatory. The Celsius temperature scale originated with a mercury thermometer, described by him in 1742 before the Swedish Academy of Sciences. Two fixed points had been chosen: one (0 degrees) at the boiling point of water, the other (100 degrees) at the melting point of ice. A few years after his death, colleagues at Uppsala began to use the familiar inverted version of this centigrade scale.

CELSUS, Aulus Cornelius
(First century AD)

Roman writer and physician. He compiled an encyclopaedia on medicine, rhetoric, history, philosophy, war and agriculture. The only extant portion of the work is the *De Medicina*, rediscovered by Pope Nicholas V and one of the first medical works to be printed (1478). In it Celsus gives accounts of symptoms and treatments of diseases, surgical methods and medical history.

CHAGAS, Carlos Ribeiro Justiniano
(1879–1934)

Brazilian physician and microbiologist born in Oliveira, Minas Gerais. He studied at the Medical School of Rio de

Janeiro, where he was introduced to the concepts and techniques of bacteriology and scientific medicine. After a few years in private practice, Chagas joined the staff of the Oswaldo Cruz Institute, where its founder and leading light, Oswald Cruz, befriended him. Much of his early work was concerned with malaria prevention and control. During one of his field missions, in Lassance, a village in the interior of Brazil, he first described a disease (Chagas' disease, or 'sleeping sickness') caused by the trypanosome (he named the organism *T Cruzi* after Cruz). Chagas elucidated its mode of spread through an insect vector, established the trypanosome's virulence in laboratory animals, and described its acute and chronic course in human beings. On circumstantial evidence, some historians have suggested that Charles Darwin suffered from Chagas' disease.

CHAMBERLAIN, Owen
(1920–)

American physicist, born in San Francisco. He was educated at Dartmouth College and at the University of Chicago, where he received his doctorate in 1949. He became professor at the University of California after working on the Manhattan atomic bomb project (1942–1946) and at the Argonne National Laboratory. The existence of antimatter had been predicted by Dirac's theory, and the first antiparticle (the anti-electron, or positron) had been discovered in 1932 by Carl Anderson. In 1955, Chamberlain, Segrè and their colleagues set up an experiment in which the anti-protons could be identified by their time of flight between two scintillators and by Cherenkov counters. They discovered a negatively charged particle with a mass very close to that of the proton. In a later experiment with collaborators from Rome, they were able to prove that these particles annihilated with protons, confirming that they were indeed anti-protons. In 1959 Chamberlain and Segrè were awarded the Nobel Prize for Physics for the discovery of the anti-proton.

CHANDRASEKHAR, Subrahmanyan
(1910–)

Indian-born American astrophysicist, born in Lahore (now in Pakistan), nephew of Raman. He was educated at the Presidency College, Madras before going to Cambridge University where he studied under Dirac. In 1936 he moved to the USA to work at the University of Chicago and Yerkes Observatory. He showed that in the final stages of stellar evolution the fate of a star depends on its mass. Low-mass stars may collapse to form white dwarfs, small hot stars, in which material is compressed to densities millions of times that of ordinary matter. Chandrasekhar demonstrated the surprising property that the larger the mass of a white dwarf, the smaller its radius. He also concluded that stars with masses greater than about 1.4 solar masses will be unable to evolve into white dwarfs and this limiting stellar mass, confirmed by observation, is known as the Chandrasekhar limit. He suggested that if the mass of a star is greater than this, it can become a white dwarf star only if it ejects its excess mass in a supernova explosion before collapse. For his work on the late evolutionary stages of massive stars he was awarded the 1983 Nobel Prize for Physics, jointly with William Fowler.

CHAPMAN, David Leonard
(1869–1958)

English physical chemist, born in Wells, Norfolk. He studied at Christ Church

Oxford, under Augustus Vernon Harcourt, graduating in chemistry in 1893 and in physics in 1894. After a short period of school teaching, he joined the chemistry staff at Owen's College, Manchester under Dixon in 1897. In 1907 he became head of the Sir Leoline Jenkins Laboratory of Jesus College Oxford, a post he held until his retirement in 1944. He was elected FRS in 1913. Chapman's earliest contributions to physical chemistry were theoretical treatments of explosion velocities in gases (using Dixon's measurements) and of electrocapillarity. Some of his equations for the former were later derived independently by Emile Jouguet and the region behind a detonation wave is known as the 'Chapman-Jouguet layer'. The electrical double layer considered in his theory of electrocapillarity is known as the 'Gouy-Chapman layer'. Chapman's main work, however, was in gas kinetics and in much of this he was ably assisted by his wife Muriel. They made important studies of the thermal and the photochemical reactions between hydrogen and chlorine. During the photochemical work Chapman devised the rotating sector method for measuring the lifetimes of chain carriers, a technique which has been widely used. He introduced the steady-state treatment into kinetics in 1913. This was later used extensively by Bodenstein, who is often credited with its invention.

CHARCOT, Jean Martin
(1825–1893)

French pathologist and neurologist, born and educated in Paris, and appointed to a position at Salpêtrière Hospital. Of working-class stock, Charcot advanced through the ranks to become the most eminent French physician of his day. His early investigations were devoted to chronic diseases such as gout and arthritis, and the diseases of old age. Increasingly, however, he turned the Salpêtrère into an international centre for the investigation of neurological diseases during his active medical service there, himself making important observations on multiple sclerosis, amyotrophic sclerosis and familial muscular atrophy. These were summarized in his *Leçons sur les maladies du système nerveux* (5 vols, 1872–1893). His 'Tuesday lessons' attracted young doctors from all over the world. During the last two decades of his life, he became intrigued by hysteria and other functional disorders, and began using hypnosis in their diagnosis and treatment. He developed the notion that ideas can cause functional diseases and argued that normal people cannot be hypnotized. Thus, hypnosis could be useful as a diagnostic aid, as well as having therapeutic possibilities. He displayed to his classes the same young women frequently; it is now known many of them were gifted actresses who could feign signs at will. His lectures stimulated the young Sigmund Freud, who also translated some of Charcot's work into German. Charcot was a powerful medical politician who had many disciples; he was also a gifted artist whose pathological illustrations grace many of his works.

CHARNLEY, Sir John
(1911–1982)

British orthopaedic surgeon, born in Bury Lancashire, and educated at Manchester University. He served as an orthopaedic specialist during World War II, then returned to the Manchester Royal Infirmary, soon devoting himself to the technical problem associated with replacing badly arthritic hip joints. He worked on animals in a search for suitable material with which to make the artificial joints. An initial operative series using artificial joints made of Teflon gave unsatisfactory long-term results, but from 1962 he used polyethylene, and this prove highly functional. With a good cementing material, and scrupulous attention to aseptic technique, he perfected the operation which has given enhanced mobility to many people. He also pioneered other joint operations. He had great insight into the mechanical dimensions of skeletal movement, and combined the engineering and surgical approaches to his craft with much skill and enthusiasm. Elected FRS in 1975, he was knighted in 1977.

CHORLEY, Richard John
(1927–)

English geomorphologist, born in Minehead, Somerset. Educated at Exeter College Oxford, he received a Fulbright scholarship to study geology at Columbia University. After various lecturing appointments in North America and the UK, he became a reader (1970–1974) at Cambridge University, where since 1974 he has been Professor of Geography. He is a leader of the group which challenged traditional geography and led to the British phase of the so-called 'quantitative revolution'. He used general system theory in the study of landforms, advocated geography as human ecology and developed the use of models in explanation. His publications include the *History of the Study of Landforms* (2 vols, 1964, 1973), *Physical Geography* (1971) *Environmental Systems* (1978) and *Geomorphology* (1984).

CLARKE, Bryan Campbell
(1932–)

English geneticist and evolutionist, born in Gatley, Cheshire. He was educated at the University of Oxford and taught at Edinburgh University. Since 1971 he has been Professor of Genetics at Nottingham University. Clarke is distinguished for studies of the ecological genetics of terrestrial snails (especially *Cepaea nemoralis*) and the elucidation of the evolution of new species in *Partula* on the Pacific Island of Moorea. He made considerable contributions to the understanding of genetical processes in natural populations of animals, refuting that most inherited variation is neutral to its possessors. This re-established the neo-Darwinian interpretation of evolution in the 1970s, following controversy produced by the discovery (through the application of electrophoresis introduced by Lewontin to the study of protein variation in population samples) of apparently excessive amounts of inherited variation. Clarke also developed the concept of frequency-dependent natural selection. He was elected FRS in 1981.

COCKERELL, Sir Christopher Sydney
(1910–)

English radio-engineer and inventor of the hovercraft, born in Cambridge. He graduated in engineering from Cambridge in 1931, and after a short period with an engineering firm in Bedford returned to Cambridge to study radio engineering, which he had become interested in as a hobby. He joined the Marconi company in 1935, and was engaged in the design of VHF transmitters and direction finders. He continued working in that field, and later on the development of radar, during World War II. In 1950 he left Marconi and established a boat hire business on the Norfolk Broads, but it was not long before his inventive mind was working on the problem of reducing the drag on a ship's hull by means of air lubrication. It was not a new concept, but Cockerell was the first to devise a means of making it a practical proposition. After many unsuccessful attempts, in 1955 he built a balsawood model powered by a model aircraft engine which could reach a speed of 12 miles per hour over land or water. Two years later he realized that a flexible skirt would be the answer to the problem of keeping a stable cushion of air under a hovercraft riding the waves of the open sea. After many difficulties in generating the necessary backing, the prototype SR-N1, built by Saunders-Roe of flying-boat fame, and weighing only 7 tonnes, made the crossing of the English Channel in July 1959. Since 1974 Cockerell has also been actively interested in the commercial development of wavepower devices. He was elected FRS in 1967 and knighted in 1969.

COHEN, Stanley
(1922–)

American biochemist, born in Brooklyn, New York City. Educated at Brooklyn College, Oberlin College in Ohio, and the University of Michigan, he held posts at the universities of Colorado and Washington before moving to Vanderbilt University in 1959. From 1967 to 1986 he was Professor of Biochemistry there. Following Levi-Montalchi's discovery of the substance now known as nerve growth factor that promotes the development of sympathetic nerves, Cohen contributed to the isolation of the compound. As a development of this work he went on to isolate a further cell growth factor, named epidermal growth factor (EGF), which was found to accelerate some aspects of natural development in newborn mice. In further studies he demonstrated the wide range of effects of this compound on various developmental processes in the body, and elucidated the mechanisms by which it is absorbed by and interacts with individual cells. In 1986 he was awarded the Nobel Prize for Physiology or Medicine jointly with Levi-Montalcini for their wok on growth factors.

COPERNICUS, Nicolas
(1473–1543)

Polish astronomer, born in Toru, Prussia (now in Poland), and brought up after his father's death (1483) by his uncle, later bishop of Ermeland. After studying mathematics at the University of Cracow (1491–1494) he went to Italy (1496) where he studied canon law and heard lectures on astronomy at the University of Bologna, while at Padua he studied medicine (1501–1505). He was made a doctor of canon law by the University of Ferrara (1503), and though nominated a canon at the cathedral of Frombork (1497), he never took holy orders. Home in Poland he was his uncle's medical adviser and had various administrative duties at Frombork, where he spent the rest of his life. Beginning while in Italy he pondered deeply on what he considered the unsatisfactory Ptolemaic description of the world, in which the Earth was the stationary centre of the universe, and became converted to the idea of a Sun-centred universe. He set out to describe this mathematically in 1512. Copernicus hesitated to make his work public, having no wish to draw criticism from Aristotelian traditionalists or from theologians such as Martin Luther who had ridiculed him, but was eventually persuaded by his disciple Rheticus to publish his complete work, *De Revolutionibus Orbium Coelestium* (1543), which he dedicated to Pope Paul III. In the new system, the Earth became merely one of the planets, revolving around the Sun and rotating on its axis. The absence of any apparent movement of the stars caused by the Earth's annual motion was interpreted as due to the great size of the sphere of the stars. The transfer of the centre of the system from the Earth to the Sun in the new arrangement greatly simplified the geometry of the planetary system, although it did not get rid of all the epicycles of Ptolemy's model, a step which had to await Kepler. Copernicus was already old and ill by the time the book was printed, and was unaware that it carried an anonymous and unauthorized 'Preface to the Reader', presenting the work as an hypothesis rather than a true physical reality, written by Andreas Osiander, a Lutheran pastor of Nuremberg who supervised the last stages of the printing Osiander's misguided intention was to forestall criticism of the heliocentric theory. The first printed copy of Copernicus's treatise, a work which fundamentally altered man's vision of the universe, reached its author on his death bed. Later banned by the Catholic church, *De Revolutionibus* remained on the list of forbidden books until 1835.

CORI, Carl Ferdinand
(1806–1984)

Czech-born American biochemist born in Prague, who shared the 1947 Nobel Prize for Physiology or Medicine with his wife Gerty Cori (and Houssay) – only the third husband and wife team to do so after the Curies in 1903 and the Joliot-Curies in 1935. They married in Prague after he graduated in medicine there in 1920. In 1922 they emigrated to the USA, and both became professors at Washington University in St Louis from 1931. As with Hans Krebs, accurate quantitating was the key to Carl Cori's success. In 1936 he isolated glucose 1-phosphate from frog muscle and showed that this ester was formed by reaction of inorganic phosphate with the storage polysaccharide, glycogen. He found the enzyme that catalysed this reaction, glycogen phosphorylase, in muscle, heart, brain and liver. He obtained it in crystalline form in 1942, and recognized that it had both inactive and active forms and required a prosthetic group, adenylic acid (later shown by Emil Fischer, Edwin Krebs and Earl Sutherland to be in allosteric activator). Cori's theory that the same enzyme catalysed glycogen synthesis was proved incorrect, however, by Leloir. Cori also identified the isomerase that converts glucose 6-phosphate to glucose 1-phosphate, and in 1951 described an alpha-1, 6-glucosidase, important in removing side chains in glycogen breakdown, and used it to determine the length of the main chain and side branches of several polysaccharides. After Gerty's death, Carl Cori worked at the Massachusetts General Hospital from 1967.

CORNFORTH, Sir John Warcup
(1917–   )

Australian chemist, born in Sydney. He entered Sydney University in 1933 to study chemistry. One year of postgraduate work in Sydney followed; he won a scholarship for further work at Oxford with Robinson and obtained his doctorate in 1941. He took part in the wartime effort to synthesize the new drug penicillin. He also studied the biosynthesis of cholesterol and other steroids. In 1962 he and George Popják were appointed co-directors of the Milstead Laboratory of Chemical Enzymology of Shell Research Ltd; Cornforth was also appointed professor at Warwick University. Cornforth and Popják studied in detail the stereochemistry of the interaction between an enzyme and its substrate. They also studied biological oxidation-reduction reactions. In 1968 Popják moved to the University of California and an extremely fruitful collaboration was terminated. Cornforth has also collaborated extensively with Hermann Eggerer on the stereochemistry of enzyme action, developing the chiral methyl group. In 1975 he was awarded the Nobel Prize for Chemistry, which he shared with Prelog for all his work on the chemistry of enzyme action. During the period 1974–1982 he was Royal Society Fellow at the University of Sussex, and he was knighted in 1977. At the age of 10 he had developed otosclerosis and within a decade, became totally deaf. The triumph of his work, in spite of this handicap is due in no small measure to the support and assistance of his wife.

COULOMB, Charles Augustin de
(1736–1806)

French physicist, born in Angoulême into a wealthy family. During his youth the family moved to Paris. An argument over career plans with his mother caused Coulomb to follow his father to Montpellier where his father had become penniless through financial speculations, but he later returned to Paris to complete his education at the École du Génie. After a long period of service in the Corps du Génie in Martinique, he returned to France in 1779, and held several public positions which he gave up at the outbreak of the Revolution. He was forced to leave Paris but returned in 1795 when he was elected a member of the new Institut de France and was appointed Inspector-General of Public Instruction (1802–1806). His experiments on mechanical resistance resulted in 'Coulomb's law' of the relationship between friction and normal pressure (1779), but he has become best known for the torsion balance for measuring the force of magnetic and electrical attraction (1784–1785). With 'Coulomb's law' he observed that the force between two small charged spheres is proportional to the product of the charges divided by the square of the distance between them. The SI unit of quantity of charge is named after him.

COURNAND, André Frédéric
(1895–1988)

French-born American physician, born in Paris. He was educated at the Sorbonne emigrating to the USA in 1930, where he became a citizen in 1941. A specialist in cardiovascular physiology, he was awarded the Nobel Prize for Physiology or Medicine in 1956 jointly with Forssman and Dickenson Richards for developing cardiac catheterization. This consisted of threading a catheter through an arm or leg vein into the right atrium of the heart, so that blood samples could be taken, and recordings made. The technique made it possible to study heart functions in health and disease, and modifications of it are now important in treating heart disease. From 1934, he was on the academic staff of Columbia University.

CRICK, Francis Harry Compton
(1916–)

English molecular biologist, born near Northampton. He received a BSc in physics at University College London. During World War II he worked on the construction of mines for the British Admiralty, and after the war he joined the laboratory of Perutz, to work on the structure of proteins. In the early 1950s, in Cambridge, he met James Watson and together they worked on the structure of DNA. 1953 saw the publication of their model of a double-helical molecule, consisting of two chains of nucleotide bases (adenine, thymine, guanine and cytosine) wound round a common axis in opposite directions. The structure they proposed suggested a mechanism for the reproduction of the genetic material and the genetic code, on which Crick continued to work for the next decade. He worked at the Laboratory of Molecular Biology, Cambridge, from 1949 to 1977, when he became Kieckhefer Professor at the Salk Institute, California. There he carried out research into the visual systems of mammals, and the connections between brain and mind. With Watson and Wilkins he was awarded the Nobel Prize for Physiology or Medicine in 1962. His autobiography *What Mad Pursuit*, was published in 1988.

CRILE, George Washington
(1864–1943)

American surgeon and physiologist, born in Chili, Ohio. He received a BA from Northwestern Ohio Normal School in 1884 and an NM from the University of Wooster in 1887. He became interested in surgical shock (abnormally low blood pressure) when a friend died following an emergency operation after an accident. While building up a busy practice he began animal experiments on surgical shock and published his first monograph on the subject in 1899. In 1900 he became Clinical Professor and in 1911 Professor of Surgery at the Western Reserve School of Medicine. He was founder and first director of the Cleveland Clinic Foundation (1921–1940). Crile continued working on surgical shock, publishing *Blood Pressure in Surgery* (1903) and *Anemia and Resuscitation* (1914). He viewed the prevention of shock as the most important principle and advocated atraumatic surgery using safe anaesthetics. He emphasized the need to monitor blood pressure during surgery, popularizing the apparatus used for this, and was one of the first to regularly use blood transfusions and adrenaline as means of combating shock. He devised a 'shockless' method of anaesthesia (which he called 'anoci-association') which aimed to separate the operative site from the nervous system. This gave excellent results although based on the erroneous 'kinetic theory' – the idea that surgical shock originates in the nervous system. He developed several operations for the endocrine glands, although some of his later physiological speculations were rather eccentric.

CURIE, Marie (originally Marya), née Sklodowska
(1867–1934)

Polish-born French radiochemist, born in Warsaw. She was brought up in poor surroundings after her father, who had studied mathematics at the University of St Petersburg, was denied work for political reasons. After brilliant high school studies, she worked as a governess for eight years, during which time she saved enough money to send her sister to Paris to study. In 1891 she also went to Paris where she graduated in physics from the Sorbonne (1893) taking first place; she then received an Alexandrovitch Scholarship from Poland which allowed her to study mathematics. Marie met Pierre Curie in 1894 and they married the following year. In 1896, Antoine Henri Becquerel had discovered the radioactive properties of uranium; Marie Curie decided to study this phenomenon for her doctoral thesis topic. She used an apparatus for measuring very small electrical currents, built by her husband, to search for elements that emitted ionizing radiations. In this way she discovered that thorium is also radioactive, and she showed that the radioactivity of uranium was an atomic property, rather than the result of interactions between the elements and another substance. In subsequent research she discovered that the radioactivity of the minerals pitchblende and chalcolite was more intense than could be explained by the uranium and thorium content alone, and from this deduced that these minerals must contain new radioactive elements. Pierre left his work on piezoelectricity to help in the laborious process of isolating the new elements by fractional crystallization. No precautions against radioactivity were taken, as the harmful effects were not known at that time. In 1898 Pierre and Marie announced the discovery of a new element, which they named polonium in honour of Marie's native country, and later the same year they announced the discovery of radium. In 1903 Marie presented her doctoral thesis (the first advanced scientific research degree to be awarded to a woman in France), and in the same year she was awarded the Nobel Prize for Physics with Pierre and Becquerel for their work on radioactivity. It was around this time that the Curies began to suffer from symptoms later ascribed to radiation sickness. In 1904 Pierre was awarded a new chair in physics at the Sorbonne, but in 1906 he was killed in a street accident; Marie succeeded him as Professor of Physics. She continued her work with radioactivity and in 1911 was awarded the Nobel Prize for Chemistry for her discovery of polonium and radium. During World War I she developed X-radiography and then became director of the research department at the newly established Radium Institute in Paris (1918–1934). She died of leukaemia, probably due to her long exposure to radioactivity. Her daughter Irine Jolio-Curie and son-in-law Fridiric Joliot-Curie followed in her footsteps in radiochemistry and also received the Nobel Prize for Chemistry.

CURIE, Pierre
(1859–1906)

French physicist, born in Paris and educated there at the Sorbonne, where he became an assistant teacher in 1878. He was appointed laboratory chief at the School of Industrial Physics and Chemistry in 1882, and remained there until in 1904 he was appointed to a new chair in physics at the Sorbonne. In studies of crystals with his brother Jacques, he discovered piezoelectricity in 1880; they observed a small electrical current being produced when certain crystals were mechanically deformed, and vice versa. Such crystals have found many uses in modern technology. The Curies used a piezoelectric crystal to construct an electrometer, capable of measuring very small electric currents, and this was later used by Pierre's wife Marie Curie in her investigations of radioactive minerals. In studies of magnetism, Pierre showed that a ferromagnetic material loses this property at a certain temperature (the Curie point) specific to the substance involved; for this work on magnetism he was awarded his doctorate in 1895. Another of his important results in magnetism was 'Curie's law', that the magnetic susceptibility of a paramagnetic material is inversely proportional to the absolute temperature. From 1898 he worked with his wife on radioactivity, and showed that the rays emitted by radium contained electrically positive, negative and neutral particles. With his wife and Antoine Henri Becquerel he was awarded the Nobel Prize of Physics (1903). He was killed in a street accident shortly after he accepted a new chair in physics at the Sorbonne.

D

DAGUERRE, Louis Jacques Mandé
(1789–1851)

French photographic pioneer, inventor of the 'daguerreotype'. Born in Cormeilles, he was a scene painter for the opera in Paris. From 1826 onwards, and partly in conjunction with Niepce, he perfected his daguerreotype process in which a photographic image is obtained on a copper plate coated with a layer of metallic silver sensitized to light by iodine vapour. This greatly reduced the exposure time required to produce an image from around eight hours for Niepce's original method to around 25 minutes.

DALE, Sir Henry Hallett
(1875–1968)

English physiologist and pharmacologist, born in London, Dale studied natural sciences at Trinity College, Cambridge, and after a brief period in physiological research with John Langley, qualified in medicine from St Bartholomew's Hospital in 1902. He worked for Starling and Bayliss at University College London before accepting a position in 1904 at the Wellcome Physiological Research Laboratories, private laboratories associated with the pharmaceutical firm Burroughs, Wellcome and Co. Around 10 years later, Dale joined the newly created Medical Research Committee (later Council) as Head of the Department of Biochemistry and Pharmacology at the National Institute for Medical Research (NIMR). Retaining his research commitments he was also appointed first Director of the NIMR in 1928, and retired in 1942. He was President of the Royal Society during 1940–1945, when he also served as Secretary to the Scientific Advisory Committee to the War Cabinet. His work focused on the physiology and pharmacology of naturally occurring chemicals, work which had started with the chance observation in 1904 that an extract of ergot of rye, a fungus, reversed the effects of adrenaline. Adrenaline had recently been shown to mimic the effects of the sympathetic nervous system, and in collaboration with his colleague Barger, Dale analysed several compounds for such 'sympathomimetic' activities. Further analysis of ergot with Ewins showed that it contained pharmacologically active compounds, including histamine, tyramine and acetylcholine. Dale's later experiments provided evidence that acetylcholine occurred naturally in animals, and played an important role in the transmission of nerve impulses across the synapse. For this work he shared the 1936 Nobel Prize for Physiology or Medicine with Loewi. Dale also did pioneering work in endocrinology, and closely associated with his research were his wider concerns with the standardization of drugs. He served on many advisory and regulatory committees at both national and international levels, and was knighted in 1932.

DALTON, John
(1766–1844)

English chemist and natural philosopher born in Eaglesfield, near Cockermouth, Cumberland. The son of a Quaker weaver, he received his early education from his father and at the Quaker school at Eaglesfield. Dalton started teaching in the village at the age of 12, but in 1781 he joined his elder brother as a school assistant. He published 'New System of Chemical Philosophy' in 1808 and enunciated that all matter was a combination of atoms which are indivisible. In 1794, he described colour blindness.

DARWIN, Charles Robert
(1809–1882)

English naturalist, born in Shrewsbury, the originator (with Wallace) of the theory of evolution by natural selection. The grandson of Erasmus Darwin and of Josiah Wedgwood, he was educated at Shrewsbury grammar school, studied medicine at Edinburgh University (1825–1827), and then with a view to the church, entered Christ's College, Cambridge, in 1828. Already at Edinburgh he was a member of the local Plinian Society; he took part in its natural history excursions, and read before it his first scientific paper – on Flustra or sea-mats. His biological studies seriously began at Cambridge, where the botanist John Stevens Henslow encouraged his interest in zoology and geology. He was recommended by Henslow as naturalist to HMS *Beagle*, then about to start for a scientific survey of South American waters (1831–1836) under its captain, Fitzroy. He visited Tenerife, the Cape Verde Islands, Brazil, Montevideo, Tierra del Fuego, Buenos Aires, Valparaiso, Chile, the Galapagos, Tahiti, New Zealand, Tasmania and the Keeling Islands: it was there that he started his seminal studies of coral reefs. During this long expedition he obtained the intimate knowledge of the fauna, flora and geology of many lands which equipped him for his later many-sided investigations. By 1846 he had published several works on his geological and zoological discoveries on coral reefs and volcanic islands – work that placed him at once in the front rank of scientists. He formed a friendship with Lyell, was Secretary of the Geological Society from 1818 to 1841, and in 1839 married his cousin, Emma Wedgwood (1808–1896). From 1842 he lived at Downe, Kent, among his garden, conservatories, pigeons and fowls. The practical knowledge thus gained (especially as regards variation and interbreeding) proved invaluable; private means enabled him to devote himself unremittingly, in spite of continuous ill health, to science. At Downe he addressed himself to the great work of his life – the problem of the origin of species. After five years collecting the evidence, he 'allowed himself to speculate' on the subject and in 1842 drew up some short notes, enlarged in 1844 into a sketch of conclusions for his own use. These embodied in embryo the principle of natural selection, the germ of the Darwinian theory; but with constitutional caution, Darwin delayed publication of his hypothesis, which was only precipitated by accident. In 1858 Wallace sent him a memoir on the Malay Archipelago, which, to Darwin's alarm, contained in essence the main idea of his own theory of natural selection. Lyell and Joseph Dalton Hooker persuaded him to submit a paper of his own, based on his 1844 sketch, which simultaneously with Wallace's went before the Linnaean Society on 1 July 1858, neither Darwin nor Wallace being present at that historic occasion. Darwin now set to work to condense his vast mass of notes, and put into shape his great work on *The Origin of Species by Means of Natural Selection*, which was published in November 1859. That epoch-making work, received through Europe with the deepest interest, was violently attacked and energetically defended, but in the end succeeded in obtaining recognition (with or without certain reservations) from almost all competent biologists. From the day of its publication, Darwin continued to work at a great series of supplemental treatises: *The Fertilisation of Orchids* (1862), *The Variation of Plants and Animals under Domestication* (1867), and *The Descent of Man and Selection in Relation to Sex* (1871) which derived the human race from hairy quadrumanous animals belonging to the great anthropoid group, and related to the progenitors of the orang-utan, chimpanzee and gorilla. In it Darwin also developed his important supplementary theory of sexual selection. Later works were *The Expression of the Emotions in Man and Animals* (1873), *Insectivorous Plants* (1875), *Climbing Plants* (1875), *The Effects of Cross and Self Fertilisation in the Vegetable Kingdom* (1876), *Different Forms of Flowers in Plants of the same Species* (1877), *The Power of Movement in Plants* (1880) and *The Formation of Vegetable Mould through the action of Worms* (1881). Though not the sole originator of the evolution hypothesis, nor even the first to apply the conception of descent to plants and animals, Darwin was the first thinker to gain for that concept a wide acceptance among biological experts. By adding to the crude evolutionism of Erasmus Darwin, Lamarck and others his own specific idea of natural selection, he supplied to the idea a sufficient cause, which raised it at once from an hypothesis to a verifiable theory. He also wrote a biography of Erasmus Darwin (1879). He was buried in Westminster Abbey. His son, Sir Francis (1848–1925), also a botanist, became a reader in botany at Oxford (1888) and produced Darwin's *Life and Letters* (1887–1903). Another son, Sir George Howard (1845–1913) was Professor of Astronomy at Cambridge (1883–1912), and was distinguished for his work on tides, tidal friction and the equilibrium of rotating masses.

DAVY, Sir Humphry
(1778–1829)

English chemist, one of the leading scientific figures of his day, noted for his pioneering work in electrochemistry and many other fields. The son of a woodcarver, he was born in Penzance, Cornwall, and was apprenticed to a surgeon and apothecary. In 1797 he began to teach himself physics and chemistry. His talent for both experiment and speculative thought soon became obvious and in 1798 Beddoes employed him as an assistant at the Pneumatic Institute in Bristol. Here Davy experimented with several newly discovered gases and discovered the anaesthetic effect of laughing gas (nitrous oxide). He also showed that heat can be transmitted through a vacuum and suggested that it is a form of motion. In 1799 he published *Researches, Chemical and Physical*, which led in 1801 to his appointment as assistant lecturer in chemistry at the Royal Institution. At once his lectures became hugely popular, his eloquence and the novelty of his experiments attracting large audiences. As a result he gained sufficient funding for research into electrochemistry, a new branch of chemistry made possible by the invention of the electric battery by Volta. Davy decomposed chemical compounds by passing an electrical current through their solutions. This research excited much attention and was later continued by Davy's successor Faraday. Davy also confirmed that water is composed of hydrogen and oxygen by electrolyzing distilled water in a special apparatus which prevented byproducts from forming. He suggested that all chemical activity is electrical, but it was Berzelius who synthesized Davy's work and his own work into a coherent system. In 1807 Davy discovered that the alkalis and alkaline earths are compound substances, formed by oxygen united with metallic bases. He isolated the metals sodium and potassium by electrolysis of fused salts, and barium, strontium, calcium and magnesium by distilling off the mercury from their amalgams. Following up the work of Courtois, he showed that fluorine and chlorine are elements related to iodine. His work on the compounds of chlorine helped to refute Lavoisier's theory that all acids contain oxygen. He also proved that diamond is a form of carbon. Another important area of his work was agriculture. His *Elements of Agricultural Chemistry* (1813) was the first book to apply chemical principles systematically to farming. Davy resigned from the Royal Institution in 1812; in the same year he was knighted and married a wealthy Scottish widow. From 1813 to 1815 the Davys travelled on the Continent taking the young Michael Faraday with them as chemical assistant and valet. In 1815, after investigating the causes of explosions in mines and finding that firedamp (methane) only ignites at high temperatures, Davy invented the celebrated safety lamp which bears his name. A wire gauze surrounding the flame conducts the heat away, thus preventing the danger point from being reached. Coal production thereafter increased markedly as deeper, more gaseous, seams could be mined. Davy's reputation in his lifetime exceeded even his considerable achievements, not the least of which were to popularize science and to interest industrialists in scientific research. He was one of the founders of the Zoological Society which in its turn founded London Zoo. He was made a Baronet in 1812. His last two years were spent on the Continent in failing health and he died in Geneva.

DAWKINS, (Clinton) Richard
(1941–)

British ethologist, born in Nairobi, Kenya. Educated at Oxford, he taught at the University of California at Berkeley before returning to Oxford (1970) where he is a Fellow of New College. His main research interests have been in theoretical modelling of ethology. His major contribution to date is his ability to expound complex evolutionary ideas so as to make them explicable to fellow biologists and laypersons alike. In *The Selfish Gene* (1976), he shows how natural selection can act on individual genes rather than at the individual or species level. He also describes how apparently altruistic behaviour in animals can be of selective advantage by increasing the probability of survival of genes controlling this behaviour. The ways in which randomly occurring small genetic changes or mutations accumulate to form the basis for evolution are discussed in *The Blind Watchmaker* (1986). *The Extended Phenotype* (1982), a more advanced book, argues that genes can have effects outside the bodies that contain them, and explores the implications of this idea for parasitology.

DE DUVE, Christian René
(1917–)

English-born Belgian biochemist, born in Thames Ditton, Surrey. He studied medicine at Louvain, worked with Theorell to determine haemoglobin and myoglobin concentrations in muscle, and worked with Carl Cori on glucose 6-phosphatase and blood sugar regulation. Returning to Louvain (1947) he became Professor of Biochemistry in 1951, and he also held a Chair of Biochemistry at Rockefeller University, New York, from 1962. From 1947 he explored the new technique of differential centrifugation, separating a tissue homogenate into its separate organelles by centrifugation at different speeds. Fortuitously using the enzyme acid phosphatase as a control market while attempting to locate the site of carbohydrate metabolism in the cell, he found that its activity, initially low, increased on storage by being released from a degrading cell organelle. Thus he discovered lysosomes, small organelles which contain acid phosphatase and other degradative enzymes. Lysosome malfunction may result in a metabolic disease, such as cystinosis. De Duve discovered peroxisomes (oxidative organelles) in a similar way. For discoveries relating to the structure and biochemistry of cells he shared the 1974 Nobel Prize for Physiology or Medicine with Albert Claude and Palade. He published a *Guided Tour of the Living Cell* in 1984.

DESCARTES, René
(1596–1650)

French philosopher and mathematician, undoubtedly one of the great figures in the history of Western thought and usually regarded as the father of modern philosophy. He was born near Tours in a small town now called la-Haye-Descartes, and was educated from 1604 to 1614 at the Jesuit College at La Flèche. He did in fact remain a Catholic all his life, and he was careful to modify or even suppress some of his later scientific views, for example his sympathy with Copernicus, following Galileo's condemnation by the Inquisition in 1634. He studied law at Poitiers, graduating in 1616; then from 1618 he enlisted at his own expense for private military service, mainly in order to travel and to have the leisure to think. He later wrote that when in Germany with the army of the Duke of Bavaria one winter's day in 1619 he had an intellectual vision in a 'stove-heated room': he conceived a reconstruction of the whole of knowledge, into a unified system of certain truth modelled on mathematics, based on physics and reaching via medicine to morality, all supported by a rigorous rationalism. From 1618 to 1628 he travelled widely in Holland, Germany, France and Italy; then in 1628 returned to Holland where he remained, living quietly and writing until 1649. Few details are known of his personal life, but he did have an illegitimate daughter called Francine, whose death in 1640 at the age of five was apparently a terrible blow for him. He published most of his major works in this period, the more popular in French and the more scholarly ones first in Latin. The *Discourse de la Méthode* (1637), the *Mediationes de prima Philosophia* (1641) and the *Principia Philosophiae* (1644) set out the fundamental Cartesian doctrines: the method of systematic doubt; the first indubitably true proposition, *cogito ergo sum* ('I think therefore I am): the idea of God as the absolutely perfect being; and the dualism of mind and matter. His theory of astronomy, which explained planetary motion by means of vortices surrounding the Sun, was eventually refuted by Isaac Newton. In mathematics he made his most lasting contribution: he reformed algebraic notation and helped found coordinate geometry. This enables geometrical problems to be reformulated and even solved algebraically. In 1649 he left Holland for Stockholm on the invitation of Queen Christina, who wanted him to give her tuition in philosophy. These lessons took place three times a week at 5 a.m. and were especially taxing for Descartes whose habit of a lifetime was to stay in bed meditating and reading until about 11 a.m. He contracted pneumonia and died. He was buried in Stockholm, but his body was later removed to Paris and eventually transferred to Saint-Germain-des-Prés.

DE VRIES, Hugo Marie
(1848–1935)

Dutch botanist and geneticist born in Haarlem, the son of a Dutch prime minister. The first instructor in plant physiology in the Netherlands, he studied at Leiden, Heidelberg and Würzburg, and became Professor of Botany at Amsterdam (1878–1918). From 1890 he devoted himself to the study of heredity and variation in plants, significantly developing Mendelian genetics and evolutionary theory. His major work was *Die Mutationstheorie* ('The Mutation Theory' 1901–1903). He described and correctly interpreted the phenomenon of plasmolysis, the process whereby the cytoplasm in plant cell shrinks away from the walls as a result of water loss from the cells. He introduced methods for studying osmotic and turgor properties of plant cells which have become standard techniques. In 1885 he showed that the plasmalemma bounding the vacuole in plant cells is semi-permeable, allowing the passage only of small molecules.

DIOPHANTIUS
(fl. third century)

Greek mathematician who lived at Alexandria. Little of his work has survived; the largest work is the *Arithmetica*, which deals with the solution of problems about numbers and, in contrast to earlier Greek work, uses a rudimentary algebraic notation instead of a purely geometric one. For example, one problem asks for two numbers so that in each case the square of the first added to the second gives a square. In a modern version of his notation, one number would be x and the other $2x + 1$. In many problems the solution is not uniquely determined, and these have become known as Diophantine problems. Typically, a Diophantine problem yields an algebraic equation which has only finitely many integer solutions, if any. Diophantus's work was rediscovered in the sixteenth century, and later editions of it inspired Vieta to offer an algebraic account of mathematics which, if historically implausible, was nonetheless significant in early seventeenth-century mathematics. The study of Diophantus's work inspired Fermat, who, well versed in Vieta's ideas, took up number theory in the 1600s with remarkable results.

DOLLOND, John
(1706–1761)

English optician, born in London of Huguenot parentage. A silk weaver by trade, in 1752 he turned to optics, and devoted himself with the help of his son Peter (1738–1820) to the development of an achromatic telescope. Achromatic lenses were first made by Chester More Hall in the period 1729–1733, apparently for his own use, but the first achromatic telescope was made by the Dollonds in 1758, following up a suggestion made to them by Klingenstierna of Uppsala. Hall and some associates brought an action against the Dollonds on the grounds of the priority of their earlier but unexploited work; the action was thrown out by the courts. Early in 1761 John Dollond was appointed Optician to King George III, only a few months before his death.

DONDERS, Franciscus Cornelis
(1818–1889)

Dutch oculist and ophthalmologist born into a talented family from Tilburg, Donders studied at the military medical school at Utrecht and later at Utrecht University; in 1840 he obtained his MD from Leiden, returning to Utrecht where, appointed Extraordinary Professor, he pursued researches in physiological chemistry, notably on plant physiology and the problem of animal and vegetable heat. Aided in part by Helmholtz's invention of the ophthalmoscope, Donders established himself as a specialist in diseases of the eye, setting up a polyclinic for eye diseases at the university. He improved the efficiency of spectacles by the introduction of prismatic and cylindrical lenses, and wrote extensively on eye physiology. In 1862 he was appointed Professor of Physiology at Utrecht.

DOPPLER, Christian Johann
(1803–1853)

Austrian physicist, born in Salzburg the son of a stonemason, he would have followed his father's craft if he had not suffered from ill-health. He showed early mathematical ability. After studying at the Polytechnic Institute in Vienna, he was appointed Professor of Mathematics and Accounting at the State Secondary School in Prague. In 1851 he was appointed Professor of Physics at the Royal Imperial University of Vienna, the first such position to be created in Austria. 'Doppler's principle', which he enunciated in a paper in 1842 when he was Professor of Elementary Mathematics and Practical Geometry at the State Technical Academy in Prague, explains the variation of frequency observed, higher or lower than that actually emitted, when a vibrating source of waves and the observer respectively approach or recede from one another. The first experimental verification was performed in Holland in 1845, using a locomotive drawing an open car with several trumpeters. The Doppler effect applies not only to sound but to all forms of electromagnetic radiation. In the case of astronomy, the changes of the spectral wavelengths of approaching or receding celestial bodies provide important evidence for the concept of an expanding universe.

DOTY, Paul Mead
(1920–)

American biochemist and specialist in arms control. He was educated at Pennsylvania State College, Columbia University and the University of Cambridge, where he graduated BS, MA and PhD respectively. He held short-term posts at the Brooklyn Polytechnic Institute and Notre Dame University, then became assistant professor (1948–1950) and associate professor (1950–1956) at Harvard. He has been full professor there since 1956, and has been Consultant to the Arms Control and Disarmament Agency of the National Security Council since 1973. In 1961 Doty discovered DNA renaturation, establishing the specificity and feasibility of nucleic acid hybridization. A very special feature of DNA is that two of the constituent bases (adenine and guanine) form chemical bonds (hybridize) to the other two bases, thymidine and cytosine. This is the basis for the formation of the double helix and for the transcription of DNA to its mirror image RNA, and allows the diagnostic use of radioactively labelled specific nucleic acids to 'find' their counterpart sequence of DNA or RNA in *vitro*. This powerful technology is the basis for all experimental manipulation of nucleic acids.

DRIESCH, Hans Adolf Eduard
(1867–1941)

German physiologist and philosopher, born in Bad Kreuznach, the son of a wealthy merchant. From 1886 to 1889 he studied zoology at the universities of Freiburg and Jena. He spent the following 10 years travelling and working as an amateur scientist, but during this time conducted his most important experiments, mainly at the Zoological Station in Naples. He is best known for an experiment which was carried out in 1891 involving the separation of the two cells produced by the first division of a fertilized sea-urchin egg. He showed that each cell could form a whole larva. This provided evidence opposed to the theory of preformation – the idea that in the fertilized egg a whole individual already exists in miniature – and helped to open new ways forward for experimental embryology. At this time Driesch favoured mechanistic explanations of development, but by 1895 he was a convinced vitalist. He began to concentrate more on philosophy than science and performed his last experiment in 1909, the year that he joined the faculty of natural science at Heidelberg. He was later appointed Professor of Philosophy at Heidelberg (1911), Cologne (1920) and Leipzig (1921).

DU BOIS-REYMOND, Emil Heinrich
(1818–1896)

German physiologist, and discoverer of neuro-electricity. His father was a Swiss teacher who had settled in Berlin – the family was French-speaking. Talent ran in the family: his brother Paul (1831–1889) was a mathematician, who made contributions to the theory of functions. Du Bois-Reymond studied a wide range of subjects in Berlin for two years before he finally chose a medical training. Working under Johannes Müller, he graduated in 1843, and plunged into research on animal electricity and especially on electricity in fishes. All through his career he was closely associated with the leading German investigators of human physiology: Schwann Schleiden, Ludwig and also the physicist Helmholtz. He succeeded Müller as Professor of Physiology in 1858, and was appointed the head of the new Physiological Institute which first opened in Berlin in 1877. Du Bois-Reymond's importance lay in his investigations of the physiology of muscles and nerves, and in his demonstrations of electricity in animals. He was successful in introducing improved techniques for measuring such effects, first investigated by Galvani. By 1849 he had evolved a delicate multiplier for measuring nerve currents. Thanks to his highly sensitive apparatus, he was able to detect an electric current in ordinary localized muscle tissues, notably contracting muscles. He observantly traced it to individual fibres, finding that their interior electrical potential was negative with regard to the surface. He demonstrated the existence of electrical currents in nerves, correctly arguing that it would be possible to transmit nerve impulses chemically. Du Bois-Reymond's experimental methods provided the basis for almost all future work in electrophysiology. He held trenchant views about scientific metaphysics. He denounced the vitalistic doctrines that were especially prominent amongst German scientists, and denied that nature contained mystical life forces independent of matter.

DUBOS, René Jules
(1901–1982)

French-born American bacteriologist, born in Saint-Brice. He became an American citizen in 1938 and worked at Rockefeller University in New York City from 1927. His doctoral thesis dealt with soil micro-organisms, and this remained his field of interest throughout his career. In 1939 he isolated an antibacterial substance from *Bacillua brevis* and named it tyrothricin. This was the first commercially produced antibiotic. Later, it was found to be a mixture of several polypeptides. Although Dubos's compounds were not particularly effective in themselves, they raised interest in penicillin, leading Waksman to isolate streptomycin and others to produce the broad-spectrum tetracyclin antibiotics of the late 1950s. In 1969 he won the Pulitzer Prize for his book *So Human An Animal.*

DUCHENNE, Guillaume Benjamin Amand
(1805–1875)

French physician, born in Boulogne-sur-Mer and educated at Douai and Paris. After 11 years as a general practitioner in Boulogne, he returned to Paris and devoted himself to the physiology and diseases of muscles. A pioneer of electrophysiology and electrotherapeutics, he did important work on poliomyelitis, locomotor ataxia and a common form of muscular (Duchenne's) dystrophy. He also developed a method of taking small pieces of muscle (biopsy) from patients for microscopical examination. Although he never held a formal hospital appointment, he worked at the Salpêtrière Hospital, where Charcot was his patron.

DULBECCO, Renato
(1914–)

Italian virologist, born in Cantanzaro. He studied medicine at Turin, and worked for the Resistance movement during World War II. In 1947 he emigrated to the USA, securing appointments at Indiana University and then at Caltech. From 1972 to 1977 he was Assistant Director of Research at the Imperial Cancer Research Fund, London. He then returned to the USA as Professor of Pathology and Medicine at the University of California (1977–1981), and since 1977 he has been Research Professor at the Salk Institute, La Jolla. Dulbecco demonstrated how certain viruses can transform some cells into a cancerous state, such that those cells grow continuously, unlike normal cells. For this discovery, he was awarded the 1975 Nobel Prize for Physiology or Medicine, jointly with his former students, Baltimore and Temin. With H Ginsberg, he published *Virology* in 1980.

DYSON, Freeman John
(1923–)

English-American theoretical physicist, born in Crowthorne. He was educated in mathematics at the University of Cambridge, and worked in bomber command during World War II. In 1947 he went to Cornell where he worked with Feynman and Bethe on quantum electrodynamics, the application of quantum theory to interactions between electromagnetic radiation and particles. He was appointed professor there in 1951 and then moved to the Institute for Advanced Studies at Princeton in 1953. He became a US citizen in 1951. While at Cornell, Dyson showed that the techniques used by Feynman, Schwinger and Tomonaga in quantum electrodynamics were mathematically equivalent. His other work in physics has included studies of ferromagnetism, statistical mechanics and phase transitions. He was also involved in the design of the Orion spacecraft and the Triga nuclear reactor. He was elected FRS in 1952 and has been awarded the Royal Society's Hughes Medal and the Max Planck Medal of the German Physical Society.

E

ECCLES, Sir John Carew
(1903–   )

Australian neurophysiologist, born in Melbourne. He studied medicine there, before undertaking postgraduate studies, on a Rhodes scholarship, in Sherrington's department of physiology in Oxford in 1925. He stayed in Oxford until 1937, achieving a permanent position as a university demonstrator and also as Tutorial Fellow at Magdalen College in 1934. While in Britain, he collaborated with Sherrington in papers on neural inhibition, and proposed that the process of neurotransmission at synaptic junctions in the nervous system was an electrical phenomenon, rather than the chemical mechanism then postulated by Loewi and Dale. He returned to Australia as Director of the Kanematsu Institute of Pathology at Sydney (1937–1944), moved to New Zealand as Professor of Physiology at Otago University (1944–1951) and then to the Australian National University at Canberra (1952–1968). On reaching compulsory retirement age in 1968 he moved to the State University of New York at Buffalo, from which he retired in 1975. Amongst many contributions, he established the relationship between inhibition of nerve cells and repolarization of a cell's membrane, much of which was related to, and dependent upon, the findings of Sir Alan Hodgkin and Sir Andrew Huxley, with whom he shared the 1963 Nobel Prize for Physiology or Medicine for discoveries concerning the functioning of nervous impulses. Eccles has made several additional significant contributions to the neuroscience. He recorded the depolarization of a post-synaptic muscle fibre in response to a neural stimulus, which he termed the EPSP (excitatory post-synaptic potential); and most notably he identified inhibitory neurons and demonstrated the role of inhibitory synapses in controlling and regulating the flow of information within the nervous system. His experiments in electrophysiology were providing increasing evidence that the work of Loewi and Dale and their associates on chemical neutrotransmission was substantially correct, and that his own hypothesis of electrical transmission was flawed. In 1944 these coincided with him meeting the philosopher Karl Popper who emphasized that science was deductive, not inductive, and that a failed scientific hypothesis was successful scientifically in that it indicated that the truth lay elsewhere. Eccles has emulated the example of Sherrington with widespread interests in the arts and philosophy. He has written several books on neurophysiology, including *The Neurophysiological Basis of Mind* (1953), and has collaborated with Popper on *The Self and its*

Brain, a particularly powerful assessment of neurobiology, consciousness, and the philosophy of self. He was elected FRS in 1941 and knighted in 1958.

EDDINGTON, Sir Arthur Stanley
(1882–1944)

English astronomer and founder of modern astrophysics, born in Kendal in Westmoreland. In 1898 he entered Owens College, Manchester, where he graduated in physics. From Manchester he went to Trinity College, Cambridge, where he graduated as Senior Wrangler in the Mathematical Tripos (1904). In 1907 he was Smith's Prizeman and was elected a Fellow of Trinity College; however, he left Cambridge on being appointed Chief Assistant at the Royal Observatory Greenwich in 1906. In 1913 he was appointed Plumian Professor of Astronomy and Experimental Philosophy at Cambridge in succession to Sir George Darwin, and in the following year also became Director of the university observatory which was to be his home for the rest of his life. In the same year his first book appeared, *Stellar Movements and the Structure of the Universe*, which dealt with the kinematics and dynamics of stars in the Milky Way. In 1916 Eddington's interest shifted to the problem of the physical constitution of the stars. He showed that stars are gaseous throughout and that the state of their equilibrium is conditioned by the effects of radiation pressure as well as gas pressure in their interiors. Assuming the perfect gas law he deduced a theoretical relationship between the mass of a star and its total output of radiation. He justified that assumption by pointing out that high temperatures in the interiors of stars would have the effect of stripping away part of the electron shells from the atoms, reducing them in size and allowing them to be closer together without upsetting the assumptions of the gas laws. Eddington suggested at the same time that extreme values of density could well exist in stars like white dwarfs and that this effect had in fact been revealed in the spectrum of the companion of Sirius. This and many other investigations such as the pulsation theory of Cepheid variable stars were published in this masterpiece *Internal Constitution of the Stars* (1926). Concurrently with these researches, Eddington had become deeply interested in Einstein's theory of relativity. He has been described as 'the apostle of relativity' in Britain, being the first British scientist to appreciate the importance of Einstein's work. He led a British expedition to the island of Principe, West Africa, on the occasion of the total solar eclipse of 29 May 1919, when he verified one of the predictions of Einstein's theory, the deflection of starlight at the edge of the Sun. In 1920 Eddington published a non-mathematical account of the theory of relativity, *Space, Time and Gravitation* which he extended to his *Mathematical Theory of Relativity* (1923). He wrote a series of scientific books for the layperson; *The Nature of the Physical World* (1928) indicates his conviction that the true foundation of physics lies in the theory of knowledge, and that fundamental constants of nature can be derived from pure theory without recourse to observation or experiment. Eddington's *Fundamental Theory* expounding this topic was posthumously published in 1946. In 1947 the Royal Astronomical Society instituted the Eddington Medal to be awarded for outstanding work on theoretical astronomy; the first recipient was the cosmologist Lemaitre in 1953. Eddington was awarded the Gold Medal of the Royal Astronomical Society in 1924, was knighted in 1930 and received the Order of Merit in 1938. He was President of the Royal Astronomical Society (1921–1923) and at the time of his death was President of the International Astronomical Union.

EDELMAN, Gerald Maurice
(1929–)

American biochemist, born in New York City. He was educated at the University of Pennsylvania and at Rockefeller University, where he has spent his entire research career, which embraces a considerable diversity of interest. He became Professor of Biochemistry in 1966. Before monoclonal antibodies became available he exploited techniques pioneered by Anfinsen to separate the light and heavy peptide subunits of an antibody, purified the light chain from a human multiple myeloma, and reported the complete amino acid sequence (1969). He also analysed the repeat nature of the antibody structure, postulating its three-dimensional relationships, and the nature of the subunit interactions, subsequently investigating the number of antibody forms (chain classes) in different vertebrates. This work, together with Rodney Porter's studies in England, enabled a picture of a typical Y-shaped human immunoglobulin (IgG) antibody molecule to be established. For these discoveries they shared the 1972 Nobel Prize for Physiology or Medicine. Around this time Edelman published on the structure of beta microglobulin and isolated the first DNA ligase (1968). His DNA interest continued in the 1970s with the development of a system for studying yeast plasmid DNA replication *in vitro*. He has also worked on nerve growth factor and the molecular embryology of cell adhesion.

EDISON, Thomas Alva
(1847–1931)

American inventor and physicist, born in Milan, Ohio, the most proliferic inventor the world has ever seen. Expelled from school for being retarded, he became a railroad newsboy on the Grand Trunk Railway, and soon printed and published his own newspaper on the train, the *Grand Trunk Herald.* During the Civil War (1861–1865) he worked as a telegraph operator in various cities, and invented an electric vote-recording machine. In 1871 he invented the paper ticker-tape automatic repeater for stock exchange prices, which he then sold in order to establish an industrial research laboratory at Newark, New Jersey, which moved in 1876 to Menlo Park and finally to West Orange, New Jersey, in 1887. He was not able to give full scope to his astonishing inventive genius. He took out more than 1000 patents in all, including the gramophone (1877), the incandescent light bulb (1879), and the carbon granule microphone as an improvement for Alexander Graham Bell's telephone. Amongst his other inventions were a megaphone, the electric valve (1883), the kinetoscope (1891), a storage battery, and benzol plants. In 1912 he produced the first talking motion pictures. He also discovered thermionic emission, formerly called the 'Edison effect'. During World War I he worked on a variety of military devices such as periscopes, flame throwers and torpedoes.

EDWARDS, Robert Geoffrey
(1925–)

British physiologist who pioneered the basic experiments in reproductive physiology that enabled *in vitro* fertilization ('test-tube babies') to be developed, in collaboration with the obstetrician

Steptoe. After military service in World War II he was educated at the universities of Wales (1948–1951) and Edinburgh (1951–1957) and following a one-year research fellowship at Caltech he became a member of staff at the National Institute of Medical Research, Mill Hill (1958–1962). He moved to the University of Glasgow (1962–1963) and to Cambridge University (1963–1989), where he became Ford Foundation Reader in Physiology (1969–1985) and Professor of Human Reproduction (1985–1989). His experimental researches focused on the mechanisms of human fertility and infertility, and the process of conception. In collaboration with Steptoe, whom he met in 1968, his specific scientific expertise contributed substantially to the successful development of the *in vitro* fertilization programme. Edwards was able to analyse and then recreate the conditions necessary for the egg and sperm to survive outside the womb, by achieving the rapid transfer of the oocyte to an optimally developed culture medium. He discovered the factors that would facilitate the ripening of immature eggs, and he provided the appropriate artificial conditions to facilitate successful fertilization and subsequent maturation of the embryo to the 8–16 cell stage, before its re-implantation into the uterus. In 1971 Edwards and Steptoe first attempted to re-implant a fertilized egg into a volunteer patient, but it was not until July 1978 that the first healthy baby was born as a result of their research. With Steptoe he established the Bourne Hallam Clinics, of which he became Scientific Director (1988–1991).

EHRLICH, Paul
(1854–1915)

German bacteriologist, born into a Jewish family in Strehlen, Silesia (now Strzelin, Poland). He entered the University of Breslau in 1872, and a year later transferred to the University of Strasburg, completing his medical degree at the University of Leipzig in 1878. He developed new dyes with specific affinities for different cell types, including the first staining procedure for the tubercle bacillus which made use of its acid-fast characteristics. He was appointed Head Physician in Friedrich von Frerichs's clinic at the Charity Hospital in Berlin. As a result of his experiments, he contracted tuberculosis, and spent two years with his family in Egypt to recover. Following his recovery, he went to work under Koch. He studied tuberculosis and cholera, and played a prominent role in the introduction of a diphtheria antitoxin. A pioneer in haematology and chemotherapy, Ehrlich's unique contribution was to conceptualize the interactions between cells, antibodies and antigens as essentially chemical responses. He recognized the need to look systematically for chemicals which attack and destroy disease-causing micro-organisms, without harming human cells. In an intensive search for such 'magic bullets', he tested hundreds of compounds to try to find one which would kill the trypanosomes which cause diseases such as sleeping sickness, discovering one partially effective compound, trypan red. He was joint winner, with Metchnikoff, of the 1908 Nobel Prize for Physiology or Medicine for his contributions to immunity and serum therapy. However, his greatest discovery was yet to come. One of his arsenic compounds (number 606) was later found to be extremely effective against spirochetes, the micro-organism which causes syphilis, at that time a much feared disease; in 1910 he announced the discovery of 'salvarsan', the complete cure.

EINTHOVEN, Willem
(1860–1927)

Dutch physiologist, born in Semarang, Dutch East Indies (now Indonesia), the son of a physician. Einthoven became a medical student at Utrecht in 1879, received a PhD in medicine in 1885 and the following year became Professor of Physiology at Leiden. He developed the string (or Einthoven) galvanometer, a sensitive current-measuring device and other apparatus for the recording of electrocardiograms. He made electrocardiograms of many patients with heart disease which he compared with records of heart sounds and murmurs. This was done via a cable 1.5 kilometres long connecting his laboratory to the university hospital. He was awarded the 1924 Nobel Prize for Physiology or Medicine.

EMPEDOCLES
(fl. c.450 BC)

Greek philosopher and poet, from Acragas (now Agrigento) in Sicily, who by tradition was also a doctor, statesman and soothsayer. He attracted various colourful but apocryphal anecdotes – such as the story that he jumped into Mount Etna's crater to support his own prediction that he would one day be taken up to heaven by the gods. His philosophy reflects both the Ionian and the Eleatic traditions, but we have only fragments of his writings from two long poems: *On Nature* describes a cosmic cycle in which the basic elements of earth, air, fire and water periodically combine and separate under the influence of dynamic forces akin to what humans might call 'love' and 'hate'. This notion was taken up and developed by Aristotle and continued to influence chemical theories for more than 2000 years. *Purification* has a Pythagorean strain and describes the fall of man, and the transmigration and redemption of souls. Empedocles was noted for his keen observation and was the first to demonstrate that air has weight. He was also aware of the possibility of an evolutionary process, and believed that some creatures less well adapted to life on Earth had perished in the past.

ERATOSTHENES
(c.276–194 BC)

Greek mathematician, astronomer and geographer, born in Cyrene. He became the head of the great library at Alexandria, and was the most versatile scholar of his time, known as '*pentathlos*' or 'all-rounder'. He measured the obliquity of the ecliptic and the circumference of the Earth with considerable accuracy. In mathematics he invented a method, the 'sieve of Eratosthenes', for listing the prime numbers less than any given number, and a mechanical method of duplicating the cube. He also wrote on geography, chronology and literary criticism, but only fragments of this work remain.

EUCLID
(fl.300 BC)

Greek mathematician. He taught in Alexandria, where he appears to have founded a mathematical school. His *Elements* of geometry, in 13 books, is the earliest substantial Greek mathematical treatise to have survived, and is probably better known than any other mathematical book, having been printed in countless editions; with modifications and simplifications it was still being used as a school textbook in the earlier part of the twentieth century. It was the first mathematical book to be printed and has stood as a model of rigorous mathematical exposition for centuries, though this aspect of it has been severely criticized by Bertrand Russell among others. The *Elements* begins with the geometry of lines in the plane, including Pythagoras's theorem, goes on to discuss circles, then ratio and incommensurable magnitudes (conjecturally the work of Eudoxus). Older material on numbers is then discussed, including the proof that there are infinitely many primes. Book X, the longest, deals with an analysis of certain kinds of lengths, and the geometry of three dimensions is explored, culminating in a proof that there are only five regular solids. He wrote other works on geometry, including the theory of conics, and on astronomy, optics and music, some of which are lost.

EULER, Leonhard
(1707–1783)

Swiss mathematician, born in Basle, where, though destined by his father for theology, he studied mathematics under Jean Bernoulli. In 1727 he went to St Petersburg to join Bernoulli's sons at the Academy of Sciences newly founded by Catherine II where he became Professor of Physics (1731) and then Professor of Mathematics (1733). In 1738 he lost the sight of one eye. In 1741 he moved to Berlin at the invitation of Frederick the Great to be Director of Mathematics and Physics in the Berlin Academy, but he returned to St Petersburg in 1706 after a disagreement with the king, and remained in Russia until his death. He was a giant figure in eighteenth-century mathematics, publishing over 800 different books and papers, mostly in Latin, on every aspect of pure and applied mathematics, physics and astronomy. In analysis he studied infinite series and differential equations, introduced or established many new functions, including the gamma function and elliptic integrals, and created the calculus of variations. His *Introductio in analysin infinitorum* (1748) and later treatises on differential and integral calculus and algebra remained standard textbooks for a century, and his notations such as e and $i = \sqrt{-1}$ have been used ever since. In mechanics Euler studied the motion of rigid bodies in three dimensions, the construction and control of ships, and celestial mechanics. For the princess of Anhalt-Dessau he wrote *Lettres une princesse d'Allemange* (1768–1772) giving a non-technical outline of the main physical theories of the time. He had an amazing skill with complicated formulae and an almost unerring instinct for the right answer, though he was less concerned with the questions of rigour which would occupy later generations. He had a prodigious memory, which enabled him to continue mathematical work though nearly blind, and he is said to have been able to recite the whole of Virgil's *Aeneid* by heart.

F

FAHRENHEIT, (Gabriel) Daniel
(1686–1736)

German instrument-maker, born in Danzig (now Gdansk, Poland). Fahrenheit was born into a merchant family. After the death of his parents in 1701 he was sent to Amsterdam, where he learned the trade of instrument maker. From 1707 he travelled widely in Europe, but by 1717 he had settled in Amsterdam. There he produced high-quality meteorological instruments, supplying eminent Dutch scholars. He devised an accurate alcohol thermometer (1709) and a commercially successful mercury thermometer (1714). In 1708 he had visited the astronomer Roemer in Copenhagen and adopted what he believed to be Roemer's practice of taking thermometric fixed points as the temperatures of melting ice and of the human body. Fahrenheit eventually chose a scale with these points calibrated at 32 and 96 degrees. The zero was at the freezing point of ice and salt. He did not use the boiling point of water as a fixed point: experiments revealed a variation of its temperature with pressure, and he suggested this as a principle for the construction of barometers. In 1724 he noticed (by chance) the supercooling of water.

FAIRCLOUGH, John Whitaker
(1930–)

English computer scientist. He graduated from Manchester University with a BSc in technology in 1954, when the institution's pioneering computer work led by Sir Frederic Calland Williams and Kilburn was being turned to commercial purposes by Ferranti Ltd. In 1954 he joined Ferranti and moved to the USA to develop and sell computer components. Finding this unsatisfying, in 1957 he joined IBM's Poughkeepsie laboratory to work on the Stretch project. In 1958 he returned to England to join IBM's Hursley laboratory, where he managed what was known as the SCAMP project. Although this project was later scrapped, Fairclough's work led to an important technological feature in IBM's highly successful system/360 machines: the control store, the theory of which had been outlined by Wilkes. Fairclough later managed the development of IBM Model 40, which was the first System/360 model, tested and first shipped to a customer in 1965. He became chairman of IBM's UK laboratories in 1983, and was Chief Scientific Adviser to the Cabinet from 1986 to 1990.

FARADAY, Michael
(1791–1867)

English chemist and physicist, creator of classical field theory. Born in Newington near London, the son of a blacksmith, he was apprenticed to a bookbinder whose books sparked his interest in science. In 1813, after applying to Humphry Davy for a job, he was taken on as his temporary assistant, accompanying him soon after on an 18-month European tour during which he met many eminent scientists and gained an irregular but invaluable scientific education. In 1827 he succeeded Davy in the Chair of Chemistry at the Royal Institution, in the same year publishing his *Chemical Manipulation.* His early publications on physical science include papers on the condensation of gases, limits of vaporization and optical deceptions. He was the first to isolate benzene, and he synthesized the first chlorocarbons. His great life work, however, was the series of *Experimental Researches on Electricity* published over 40 years in *Philosophical Transactions of the Royal Society*, in which he described his many discoveries, including electromagnetic induction (1831), the laws of electrolysis (1833) and the rotation of polarized light by magnetism (1845). He received a pension in 1835 and in 1858 was given a house in Hampton Court by Queen Victoria. As adviser to the Trinity House in 1862 he advocated the use of electric lights in lighthouses. Greatly influential on later physics, he nevertheless had no pupils and worked with only one long suffering assistant. He is generally considered the greatest of all experimental physicists.

FERMAT, Pierre de
(1601–1665)

French mathematician, born in Beaumont. He studied law at Toulouse, where he became a councillor of parliament. His passion was mathematics, most of his work being communicated in letters to friends containing results without proof. His correspondence with Pascal marks the foundation of probability theory. He studied maximum and minimum values of functions in advance of the differential calculus, and wrote an unpublished account of the conic sections, extending Vieta's notation to two variables. He is best known for his work in number theory, proofs of many of his discoveries being first published by Leonhard Euler a hundred years later. His 'last theorem' is the most famous unsolved problem in mathematics; it states that there are no positive integers x, y and z with $Xn + yn = Zn$ if n is greater than 2. It is not known what proof, if any, Fermat had of this result, but he did discover a valid proof for the case n = 4. In optics Fermat's principle was the first description of a variational principle in physics; it states that the path taken by a ray of light between two given points is the one in which the light takes the least time compared with any other possible path.

FERMI, Enrico
(1901–1954)

Italian-American physicist, born in Rome. He studied at the Scuola Normale Superiore in Pisa and later obtained a scholarship to study at Göttingen and Leiden. In Göttingen he studied under Born and met Heisenberg and Pauli, and in Leiden he worked with Ehrenliest. At Rome in 1927 he was appointed to the first Chair of Theoretical Physics in Italy.

He worked on modifying the classical theory of statistical mechanics developed by Einstein and Satyendra Nath Bose to take into account the Pauli exclusion 1011 principle. This was later further modified by Dirac to take into account ideas of quantum mechanics, and the resulting Fermi-Dirac distribution is a fundamental part of statistical physics and can explain a wide variety of phenomena from semiconductors to neutron stars. In 1934 Fermi presented a theory to describe the beta decay of nuclei using the neutrino proposed by Pauli. This was able to explain the energy spectrum of the emitted beta particles and the lifetimes of the nuclei. With Segrè and a research group he also studied induced radioactivity by bombarding elements with neutrons. This led to the discovery that slow neutrons, neutrons that have been passed through an absorber such as paraffin to reduce their energy, are much more efficient than high-energy neutrons in initiating nuclear reactions. Their discovery was an important step in the development of nuclear power and weapons. Fermi was awarded the 1938 Nobel Prize for Physics. Fearing for the safety of his Jewish wife in the light of Italy's anti-semitic legislation, he went straight from the Nobel prize presentation in Stockholm to the USA, where he became professor at Columbia University (1939). In 1942 Fermi built the world's first nuclear pile in a disused squash court and produced the first controlled chain reaction. After World War II he continued in nuclear research. In 1952 he discovered 'resonances' in the invariant mass peak when he scattered pions off protons, indicating that very short-lived fundamental particles had been produced. The element fermium was named after him.

FISCHER, Emil Hermann
(1852–1919)

German chemist, born in Euşkirchen, Prussia. He was the son of a Protestant merchant and studied chemistry in Bonn. He worked with Baeyer in Strasburg and Munich. In 1882 he became professor in Erlangen, then in Würzburg (1882) and finally he succeeded Hofmann in Berlin (1892), where he remained until his death. He made important studies of the chemistry of sugars. After his discovery of phenylhydrazine in 1875, he found that it reacted with aldehydes to give phenylhydrazones. By a series of related reactions, phenylhydrazine reacts with a simple sugar to give an osazone and this permits interconversion of simple sugars. This, together with studies of optical activity, led to the elucidation of the structures of the 16 possible aldohexoses (which include glucose). Textbook diagrams of the 16 isomers are known as Fischer projections. The frequent use of phenylhydrazine impaired his health and probably shortened his life, but it was for this work that he was awarded the Nobel Prize for Chemistry in 1902. With his cousin Otto Fischer he elucidated the structure of rosaniline dyes. He also made significant discoveries concerning the structures of caffeine and related compounds. From 1899 he turned his attention to proteins and later to tannins. At the height of his powers he was considered the greatest living organic chemist, but he led a very simple and uneventful life.

FISCHER, Hans
(1881–1945)

German chemist, born in Frankfurt. He received his PhD in chemistry from Marburg University in 1904 and qualified in medicine from Munich University in

1908. Medical work in Munich was followed by chemical research in Berlin at Emil Fischer's institute. From Berlin he moved to Innsbruck (1916) to become Professor of Medical Chemistry and then back to Munich as Professor of Organic Chemistry in 1921. He remained there actively pursuing research until his death. His most important researches concerned the structure of the naturally occurring pigments haemin and chlorophyll. The crowning glory of his work was the synthesis of haemin, and for this he was awarded the Nobel Prize for Chemistry in 1930.

FLEMING, Sir Alexander
(1881–1955)

Scottish bacteriologist, and the discoverer in 1928 of penicillin, born on a farm in Loudoun, Ayrshire. He was educated at Kilmarnock, and became a shipping clerk in London for five years before matriculating (1902) and embarking on a brilliant medical studentship, qualifying as a surgeon at St Mary's Hospital, Paddington, where he spent the rest of his career. It was only by his expert marksmanship in the college rifle team, however, that he managed to find a place in Sir Almroth Wright's bacteriological laboratory there. As a researcher he became the first to use anti-typhoid vaccines on human beings, and pioneered the use of salvarsan against syphilis, a treatment introduced by Ehrlich in 1910. In 1922, while trying unsuccessfully to isolate the organism responsible for the common cold, he discovered lysozyme, an enzyme present in tears and mucus that kills some bacteria without harming normal tissues. While this was not an important antibiotic in itself, as most of the bacteria killed were non-pathogenic, it inspired his search for other antibacterial substances. In 1928 by chance exposure of a culture of staphylococci he noticed a curious mould, penicillin, which he found to have unsurpassed antibiotic powers. Unheeded by colleagues and without sufficient chemical knowledge, he had to wait 11 years before two brilliant experimentalists at the William Dunn School of Pathology at Oxford, Florey and Chain, with whom he shared the 1945 Nobel Prize for Physiology or Medicine, perfected a method of producing the volatile drug. Fleming was appointed Professor of Bacteriology at London in 1938. He was elected FRS in 1943, and knighted in 1944.

FLEMMING, Walther
(1843–1905)

German biologist, born in Sachsenberg. Flemming studied medicine in five German universities, going on to become Professor of Anatomy at Kiel. Flemming is renowned for his investigations of cell division (mitosis), although he also made significant advances in microscope techniques. Developing the new aniline dyes as microscopic stains, and deploying improved microscopes, he found that dispersed fragments in an animal cell nucleus became strongly coloured; he called this substance chromatin. He noted that, in cell division, the chromatin granules combined to constitute larger threads: in 1888 Waldeyer-Hartz was to name these chromosomes. Flemming further demonstrated that elementary nuclear division as described by Remak was not universal; the more typical type of cell division Flemming named mitosis. In this process, the chromosomes divided lengthwise, and the indistinguishable halves moved to opposite sides of the cell. The cell then divided, giving two daughter cells with as much chromatin as the original. Flemming provided a superb account of the process in 1882. He was unaware of Mendel's work. The application of Flemming's idea to genetics was not to come for another 20 years.

FLOREY, Howard Walter, Baron Florey of Adelaide and Marston
(1898–1968)

Australian pathologist, born and educated in Adelaide. He studied physiology as a Rhodes scholar under Sherrington at Oxford, and pathology in Cambridge, becoming a lecturer there in 1927. His early researches included an analysis of the effectiveness of lysozyme, a naturally occurring antibacterial component of tears, saliva and nasal secretions. He was appointed Professor of Pathology at Sheffield (1931–1935) and then at Oxford (1935–1962), where he headed the Sir William Dunn school of pathology. Here his work on lysozyme had developed to such an extent that he required biochemical assistance, and he appointed a refugee from Nazi Germany, Chain. With Heatley and Abraham, they examined and synthesized a wide range of antibacterial compounds. In 1938 they began work on penicillin, an antibacterial substance produced by a mould, and first reported in 1929 by Alexander Fleming, but which was then regarded as unsuitable and too unstable for routine clinical use. Together the Oxford scientists succeeded in isolating sufficient penicillin to enable them to report on its biological properties, especially its marked effect on staphylococcus bacteria, and its low toxicity. These features indicated its practical therapeutic potential and with the help of a small grant from the Medical Research Council they started, during the early months of World War II, to assay and purify penicillin for clinical use. By 1941 they had carried out successful tests on nine patients, and hampered by the paucity of resources in war-torn Britain, Florey and Heatley travelled to the USA to persuade pharmaceutical companies to assist in the development of large-scale production methods for penicillin. Enough penicillin was ready in time to treat casualties in the D-Day battles in Normandy, where it proved highly successful in combating previously fatal bacterial infections. For this vital work, Florey shared the 1945 Nobel Prize for Physiology or Medicine with Fleming and Chain. He was provost of Queen's College, Oxford, from 1962, and was made a life peer in 1965. He never forgot his Australian heritage, and did much to found the Australian National University at Canberra.

FLORY, Paul John
(1910–1985)

American physical chemist, born in Sterling Illinois. He was educated at Ohio State University, and in 1934 obtained a research post under Carothers at Du Pont, Wilmington. In 1938 he moved to a basic science research laboratory at the University of Cincinnati, but in 1940 returned to industry at the Linden Laboratory of Standard Oil. In 1943 he became leader of the Goodyear fundamental research group, in 1948 professor at Cornell University (after giving the Baker Lectures there), and in 1957 Executive Director of the Mellon Institute. From 1961 until his retirement in 1976 he held chairs at Stanford University. He was awarded the Nobel Prize for Chemistry in 1974. His many other honours and awards included the Debye Award (1969) and Priestley Medal (1974) of the American Chemical Society. Flory's main research area was the physical chemistry of polymers. By the 1930s many polymers were known, but their nature as macromolecules had only just been recognized. Flory's work brought them within the scope of kinetics, thermodynamics and statistical mechanics. This led to detailed understanding of their properties and those of their solutions, and of the chemical reactions involved in their formation and breakdown. From the 1950s Flory also worked on liquid crystal behaviour.

FOUCAULT, Jean Bernard Léon
(1819–1868)

French physicist, born in Paris. He first entered on a career as a physician, but found this to be impossible as he came to detest the sight of blood. He turned to experimental physics and determined the velocity of light by the revolving mirror method originally proposed by Arago, and also proved that light travels more slowly in water than in air (1850); subsequently he showed that the ratio of the speeds in the two media is the inverse of the ratio of their respective refractive indices. This was convincing evidence of the wave nature of light as opposed to the corpuscular theory, and earned Foucault his doctorate. In 1851, by means of a freely suspended pendulum, more than 200 feet long, he convincingly demonstrated the rotation of the Earth to a large crowd in a Paris church. In 1852 he constructed the first gyroscope and in 1857 the Foucault prism. In 1858 he improved the mirrors of reflecting telescopes.

FOURIER, Jean Baptiste Joseph, Baron de
(1768–1830)

French mathematician, born in Auxerre. The son of a tailor, Fourier was orphaned at the age of eight. He revealed his talent for mathematics during his education at a military school and an abbey, and later took an active part in promoting the Revolution. In 1795 he joined the staff of the École Normale in Paris, newly formed to train senior teachers, where his success led to the offer of the Chair of Analysis at the École Polytechnique. He accompanied Napoleon during the invasion of Egypt in 1798, and on his return in 1802 was made Prefect of Isère in Grenoble and created baron in 1808. After 14 years at Grenoble, he resigned to rejoin Napoleon during the Hundred Days. He was later made a member of the Academy of Sciences of Paris (1817), becoming joint secretary with Cuvier in 1822. He died in 1830 of a disease contracted in Egypt. Fourier introduced the expansion of functions in trigonometric series, now known as Fourier series. This ended a long period of controversy on the subject and it became generally accepted that almost any function of a real variable can be expressed as a series containing the sines and cosines of integral multiples of the variable: for example, any complex musical sound can be represented as the sum of many individual pure frequencies. This method has become an essential tool in mathematical physics and a major theme of analysis. His *Thèorie analytique de la chaleur* (1822) applied the technique to the solution of partial differential equations to describe heat conduction in a solid body. His work did receive some criticism, however, as he failed to produce a general proof that the Fourier series actually converges to the value of the function involved; this difficulty was not satisfactorily resolved until almost a century later.

FRANKLIN, Benjamin
(1706–1790)

American statesman and scientist, youngest son and 15th child of a family of 17, born in Boston. He was apprenticed at 12 to his brother James, a printer, who started a newspaper, the *New England Courant*, and later Benjamin assumed the paper's management. He later established his own successful printing house in Philadelphia, and in 1729 he purchased the *Pennsylvania Gazette*. In 1732 he commenced the publication of *Poor Richard's Almanac*, which attained an unprecedented circulation. In 1736 Franklin was appointed Clerk of the Assembly, in 1737 Postmaster of Philadelphia, and in 1754 Deputy Postmaster-General for the colonies, a post which took him to London on diplomatic service for a number of years. He participated actively in the deliberations which resulted in the Declaration of Independence on 4 July 1776. Franklin was US minister in Paris till 1785, when he returned to Philadelphia, and was elected President of the state of Pennsylvania. In 1788 he retired from public life. Despite this highly eventful political career, Franklin made many important contributions to science. In 1746 he commenced his famous researches in electricity which earned his election to the Royal Society. He brought out fully the distinction between positive and negative electricity in this theory of a single electric fluid; he suggested a method of proving that lightning and electricity are identical (first performed by Thomas Franois Dalibard at Marley, France, in 1752); and he suggested the protecting of buildings by lightning-conductors. Further, he discovered the course of storms over the North American continent; the course of the Gulf Stream, its high temperature, and the use of the thermometer in navigating it; and the various powers of different colours to absorb solar heat.

G

GABOR, Dennis

(1900–1979)

Hungarian-born British physicist, born in Budapest. After receiving a doctorate in engineering in Berlin (1927) he worked there as a research engineer, but left Germany in 1933. After spending some years with the British Thompson Houston Company in Rugby – during which time he wrote the first book on the electron microscope – he was appointed to a readership at Imperial College, London (1948), and later became Professor of Applied Electron Physics (1958–1967). An inventor rather than a discoverer, he did important work on communications theory and information theory throughout his career, but he is best remembered for conceiving in 1947 the technique of (and the name) holography; this is a method of photographically recording and reproducing three-dimensional images, and for this he was awarded the Nobel Prize for Physics in 1971. The invention came about during work on another of his long-term preoccupations; the improvement of the resolution of the electron microscope. Gabor also worked on television technology, optical processing methods for improving image quality, and an acoustical technique analogous to holography for detecting dense objects in water or under ground. In the late 1960s he developed an acute interest in the socio-political and environmental questions raised by the Club of Rome, an international group which first raised the alarm about the clash between economic expansionism and limitations imposed by the Earth's finite resources. He visited many countries to lecture on these topics.

GAJDUSEK, Daniel Carleton

(1923–   )

American virologist, born in Yonkers, New York. He studied physics at Rochester and medicine at Harvard. He spent much time in Papua New Guinea, studying the origin and dissemination of infectious diseases amongst the Fore people, especially a slowly developing lethal viral disease called *kuru.* He showed that it was spread through cannibalism, as some women and children ritually ate the brains of dead *kuru* victims. The causative agent of *kuru* was the first of the 'slow viruses' to be identified; these viruses have since been implicated in other diseases, such as sheep scrapie, 'mad cow' disease (bovine spongiform encephalopathy), and Creutzfeldt-Jacob dementia of human beings. Gajdusek shared the 1976 Nobel Prize for Physiology or Medicine with Blumberg.

GALEN (Claudius Galenus)
(c.130–c.201 AD)

Greek physician, born in Pergamum in Mysia, Asia Minor, where his father was an architect. He studied medicine there and at Smyrna, Corinth and Alexandria. He was chief physician to the gladiators in Pergamum from AD 157, then moved to Rome and became friend and physician to the emperor Marcus Aurelius. He was also physician to emperors Commodus and Severus. Galen was a voluminous writer on medical and philosophical subjects. The work extant under his name consists of more than 80 genuine treatises and some 15 commentaries on Hippocrates. Galen saw himself as completing and perfecting the medical ideas of Hippocrates, whom he greatly admired. He had, however, a low opinion of most other doctors past and contemporary, and much of his writing is polemical in character. Despite the veneration he reserved for Hippocrates, Galen was an active experimentalist, dissecting animals and using the anatomical and physiological information thus obtained in constructing his theories of human bodily structures and functions. This occasionally led him to describe structures, such as the rete mirabile (a vascular network in the brain of some animals), which later doctors assumed to be present in human beings. In addition to the humoral theory of Hippocrates (which Galen accepted), Galen elaborated a physiological system whereby the body's three principal organs – heart, liver and brain – were central to living processes. The liver took ingested food and converted it to blood. The blood then went via the vena cava to the heart, where it was impressed with vital spirits. Some of it also went to the brain, having passed through the septum of the heart; in the brain further refinement into animal spirits took place. Galen was a shrewd and successful practitioner, who, if he is to be believed, cured many patients. He put great store on the pulse, which he used as a diagnostic aid. Galen also admired Plato's philosophy. Although not a Christian, he was a monotheist and thus his work was easily assimilated into Christian orthodoxy in the centuries after his death. His *De usu partium* ('The uses of the parts') was in essence a hymn to the creator, whereby the organs of the body were seen as perfectly adapted to the functions which they served. For many centuries he was venerated as the standard authority on medical matters, the man who had perfected the medical systems of antiquity.

GALILEO (Galileo Galilei)
(1564–1642)

Italian astronomer, mathematician and natural philosopher. Born, in Pisa, the son of a musician, he matriculated at Pisa University (1581) where he accepted the Chair of Mathematics in 1589. His discovery (1582) of the isochronism of the pendulum, which indicated the value of the pendulum as a timekeeper, was made while watching the swinging of a lamp in the cathedral of Pisa. In the study of falling bodies, Galileo showed that contrary to the Aristotelian belief that the rate at which a body falls is proportional to its weight, all bodies would fall at the same rate if air resistance were not present. He also showed that a body moving along an inclined plane has a constant acceleration, and demonstrated the parabolic trajectories of projectiles. In 1592 he moved to the University of Padua, where his lectures attracted pupils from all over Europe. He made his first contribution to astronomy in 1604 when a bright new star appeared in the constellation Ophiuchus and was shown by him to be more distant than the planets, thus confirming Tycho Brahe's conclusion that changes take place in the celestial regions beyond the planets. In 1609 he reinvented the telescope, having heard an account of that device recently constructed in Holland, and perfected a refracting telescope which led to many astounding astronomical revelations published in his *Sidereus Nuncius* ('Sidereal Messenger' 1610). These included the mountains of the Moon, the multitude of stars in the Milky Way, and quite particularly, the existence of Jupiter's four satellites, the 'Medicean stars', named in honour of his future patron. The book was received with great acclaim, and Galileo was appointed 'Chief Mathematician and Philosopher' by the Grand Duke of Tuscany. On a visit to Rome in 1611 he was elected a member of the Accademia dei Lincei and feted by the Jesuit mathematicians of the Roman College. Further discoveries included the phases of Venus, spots on the Sun's disc, the Sun's rotation, and Saturn's appendages (although not then recognized as a ring system). These brilliant researches made Galileo confident that he might express his conviction in the truth of the heliocentric Copernican system which had been proscribed as heretical by the ecclesiastical authorities and which Cardinal Bellarmine had formally asked him in 1616 to abstain from advocating. In his second book, *The Assayer* (1623), Galileo apparently followed that advice, but in his later controversial book *A Dialogue on the Two Principal Systems of the World* (1632), he defended the Copernican system, in disregard of Bellarmine's admonition. The sale of the book was prohibited and Galileo was cited to Rome by the Inquisition. He was detained, finally examined by the Inquisition and under threat of torture recanted. After several days in the custody of the Inquisition he was relegated first to Villa Medici, then to the house of his friend the archbishop of Sienna, and finally allowed to live under house arrest in his own home at Arcetri, near Florence where he continued his researches and completed his *Discourses on the Two New Sciences* (1638), in many respects his most valuable work, in which he discussed at length the principles of mechanics. His last telescopic discovery, that of the Moon's librations, was made in 1637 only a few months before he went blind. He continued working until his death on 8 January 1642. The sentence passed on him by the Inquisition was formally retracted by Pope John Paul II on 31 October 1992.

GALTON, Sir Francis
(1822–1911)

English scientist, cousin of Charles Darwin. Born in Birmingham to wealthy parents Galton studied medicine at the Birmingham Hospital and King's College, London, and graduated from Trinity College, Cambridge in 1844, when his father's death left him with an ample fortune. In 1846 he travelled North Africa, undertaking a major expedition in 1850 in South Africa, and publishing his *Narrative of an Explorer in Tropical South Africa* and the *Art of Travel* (1855). In 1856 he was elected FRS, and for the rest of his long life he adopted the style of a London-based gentleman scientist. Galton's investigations covered many domains. He was, for instance, a pioneer in meteorology, discovering and naming the anticyclone and publishing, in *The Times* in 1875, the first newspaper weather map. He was also an early enthusiast for the use of fingerprinting in crime detection; colour blindness intrigued him. What unified his researches was an accent upon quantification. He trailblazed the use of statistical techniques, in 1888 presenting to the Royal Society a method for calculating correlation coefficients. An early convert to the evolutionary thinking of his cousin Darwin, Galton considered the respective roles of environment and heredity in shaping human and animal populations. To resolve the nature/nurture conundrum, he selectively bred plants and animals, and made special studies of the medical histories of identical twins. Growing convinced of the key role of heredity, Galton coined the term eugenics to designate the science of creating superior offspring and of preventing inferior populations. He expounded his hereditarian and eugenic ideas in *Hereditary Genius* (1869), *English Men of Science: Their Nature and Nurture* (1874) and *Natural Inheritance* (1889), and endowed a eugenics chair at London University. He was knighted in 1909.

GALVANI, Luigi
(1737–1798)

Italian physiologist, born in Bologna. Galvani studied in his home town, in 1768 becoming a lecturer in anatomy, and from 1782 Professor of Obstetrics. He is famous for the discovery of animal electricity. This he did by chance. He noted that dead frogs undergoing drying by being fixed to an iron fence with brass skewers suffered convulsions. He then showed that paroxysms followed if a frog was part of a circuit involving metals. Galvani believed that electricity of a hitherto unknown sort (animal electricity, or galvanism) was generated in the material of the muscle and nerve. This was later proved erroneous by Volta who in 1800 devised the voltaic pile and resolved the problem by showing that the current arose from the metals, not the frog. Galvani's names lives on in the word 'galvanized' meaning stimulated as if by electricity, and in the galvanometer, used from 1820 to detect electric current.

GARROD, Sir Archibald Edward
(1857–1936)

English physician, born in London into a prominent medical and scientific family. He was educated at Oxford and St Bartholomew's Hospital, London. He held appointments at Several London hospitals – including, eventually, St Bartholomew's where he did important work on arthritis, co-authored a textbook on children's diseases and, above all investigated what he first called *Inborn Errors of Metabolism* (1909). The model for this concept was alkaptonuria, an inherited metabolic disorder in which an acid is excreted in quantities in the urine, causing it to blacken on standing. The other rare hereditary conditions which Garrod discussed were albinism, cystinuria and pentosuria. Garrod's work was an early application of the new Mendelian genetics in the study of human disease. In his *The Inborn Factors in Disease* (1931), he developed his ideas of biochemical individuality. During World War I, he served in Malta as a colonel in the Royal Army Medical Corps, where he developed an interest in classical archaeology. A pioneer of clinical investigation, Garrod succeeded Osler as Regius Professor of Medicine at Oxford in 1920.

GAY-LUSSAC, Joseph Louis
(1778–1850)

French chemist and physicist, born in Saint Leonard in Haute Vienne. He was educated at the École Polytechnique and the École des Ponts et Chaussées. He became assistant to Berthollet in 1800 and subsequently held various posts including Professors of Chemistry at the École Polytechnique (from 1810), Professor of Physics at the Sorbonne (1808–1832), Professor of Chemistry at the National Museum of Natural History (from 1832), superintendent of the government gunpowder factory (from 1818) and Chief Assayer to the Mint (from 1829). He became an Academician (1806), member of the chamber of deputies (1831) and member of the upper house (1839). Gay-Lussac was elected an Honorary Fellow of the Chemical Society in 1849. His earliest research work was on the expansion of gases with temperature increases, and he discovered independently the law which in Britain is commonly known as Charles's law. In 1804 he made balloon ascents in association with Biot to make magnetic and atmospheric observations, and in 1805–1806 he travelled with Humboldt, making measurements of terrestrial magnetism. In 1808 he published his important law which states that when chemical combination occurs between gases, the volumes of those consumed and of those produced, measured under standard conditions of temperature and pressure, are in simple numerical relation. This was based on work which he had begun with Humboldt in 1805. From around 1808 Gay-Lussac's work became purely chemical, and much of it was done in collaboration with Thènard. Their work included the isolation and investigation of sodium, potassium, boron and silicon, extensive studies of the halogens (involving controversy with Sir Humphrey Davy), and the improvement of methods of organic analysis. His last great pure research was on prussic acid and cyanogen, and their derivatives. During the later part of his career, Gay-Lussac did much work as a technical adviser to industry.

GEER, Baron Gerhard Jacob de
(1858–1943)

Swedish quaternary geologist, born in Stockholm, the son of a prime minister of Sweden. Educated at Uppsala, he was Professor of Geology there (1897–1924) and founded the Geochronological Institute of the University of Stockholm, serving as its first director from 1942. He was himself a member of parliament from 1900 to 1905. He first took an interest in quaternary geology as a student, and later, following fieldwork in Spitzbergen, he turned his attention to local quaternary deposits and devised a novel and valuable method for dating by comparing sequences of varves (the annual deposits of sediment under glacial meltwater). He was able to decipher an annual chronology reaching back some 15,000 years from the present day. He successfully correlated the Swedish varve sequence with others from the Himalayas, Iceland, Newfoundland, Canada, Argentina, New Zealand and elsewhere, demonstrating global climatic events and greatly advancing knowledge of the later geological history of the last Ice Age. The method was eventually enhanced by the use of increasingly sophisticated radioisotope methods.

GEHRING, Walter Jacob
(1939–)

Swiss geneticist, born in Zürich. He was educated at the Realgymnasiuim, Zürich, the University of Zürich and Yale University. He was Associate Professor of Anatomy and Molecular Biophysics at Yale University from 1969 to 1972, and has been Professor of Genetics and Developmental Biology at the University of Basle since 1972. His career for many years was concerned with the genetics of *Drosophila melanogaster*, the fruit fly, and in the mid-1980s he discovered a short regulatory DNA sequence which he called the homeobox. Certain mutations in the fruit fly affect whole developmental pathways; for example, antennae can be converted to limbs by a mutation called antennapoedia. Such mutations are called homeotic mutants and arc controlled by homeobox sequences. It was later found that similar control sequences occur in mammals, thus encouraging new lines of research to understand the biology and molecular biology of such major developmental pathways.

GEIGER, Hans Wilhelm
(1882–1945)

German physicist, born in Neustadt-an-der Haardt and educated at the University of Erlangen where he received his PhD in 1906. He then worked under Rutherford at Manchester (1906–1912). With Rutherford, he devised a means of detecting alpha particles in 1908. The instrument consisted of a gas-filled tube with a wire at high electric potential down the centre. Passage of an a-particle produced ionization in the gas, and this would result in a short burst of current which could be measured. They used the device to show that a-particles carry two units of charge. Soon after, Geiger and Ernest Marsden demonstrated that when a-particles are incident on a thin metal foil (gold, silver and copper were used), they are occasionally deflected through large angles. This led Rutherford to propose that the atom has a compact nuclear core surrounded by electrons. With Rutherford, Geiger also showed that two a-particles are emitted in the radioactive decay of uranium, and with J. M. Nuttall he demonstrated the linear relationship between the logarithm of the range of a-particles and the radioactive time constant of the emitting nucleus, now called the Geiger-Nuttall rule. In 1912 he became head of the Physikalisch Technische Reichsanstalt in Berlin, where he obtained confirmatory evidence of the Compton effect. He later became professor at Kiel Linkersity (1925), where he and Walther Müller made improvements to the particle counter, resulting in the modern form of the Geiger-Müller counter, which also detects electrons and ionizing radiation.

GILBERT, or GYLBERDE, William
(1544–1603)

English physician and geophysicist, born in Colchester. He graduated from St John's College, Cambridge (BA 1561), where he was elected to a fellowship in 1561, and received an MA in 1564 and MD in 1569. In 1573 he moved to a practice in London and became a Fellow of the Royal College of Physicians, for whom he was censor (1581–1588), treasurer for nine years and president from 1600. He was appointed physician to Queen Elizabeth (1601) and briefly before his death to King James VI and I (1603). He found time for amateur scientific experiments and took a particular interest in magnetism. He was the first to consider the Earth as one vast spherical magnet and studied the attraction of magnets, the direction relative to the Earth's magnetic field declination and the use of declination to determine latitude at sea. He published *De Magnete, Magneticisque Corporibus, et de Magno Tellure Physiologia. Nova* in 1600, the first great physical book to be published in England. In it he conjectured that terrestrial magnetism and electricity (produced by rubbing amber) were two allied emanations of a single force. Having distinguished the magnetic and amber effects, he demonstrated that many other substances when rubbed exhibited the same phenomenon, and these he called 'electrics'; all others were 'non-electrics'. From this Sir Thomas Browne derived the term 'electricity' in 1646. Gilbert devised his versorium (a light metallic needle turning on a vertical axis) to test for the amber effect. He also demonstrated how the magnetic strength of lodestones (natural magnets) could be increased by 'arming' them with soft-iron pole pieces. The gilbert unit of magnetomotive power is named after him. He bequeathed all his books, apparatus, instruments and mineral collections to the Royal College physicians, although all perished during the Great Fire of London in 1666. Some of his papers were published posthumously including *De Mundo Nostri Sublunari Philosophia Nova* (1651).

GMELIN, Leopold
(1788–1853)

German chemist and physiologist, son of Johann Friedrich Gmelin, born in Göttingen. He studied in Germany and Italy and taught at the university of Heidelberg, where he became Professor and Director of the Chemical Institute in 1817, posts that he held until 1851. Gmelin's interests ranged from physiology – he wrote his thesis on the black pigmentation in the eyes of cattle and later studied the chemistry of digestion – to chemistry and mineralogy. He worked towards a definition of organic chemistry suggesting that an organic compound must contain carbon and hydrogen, and he introduced the terms 'ester' and 'ketone'. One of the pioneers of physiological chemistry, he prepared uric acid, formic acid and potassium ferricyanide (Gmelin's salt). He also developed a test which shows the presence of bile pigments (Gmelin's test). Renowned as teacher, he is even more famous as author of the mighty *Handbuchen der theoretisch Chemie* (1817–1819) which had grown to 10 volumes by 1870. The organic section was subsequently dropped, but the inorganic section continued as *Gmelin's Handbuch anorganische Chemie*. Gmelin died in Heidelberg.

GOLDSTEIN, Joseph Leonard
(1940–)

American molecular geneticist, born in Sumter, South Carolina. He graduated with a medical degree from the University of Texas, Dallas, in 1966. As an intern at Massachusetts General Hospital, Boston, he met Michael Brown at the start of a highly productive collaboration. In 1972 he returned to the University of Texas to become Head of the Division of Medical Genetics, Professor of Internal Medicine (from 1976), and Professor of Medical Genetics (from 1977). With Brown, he has worked on cholesterol metabolism in the human body, studying individuals with the inherited genetic disease familial hypercholesterolemia, which causes high levels of cholesterol in the blood. Cholesterol is carried in the blood forming a complex molecule with low-density lipoproteins (LDLs). Normally, cells in the liver pick up the cholesterol from this complex thereby decreasing the level of cholesterol; Goldstein found that in patients with hypocholesterolemia, the liver cells cannot bind to the complex because they are missing a receptor site for the LDLs. In 1984 Goldstein and Brown described several mutations in the gene that codes for the LDL receptor and opened up possibilities for new drugs to combat this disease. For this work, they were jointly awarded the 1985 Nobel Prize for Physiology or Medicine.

GOLGI, Camillo
(1843–1926)

Italian histologist, born in Corteno, Lombardy. He studied medicine at the University of Pavia and after graduating in 1865, studied psychiatry for a while. In 1872 he became Medical Director of a clinic for the incurable at Abbiategrasso, where he tried to continue the research he had already started on the microscopic anatomy of the nervous system, and to develop new staining methods for his experimental tissues. He returned to Pavia as a lecturer (1875) and then as professor (1876–1918), with a brief intermission at the University of Sienna as Professor of Anatomy (1879). Many of this most important studies were on the anatomy of the brain and other nervous tissues, and he developed a technique, still known by his name, of impregnating particular nerve cells with silver salts. Until that time the structural complexity of the central nervous system had deterred investigators, but Golgi's method, which selectively stained only a few of the nerve cells meant that they stood out from the unstained background of other cells, and could be closely examined under the microscope. He believed that the processes of the nerve cells he described formed a continuous network, and that it was through this continuum that communication within the nervous system was effected. It was Ramón y Cajal, using an adaptation of Golgi's method, who demonstrated that nerve cells were not continuous but were discrete entities separated one from the other. The two men shared the 1906 Nobel Prize for Physiology or Medicine, although Golgi used the award ceremony as an occasion to revile Ramón y Cajal's interpretations. Golgi's work opened up a new field of research into the fine structure of the central nervous system, sense organs, muscles and glands.

GOODALL, Jane
(1934–)

English primatologist and conservationist born in London. She worked in Kenya with the anthropologist Louis Leakey who in 1960 raised funds for her to study chimpanzee behaviour at Gombe in Tanzania. She obtained her PhD from Cambridge in 1965 and subsequently set up the Gombe Stream Research Centre. She has been a visiting professor at the Department of Psychiatry and Progam of Human Biology at Stanford University (1971–1975) and visiting professor of zoology at Dar es Salaam since 1973. Since 1967 she has been Scientific Director of the Gombe Wildlife Research Institute. With her co-workers at Gombe, she has carried out a study of the behaviour and ecology of chimpanzees which at over 30 years is the longest unbroken field study of a group of animals in their natural habitat. This research has transformed the understanding of primate behaviour by demonstrating its complexity and the sophistication of inter-individual relationships. Among her major discoveries was the ability of chimpanzees to modify a variety of natural objects such as the stems of plants to use as tools to collect termites, and sticks and rocks as missiles for defence against possible predators. She also showed that they hunt animals for meat and that the adults share the proceeds of such kills. She has been active in chimpanzee conservation in Africa and their welfare in those countries where they are extensively used in medical research. Her books include *In the Shadow of Man* (1971) and *The Chimpanzees of Gombe: Patterns of Behavior* (1986). She has received many awards for conservation and for her scientific research including the Albert Schweitzer Award (1987), the Encyclopaedia Britannica Award (1989) and the Kyoto Prize for Science (1990).

GOULD, Stephen Jay
(1941–   )

American palaeontologist, born in New York City. Educated at Antioch College, Ohio, and Columbia University, he has been Professor of Geology since 1973 and Alexander Agassiz Professor of Zoology since 1982 at Harvard. His research has extended from studies of the Irish elk (a giant fossil deer from the Pleistocene epoch) to the evolution of Caribbean snails, but he is primarily known for his theoretical contributions. In an influential paper published in 1972, Gould, together with the palaeontologist Niles Eldgredge, posited the theory of 'punctuated equilibrium'. This proposed that most evolutionary change occurs rapidly during the process of allopatric speciation in small populations, when genetic change leads to the evolution of new species, and that species then persist for long periods with little or no change. Gould has also championed the idea of 'hierarchical evolution', suggesting that natural selection operates at many levels, including genes and species as well as at the traditional level of individuals. He has been critical of the 'adaptationist program', emphasizing that many characters of organisms are not 'adaptive' in the strict sense. He has popularized his ideas in a monthly column in *Natural History* magazine, and a series of collected essays including *Ever Since Darwin* (1977), *The Panda's Thumb* (1980), *Hens' Teeth and Horses' Toes* (1983), *The Flamingo's Smile* (1985), *Bully for Brontosaurus* (1991) and *Eight Little Piggies* (1993). His books have won many awards, including the 1990 Science Book Prize for *Wonderful Life* (1989), a reinterpretation of the Cambrian Burgess Shale fauna. Gould has admitted to Marxist influence in his scientific work, and has been a forceful speaker against pseudo-scientific racism and biological determinism; *The Mismeasure of Man* (1981) is a critique of intelligence testing. He was also a witness in a courtroom trial concerning the teaching of evolution in American public schools.

GRAAF, Regnier de
(1641–1673)

Dutch physician and anatomist, born in Schoonhoven. He studied medicine at Utrecht and Leiden, where he was a student of Franciscus Sylvius and a contemporary of Swammerdam (with whom he was later involved in violent priority disputes). He went into practice at Delft, but pursued researches and as early as 1664 published his notable treatise on the pancreatic juice. In 1672, on the basis of human and animal dissection, he discovered the Graafian vesicles of the female gonad, coining the term 'ovary' for the organ. Not noticing the rupture of the ovarian follicles, Graaf was not able to develop a satisfactory theory for explaining the role of the ovary; the mammalian egg was not discovered until the work of Baer in 1827. Graaf is rightly credited with having been one of the founders of experimental physiology.

GRANIT, Ragnar Arthur
(1900–   )

Finnish-born Swedish physiologist, born in Helsinki to a family of Swedish origin. He was a volunteer in the Finnish army fighting for independence from Russia after the Russian Revolution, and in 1919 enrolled at Helsinki University. He studied psychology and medicine, and becoming interested in vision and visual perception, decided to specialize in neurophysiology. After graduating he was appointed instructor in physiology in 1927 and the following year visited England to work for a short period in Sherrington's laboratory at Oxford, during which he also met Adrian, and learned of the latter's success in amplifying the small electrical signals generated by neural impulses. Granit perceived that similar techniques could be used to study the physiology of the retina, and in 1920 won a fellowship in medical physics to work at the Johnson Foundation at the University of Pennsylvania. There he worked under the supervision of Bronk and met both Wald and Hartline with whom he was to share the 1967 Nobel Prize for Physiology or Medicine. After two years he returned to Helsinki, although he continued to spend periods in Oxford, and became Professor of Physiology in 1937. He entered the military medical service during the Soviet invasion of Finland in 1939 and the following year he escaped to Sweden, where he became Professor of Neurophysiology at the Karolinska Institute in Stockholm (1940–1967). His analyses of retinal processing revealed that visual mechanisms were complex responses to light and dark, and he pioneered the recording of the mass response of the retina, the electroretinogram (ERG). He studied the ERGs of animals that lived in a variety of different light conditions, and the technique was widely adopted for the physiological investigation of clinical conditions. Granit's micro-electrode studies enabled him to study the detailed responses of isolated retinal cells to light of different wavelengths and intensities, from which he was able to explain the mechanisms of colour discrimination. He summarized his work in visual physiology in *Sensory Mechanisms of the Brain* (1963). He has also contributed to research on the spinal cord, on pain mechanisms and on the philosophy of science, and has written a biography of Sherrington.

GREEN, Michael Boris
(1945–   )

English theoretical physicist, born in London. He was educated at Cambridge University before obtaining a postdoctoral fellowship at the Institute for Advanced Study at Princeton (1970–1972). From 1972 to 1977 he worked at Cambridge before becoming a Science and Engineering Council Advanced Fellow at Oxford University from 1977 to 1979, when he was appointed lecturer at Queen Mary and Westfield College, London. He became professor there in 1985, and returned to Cambridge in 1993 as John Humphrey Plummer Professor of Theoretical Physics. With John H. Schwarz and Edward Witten, he was the founder of superstring theory. This is based on the idea that the ultimate constituents of nature, when inspected at very small scales, do not exist as point-like particles but as 'strings' in more than three dimensions. It was first introduced as a possible way to avoid the difficulties encountered by early unification schemes involving gravity, but now string theories are considered very good candidates for the actual laws of physics at the ultimate small scale. For this work he was awarded the Maxwell Medal by the Institute of Physics (1987), the William Hopkins Prize by the Cambridge Philosophical Society (1987) and the Dirac Medal of the International Centre for Theoretical Physics (1989). He was elected FRS in 1989.

GROTHENDIECK, Alexandre
(1928–   )

French mathematician, born in Berlin. He became a French citizen after fleeing Germany in 1941 and being interned for a while during World War II. After early important work on infinite-dimensional vector spaces, he switched to algebraic geometry, where he revolutionized the subject. His work led to a unification of geometry, number theory, topology and complex analysis, based on the complex but profound concept of the scheme. His work led to a resolution of the important Weil conjectures, of which the last and most troublesome was finally solved by his pupil Pieffe Deligne in 1972. His work has also had profound implications for the theory of logic. He was awarded the Fields Medal (the mathematical equivalent of the Nobel Prize) in 1966. Later he became deeply involved in a pacifist anti-militarist movement, and then in the teaching of mathematics. He has engaged in lengthy polemics about the way his work has been taken up by others, but his remarkably powerful introduction of the ideas of category theory at the very basis of algebraic geometry have had the effect of extending the language of geometry from fields to rings. In particular, it has enabled questions about the integers to be treated using the geometrical techniques hitherto available only when dealing with the rational or real numbers, thus opening up many important but previously intractable problems.

GUERICKE, Otton von
(1602–1686)

German engineer and physicist, inventor of the vacuum-pump, born in Magdeburg. He worked in Leipzig, Helmstadt, Jena and Leiden, studying law and mathematics, mechanics and the art of fortification. An engineer in the Swedish army, he became one of the four burgomasters of Magdeburg from 1646 to 1681, elected for his service to the town as an engineer and diplomat during its siege in the Thirty Years' War. His interest in the possibility of a vacuum led him to modify a water-pump so that it would remove most of the air from a container. Such primitive vacuum pumps enabled the natural philosophers of the day to study new areas of physics. He arranged a dramatic demonstration of the effect of atmospheric pressure on a near vacuum in 1654 at Regensburg before the emperor Ferdinand III. Two large metal hemispheres were placed together and the air within pumped out; they could not then be separated by two teams of eight horses, but fell apart when the air was allowed to re-enter. He showed that in a vacuum candles cannot remain alight and small animals die, and he devised several experiments that demonstrated the elasticity of air. He also carried out some experiments in electricity and magnetism.

GURDON, John Bertrand
(1933–)

English geneticist, born in Dippenhall, Hampshire. He was educated at Christ Church, Oxford, where he graduated BA in 1956 and DPhil in 1960. He became a lecturer in the zoology department of the University of Oxford (1965–1972) and a staff member at the Medical Research Council Laboratory of Molecular Biology, Cambridge (1972–1983). Since 1983 he has been John Humphrey Plummer Professor of Cell Biology at the University of Cambridge, and since 1991, Chairman of the Wellcome Cancer Research Campaign Institute. He was elected FRS in 1971, and became a Foreign Member of the US National Academy of Sciences in 1980. A central question in biology this century has concerned the mechanism by which one cell type (the fertilized egg) gives rise to all the different cell types in the adult animal. For example, does a fully differentiated cell type such as intestine lose the information to become any other cell type, or does this information remain in the quiescent state? In 1968 Gurdon showed that transplantation of a nucleus derived from frog gut epithelium into an enucleated fertilized egg gave rise to a normal tadpole. This demonstrated that fully differentiated animal cells retain the genetic information to become any cell type under the correct environmental stimuli.

H

HABER, Fritz
(1868–1934)

German physical chemist, born in Breslau (now Wroclaw, Poland). The son of a dyestuffs merchant, he took up chemistry initially with a view to entering the family business. After study at the universities of Berlin and Heidelberg, he obtained his doctorate at the Technische Hochschule, Charlottenberg. There followed a period of uncertainty in which he attempted to find a satisfying career in the organic chemical industry and in the family business. In 1894, however, he became an assistant at the Technische Hochschule in Karlsruhe, and began the study of physical chemistry and its technical applications. Haber became *Privatdozent* in 1896, Extraordinary Professor in 1898, and in 1906 Professor of Physical Chemistry and Electrochemistry. In 1911 he moved to Berlin to direct the Kaiser Wilhelm Institute for Physical Chemistry and Electrochemistry, from which he resigned in 1933 in protest at the anti-Jewish policies of the Nazi regime. He accepted an invitation to work in Cambridge, but decided to winter first in Italy, and while travelling south he died at Basle in January 1934. At Karlsruhe much of his research was in electrochemistry, for example his study of the course of the electrolytic reduction of nitrobenzene, but he also worked in several other areas of physical chemistry. In 1904 he began to study the direct synthesis of ammonia from nitrogen and hydrogen gases, work which continued after his move to Berlin and which, in association with Bosch, led to the large-scale production of ammonia. This was important in maintaining an explosives supply for the German war effort in 1914–1918. It also led to Haber receiving the Nobel Prize for Chemistry in 1918. This occasioned some criticism because Haber had been involved in the organization of gas warfare. In the 1920s he made abortive attempts to extract gold from sea-water, with a view to financing Germany's war reparations.

HAFFKINE, Waldemar Mordecai Wolff
(1860–1930)

Russian-born British bacteriologist, born in Odessa. He worked as an assistant to Pasteur (1889–1893), and as bacteriologist to the government of India (1893–1915) he introduced his method of protective inoculation against *Vibrio cholerae*, the bacteria which cause cholera, using a heat-killed culture prepared from a highly virulent strain. In 1902 he was wrongly accused of having sent contaminated vaccine to the Punjab, but he was exonerated in 1907. He became a British subject in 1899.

HAHNEMANN, (Christian Friedrich) Samuel
(1755–1843)

German physician and founder of homeopathy, born in Meissen. He studied at Leipzig, and for 10 years practised medicine. After six years of experiments on the curative power of bark (the source of quinine), he came to the conclusion that drugs produce a very similar condition in healthy persons to that which they relieve in the sick. This was the origin of his famous principle, *similia similibus curantur* ('like cures like'), which he contrasted to the ordinary belief of allopathic (i.e. ordinary) practitioners. His own infinitesimal doses of medicine provoked the apothecaries, who refused to dispense them; accordingly he illegally gave his medicines to his patients, free of charge, and was prosecuted in every town in which he tried to settle from 1798 until 1810. He then returned to Leipzig, where he taught his system until 1821, when he was again driven out. He retired first to Köthen, and then in 1835 to Paris. He spent much time undertaking 'proving' of a number of drugs, which then entered the homeopathic pharmacopoeia. Many of these were herbal in origin, and subsequent homeopathists have continued to emphasize natural remedies. In his later years, he developed his idea that most chronic diseases are caused by the 'psora', a material he believed to be present on the surface of the skin. By the time of his death, his system had been taken up by practitioners throughout Europe and North America, although their relations with ordinary doctors were often bitter.

HALLEY, Edmond
1656–1742)

English astronomer and mathematician, born in London. Educated at St Paul's School and Queen's College, Oxford, he published three papers on the orbits of the planets, on a sunspot, and on the occultation of Mars while still an undergraduate at Queen's. In 1676 he left for St Helena to make the first catalogue of the stars in the southern hemisphere (*Catalogus Stellarum Australium*, 1679). In 1680 he was in Paris with Cassini, observing comets. It was in cometary astronomy that he made his greatest mark. His calculation of the orbital parameters of 24 comets enabled him to predict correctly the return (in 1758, 1835 and 1910) of a comet that had been observed in 1583, and is now named after him. He as the first to make a complete observation of the transit of Mercury, and the first to recommend the observation of the transits of Venus with a view to determining the Sun's parallax. He established the mathematical law connecting barometric pressure with heights above sea-level (on the basis of Boyle's law). He published studies on magnetic variations (1683), and trade winds and monsoons (1686), investigated diving and under-water activities, and voyaged in the Atlantic to test his theory of the magnetic variation of the compass which he embodied in a magnetic sea-chart (1701). Halley predicted with considerable accuracy the path of totality of the solar eclipse that was observed over England in 1715. He was the first to realize that the Moon's mean motion had a secular acceleration. He also noticed that stars such as Aldebaran, Acturus and Sirius had a proper motion, and that they had gradually changed their positions over the previous two millennia. In map-making Halley was the first to use an isometrical representation. He was also the first to predict the extraterrestrial nature of the progenitors of meteors. He encouraged Isaac Newton to write his celebrated *Principia* (1687), and paid for the publication out of his own pocket. With his *Breslau Table of Mortality* (1693), he laid the actuarial foundations for life insurance and annuities. In 1703 he was appointed Savilian Professor of Geometry at Oxford, where he built an observatory on the roof of his house which is still to be seen, and in 1720 he succeeded Flamsteed as Astronomer Royal of England.

HAMILTON, Sir William Rowan
(1805–1865)

Irish mathematician, the inventor of quaternions, born in Dublin. At the age of nine he had a knowledge of 13 languages, and at 15 he read Isaac Newton's *Principia* and began original investigations. In 1827, while still an undergraduate, he was appointed Professor of Astronomy at Dublin and Irish Astronomer Royal; he was knighted in 1835. His first published work was on optics, and led his colleague Bartholomew Lloyd to discover the unexpected phenomenon of conical refraction of light in certain crystals. He then developed a new approach to dynamics, later and independently proposed by Jacobi, which found favour only with the work of Lie and Poincaré, and which became of considerable importance in the twentieth-century development of quantum mechanics. In 1843 he introduced quaternions after realizing that a consistent algebra of four dimensions was possible if the requirement of commutativity was dropped, and interpreted them in terms of rotations of three-dimensional space. The discovery led to much work on other abstract algebras and so proved to be the seed of much modern algebra. Because quarternions split naturally into a one- and a three-dimensional part, their discovery allowed the successful introduction of vectors into physical problems.

HARDY, Godfrey Harold
(1877–1947)

English mathematician born in Cranleigh. Educated at Winchester and Cambridge, he became a Fellow of Trinity College in 1900. In 1919 he became Savilian Professor of Geometry at Oxford, and he later returned to Cambridge as Sadleirian Professor of Pure Mathematics (1931–1942). Hardy was an internationally important figure in mathematical analysis, and was chiefly responsible for introducing English mathematicians to the great advances in function theory that had been made abroad. In much of his work in analytic number theory, the Riemann zeta function Fourier series and divergent series, he collaborated with Littlewood. He brought the self-taught Indian genius Ramanujan to Cambridge and introduced his work to the mathematical world, rating it far above his own. Their greatest joint achievement was an exact formula for the partition function, which expresses the number of ways a number can be written as a sum of smaller numbers. His mathematical philosophy was described for the layperson in his book *A Mathematician's Apology* (1940), in which he claimed that one of the attractions of pure mathematics was its lack of practical use. Cricket was the other great passion of his life.

HARINGTON, Sir Charles Robert
(1897–1972)

British chemist, born in north Wales. He studied at Cambridge and then in Edinburgh (1919–1920), inspired by Barger with whom he later reported the constitution and synthesis of thyroxine (1927). Moving to the Royal Infirmary in Edinburgh, he studied protein metabolism and then spent a year in the USA with

D. D. Van Slyke and Dakin before becoming a lecturer (1922–1931) and Professor of Pathological Chemistry (1931–1942) at University College London. From 1942 to 1946 he was Director of the National Institute for Medical Research in London. In 1926 Harington published a provisional structure for thyroxine as a tetra-iodo derivative of tyrosine and, with Barger, confirmed this the following year. In 1927 the hormone was also tested on two myxoedema patients with modest, but not lasting, success. He reported an improved method of synthesis with better yield in 1940. His numerous publications included *The Thyroid Gland: Its Chemistry and Physiology* (1933). Harington's later years were mainly involved in administration. He was elected FRS in 1931, and knighted in 1948.

HARVEY, William
(1578–1657)

English physician, and discoverer of the circulation of the blood. Born in Folkestone the eldest of seven sons in the family of yeoman farmer, Harvey went to school in Canterbury, proceeding to study medicine at Caius College, Cambridge. After graduating in 1597, he moved to Padua, working under Fabrizio. In 1602 he set up in practice in London as a physician. Elected a Fellow of the Royal College of Physicians in 1607, two years later he was appointed physician to St Bartholomew's Hospital, and in 1615 he was Lumleian Lecturer at the College of Physicians. In 1628 he published his celebrated treatise, *Exercitatio Anatomica de Motu Co dis et Sanguinis* ('An Anatomical Exercise on the Motion of the Heart and the Blood in Animals'), in which he expounded his views on the circulation of the blood. Successively physician to James I (from 1618) and to Charles I (from 1640), he accompanied the Earl of Arundel in his embassy to the emperor in 1636, publicly demonstrating his theory at Nuremberg. A convinced royalist he was present at the Battle of Edgehill in 1642, attending on the King; he then accompanied Charles to Oxford, becoming warden of Merton College. In July 1646 on the surrender of Oxford to the parliamentary forces, he returned to London, retired from professional life, and devoted himself entirely to his researches. His book on animal reproduction, *Exercitationes de Generatione Animalium*, appeared in 1651. He was buried at Hempstead near Saffron Walden. The key claim of Harvey's distinguished work on the cardiovascular system was that 'the blood performs a kind of circular motion' through the bodies of men and animals. Previously the movement of the blood had been seen as a kind of bodily irrigation. After actually viewing it experimentally in animals, Harvey concluded that the heart was a muscle functioning as a pump, and that it effected the movement of the blood through the body via the lungs by means of the arteries, the blood then returning through the veins to the heart. Harvey upheld the difference between venous and arterial blood. Through experiment and dissection he demonstrated the one-way nature of the valves in the arteries and veins. He showed that in systole the heart contracted, expelling blood; the right ventricle supplied the lungs and the left ventricle provided blood for the arterial system. Blood, he insisted, flowed through the veins towards the heart. Because Harvey's views contradicted ideas central to medicine since Galen, he was widely ridiculed by traditionalists, notably in France. Harvey was not able to show how blood passed from the arterial to the venous system, there being no connections visible to the naked eye. He rightly supposed that the links must be too minute to see. Malpighi observed them with a microscope, shortly after Harvey's death. Harvey's notable *Essays on Generation in Animals* (1651) made public his embryological researches, in which he espoused epigenetic rather than preformationist views. Harvey affirmed the doctrine that every living being has its origin in an egg. Harvey was a gifted experimenter and a master of patient reasoning, and modern animal physiology may be said to have begun with his labours.

HASSEL, Odd
(1897–1981)

Norwegian physical chemist, born in Oslo. His scientific education was at the universities of Oslo, Munich (in the laboratory of Fajans) and Berlin (DPhil 1924). From 1925 until his retirement in 1964 he was on the staff of the Department of Physical Chemistry of the University of Oslo as professor and director from 1934. He received the Nobel Prize for Chemistry jointly with Barton in 1969 and was an Honorary Fellow of the Royal Society of Chemistry. Hassel's most distinguished researches were carried out in the 1930s and involved the application of X-ray and electron diffraction, and the measurement of dipole moments. He elucidated the details of the molecular structure of cyclohexane and related compounds, and thereby helped to establish the concepts and procedures of 'conformational analysis'. Due largely to World War II (he was imprisoned during the German occupation of Norway) much of his work was not well known until the 1950s. Conformational analysis was extensively developed by Barton in the 1950s and 1960s, and has become a very important feature of organic chemistry. Hassel's later work was on charge-transfer complexes.

HAUPTMAN, Herbert Aaron
(1917–)

American mathematical physicist, born in New York City. He was educated at the City College of New York, Columbia University and the University of Maryland. Working at the US Naval Research Laboratories in Washington with Karle during the 1950s and 1960s, he helped develop a statistical technique that radically increased the speed of methods by which X-ray crystallography mapped structures of molecules. Using this 'direct method' the time taken to establish a molecular structure from the pattern of visible dots obtained by exposing a crystal to an X-ray beam was reduced from years to days. Hauptman and Karle published their key monographs in 1953, but its importance remained unacknowledge for years. By the late 1960s, however, the repeated success of their method had ensured its establishment as a standard crystallographic technique. It has so far been applied to molecules of up to 200 atoms, including many important biological molecules such as steroids and other drugs. In 1985, 22 years after publishing the work, Hauptman and Karle were jointly awarded the Nobel Prize for Chemistry. Hauptman became professor at the University of Buffalo in 1970 and since 1972 has continued his research on X-ray crystallography at the Medical Foundation at Buffalo.

HAWKING, Stephen William
(1942–   )

English theoretical physicist, born in Oxford. He graduated from the University of Oxford and received his PhD from Cambridge. He was elected FRS in 1974, and became Lucasian Professor of Mathematics at Cambridge in 1980. His early research on relativity led him to study gravitational singularities such as the 'Big Bang' when the universe originated, and the 'black holes' where space-time is curved due to enormous gravitational fields. The theory of black holes, which result when stars collapse at the end of their lives, owes much to his mathematical work. In the early 1970s, Hawking and colleagues proved mathematically that the only properties conserved when an object becomes a black hole are its mass, angular momentum and electric charge. Since 1974 he has shown that black holes could actually emit thermal radiation and could evaporate. By a process predicted by theories of quantum mechanics, mass can be lost from the black hole and escape entirely from its gravitational pull. This is known as the Hawking process, and he showed that the rate of mass loss would be inversely proportional to the mass of the hole. A galactic mass would take around 1090 years to evaporate. His book *A Brief History of Time* (1988) is a bestselling popular account of modern cosmology. His achievements are especially remarkable because from the 1960s he has suffered from a highly disabling and progressive neuromotor disease.

HAWORTH, Sir (Walter) Norman
(1883–1950)

English chemist, born in Chorley. He learned most of his early chemistry from working in his father's linoleum factory and it was not until 1903 that he enrolled at the University of Manchester, where he studied under William Henry Perkin Jr. A scholarship enabled him to study with Wallach at Göttingen, and he then returned to Manchester to investigate terpenes. In 1911 he moved to Imperial College, London, but was there for only one year before moving to a post at the University of St Andrews. He soon discontinued his work on terpenes and joined the group at St Andrews founded by Irvine to study the chemistry of sugars. In 1920 he moved to King's College, Newcastle, and in 1925 took up the Chair of Organic Chemistry at the University of Birmingham, where he was joined by Hirst for his most productive period. Scurvy had been shown to be a vitamin deficiency disease and Szent-Györgvi had isolated from the adrenal glands and from fruit juice a substance named hexuronic acid which was later shown to be the antiscorbutic factor vitamin C. A sample was sent to Haworth and, with a group including Hirst, he elucidated its chemical structure and confirmed this by chemical synthesis. He called it ascorbic acid. For this work he shared the Nobel Prize for Chemistry with Karrer in 1937. He was elected FRS in 1928, and knighted in 1947.

HELMHOLTZ, Hermann von
(1821–1894)

German physiologist and physicist, born in Potsdam, one of the last of the scientific polymaths. As his father could not afford an expensive university education, Helmholtz studied medicine for which state aid was available but only if the beneficiary was committed to eight years service as an army surgeon. During his medical studies he followed courses on physics and chemistry, and studied mathematics privately. After qualifying he was appointed surgeon to the regiment at Potsdam. His scientific career began when he was released from military duty in 1848. He was successively appointed as Professor of Physiology at Königsberg (1849), Bonn (1855) and Heidelberg (1858). In 1871 he became Professor of Physics in Berlin, and in 1887 president of the newly founded Physikalisch Technische Reichsanstalt for research in the exact sciences and precision technology, to which Werner von Siemens had donated a large sum of money. Helmholtz was equally distinguished in physiology, mathematics, and experimental and mathematical physics. His physiological works are principally connected with the eye, the ear and the nervous system. His work on vision (e.g. on the perception of colour) is regarded as fundamental to modern visual science. He invented an ophthalmoscope (1850) independently of Babbage. He was also important for his analysis of the spectrum, his explanation of vowel sounds, his papers on the conservation of energy with reference to muscular action, his paper on *Conservation of Energy* (1847), his two memoirs in Crelle's *Journal* on vortex motion in fluids and on the vibrations of air in open pipes, and for researches into the development of electric current within a galvanic battery. He was elected a Foreign Member of the Royal Society and in 1873 was awarded the society's Copley Medal.

HERACLEIDES OF PONTUS AND EKPHANTUS
(c.338–c.315 BC)

Greek philosopher and astronomer, born in Heraklea, near the Black Sea. He migrated to Athens to become a pupil of Speusippus and Plato. Although all his writings are lost, Aetius reports that Heracleides was the first to propose that the Earth is spinning like a wheel from west to east, thinking that it was highly improbable that the immense spheres of the stars and planets could rotate once every 24 hours. It is also suggested that he took the first steps towards a heliocentric solar system, concluding that Mercury and Venus orbited the Sun rather than the Earth. He also considered the cosmos to be infinite and that the other planets were Earth-like with atmospheres.

HERSCHEL, Caroline Lucretia
1750–1848)

German-born British astronomer, sister of William Herschel, born in Hanover. In 1772 her brother brought her to England as assistant with his musical activities, and she became his devoted collaborator when he abandoned his first career for astronomy (1782). Between 1786 and 1797 she discovered eight comets. Among her other discoveries was the comparison of the Andromeda nebula (1783). In 1787 she was granted a salary of £50 a year from the king as her brother's assistant at Slough. Her part in William's observational work made her thoroughly familiar with the heavens: her *Index to Flamsteed's Observations of the Fixed Stars* and a list of errata were published by the Royal Society (1798). Following her bother's death she returned at the age of 72 to Hanover where she worked on the reorganization of his catalogue of nebulae. For this *Reduction and Arrangement in the Form of a Catalogue in Zones of all the Star Clusters and Nebulae Observed by Sir William Herschel*, though unpublished, she was awarded the Gold Medal of the Royal Astronomical Society (1828). She was elected (with Somerville) an honorary member of the Royal Astronomical Society (1835) and a member of the Royal Irish Academy (1838). On her

96th birthday she received a gold medal from the King of Prussia.

HERSCHEL, Sir John Frederick William (1792–1871)

English astronomer, only child of Sir William Herschel, born in Slough and educated briefly at Eton, afterwards at home, and at St John's College, Cambridge, where he was Senior Wrangler and Smith's Prizeman (1813), and was made a Fellow of his college. His first award was the Copley Prize of the Royal Society for his mathematical researches in 1821. In collaboration with Sir James South, he re-examined his father's double stars (1821–1823) and produced a catalogue which earned him the Lalande Prize (1825) and the Gold Medal of the Royal Astronomical Society (1826). He reviewed his father's great catalogue of nebulae in Slough (1825–1833), adding 525 new ones, for which he received the Gold Medals of the Royal Astronomical Society (1826) and the Royal Society (1836). To extend the survey to the entire sky he went to South Africa and set up the seven-foot Slough reflector at Feldhausen near Capetown. In four years (1834–1838) there he completed a survey of nebulae and clusters in the southern skies, observing 1,708 of them, the majority previously unseen. He also discovered over 1,200 pairs of double stars, catalogued over a thousand objects in the Magellanic Clouds, and extended his father's star-gauging exercise to southern fields. The preparation for publication of his massive southern observations occupied him for many years. These *Cape Observations* (1847) earned him the Copley Medal of the Royal Society (1847). John Herschel dominated British science for 50 years and excelled in many branches. His preferred interest was chemistry; he was a pioneer photographer, the inventor of the fixing process using thiosulphate of soda (1819) and independently of sensitized paper (1839), as well as the originator of the terms positive and negative in photography. He never occupied an academic post, supporting his researches from his private means, his one official appointment being Master of the Mint (1850–1855). He was made a baronet at Queen Victoria's coronation. Among his numerous honours were the Prussian *Pour le Mérite* and Membership of the French Institute. He is buried in Westminster Abbey close to the grave of Isaac Newton.

HERSCHEL, Sir (Frederick) William (1738–1822)

German-born British astronomer, born in Hanover, the son of a musician who instructed his sons in the same profession. William joined the Hanoverian Guards band as an oboist and moved in 1755 to England where he built up a successful career in music eventually settling in Bath in 1766. It was here that his interest in astronomy began. He built his own telescopes, learning to cast his own metal discs for his mirrors. In 1781 he discovered the planet Uranus, the first to be found telescopically, which he named *Georgium Sidus* in honour of King George III who a year later appointed him his private astronomer. At Slough, near Windsor, assisted by his sister Caroline Herschel, he continued his researches and built ever larger telescopes, up to his 40-foot long reflector (completed in 1789). He lived and worked in Slough for the rest of his life. Herschel's discoveries included two satellites of Uranus (1787) and two of Saturn (1789), but his epoch-making work lay in his studies of the stellar universe. He drew up his first catalogue of double stars (1782), later demonstrating that such objects constitute bodies in orbit around each other (1802), and observed the Sun's motion through space (1783). His famous paper *On the Construction of the Heavens* (1784), based on star counts in thousands of sample portions of the sky, produced a model of the Milky Way as a non-uniform aggregation of stars; such studies occupied him for the rest of his life. Following the publication of Messier's catalogue of nebulae and star clusters (1781), he began a systematic search for such non-stellar objects which revealed a total of 2,500, published in three catalogues (1786, 1789 and 1802). He distinguished different types of nebulae, realizing that some were distant clusters of stars while others were nebulosities. Herschel was knighted in 1816. The epitaph on his tomb sums up his immense influence on the course of astronomy: *Coelorum perupit claustra* – he broke the barriers of the heavens.

HERSHEY, Alfred Day
(1908–)

American biologist, born in Owosso, Michigan. He studied at Michigan State College, and from 1950 to 1974 worked at the Carnegie Institution in Washington, where from 1962 he was Director of the Genetics Research Unit. He became an expert on the viruses which infect bacteria (bacteriophage or 'phage'), and set up the Phage Group with Luria and Delbrück in the late 1940s, to encourage the use of phage as an experimental tool. At that time, it was not known whether DNA or protein was the genetic material; Avery had suggested in 1944 that DNA was the genetic material but, experimental evidence was lacking. Working with Martha Chase in the early 1950s, Hershey radioactively labelled the protein and the DNA of the phage particles with different markers.

HERZBERG, Gerhard
(1904–)

German Canadian physicist, born in Hamburg. He was educated at the TechnischeHochscule, Darmstadt and at the universities of Güttingen and Bristol. From 1930 he taught at Darmstadt, but he emigrated to Canada in 1935. He was Research Professor of Physics at the University of Saskatchewan from 1935 to 1945, and after a brief appointment at Yerkes Observatory, he became Director of the Division of Pure Physics of the National Research Council, Ottawa, from 1949 to 1969. Herzberg is distinguished for his applications of spectroscopic methods in astrophysics, atomic and molecular physics, and physical chemistry. Of particular importance have been his extremely precise determinations of the energy levels of atomic and molecular hydrogen and isotopes thereof, and of radicals such as CH, CH2, CH3 and N112. His several books have been very influential, particularly *Molecular Spectra and Molecular Structure*, which appeared in four volumes over a period of 40 years (1939–1979). Hemberg was awarded the Nobel Prize for Chemistry in 1971. He was elected a Fellow of the Royal Society in 1968 and received its Royal Medal in 1971. He is also an Honorary Fellow of the Royal Society of Chemistry and was awarded its Faraday Medal in 1970.

HEYWOOD, Vernon Hilton
(1927–)

British botanist, born in Edinburgh and educated at Edinburgh University and Pembroke College, Cambridge. After a distinguished early career at the botany department of the University of

Liverpool (1955–, professor 1964–1968) he was Professor of Botany at the University of Reading from 1964 to 1988. Since 1987 he has been Director of the Botanical Gardens Conservation Secretariat, and from 1988 to 1992 he was Chief Scientist (Plant Conservation) and Director of Plant Science for the International Union for the Conservation of Nature and Natural Resources (IUCN), based at Kew. Heywood is a world authority on the families Compositae and Umbelliferae, and has had long-standing interests in the flora of Spain and in plant conservation. He has published nearly 200 papers and is also the author or editor of over 36 books, including *Principles of Angiospern: Taxonomy* (1963, with Peter Davis), one of the most important twentieth-century works on the theory and practice of plant taxonomy; *Plant Taxonomy* (1967); *The Biology and Chemistry of the Umbelliferae* (1971, editor); *Flowering Plants of the World* (1978, editor); *Our Green and Living World* (1984, joint author); *The Botanic Gardens Conservation Strategy* (1989); and *International Directory of Botanic Gardens* (1990).

HILBERT, David
(1862–1943)

German mathematician, born in Konigsberg (now Kaliningrad, Russia). He studied and taught at the university there until he became professor at Göttingen (1895–1930). His definitive work on invariant theory, published in 1890, was so novel that it brought to a halt for a generation the work on a subject that had occupied so many nineteenth-century mathematicians, and laid the foundations for modern algebraic geometry. In 1897 he published a lengthy report on algebraic number theory which was the basis of much later work. In this field he established a synthesis of the ideas of Dedekind, Kummer and Kronecker that opened the way to a structural analysis of the subject while playing down the explicit, if formal side. In 1899 he was the first to give abstract axiomatic foundations of geometry which made no attempt to define the 'meaning' of the basic terms but only to prescribe how they could be used. This was later to inspire him to look for a foundation of all mathematics in similar terms. With his students, he then worked on the theory of integral equations, the calculus of variations, and theoretical physics (which he claimed was too complex to be left to physicists). The concept of the 'Hilbert space' (an infinite-dimensional space) was implicit in this work, and was made explicit by Erhard Schmidt in 1906. Unease with the contemporary foundations of mathematics, and the specific proposals of Brouwer and Weyl, led him to take up mathematics logic. He proposed that all mathematics should be based on a finite set of axioms and that a fully rigorous theory of proofs in mathematics should be given along the lines of elementary logic (strictly speaking, using a first-order language, for which a rigorous theory was available). This aim was later shown to be unattainable by Gödel. At the International Congress of Mathematicians in 1900 he listed 23 problems which he regarded as important for contemporary mathematics; the solutions of many of these have led to major advances, while others remain unsolved.

HIPPARCHOS
(c.180–125 BC)

Greek astronomer, born in Nicaea in Bithynia, the most outstanding astronomer of the ancient world. He made his observations from Rhodes where he spent a long time, and may also have lived in Alexandria. He compiled a catalogue of 850 stars (completed in 129 BC) giving their positions in celestial latitude and longitude, the first such catalogue ever to exist, which remained of primary importance up to the time of Halley. In comparing his observed star positions with earlier records he made his great discovery of the precession of the equinoxes, i.e. the shifting of the point of intersection of the Sun's annual path with the celestial equator. He observed the annual motion of the Sun through the sky, developed a theory of its eccentric motion, and measured the unequal durations of the four seasons. He made similar observations of the Moon's more complex motion. Following the method of eclipse observations used by Aristarchos, he estimated the relative distances of the Sun and Moon, and improved calculations for the prediction of eclipses. He developed the mathematical science of plane and spherical trigonometry required for his astronomical work. In the field of geography Hipparchos was the first to fix places on the Earth by latitude and longitude.

HIPPOCRATES
(c.460–377 or 359 BC)

Greek physician, known as the 'father of medicine', and associated with the medical profession's 'Hippocratic Oath'. The most celebrated physician of antiquity, he was born and practised on the island of Cos, but little is known of him except that he taught for money. The so-called 'Hippocratic Corpus' is a collection of more than 70 medical and surgical treatises written over two centuries by his followers, and only one or two can be fairly ascribed to him. He seems to have tried to distinguish medicine proper from the traditional wisdom and magic of early societies, and laid the early foundations of scientific medicine; he was said to be good at diagnosis and prognosis, and his followers developed the theories that the four fluids or humours of the body (blood, phlegm, yellow bile and black bile) are the primary seats of disease. They conceived that excesses or deficiencies of the humours caused diseases, which were to be treated by measures (such as drugs, diet, change of life, blood-letting) which countered them. They thought that the doctor must be the servant of nature, believing that the body has natural tendencies to correct the humoral imbalance (the 'healing power of nature'). The humoral doctrine also included a notion of temperaments, whereby each person had his or her own ideal balance, which the doctor needed to take into account when planning therapy. The balance also naturally changed during the course of the life cycle, with blood the dominant humour during early life, and phlegm the one of old age. The Hippocratic writings contain many treatises which long exerted great influence. The Oath has been seen as the foundation of Western medical ethics, and is still occasionally used in a Christianized version. However, it was as much a guild agreement as an ethical statement, although some of its prohibitions, such as the injunction against procuring abortions, appealed to later doctors. Of his works, *Airs, Waters, Places* contained shrewd observations about the geography or disease and the role of the environment in shaping the health of a community. *Epidemics X* examined epidemics in a population and offered case histories of patients with acute diseases. *The Sacred Disease* elaborated a rigorous defence of the naturalistic causes of diseases, in the context of a monograph on epilepsy, attributed to supernaturalistic influences by many in early Greek society. *Aphorisms* consisted of a series of short pithy statements, mostly about clinical situations, but beginning with the most famous, 'life is short, the art is long'.

HIRST, Sir Edmund Langley
(1898–1975)

English chemist, born in Preston. He attended Madras College while his father was a minister of the Baptist Church in St Andrews, and then proceeded to St Andrews University, where he took a degree in chemistry, and a PhD on the chemistry of sugars. He joined the staff of St Andrews, but left in 1923 and worked successively at the universities of Manchester, Durham and Birmingham. His first chair was in Bristol (1936) and he became professor at the University of Manchester in 1944. He ended his distinguished academic career as Forbes Professor of Organic Chemistry at the University of Edinburgh. Hirst was a gracious and charming scholar who made important contributions to the chemistry of carbohydrates. His most famous work was his collaboration with Norman Haworth in the laboratory synthesis of vitamin C. He was elected FRS in 1934 and knighted in 1964.

HODGKIN, Dorothy Mary, née Crawford
(1910–)

British crystallographer, born in Cairo, Egypt. She studied chemistry at Somerville College, Oxford, moved to Cambridge to study for her PhD, and became a Fellow and Tutor at Somerville in 1934. After various appointments within the university, she became the first Royal Society Wolfsan Research Professor at Oxford in 1960. She received the Nobel Prize for Chemistry in 1964, only the third woman to do so. In 1965 she was admitted to the Order of Merit, the first woman to be so honoured since Florence Nightingale. Hodgkin was elected FRS in 1947 and gave the Bakerian Lecture in 1972. She is an Honorary Fellow of the Royal Society of Chemistry and was awarded the Longstaff Medal in 1978. She was President of the International Union of Crystallography from 1972 to 1978. In X-ray crystallography studies at Cambridge, Bernal introduced her to the study of biologically interesting molecules, which was extremely difficult and tedious in the 1930s. With Bernal she began work on sterols and continued this after her return to Oxford. Her detailed X-ray analysis of cholesterol was a milestone in crystallography, but an even greater achievement was the determination of the structure of penicillin (1942–1945). After World War II computational facilities increased; even so, the determination of the structure of vitamin B12, which was her real triumph, occupied eight years (1948–1956). her later work on insulin, an even more complicated molecule, was able to use sophisticated computers.

HOFFMANN, Roald
(1937–)

American chemist, born in Zloczow, Poland. Having survived the Nazi occupation by hiding in an attic for almost a year, he and his mother (his father was killed in a breakout from a labour camp) became refugees when Poland was liberated by the Russians. After moving around Europe for some years, they and Roald's stepfather (Paul Hoffmann) migrated to the USA, arriving in New York City in 1949. In 1955 he entered Columbia University and took a number of arts courses as well as chemistry. He obtained a PhD from Harvard in chemical physics in 1962, having spent nine months at the University of Moscow. He was elected a Junior Fellow at Harvard and in 1964 began a collaboration with Robert Woodward, in which factors controlling the way in which cyclization reactions occur when bond breaking and making occur simultaneously were established. The results of these considerations became known as the Woodward-Hoffman rules for the conservation of orbital symmetry and have been of remarkable predictive value, stimulating much productive experimental work. It was for this work that, along with Fukui, he received the 1981 Nobel Prize for Chemistry. In 1965 Hoffmann moved to Cornell University, and in 1974 he was appointed John A. Newman Professor of Physical Science. His most recent work concerns the synergism of molecular orbital calculation and experiment in a number of areas of inorganic chemistry. These studies have led to an increased understanding of bonding, and to the prediction of chemical species subsequently synthesized by others. Another area of interest is the bonding of species absorbed on surfaces. He also writes popular articles on science and has hosted a TV programme on chemistry. His second book of poems was published in 1990. In addition to the Nobel Prize he has received many awards and honours, including the Priestley Medal of the American Chemical Society.

HOLLEY, Robert William
(1922–)

American Biochemist, born in Urbana, Illinois. Working mainly at Cornell Medical School and at the Salk Institute in California (from 1968), he was a member of the team which first synthesized penicillin in the 1940s. In 1962, with Benzer, using poly UG and poly UC, he identified two distinct leucyl transfer RNAs (leu tRNS) characterized by different codons, and suggested that this provided the physical basis for the degeneracy of the amino acid code suggested by Crick. In the bacterium *Escherichia coli* he found five differently coding leu tRNAs (1965). He secured the first pure sample of a tRNA (1 gram of leu tRNA from 90 kilograms of yeast) and in 1965 published the full molecular structure of this nucleic acid – Crick's 'adaptor molecule', which plays a central role in the cellular synthesis of proteins. He shared the 1968 Nobel Prize for Physiology or Medicine with Khorana and Nirenberg.

HOPWOOD, David Alan
(1933–)

British geneticist, born in Kinver, Staffordshire. He was educated in Cheshire and at St John's College, Cambridge, where he received his PhD. After some years as a university demonstrator and Fellow of Magdalene and St John's colleges in Cambridge, he moved to a lectureship at the University of Glasgow in 1961. In 1968 he became the first John Innes Professor of Genetics at the University of East Anglia and head of the genetics department at the John Innes Institute. Elected FRS in 1979, he is a member of scientific academies of India, Hungary and China, and has honorary fellowships from the Federal Institute of Technology (ETH), Zürich, and the University of Glasgow. Hopwood's major contribution has been in the understanding of the genetics of bacteria belonging to the genus *streptomyces*; these organisms produce the great majority of the antibiotics used in human and veterinary medicine and in agriculture. His work has led to the ability to manipulate genes for antibiotic production – this has stimulated the development of novel methods to improve antibiotic production and to generate new antibiotics by genetic engineering.

HORROCKS, Jeremiah
(1619–1641)

English astronomer, born in Toxteth, Liverpool. In 1632 he entered Emmanual College, Cambridge, and in 1639 he became Curate of Hoole, Lancashire, where he made the first observation of the transit of Venus (24 November 1639 according to the Julian calendar), deduced the solar parallax corrected the solar diameter, and made tidal observations. Horrocks was an enthusiastic admirer of Kepler and made considerable improvements to the equation of motion of the Moon, noticing that the line of apsides was librating and that the eccentricity was changing. Both these were attributed to the effect of the solar gravitational field. He also noticed irregularities in the motion of Jupiter and Saturn. These have now been shown to be the clue to their mutual galvinational attraction. Erroneously Horrocks believed that comets were blown out of the Sun, their velocities decreasing as they receded but increasing again when they started to fall back. He was the first person to undertake a continuous series of tidal observations, hoping eventually to understand the underlying causes of these variations. His 19 March 1637 observations of the lunar occulation of the stars in the Pleiades indicated that they disappeared instantaneously. He concluded that the stars had apparent diameters that were negligible and thus not capable of measurement.

HOUNSFIELD, Sir Godfrey Newbold
(1919–)

English electrical engineer, born in Newark. He studied in London at the City and Guilds College and Faraday House College, worked as an instructor in electronic and radar communications in the RAF during World War II, joined Thorn/EMI in 1951 and became head of medical systems research there in 1972. He led the team which (independently of Cormack) developed the technique of computer-assisted tomography (CAT scanning), which produces detailed X-ray pictures of the human body, including the soft tissues which are normally almost transparent to X-rays. The images are built up by computer from large numbers of measurements of the absorption of X-rays transmitted in different directions through the body. The EMI scanner system which resulted from this research represented a major breakthrough in the non-invasive diagnosis of disease, and won the MacRobert Award in 1972. In 1978 Hounsfield was appointed Professorial Fellow in Imaging Sciences at the University of Manchester, and he shared the 1979 Nobel Prize for Physiology or Medicine with Cormack. He was knighted in 1981.

HOUSSAY, Bernardo Alberto
(1887–1971)

Argentine physiologist, born in Buenos Aires. A precocious child, Houssay graduated from the School of Pharmacy of the University of Buenoz Aires in 1904 aged 17. He went on to study medicine while working as a hospital pharmacist and holding teaching appointments. From 1909 he was Professor of Physiology at the veterinary school, and after graduating in medicine in 1911 he took on additional duties, including a private practice and the directorship of a municipal hospital service. From 1919 he was Professor of Physiology at the medical school, until dismissed by the military dictatorship of 1943 for allegedly being too pro-American. He was reinstated under a general amnesty in 1945, but was dismissed again after Juan Perón became president in 1946. He continued his research privately until reinstated in 1955. From his student days, Houssay was interested in the pituitary gland and later studied interactions between the pituitary gland and insulin. He shared the 1947 Nobel Prize for Physiology & Medicine with Carl & Gerty Cori.

HOYLE, Sir Fred
(1915–)

English astronomer and mathematician, born in Bingley, Yorkshire. Educated at Bingley Grammar School and Emmanuel College, Cambridge, he taught mathematics at Cambridge (1945–1958), was Plumian Professor of Astronomy and Experimental Philosophy there (1958–1972) and Professor-at-Large at Cornell University (1972–1978). In 1948, with Bondi and Gold, he propounded the 'steady state' theory of the universe, which proposes that the universe is uniform in space and unchanging in time, its expansion being fuelled by constant spontaneous creation of matter. The theory was later to be displaced through new evidence, but it was a valuable contribution to cosmology. He also suggested the currently accepted scenario of the build up to supernova explosion in stars, in which the nuclear reactions fuelled by hydrogen and helium in stars are followed by the production and nuclear 'burning' of heavier elements. A chain of reactions then leads to supernova explosions, in which the matter is ejected into space, and recycled in second-generation stars which form from the remnants. He has been a successful writer on popular science and science fiction; his books include *Nature of the Universe* (1952), *Frontiers of Astronomy* (1955) and *The Black Cloud* (1957). He was elected FRS in 1957 and knighted in 1972.

HUBBLE, Edwin Powell
(1889–1953)

American astronomer, born in Marshfield, Missouri. After studying law at the University of Chicago and as a Rhodes scholar at Oxford (1910–1913), he returned to the USA but soon abandoned law to take up the study of astronomy. He held a research position at the Yerkes Observatory (1914–1917), and following military service during World War I, moved in 1919 to the Carnegie Institution's Mount Wilson Observatory where he began his fundamental investigations of the realm of the nebulae. Using Cephied variable stars as distance indicators, he succeeded in determining the distance to the Andromeda nebula (1923). He found that spiral nebulae are independent stellar systems, and that the Andromeda nebula in particular is very similar to our own Milky Way galaxy. In 1929 he announced his discovery that galaxies recede from us with speeds which increase with their distance. This was the phenomenon of the expansion of the universe, the observational basis of modern cosmology. The linear relation between speed of recession and distance is known as Hubble's law. Hubble remained on the staff of Mount Wilson until his death. The 2.4-metre aperture Hubble Space Telescope launched in 1990 was named in his honour.

HUDSON, William
(1734–1793)

English botanist and apothecary, born and educated in Kendal. He was apprenticed to a London apothecary and received as the prize for botany a copy of Ray's Synopsis. From 1757 to 1758 he was sub-librarian at the British Museum and his studies of the Sloane Herbarium formed the basis of his adaptation of Linnaean nomenclature to the plants that Ray had described – work which was much more accurate than that of Sir John Hill's *Flora Britannica* of 1760. In the early 1760s, Hudson practised as an apothecary in Haymarket, London. During this period, the first edition of his *Flora Anglica* was published in 1762. This was the first British botanical work to adopt the Linnaean classification system and its binomial nomenclature, and contained much original work. Hudson was later praefector horti to the Apothecaries' Company at Chelsea Physic Garden (1765–1767). An enlarged second edition of *Flora Anglica* was published in 1778. He also studied insects and molluscs. His insect collection, and most of his herbarium, were destroyed in a fire at his Panton Street home in 1783. Thereafter he moved to Jermyn Street, and died there of paralysis.

HUGGINS, Charles Brenton
(1901–)

Canadian-born American surgeon, born in Halifax, Nova Scotia. He worked at Chicago University from 1927, where he became Professor of Surgery in 1936 and was head of the Ben May Laboratory for Cancer Research from 1951 to 1969. He has been a pioneer in the investigation of the physiology and biochemistry of the male urogenital tract, including the prostate gland. Research on dogs (a species which also can develop benign and malignant tumours of the prostate) led him to the possibility of using hormones in treating such tumours in human beings. He also worked on the use of hormones in treating breast cancer in women. He shared the 1966 Nobel Prize for Physiology or Medicine with Rous for their cancer research, in Huggin's case, his discovery of hormonal treatment for cancer of the prostate gland.

HUNTER, John
(1728–1793)

Scottish physiologist and surgeon, born in Long Calderwood, East Kilbride. After working as a Glasgow cabinet-maker, he came to London, assisting between 1748 and 1759 at the anatomy school run by his elder brother, William Hunter, where he learned his dissecting skills. He then studied surgery at St George's and St Bartholomew's hospitals. In 1760 he entered the army as Staff Surgeon; service on the Belleisle and Portugal expedition provided the basis for his later expertise in gunshot wounds. In 1768 he became surgeon at St George's, and in 1776 was appointed Surgeon-Extraordinary to George III. In 1790 he was made Surgeon-General to the army. An indefatigable biological and physiological researcher, he built up huge collections of specimens to illustrate the processes of plant and animal life and elucidate comparative anatomy. His museum grew to contain an astonishing 13,600 preparations; on his death, it was bought by the government and subsequently administered by the Royal College of Surgeons in London. In the field of human pathology, Hunter investigated a wide range of subjects, from venereal disease and embryology to blood and inflammation. He developed new methods of treating aneurysm, and was the first to apply pressure methods to the main trunk blood arteries. He also succeeded in grafting animal tissues. His *Natural History of Human Teeth* (1771–1778) gave dentistry a scientific foundation. His biological studies included work on the habits of bees and silkworms, on hibernation, egg incubation and the electrical discharges of fish. Hunter trained many of the leading doctors and natural historians of the next generation, including Jenner. he was buried in the church of St Martin-in-the-Fields. He has widely been dubbed the founder of scientific surgery.

HUXLEY, Sir Andrew Fielding
(1917–)

English physiologist, grandson of T. H. Huxley and half-brother of Aldous and Julian Huxley, born in London. He studied natural sciences at Trinity College and at the Physiological Laboratory, Cambridge. He graduated in 1938 and began neurophysiological research with his tutor, Alan Hodgkin. But during World War II (1940–1945) he worked on operational anti-aircraft research. He became a Fellow (1941–1960) and Assistant Director of Research in Physiology (1952–1960) at Trinity College, Cambridge in the Physiological Laboratory he was demonstrator (1946–1950), Assistant Director of Research (1951–1959) and Reader in Experimental Biophysics (1959–1960). He moved to University College London, as Joddrell Professor of Physiology (1960–1969) and as Royal Society Research Professor (1969–1983), before returning to Cambridge as Master of Trinity College (1984–1989). With Hodgkin he provided a physico-chemical explanation for the conduction of impulses in nerve fibres, using the giant axon preparation of the squid described by John Young. In 1950 he changed direction, to muscle physiology, and devised a special interference microscope with which to study the contraction and relaxation of muscle fibres. From his observations and measurements of different components of the cell fibre, he postulated a 'sliding filament theory' to account for the functional and morphological changes that occur during muscular contraction. The theory suggested that different sections of the muscle fibre overlapped, thus shortening the overall length of the fibre and generating force in the direction of the shortening. Huxley served as President of the Royal Society (1980–1985) and was awarded the Order of Merit in 1981. In 1963 he shared the Nobel Prize for Physiology or Medicine with Hodgkin and Eccles.

HUXLEY, Sir Julian Sorell
(1887–1975)

English biologist and humanist, grandson of T. H. Huxley, brother of novelist Aldous Huxley and half-brother of Andrew Huxley. Educated at Eton and Balliol College, Oxford, where he won the Newdigate Prize (1908), he was professor at the Rice Institute, Texas (1913–1916), and after World War I became Professor of Zoology at King's College, London (1925–1927), Fullerian Professor at the Royal Institution (1926–1929), and Secretary to the Zoological Society of London (1935–1942). He extended the application of his scientific knowledge to political and social problems, formulating a pragmatic ethical theory of 'evolutionary humanism', based on the principle of natural selection. This lead to his enthusiastic, although later regretted, welcome for the ideas of the Jesuit mystic and palaeontologist, Teilhard de Chardin. He was the first Director-General of UNESCO (1946–1948). His influence was based on his capacity for synthesis stimulation, inspiring the work of Biologists such as Elton and Ford, rather than his own fairly meagre scientific accomplishments in animal behaviour and experimental embryology. His *Evolution: The Modern Synthesis* (1942) was undoubtedly the key work in the acceptance of the neo-Darwinian synthesis of evolution, although it was chiefly a presentation of the ideas of Fisher, Dobzhansky, Mayr, George Gaylord Simpson, J. B. S. Haldane and others. His writings also included *Essays of a Biologist* (1923). *Religion without Revelation* (1927), *Animal Biology* (with Haldane, 1927), *The Science of life* (with H. G. Wells, 1931), *Problem of Relative Growth* (1932), *Evolutionary Ethics* (1943), *Biological Aspects of Cancer* (1957) and *Towards a New Humanism* (1957). Huxley's life is described in his two-volume autobiography *Memories* (1970, 1973), and in *Evolutionary Studies* (1989).

HYPATIA
(c.370–415 AD)

Greek philosopher, the first notable female astronomer and mathematician, who taught in Alexandria and became head of the neoplatonist school there. She was the daughter of Theon, a writer and commentator on mathematics, with whom she collaborated, and was herself the author of commentaries on mathematics and astronomy, although none of these survives. Hypatia was renowned for her beauty, eloquence and learning, and drew pupils from all parts of the Greek world, Christian as well a pagan. Cyril, archbishop of Alexandria, came to resent her influence, and she was brutally murdered by a Christian mob he may have incited to riot.

J

JACKSON, Benjamin Daydon
(1846–1927)

English botanist and bibliographer, born in Stockwell, London and educated at private schools. From an early age Jackson had strong botanical and especially bibliographical instincts. A born indexer, his greatest work was the compilation of *Index Kewensis* an index of all names of flowering plants hitherto described; the first volume was published in 1892. This work began in 1882, as the result of a suggestion of Charles Darwin to Sir Joseph Dafton Hooker; the manuscript, when completed in 1895, weighed over one ton. An absolutely essential tool for the plant taxonomist, *Index Kewensis* continues to be updated regularly. Jackson also compiled an indispensable *Glossary of Botanical Terms* (1900), and the *Catalogue of the Library of the Linnaean Society* (1925), and wrote two botanical bibliographies, *Guide to the Literature of Botany* (1881) and *Vegetable Technology* (1882). He edited and annotated facsimiles of early botanical works including Gerard's *Catalogue of Plants Cultivated in His Garden in the Years 1596–1599* (1876) and William Turner's *Libellus de Re Herbario Novus* of 1538 (1877). He was an authority on Linnaeus, and at celebrations in Sweden to mark the bicentenary of Linnaeus's birth in 1907, he received the Order of Knighthood of the Polar Star.

JACOB, François
(1920–   )

French biochemist, born in Nancy. Educated at the University of Paris and the Sorbonne, he worked at the Pasteur Institute in paris from 1950. In 1960 he became Head of the Cellular Genetics Unit at the Pasteur, and in 1964 he was appointed Professor of Cellular Genetics at the College de France. During the 1950s, he worked on the nutritional requirements of the bacterium *Escherichia coli*. He found that the enzyme B-galactosidase is only produced when this bacterium is grown in the presence of the sugar lactose, and suggested that genes are turned on and off by other genes which regulate the genes that code for the enzyme. He formulated the 'operon system', which consists of the structural genes that code for a protein, and a regulator gene which controls the structural genes by producing a 'repressor' molecule which binds to a specific section of DNA known as the 'operator'. The theoretical 'repressor' molecule was identified by Walter Gilbert some years later. With Lwoff and Monod, Jacob was awarded the 1965 Nobel Prize for Physiology or Medicine for research into cell physiology and the structure of genes. His autobiography, *The Statue Within*, was published in 1988.

JANSSEN, (Pierre) Jules César
(1824–1907)

French astronomer, born in Paris. After a time working in a bank he gave up in order to study mathematics and physics at the Sorbonne. He had a highly adventurous life in which he undertook many scientific missions on behalf of his country: to Peru in 1857 to determine the magnetic equator, to the Azores in 1867 to make magnetic observations, and to Japan and Algeria in 1874 and 1882 respectively to observe the century's two transits of Venus. he took part in eclipse expeditions including that of 1870 when he escaped from the beleaguered Paris in a balloon. His spectroscopic observations during the total eclipse of the Sun in India (1868) revealed the gaseous nature of solar prominences; the following day came his discovery of a method of observing prominences outside total eclipse. The same method, independently arrived at by Lockyer, brought both men to fame and the French Academy of Sciences commemorated their discovery by striking a medallion bearing their names and profiles. In 1875 the French government set up a new astrophysical observatory in Meudon near Paris with Janssen in charge from 1876. There he concentrated on solar physics, producing the first photographs of a fine grain-like structure (granulation) on the Sun's surface. In 1890 he set up an observing station on Mont Blanc where he was able to study the solar spectrum through a minimum of the Earth's interfering atmosphere. Janssen remained Director of Meudon Observatory until his death.

JOHANSON, Donald Carl
(1943–)

American palaeoanthropologist, born in Chicago of Swedish immigrant parents. Johanson obtained a BA at the University of Illinois in 1966 and conducted graduate studies in Chicago. In 1972 he became Assistant Professor of Anthropology at the Case Western Reserve University in Cleveland, and Associate Curator of Anthropology at the Cleveland Museum of Natural History. Johanson made spectacular finds of fossil hominids 3–4 million years old at Hadar in the Afar triangle of Ethiopia (1972–1977, 1990–1992). They included 'Lucy', a female specimen, and the 'first family', a scattered group containing the remains of 13 individuals. Lucy was named after the song 'Lucy in the sky with diamonds', played at the camp celebration following her discovery. Johanson suggested that these remains belong to a previously undiscovered species, which he named *Australopithecus afarensis* (Afar ape-man), more primitive than Australopithecus africanus which had been named by Dart. Johanson put forward the theory of 'mosaic evolution' – the idea that some parts of the body became human before others which challenged the idea that bipedalism, large brains and tool-making developed together. He suggested that bipedalism came first by about two million years. In 1981 Johanson was founding director of the Institute of Human Origins, Berkeley, California.

JOHNUBIPRASAD, Ashoka
(1955–)

Indian polymath, who used a shortened form of his surname for publications, born in Gorakhpur, India, son of a meritorious judge in a highly distinguished family; his great-grandfather was one of the Fathers of the Indian Constitution and later First President of the Indian Republic and his great-uncle led the revolt resulting in the downfall of Mrs

Gandhi's dictatorship in the mid-1970s. He was educated at the ultra elite Kolvin School in Lucknow where he had a distinguished academic record despite dyslexia and later at the Cawnpore University, Edinburgh, Leeds and London. Abdus Salam, after his only meeting with him, remarked that he would be best remembered for 'establishing it was still possible to be a polymath in the twentieth century'. Sir Derek Barton, described him as a 'perfect example of twentieth-century genius'. By the time he was 35, he had earned top accolades in medicine, paediatrics, geography, pathology, clinical genetics, psychiatry, surgery, public health, history, anthropology, biology, mathematics and qualifications in aviation-medicine. He rose to hold full professorships in several centres and an adjunct full professor of medical anthropology at Columbia University, New York. His major contribution was establishing a link between GABA, Valproate and mania which led to a safe alternative to toxic lithium. He also has a syndrome named after him which links Hashimoto's with mania. His lectures on 'Philosophical Bankruptcy in Modern Day Science' in New York and Philadelphia won wide acclaim. He steadfastly refused any honour and maintained a low profile. Several disputes with colleagues led to retirement to his home town where he is leading the life of a recluse.

JOLIOT-CURIE, Frédéric, originally Jean-Frederic Joliot
(1900–1958)

French physicist, born in Paris and educated at the École Primaire Supérieure Lavoiser in Paris. After military service and a job at a steel mill, he joined the Radium Institute under Marie Curie (1925), there he studied the electro-chemical properties of polonium. He married Marie's daughter Irène (see Irène Joliot-Curie) in 1926, and in 1931 they began collaborating in research. In 1933–1934 they produced the first artificial isotope and for this work they were awarded the Nobel Prize for Chemistry in 1935. He became professor at the College de France in 1937. Aware of the consequences of nuclear fission, in 1939 he persuaded the French government to purchase the world's major stocks of heavy water (used as a moderator in chain reactions) from Norway. When France was invaded by the Germans in 1940, he arranged for it to be secretly shipped to England. During World War II he became a strong supporter of the Resistance movement and a member of the Communist party. After the liberation he became High Commissioner for Atomic Energy (1946–1950), a position from which he was dismissed for his political activities. He succeeded his wife as head of the Radium Institute. He was awarded the Stalin Peace Prize (1951) and was given a state funeral by the Gaullist government when he died from cancer, caused by lifelong exposure to radioactivity.

JOLIOT-CURIE, Irène, née Curie
(1897–1956)

French physicist, born in Paris, daughter of Pierre and Marie Curie. She was educated at home by her mother, and during World War I served as a radiographer in military hospitals. In 1918 she joined her mother at the Radium Institute in Paris and began her scientific research in 1921. In 1926 she married Frédéric Joliot, and they collaborated in studies of radioactivity from 1931. In work on the emissions of polonium, they studied the highly penetrating radiation observed by Bothe and demonstrated its ability to eject Protons from paraffin wax; the radiation emitted was in fact neutrons, but they misinterpreted their results and attributed it to a consequence of the Compton effect. Chadwick read their paper and built on this work in his discovery of the neutron. In 1933–1934 the Joliot-Curies made the first artificial radioisotope by bombarding aluminium with alpha particles to produce a radioactive isotope of phosphorus. It was for this work that they were jointly awarded the Nobel Prize for Chemistry in 1935. Similar methods led them to make a range of radioisotopes, some of which have proved indispensable in medicine, scientific research and industry. During World War II Irène Joliot-Curie escaped to Switzerland. Back in Paris after the war, she became director of the Radium Institute in 1946 and a director of the French Atomic Energy Commission. She died from leukaemia due to long periods of exposure to radioactivity.

JOULE, Jarnes Presscott
(1818–1889)

British natural philosopher, born in Salford. Shy and often ill as a child, Joule was educated by private tutors, notably the chemist Dalton. A laboratory in the parental home was the site of Joule's first electrical experiments. In 1838 he described his new electromagnetic engine in Sturgeon's *Annals of Electricity*. Joule's belief that such machines would supersede the steam engine was revised in the light of systematic investigations into engine 'duty' or efficiency. The 'Joule effect' (1840) asserted that the heat produced in a wire by an electric current was proportional to the resistance and to the square of the current. He came to believe that heat was derived from work, whether chemical, mechanical or electrical in form: contradicting the caloric theory. Heat and work were interconvertible. Painstaking experimental researches provided successive determinations of the mechanical equivalent of heat, the first announced to the peripatetic British Association for the Advancement of Science (BAAS) in 1843 and the last in 1878 carried out for the BAAS Committee on Standards. In 1845 Joule demonstrated the conversion of work into heat by agitating water with a rotating paddle wheel apparatus. This famous experiment was repeated at the 1847 BAAS meeting in Oxford where Joule met and greatly impressed Kelvin. Between 1853 and 1862 they collaborated on the 'porous plug' experiments showing that when a gas expands without doing external work its temperature falls (the Joule-Thomson effect). Recognition came with the award of the Royal Society's Royal (1852) and Copley (1870) medals. During the 1850s with the promotion of the kinetic theory of gases (Joule calculated the average velocity of a gas molecule in 1848), the new thermodynamics, and energy physics, his ideas were recast in terms of the principle of the conservation of energy.

K

KARRER, Paul
(1889–1971)

Swiss chemist, born in Moscow. He was the son of a Swiss dentist who practised in Russia, but he was educated in Switzerland and studied chemistry under Alfred Werner at the University of Zürich. After graduating he worked with organo-arsenic compounds, and this interest led him to move to Frankfurt in 1912 to work with Ehrlich. In 1915, after Ehrlich's death, he became director of the chemical division of Georg Speyer Haus. He remained in this position for only three years before returning to Zürich where in 1919 he succeeded Werner as professor. He remained there for the rest of his life. All Karrer's chemical studies involved natural products. He began with amino acids and proteins, and continued with polysaccharides such as starch and cellulose. During the 1920s he developed an interest in plant pigments, a topic which was to occupy him for the rest of his career. An early triumph was his elucidation of the structure of carotene, which led to important discoveries concerning vitamin A. He also elucidated the structures of vitamins E, K and B2 (riboflavin). For these achievements and for important studies of the chemistry of vitamin C and biotin, he shared the 1937 Nobel Prize for Chemistry with Norman Haworth. Karrer's chemical studies continued during the 1940s and 1950s, and included important work on the coenzyme nicotinamide-adenine dinucleotide, carotenoids and the curare-like alkaloids. He was rector at the University of Zürich from 1950 to 1952, and published the world renowned textbook *Lehrbuch der organischen Chemie* in 1928.

KATZ, Sir Bernard
(1911–)

British biophysicist, born and educated in Leipzig, Germany. After qualifying in medicine he left Nazi Germany in 1935, and began physiological research at University College London, initially in Hill's laboratory. He started working on problems associated with the electrical stimulation of nerves and the processes of neuromuscular transmission, and received his PhD in 1938. Just before the outbreak of World War II he joined Eccles at the Kanematsu Institute in Sydney, Australia, where he collaborated in further neurophysiological experiments with both Eccles and Kuffler, before enlisting as a radar officer in the Australian Royal Air Force. In 1946 he returned to University College, where he remained for most of his career (Lecturer 1946–1950; Reader in Physiology 1950–1952; Professor of Biophysics 1952–1978) although he spent a substantial part of the late 1940s in Cambridge and

Plymouth, working with Sir Alan Hodgkin and Sir Andrew Huxley on the mechanisms by which the nerve impulse is transmitted, using the giant axon preparation of the squid, first reported by John Young. For the nest three decades Katz's work focused on the mechanisms of neural transmission, in particular how the chemical neurotransmitter acetylcholine is stored in nerve terminals and released by neural impulses. In collaboration with Paul Fatt and Jose del Castillo, he showed that the chemical was stored in nerve terminals, in small packets called vesicles that could be observed using an electro microscope, and was released in specific portions called quanta when stimulated by the arrival of the neural impulse. For this work he shared the 1970 Nobel Prize for Physiology or Medicine with Axelrod and Ulf von Euler. He was elected FRS in 1952, and knighted in 1969.

KELVIN, William Thomson, 1st Baron of Largs
(1824–1907)

British natural philosopher, born in Belfast. He came to Glasgow when his father was made Professor of Mathematics there in 1832. At the age of 10, William Thomson matriculated at the university and in 1841 he entered Peterhouse College, Cambridge. He completed the exacting Mathematics Tripos in 1845 as Second Wrangler, first Smith's Prizeman, and winner of the Silver Sculls rowing trophy. After acquiring experimental skills in Regnault's Paris laboratory, Thomson was appointed Professor of Natural Philosophy at Glasgow (1846), a post he held until 1899 when he retired, only to re-enrol as a research student. In a career of prodigious versatility and international distinction he harmonized physical theory and engineering practice, emphasizing accurate measurement, economy and the minimization of waste. The application of continental mathematics, combined with the use of mechanical models in physical theory was a hallmark of his scientific style. While a student he read Fourier's *Analytic Theory of Heat*, praised it as a 'mathematical poem', and in his first articles defended its adventurous mathematical techniques. From 1842 he addressed problems in electromagnetism, developing an analogy with Fourier's macroscopic heat flow analysis and stressing contiguous action rather than forces acting at a distance. Through the British Association for the Advancement of Science, Thomson worked to establish national and international electrical standards. When the first transatlantic telegraph cable became operational in 1866, he was knighted for the major consultative role he had played. His mirror galvanometer and siphon recorder were designed as receivers, for rapid transmission on such long-distance cables. Numerous electrical instruments were manufactured under patent, bringing substantial returns. In 1848 Thomson proposed his absolute (Kelvin) scale of temperature, independent of any physical substance. The following year he gave a rigorous account of Carnot's theory of heat-engines, and in 1851 he reconciled Carnot's theory and the mechanical theory of heat given credence by Joule. Exhaustive memoirs on the 'dynamical theory of heat', contemporaneously with Clausius and Rankine, established the second law of thermodynamics. Thomson promoted the new unifying concept of 'energy' in physics, and considered the consequences of a general dissipation of the energy of the universe. Cosmological convictions fused with geophysics to generate estimates of the Earth's age now regarded as far too low, but which at the time provided temporal arguments against Charles Darwin's theory of evolution. Thomson also worked on hydrodynamics in the context of the burgeoning Clyde shipbuilding industry. His yacht the *Lalla Rookh*, served both as symbol of affluence and floating laboratory. Created Britain's first scientific peer in 1892, he is buried in Westminster Abbey, beside Isaac Newton.

KENDALL, Edward Calvin
(1886–1972)

American chemist, born in South Norwalk, Connecticut. After training in Canada he joined the firm of Park, Davis (1910) and was asked to isolate the active principle of the thyroid gland. He left in 1911, going first to St Luke's Hospital in New York and then to the Mayo Foundation, Rochester (1914), where he was appointed professor and head of biochemistry. In December 1914 he isolated thyroxine (33 grams of thyroxine from three tons of pig thyroid gland), and in 1917 he described its properties and physiological activity in detail. Its structure was elucidated partly by Kendall and partly by Harington. In collaboration, Kendall isolated cortisone (around 1936) and 29 related steroids from the adrenal cortex. In particular he studied and determined the metabolic effects of cortisone, corticosterone, 17-hydroxycorticosterone, dihydrocorticosterone and cortisol, and observed the loss of resistance to toxic chemicals caused by adrenalectomy. He prepared synthetic corticosterone in 1944 and cortisone (active against Addison's disease) in 1947. With Hench, he found that cortisone was effective against rheumatic fever and that cortisone plus ACITI (adrenocortical trophic hormone) was effective against rheumatoid arthritis. Kendall, Hench and Reichstein shared the 1950 Nobel Prize for Physiology or Medicine.

KENDREW, Sir John Cowdery
(1917–1997)

English molecular biologist, born in Oxford. He was educated at Clifton College, Bristol, and Trinity College, Cambridge. Elected a Fellow of Peterhouse College, Cambridge (1947–1975), he was a co-founder (with Perutz) and Deputy Chairman of the Medical Research Council Unit for Molecular Biology at Cambridge (1946–1975). During World War II he was seconded to the Ministry of Aircraft Production and Scientific Advisor to the Allied Air Commander in Chief. He was also Scientific Advisor to the Ministry of Defence (1960–1964). Kendrew carried out researches in the chemistry of the blood and determined by X-ray crystallography the structure of the muscle protein myoglobin, giving its outline in 1957 and detailed three-dimensional structure in 1959. By observing the alpha helical nature of the polypeptide chain, he was the first to confirm the structure proposed by Pauling. He was awarded the 1962 Nobel Prize for Chemistry jointly with Perutz. He wrote *The Thread of Life* (1966), was elected FRS in 1960 and knighted in 1974.

KEPLER, Johannes
(1571–1630)

German astronomer, born in Weilderstadt, Württemberg, the son of a mercenary soldier, and educated at the University at Tübingen where he obtained a master's degree in theology in 1591. Being a deeply religious Lutheran he originally intended to enter the church ministry, but when the Protestant Estates of Styria were looking for a Professor of

Mathematics at Graz, Kepler accepted the post (1594), having been strongly influenced by the teaching of Michael Maestlin, his Professor of Mathematics at Tübingen and an enthusiastic supporter of the ideas of Copernicus. Among Kepler's duties at Graz was the publication of almanacs to forecast the weather and to predict favourable days for various undertakings with reference to the rules of astrology: in fact Kepler was for a time astrologer to the Duke Albrecht of Wallenstein, the great soldier of the Thirty Years' War. By temperament inclined to mysticism, Kepler was overjoyed when he thought he had found, as recorded in his first major publication, the *Mysterium Cosmographium* (1596), that the distances from the Sun of the six planets including the Earth could be related to the five regular solids of geometry, of which the cube is the simplest. He sent copies of his book to Galileo and Tycho Brahe, the greatest astronomers of the day, who responded in a friendly manner. When Kepler was in difficulties in Graz, the latter invited him to join him in Prague. Kepler arrived in Prague in 1600, and when Tycho Brahe died in 1601 he was appointed to succeed him as imperial mathematician by the emperor Rudolf II. Among Kepler's early work at Prague was a treatise on optics (1604) and a paper on the nova of the same year which appeared in the constellation Ophiuchus ('Kepler's supernova'). His chief interest was the study of the planet Mars, for which Tycho's numerous observations provided very valuable material. He found that the movement of Mars could not be explained in terms of the customary cycles and epicycles, and published his discovery in his *Astronomia Nova* (1609), which he dedicated to the emperor. In this fundamental book Kepler broke with the tradition of more than 2,000 years by demonstrating that the planets do not move uniformly in circles but in ellipses with the Sun at one focus and with the radius vector of each planet describing equal areas of the ellipse in equal times (Kepler's first and second laws). He completed his researches in dynamical astronomy 10 years later by formulating, in his *De Harmonica Mundi* (1619), his third law which connects the periods of revolution of the planets with their mean distances from the Sun. In 1627 he published the *Tabulae Rudolphinae* named after the emperor, which contained the ephemerides of the planets according to the new laws, and also an extended catalogue of 1,005 stars based on Tycho's observations. Kepler died of fever at Regensburg where he had journeyed to the Diet in quest of funds.

KERR, Roy Patrick
(1934–)

New Zealand mathematician, born in Kurow. After receiving an MSc in New Zealand and a PhD from Cambridge, Kerr returned to New Zealand to become Professor of Mathematics at the University of Canterbury, Christchurch. His main contribution has been in the field of astrophysics. In 1916, Schwarzschild had introduced the idea that when a star contracts under gravity, there will come a point at which the gravitational field is so intense that nothing – not even light – can escape, and a black hole is formed. Schwarzschild had derived a mathematical description of the properties of such an object and its gravitational distortion of the surrounding space and time, with the assumption that the black hole is not rotating. However, the condition of non-rotation is unrealistic as almost all stars are found to rotate, and it is thought that this rotation would be preserved during gravitational collapse. Kerr found a new solution to Einstein's equations taking account of the resulting angular momentum to give the 'Kerr metric', an expression which completely describes the properties of any black holes that physicists expect to exist. In later work, he formulated the Kerr-Schild solutions, which were very useful in exact solution of the equations of general relativity.

KHORANA, Har Gobind
(1922–)

Indian-born American molecular chemist, born in Raipur (now in Pakistan). He studied at Punjab University, receiving his BSc in 1943 and an MSc in 1945. He was awarded a PhD in organic chemistry by Liverpool University, and was a Research Fellow at Cambridge before moving to Vancouver as Head of the Department of Organic Chemistry (1952). During 1960–1970, he was professor and co-director of the Institute of Enzyme Research at the University of Wisconsin, and since 1970, he has been Professor of Biology and Chemistry at MIT. His early work was on the biochemistry of enzymes, but in the 1960s he turned to the nucleic acids and the genetic code. He determined the sequence of the nucleic acids, also known as 'bases', for each of the 20 amino acids in the human body. Each amino acid was usually found to have a pattern of three base codes, but Khorana discovered some with more than one triplet sequence of bases, and showed that some triplets code for start/stop sequences. His work on nucleotide synthesis at Wisconsin was a major contribution to the elucidation of the genetic code. In the early 1970s, he was one of the first to artificially synthesize a gene, initially from yeast, and then later from the bacterium *Escherichia coli*. He shared the 1968 Nobel Prize for Physiology or Medicine with Nirenberg and Holley.

KILBURN, Tom
(1921–)

English scientist born in Dewsbury. Educated in Sidney Sussex College, Cambridge, where he read mathematics, his wartime work at the Telecommunications Research Establishment at Malvern brought him into contact with Sir Frederick Calland Williams. When the latter moved to Manchester University in 1946, Kilburn was seconded to follow him and together they built the world's first operational computer in 1948. Kilburn helped perfect the storage device (the world's first electronic random access memory) and published the results. He was awarded his PhD in 1948. In the early 1950s, Williams handed over to Kilburn the direction of the university's computer projects, which by now included important collaborative ventures with the Manchester-based electronics firm of Ferranti. With Kilburn's help Ferranti introduced a string of technically successful computers, such as the Mark I (1951) and the Atlas (1962). Kilburn's design for the Atlas was a high water mark in British computing, pioneering many modern concepts in paging, virtual memory and multi-programming. Kilburn became Professor of Computing Science at Manchester (1964–1981) and was elected FRS in 1965.

KIRWAN, Richard
(1733–1812)

Irish chemist, born in Galway. He was educated at the University of Poitiers with the idea of becoming a Jesuit, but gave up the idea when he became heir to the family estates on the death of his brother. After practising briefly as a lawyer, Kirwan spent 10 years in London and was elected FRS in 1780. On his return to Ireland he helped to found the Royal Irish Academy, presiding over it from 1799 to his death. Kirwan did valuable work on chemical affinity and the composition of salts, publishing the first systematic work on mineralogy in English in 1784. He is best known, however, for his opposition to the discoveries of Lavoisier, instead maintaining the traditional view that when a substance burns it loses a vital essence referred to as 'phlogiston'. He also challenged the revolutionary views of the Scottish geologist Hutton, who in 1785 argued that the Earth was hot and that crystalline rocks were igneous in origin. Towards the end of his life, Kirwan turned to metaphysics. He died in Dublin.

KLINGENSTIERNA, Samuel
(1698–1765)

Swedish mathematician and scientist, born in Linköping. He studied law at Uppsala, but subsequently turned to mathematics and physics. He was appointed secretary to the Swedish Treasury, and given a scholarship to travel and study. He studied under Christian von Wolff at Marburg and Jean Bernoulli at Basle. He was appointed Professor of Mathematics at Uppsala, and in 1750 Professor of Physics there; he became tutor to the crown Prince (later Gustav III) in 1756. Klingenstierna showed that some of Isaac Newton's views on the refraction of light were incorrect, and designed telescope lenses which substantially reduced image defects arising from chromatic and spherical aberrations. By communicating his findings to Dollond, he also contributed to Dollond's success in constructing achromatic telescopes.

KLUG, Sir Aaron
(1926–   )

Lithuanian-born English biophysicist, born in Zelvas. He moved to South Africa as a young child and studied physics at the universities of Witwatersrand and Cape Town. He became a research student in the Cavendish Laboratory of Cambridge (1949–1952) before moving to London to Bernal's department at Birkbeck College (Nuffield Research Fellow, 1954–1957) where he worked with Rosalind Franklin and became particularly interested in viruses and their structure. Following Franklin's death in 1958, he became Head of the Virus Structure Research Group at Birkbeck (1958–1961) and returned to Cambridge in 1962 as a Fellow of Peterhouse and member of staff at the Medical Research Council's Laboratory of Molecular Biology, becoming its director in 1986. His studies employed a wide variety of techniques, including X-ray diffraction methods, electron microscopy, and structural modelling, to elucidate the structure of viruses such as the tomato bushy stunt virus and the polio virus. From the 1970s he also applied these successful methods to the study of chromosomes, and other biological macromolecules such as muscle filaments. He was elected FRS in 1969, awarded the Nobel Prize for Chemistry in 1982 and knighted in 1988.

KNOPOFF, Leon
(1925–   )

American geophysicist, born in Los Angeles. He was educated at Caltech, where he graduated with a BS in electrical engineering (1944), an MS (1946), and a PhD in physics (1949). His first appointment was at the University of Miami in Oxford, Ohio (1948–1950);

since 1950 he has worked at the University of California at Los Angeles (UCLA) where he became Professor of Geophysics in 1959, Professor of Physics in 1961 and research musicologist in 1963. From 1972 to 1986 he was also Associate Director of their Institute of Geophysics and Planetary Physics, Knopoff devised the first representation theorem for the full seismic wave equation (1956) and made important advances in work on the diffraction of seismic waves, e.g. by the core of the Earth (1959). He was the first to apply long-period seismic array data to the interpretation of geological structures (1966), and deduced that North America comprises a number of different tectonic provinces (1974). In 1967 he pioneered numerical models to simulate seismicity and geological faulting, having in 1956 demonstrated the relationship between seismic energy release and velocity of slip on the fault; later he described the universal power law for the spatial distribution of earthquakes (1980), which has implications for the geometry of faults. He also showed that the value in the Gutenberg-Richter magnitude-frequency law is a characteristic of seismicity on a complex fault system (1992).

KOCH, (Heinrich Hermann) Robert
(1843–1910)

German physician and pioneer bacteriologist, born in Klaustbal in the Ham. In 1862 he entered the University of Göttingen, and he received his medical degree in 1866. He settled as a physician in Rakwitz, and later practised medicine at Hanover and elsewhere. His work on wounds, septicaemia and splenic fever gained him a seat on the Imperial Board of Health in 1880. Koch proved that the anthrax bacillus was the sole cause of the disease, and demonstrated that its epidemiology was a result of the natural history of the bacterium. This work was published in 1876 and 1877; later he quarrelled with Pasteur over anthrax. Despite his severe myopia, further researches in microscopy and bacteriology led to his discovery on 24 March 1882 of the tubercle bacillus that causes tuberculosis, responsible for one in seven of all deaths in Europe during this period. In 1883 he was leader of the German expedition sent to Egypt and India in quest of the cholera germ; for his discovery of the cholera bacillus he received a gift of £5,000 from the government. In 1890 he produced a drug named tuberculin which he claimed could prevent the development of tuberculosis. It was found to be ineffective as a cure, but later proved useful in diagnosis. He became Professor at Berlin and Director of the Institute of Hygiene in 1885, and first Director of the Berlin Institute for Infectious Diseases in 1891. In 1896 and 1903 he was summoned to South Africa to study rinderpest and other cattle plagues: in disclosing the causes of disease and expounding the means of prevention, Koch was unsurpassed. He won the Nobel Prize for Physiology or Medicine in 1905 for his work on tuberculosis. His formulation of essential scientific principles known as 'Koch's postulates' for investigating the causes of infectious diseases established clinical bacteriology as a medical science in the 1890s.

KOCHER, Emil Theodor
(1841–1917)

Swiss surgeon, born and educated in Bern. He became professor at the University of Bern in 1871. In the first generation of surgeons able to exploit the new possibilities of anti- and aseptic techniques, Kocher developed general surgical treatment of disorders of the thyroid gland, including goitre and thyroid tumours. His observations of patients suffering the long-term consequences of removing the thyroid gland helped elucidate some of its normal functions; by the 1890s, the isolation of one of the active thyroid hormones made replacement therapy possible. Kocher's Bern clinic attracted many young surgeons from all over the world. His textbook *Operative Surgery* (1894) went through many editions and translations. He kept meticulous case records, and followed patients over long periods. He advocated the practice of 'physiological surgery', whereby the surgeon's aim was to keep operative trauma to a minimum, thereby conserving as much of the patient's tissues as possible. He pioneered operations of the brain and spinal cord, and during World War I did experimental work on the trauma caused by gunshot wounds. Kocher was the first surgeon to be awarded the Nobel Prize for Physiology or Medicine (1909), for his work on the physiology, pathology, and surgery of the thyroid gland.

KÖHLER, Georges Jean Franz
(1946–   )

German immunochemist, born in Munich. After receiving his doctorate at the University of Freiberg in 1974, he joined Milstein at the Medical Research Council Laboratory in Cambridge, UK, where they discovered how to produce hybridomas – cell lines grown in culture that were derived (created) by the fusion of an antibody-generating cell from mouse spleen with a cancer cell. Hybridomas possess the property of infinite life and generate a colony of cells (clone) that produce a single type of antibody (monoclonal antibody) against a specific antigen (foreign body). Mixed antibody-producing cells generate a confusing mixture of antibodies, and the use of hybridomas opened the way to a precise examination of antibody structure. Köhler moved to the Basle Institute of Immunology in 1976, and in 1984 he became one of three directors of the Max Planck Institute of Immune Biology in Freiberg. He continued his research by using markers for drug resistance to study the pattern of inheritance of hybridoma cells, and demonstrated that structural mutants of immunoglobulins could be formed by hybridomas (1980). More recently he has studied the carbohydrate component of immunoglobulins by using the antibiotic tunicamycin, which specifically inhibits the attachment of carbohydrate to certain lipids. Commercially produced monoclonal antibodies now provide an unambiguous and sensitive way of identifying and quantifying a wide range of substances, from the animal species of the protein used in processed food, pregnancy testing and the concentrations of drugs, hormones or other metabolites in body fluids to diagnosing and treating cancer and other diseases. For this outstanding contribution to human well-being, Köhler shared with Milstein and Jerne the 1984 Nobel Prize for Physiology or Medicine.

KOLFF, Willem Johan
(1911–   )

Dutch-born American physician, developer of the artificial kidney, born in Leiden. A medical student in Leiden, he received his MD in 1946 from Gröningen University. Kolff constructed his first rotating drum artificial kidney in wartime Holland and treated his first patient with it in 1943. This dialysis machine used a series of membranes to remove impurities from the blood which would ordinarily be filtered out by the healthy kidney.

From 1950, when he moved to the USA, he worked primarily at the Cleveland Clinic and Utah University, developing the artificial kidney further; he was also involved in research on the heart-lung machine used during open-heart surgery.

KOLMOGOROV, Andrei Nikolaevich
(1903–1987)

Russian mathematician, born in Tambov. He studied at Moscow State University where he graduated in 1925, and remained there throughout his career, as professor from 1931 and 2nd Director of the Institute of Mathematics from 1933. He worked on a wide range of topics in mathematics, including the theory of functions of a real variable, functional analysis, mathematical logic, and topology. He is particularly remembered for his creation of the axiomatic theory of probability in his book *Grundbegriffe der Wahrscheinlichkeitsrechnung* (1933), which he interpreted in the language of measure theory introduced by Lebesgue. His work with Khinchin on Markov processes was also of lasting significance; in this he formulated the partial differential equations which bear his name and which have found wide application in physics and chemistry. Kolmogorov also worked in applied mathematics, on the theory of turbulence, on celestial mechanics where he refined ideas of Poincaré, on information theory and in cybernetics.

KORNBERG, Arthur
(1918–)

American biochemist, born in Brooklyn, New York City. A graduate in medicine from Rochester University. He became director of enzyme research at the national Institutes of Health (1947–1952) and head of the microbiology department of Washington University (1953–1959). In 1959 he was appointed professor at Stanford University. When Crick and James Watson developed their model of the DNA Molecule, the details of how DNA synthesis occurred and how DNA gives rise to protein remained unknown. In studies of *Escherichia coli*, Kornberg discovered DNA polymerase, the enzyme that synthesizes new DNA from a mixture of the four deoxynucleoside triphosphates, and showed that synthesis required a DNA template and base-pairing, giving two helical strands of opposite polarity as predicted by Crick and Watson. For this work he was awarded the 1959 Nobel Prize for Physiology or medicine jointly with Ochoa. Around 1972 Kornberg showed that this bacterial enzyme (polymerase I) proof-read the new DNA for accuracy and could replace by DNA the RNA 'primer' to which the new DNA was initially joined in the Okazaki fragment. Kornberg became the first to synthesize viral DNA (1967) and wrote *DNA Replication* (1980). He also elucidated the synthesis of the coenzymes nicotinamide-adeninedinucleotide (NAD) and flavine-adeninedinucleotide (FAD), showed that acylCoA was the activated form of a fatty acid, and studied the *Escherichia coli* enzymes anthranilate synthetase and phosphopholipase A.

KOSSEL, Albrecht
(1853–1927)

Swiss-born German physiological chemist, born in Rostock. He studied medicine at the University of Strassburg and worked under Hoppe-Seyler. In 1895 he became Professor of Physiology and Director of the Physiological Institute, Marburg, and he was later professor at Heidelberg (1901–1923). Kossel carefully isolated nuclei, freed from cytoplasm by treatment with proteases, from the heads of spermatozoa and from avian red cells, and showed them to contain nuclein, the nucleoprotein previously found and identified as rich in phosphorus by Friedrich Miescher (1844–1895), another of Hoppe-Seyler's students. Kossel separated nuclein into its two components, protein and nucleic acid; he showed that the latter was the component rich in phosphorus and contained the four DNA bases, adenine, guanine (already known), thymine and cytosine, and carried out a partial identification of their structures. He was able to explain that in a blood leukaemia, the 'guanine' found in the blood in large amounts derived from decomposed young nucleated erythrocytes. He also discovered histidine in spermatozoa (1896). In 1910 Kossel was awarded the Nobel Prize for Physiology or Medicine.

KREBS, Edwin Gerhard
(1918–)

American biochemist, born in Lansing, Iowa. Krebs trained at the University of Illinois and with Carl Cori before joining the Howard Hughes Medical Institute and Department of Pharmacology at the University of Washington School of Medicine. Stimulated by Cori's discovery of the activation of glycogen phosphorylase by adenylic acids, Krebs and Edmond Fischer showed that phosphorylation-dephosphorylation processes are involved, catalysed by two enzymes, phosphorylase kinase and phosphorylase phosphatase, respectively (1955). In 1959 they went on to show that phosphorylase kinase is also regulated by phosphorylation dephosphorylation and that cyclic AMP (an adenylic acid derivative discovered by Earl Sutherland) was involved. These initial findings led to the discovery of the phosphorylation amplification cascade of enzymes that initiates the rapid switching on of glycogen phosphorylase and other enzymes, notably pyruvate, dehydrogenase, under the influence of hormone; such as glucagon and adrenaline (epinephrine). Similar systems controlled by other activators, such as protein kinase C by diacylglycerol, were also subsequently discovered. His more recent work relates to the homology of the kinases (of evolutionary significance) and the properties of the phosphatases. With Fischer, Krebs was awarded the 1992 Nobel Prize for Physiology or Medicine. He was elected FRS in 1947 and knighted in 1958.

KREBS, Sir Hans Adolf
(1900–1981)

German-born British biochemist, born in Hildesheim. After training in medicine and working as assistant to Warburg at the Kaiser Wilhelm Institute for Cell Physiology, Berlin he emigrated to the UK (1934) and worked with Frederick Gowland Hopkins on redox reactions. He became a lecturer in pharmacology (1935–1945) and Professor of Biochemistry at Sheffield (1945–1954), then Whitley Professor of Biochemistry at Oxford (1954–1967). In 1932, with K. Henseleit, he described the urea cycle whereby carbon oxide and ammonia form urea in the presence of liver slices and catalytic amounts of ornithine and citrulline (previously found only in citrus fruits). Leading on from his earlier work, showing that all the hydrogens of glucose passed through fumarate and that pyruvate had a central role in this process, he discovered that citrate also acted as a catalyst while citrate could be formed from oxaloacetate. In this way he elucidated the citric acid cycle (Krebs' cycle) of energy production (around 1943). He also discovered D-amino acid oxidase L-glutamine synthetase, purine synthesis in birds, that ketone bodies were formed in starvation and that rat heart muscle preferentially utilized ketone bodies (1961). In 1953 he shared with Lipmann the Nobel Prize for Physiology or Medicine for his discovery of the citric acid cycle. He was elected a Fellow of the Royal Society in 1947 and was awarded the society's Royal (1954) and Copley (1961) medals.

KROTO, Harold Walter
(1939–)

English chemist, born in Wisbech and educated at the University of Sheffield (BSc 1961, PhD 1964). He then moved to the National Research Council, Ottawa, where he developed his interest in the electronic spectroscopy of free radicals (which he had begun in his PhD days) and in microwave spectroscopy. In 1966 he moved to Bell Telephone Laboratories, New Jersey, where he carried out Raman spectroscopic studies of liquids and quantum chemistry calculations. Kroto is distinguished for his work in detecting unstable molecules, especially those containing reactive multiple bonds, using methods such as microwave and photoelectron spectroscopy. His studies extend to molecules which exist in interstellar space. He discovered, along with astronomers at the Herzberg Research Institute of Astrophysics at Ottawa, polyyne molecules in interstellar space – the most complex and heaviest interstellar molecules known. In 1985, together with co-workers at Rice University, Texas, he discovered the third allotrope of carbon C60, known as 'buckminsterfullerene' (familiarly 'buckyballs') after the American architect Buckminster Fuller. The 'football' shape of C60, built up from an array of pentagonal and hexagonal faces, has the same topology as the buildings designed by this architect. Kroto was elected FRS in 1990, and is now Royal Society Research Professor at the University of Sussex. He won the Nobel prize in 1997 for chemistry.

KUHN, Richard
(1900–1967)

German chemist, born in Vienna-Döbling. After study at the University of Vienna he worked for his doctorate with Willstätter in Munich. In 1926 he moved to Zürich, and in 1929 to the Kaiser Wilhelm Institute for Medical Research in Heidelberg, where he remained for the rest of his life. His early work on enzymes led to an interest in problems of stereochemistry that preoccupied him subsequently. Work on conjugated polyenes led to important studies on carotenoids and vitamin A. Later work on vitamins B2 and B6 and on 4-aminobenzoic acid earned him the award of the Nobel Prize for Chemistry (1938). He was forbidden by the Nazi government to accept the award, but it was presented to him after World War II. His research continued actively after the war when he worked on resistance factors effective in preventing infections in both plants and animals. He and his collaborators published over 700 scientific papers.

KUMMER, Ernst Eduard
(1810–1893)

German mathematician, born in Sorau. He studied theology, mathematics and philosophy at the University of Halle, and then taught at the gymnasium in Liegnitz (1832–1842), where Kronecker was among his students. He became known to Jacobi and Dirichlet through his work on the hypergeometric series, and was elected a member of the Berlin Academy of Sciences in 1839. He was later appointed Professor of Mathematics at Breslau (1842–1855) and subsequently at Berlin (from 1855). Kummer worked in number theory, where, in generalizing Gauss's law of quadratic reciprocity, he gained a significant insight on Fermat's last theorem and proved it rigorously for many new cases. The 'ideal numbers' he introduced here were later developed by Dedekind and Kronecker into one of the fundamental tools of modern algebra. He also worked on differential equations and in geometry, where he discovered the quartic surface named after him.

L

LACK, David
(1910–1973)

English ornithologist, born in London and educated at Magdalene College, Cambridge. He taught at Dartington Hall from 1933, and then in 1945 moved to Oxford as Director of the Edward Grey Institute. Whilst teaching at Dartington, he published a popular book, *The Life Of the Robin* (1943), which formed the basis of his reputation. He was very influential in studying the Geospiza finches of the Galapagos Islands (*Darwin's Finches*, 1947) and the effect of the diet and food supply on their adaptation and differentiation. While in Oxford, Lack initiated a series of long-term studies, notably of great tits in Wytham Woods, but also of a number of other species, including swifts; the results were published in a popular book, *Swifts in a Tower* (1956). He was a dominant influence in the transformation of ornithology from an observational to a scientific discipline. His most important books were *National Regulation of Numbers* (1954), *Population Studies of Birds* (1966), *Ecology Adaptations for Breeding in Birds* (1968), *Birds* (1971). He was elected FRS in 1951.

LAËNNEC, René Théophile Hyacinthe
(1781–1826)

French physician, born in Quimper, Brittany. An army doctor from 1799, in 1814 he became editor of the *Journal de Medicine* and physician to the Salpétrière, and in 1816 chief physician to the Hospital Necker, where he invented the stethoscope in the same year. His stethoscope consisted of a simple hollowed tube of wood, with adaptations at the end to help transmit sound more easily. The familiar binaural stethoscope, with rubber tubing going to both cars, was not developed until after Laënnec's death. He demonstrated the importance of the stethoscope in diagnosing diseases of the lungs, heart and vascular systems, and introduced the basic vocabulary to describe heart and lung sounds. In 1819 he published his *Traité de l'auscultation médiate*, the fruit of three years' intense labour and including outstanding clinical and pathological descriptions of many chest diseases, such as tuberculosis, pneumonia, pleuritis and bronchitis. He followed his patients throughout their illnesses and performed postmortem examinations in fatal cases, correlating the signs and symptoms he had noticed while the patient was alive with the pathological lesions in their dead bodies. He was a pious Catholic and royalist, and his academic career after the end of the Napoleonic Wars benefited from the political conservatism which then prevailed. He died of tuberculosis, a disease he had described so brilliantly.

LAMARCK, Jean-Baptiste Pierre Antoine de Monet, Chevalier de
(1744–1829)

French naturalist and evolutionist, born in Bazentin from a long line of military horsemen. At the age of 19 he escaped from Jesuit school to join the army, where he rapidly distinguished himself in battle against the Germans. As an officer at Toulon and Monaco he became interested in the Mediterranean flora. Resigning after an injury, he held a post in a Paris bank, and meanwhile began to study medicine and botany. In 1773 he published *Flore francaise*, the first key to French flowers, and the following year became keeper of the royal garden (later the nucleus of the Jardin des Plantes). From 1794 he was keeper of invertebrates at the newly formed Natural History Museum. He lectured on zoology, originating the taxonomic distinction between vertebrates and invertebrates. His *Histoire des animaux sans vertebres* appeared in 1815–1922. By about 1801 he had begun to think about the relations and origin of species, expressing his conclusions in his famous *Philosophie zoologique* (1809). Lamarck broke with the notion of immutable species, recognizing that species needed to adapt to survive environmental changes, and postulating a gradual process of development from the simple to the complex. In this he foreshadowed Charles Darwin's ideas, but he became best known for espousing the idea that the development or atrophy of organs by 'use or disuse' can be inherited by later generations. This 'use or disuse' was mediated by habitual behaviour relating to the animal's needs, not (except in the highest vertebrates) by conscious volition, a suggestion for which he has been unfairly chastized by evolutionists from Darwin onwards. Eventually, hard work and illness enfeebled his sight and left him blind and poor, a pathetic figure suffering in old age the taunts of the anti-evolutionist Cuvier.

LAMB, Willis Eugene
(1913–)

American physicist, born in Los Angeles and educated at the University of California. He later became professor at Columbia University. New York (1938–1951), before being appointed to similar posts at Stanford (1951–1956), Oxford (1956–1962) and Yale (1962–1974). He moved to the University of Arizona as Professor of Physics in 1974. His accurate studies of the hyperfine structures of the hydrogen spectrum showed that the two possible energy states of hydrogen, rather than being equal as predicted by Dirac, differed in energy by a very small amount. This became known as the 'Lamb shift', and prompted a revision of the theory of interaction of the electron with electromagnetic radiation. This revision led to the theory known as quantum electrodynamics. Lamb shared with Kusch the 1955 Nobel Prize for Physics for this research.

LANDAU, Lev Davidovich
(1908–1968)

Soviet physicist, born in Baku. At the age of 14 he entered Baku University and he received his PhD from Leningrad University in 1927. He studied with Niels Bohr in Copenhagen, and became Professor of Physics at Moscow in 1937. In 1932, following the discovery of the neutron, he proposed the existence of neutron stars.

Landau developed a theory to describe the properties of liquid helium. Below a critical temperature, helium has zero viscosity in one direction, a property known as superfluidity. Rather than trying to describe the properties of the liquid in terms of the individual atoms, he explained the behaviour in terms of the collective behaviour of the atoms in the liquid. This was later further developed by Feyman. Landau received the 1962 Nobel Prize for Physics for work on theories of condensed matter, particularly helium.

LANDSTEINER, Karl
(1868–1943)

Austrian-born American serologist, immunologist and pathologist, born in Vienna. He studied chemistry in Vienna, Würzburg, Munich and Zürich before turning to medicine, and received his MD in 1891. During a series of assistantships in Vienna, he became interested in serology. Blood transfusions had often been attempted as a means of treating blood loss following injuries, but rather than aiding recovery they frequently hastened death. Landsteiner showed that people could be divided into groups according to the ability of their blood serum to make the red blood cells of other people cluster together. In 1901 he published a paper describing a technique for dividing human blood into three groups: A, B and C (later O). His colleagues found a fourth group, later termed AB. These discoveries led to the development of safe blood transfusions. In 1909 Landsteiner was appointed to the Chair of Pathology at the University of Vienna. Life for scientists was very difficult in postwar Vienna; Landsteiner left to become a prosecutor at a hospital in the Hague. In 1922 he joined the Rockefeller Institute in New York where he remained for the rest of his life. In 1927 Landsteiner and his colleagues discovered the M, N and MN blood factors. He was awarded the 1930 Nobel Prize for Physiology, or Medicine. In 1940 he was involved in the discovery of the Rhesus (Rh) groups, and also conducted important work on poliomyelitis, being the first to isolate the poliomyelitis virus (1908).

LAPLACE, Pierre Simon, Marquis de
(1749–1827)

French mathematician and astronomer, born in Beaumont-en-Auge, Normandy, the son of a farmer. He studied at Caen, went to Paris and became Professor of Mathematics at the École Militaire where he gained fame by his researches on the inequalities in the motion of Jupiter and Saturn, and the theory of the satellites of Jupiter. In 1785 he became a member of the French Academy of Sciences, and prospered at a time when mathematics flourished under Napoleon, who had a strong interest in the science. In 1799 he entered the senate, becoming its Vice-President in 1803, and Minister of the Interior for six weeks, after which he was replaced because of an incapacity for administration. He was created marquis by Louis XVIII in 1817. His astronomical work culminated in the publication of the five monumental volumes of *Mécanique céleste* (1799–1825), the greatest work on celestial mechanics since Newton's *Principia*. His *Systéme du monde* (1796) was a nonmathematical exposition, of masterly clarity, of all his astronomical theories, and his famous nebular hypothesis of planetary origin occurs as a note in later editions. This proposed that the solar system originated as a massive cloud of gas, and that the centre collapsed to form the Sun, leaving outer remnants which condensed to form the planets. In his study of the gravitational attraction of spheroids he formulated the fundamental differential equation in physics which bears his name. He also founded the modern form of probability theory.

LARTET, Edouard Arman Isidore Hippolyte
(1801–1871)

French vertebrate palaeontologist, stratigrapher and prehistorian, born in St Guiraud, Gers, and educated at the college in Auch where he received a prize from Napoleon. Lartet went to Toulouse to study law, but his life changed direction when he inherited the family estate. He undertook important studies of French Tertiary and Quaternary vertebrates and discovered the fossil jawbone of an ape, Pliopithecus, in the Tertiary formations of Sansan, Gers (1836), which refuted the assertion of Cuvier that neither men nor apes could be found in a fossil state. His studies in 1860 at the prehistoric sites of Massat and Aurignac yielded conclusive proof of the contemporaneity of man and extinct animal species. His son Louis (1840–1899) was a well-known stratigrapher and palaeontologist.

LAVOISIER, Antoine Laurent
(1743–1794)

French chemist, one of the greatest names in science. He was born in Paris of a prosperous family and was educated at the College Mazarin. While training as a lawyer, he attended chemistry lectures by G. F. Rouelle and soon showed a genius for many branches of science. He was elected to the Academy of Sciences at an unprecedented young age in 1768 and later became one of its most influential members, playing a large part in the establishment of the metric system, its greatest achievement. From the mid-1760s onward, his scientific research was carried out in parallel with his involvement in finance, civic duties and politics – activities which shaped his career and which in the end cost him his life. In 1768 Lavoisier gained a lucrative position as a member of the Ferme Générale, a private consortium which collected taxes for the government. In 1775 he was made one of the commissioners in charge of the production of gunpowder at the Arsenal in Paris, and in 1787 he was elected to the provincial assembly of Orléans. During his career he campaigned for social reforms and when the Revolution broke out in 1789, he supported it. However, as the Revolution grew more extreme, the former members of the Ferme Générale became objects of public hatred. Lavoisier lost his position at the Arsenal in 1791, and even his record as a reformer and his distinction as a scientist could not save him from the Reign of Terror (1792–1794). He was guillotined in Paris on 8 May 1794. Lavoisier's greatest achievement was to discover what happens during combustion. In 1772, dissatisfied with the long-held belief that when a substance burned it lost a vital essence, termed 'phlogiston', Lavoisier began an extensive series of experiments on burning and in 1774 followed up work by Priestley and Scheele. He showed that when phosphorus, sulphur or mercury is heated, only some of the available air is used up, and that the remainder does not support life or combustion: he therefore concluded that air is a mixture of gases. He identified the portion of the air used during combustion with the 'air' driven off when Priestly's 'red precipitate of mercury' was heated. He showed that this gas combined with carbon to form the 'fixed air' (carbon dioxide) discovered by Joseph Black, and that it was a constituent of acids. Synthesizing these results he realized that combustion is a process in which the burning substance combines with a constituent of the air, a gas which he named 'oxygine' ('acid maker'). His discovery, publicized in a series of papers and demonstrations from 1779 onwards, threw the scientific world into a ferment and was not generally accepted until the 1790s. His best advocate was his book *Traité elémentaire de chimie* (1789), one of the most celebrated works in chemistry. In 1783 Lavoisier showed that water, regarded from earliest times as an element, was composed only of hydrogen and oxygen, a discovery in which he was preceded by Cavendish. He burned alcohol and other organic compounds in oxygen, then weighed the water and carbon dioxide formed, thus laying the basis of quantitative organic analysis. Lavoisier's genius for meticulous experiment distinguished all his work and his influence on quantitative chemical method was as far-reaching as his actual discoveries. He was also a pioneer of physiological chemistry, making extensive studies of fermentation, respiration (which he recognized as a form of combustion) and animal heat. He was the first person to make a clear distinction between an element and a compound. Together with Berthollet, Guyton de Morveau and Fourcroy, he introduced a new System of chemical nomenclature in *Méthode de nomenclature chimique* (1787) to replace the chaotic traditional system. It is still used today.

LAWS, Richard Maitland
(1926–)

English mammalogist and Antarctic scientist, born in Whitley Bay, Northumberland. He was educated at Cambridge University, where he received his PhD in 1953. From 1947 to 1953 he was employed by The Falkland Islands Dependencies Survey, and he later moved to the National Institute of Oceanography. He studied large mammals particularly seals and whales on basis of growth ring in the teeth and had enormous impact on ecology through his leadership of British Antarctic Survey.

LAWTON, John Hartley
(1943–)

English ecologist, born in Preston, Lancashire. He was educated at Durham University, and taught at the universities of Oxford and York. Since 1989 he has been Director of the Natural Environmental Research Council's Centre for Population Biology at Imperial College, London. His early work concentrated on ecology energetics, leading to the recognition of patterns in energy use by animal populations and thus the determinants of the structure of food-webs and the ratios of predator and prey species. One of his most illuminating contributions was the recognition of the role of plant structural or architectural complexity in the control of the species richness of insect herbivore communities. This work, and his collaborative studies on the dynamics of single species and predator-prey populations, naturally led to questions on the structuring of the insect communities on plants. His showed that there is little evidence for overt horizontal, interspecific interactions and suggested that more subtle, plant-mediated competitive effects, as well as vertical interactions such as enemy-free space, play important roles. Much of his fieldwork, which is both extensive and experimental, has been conducted on bracken and its herbivores. Besides contributions to fundamental science, Lawton has directed his findings towards the biological control of this worldwide weed. He was elected FRS in 1989.

LEAKEY, Mary Douglas, née Nicol
(1913–   )

English archaeologist and anthropologist, born in London. Wife of Louis Leakey, her excavations and fossil finds in East Africa revolutionized ideas about human origins. Her interest in prehistory was roused during childhood trips to south-west France, where she collected stone tools and visited the painted caves around Les Eyzies. She met Louis Leakey while preparing drawings for his book *Adam's Ancestors* (1934), and moved shortly afterwards to Kenya where she undertook pioneer archaeological research (1937–1942) at sites such as Olorgesailie and Rusinga Island. In 1948 at Rusinga, in Lake Victoria, she discovered *Proconsul africanus* a 1.7-million-year-old dryo-pithecine (primitive ape) that brought the Leakeys international attention and financial sponsorship for the first time. From 1951 she worked at Olduvai Gorge in Tanzania, initially on a modest scale, but more extensively from 1959 when her discovery of the 1.75-million-year-old hominid *Zinjanthropus* (subsequently reclassified as *Australopithecus*), filmed as it happened, captured the public imagination and drew vastly increased funding. *Homo habilis* – a new species, contemporary with, but more advanced than Zinjanthropus – was found in 1960 and published amidst much controversy in 1964. Perhaps most remarkable of all was her excavation in 1976 at Laetoli, 30 miles south of Olduvai, of three trails of fossilized hominid footprints which demonstrated unequivocally that our ancestors already walked upright 3.6 million years ago. Her books include *Olduvai Gorge: My Search for Early Man* (1979) and an autobiography, *Disclosing the Past* (1984). Her son Richard Leakey (b. 1944) also became a distinguished palaeoanthropologist.

LE BEL, Joseph Achille
(1847–1930)

French chemist, born in Pechelbron Rhin. Born of wealthy parents, educated at the École Polytechnique, subsequently worked with Balard and with Wurtz. In 1874 both he and van't Hoff (who was also working with Wurtz) independently published papers establishing the relationship between the three-dime arrangement of atoms in a molecule chemistry) and optical activity. After this important contribution to the progress of organic chemistry, Le Bel made a few significant discoveries. He divided his time between the family factory and a private laboratory in Paris where he worked mainly on activity. He held no academic appointments but was appointed President of the Chemistry Society in 1892.

LE CHATELIER, Hend Louis
(1850–1936)

French chemist and metallurgist, born in Paris. He was educated at the École Polytechnique and the École des Mines. After brief service in the Corps des Mines, he was invited in 1877 to become Professor of General Chemistry in the École des Mines. In 1887 he exchanged this chair for that of industrial chemistry and retained this post until his retirement in 1919. Parallel with this post he also held from 1898 to 1908 the Chair of Mineral Chemistry at the College de France and, from 1907 onwards, the chair at the Sorbonne previously occupied by Moissan. His awards and honours included the Bessemer Medal of the Iron and Steel Institute (1910) and the Davy Medal of the Royal Society (1916), of which he became a Foreign Member in 1913. La Chatelier's earlier researches were on the nature and setting of cements. This work led him to consider the fundamental laws of chemical equilibrium and in 1884 to formulate the principle named after him, which states that if a change is made in pressure, temperature or concentration of a system in chemical equilibrium, the equilibrium will be displaced in such a direction as to oppose the effect of this change. He devised the platinum/platinumrhodium thermocouple for measuring high temperatures. Le Chatelier later studied gaseous explosions in connection with safety in mines, and made many studies of the metallurgy of steel and other alloys. He invented the inverted stage metallurgical microscope.

LEDER, Philip
(1934–)

American geneticist, born in Washington DC. He was educated at Harvard University, and became a research associate at the National Heart Institute and the National Cancer Institute, and Laboratory Chief of Molecular Genetics at the National Institute for Child Health And Human Development in Bethesda, Maryland (1972–1980). In 1980 be became Professor of Genetics at Harvard University Medical School, and he is currently John Emory Andrus Professor of Genetics at Harvard and a senior investigator at the Howard Hughes Medical Institute. In the late 1970s, Leder worked extensively on the structure and function of the globin genes. He discovered how the multiple globin genes are arranged on the chromosome, and how the coding and non-coding regions of the genes are organized. More recently, he has worked on the function of oncogenes in transgenic mice. Oncogenese are normal cellular genes which can be activated in a variety of ways to become carcinogenic. To examine the relative influences of oncogene activation and environmental effects, activated oncogenes are injected into mouse eggs to create transgenic mice. Leder has the worldwide patent on the 'oncomouse' and has shown that the introduction of cancer to transgenic mice requires the cooperative action of more than one oncogene. He received the National Medal of Science in 1989.

LEDERBERG, Joshua
(1925–)

American biologist and geneticist, born in Montclair, New Jersey. He studied biology at Columbia, and became professor at Wisconsin (1947–1959) and Stanford (1959–1978), and President of Rockefeller University (1978–1990). He has been director and consultant to several biotechnology companies. Procter and Gamble, Ohio, the Celanese Corp. and Alfyniax, Palo Alto. With Tatum, he showed that bacteria can reproduce by a sexual process known as conjunction. By mating two different strains, he found that the new strain had characteristics of both parents. He also introduced the technique of 'replica plating', whereby mutants of a bacterial strain can be retrieved from the original bacterial colony. Working with Zinder, Lederberg made a further fundamental contribution through his description of 'transduction' in bacteria, whereby the bacterial virus transfers part of its DNA into the host bacterium; this led to the development of techniques for manipulation of genes. He served as a consultant to the US space programme in the early 1960s, and was also a consultant to the World Health Organisation on biological warfare. In 1958 he was awarded the Nobel Prize for Physiology or Medicine, jointly with Tatum and Beadle.

LEE, Taung-Dao
(1926–   )

Chinese-born American physicist, born in Shanghai. Educated at Kiangsi and at Chekiang University, he won a scholarship to Chicago in 1946 where he worked under Teller. He became a lecturer at the University of California, and from 1956 was professor at Columbia University, as well as a member of the Institute for Advanced Study (1960–1963). The electromagnetic and strong nuclear forces are known to conserve a quantum property known as parity, but Feynman and Martin Block suggested this may not be true for the weak nuclear force. With Yang, Lee made a thorough analysis of all the known data in particle physics and concluded that parity was unlikely to be conserved in weak interactions. They suggested a simple experiment which would prove it. In the same year (1956), a similar experiment by a group of physicists from Columbia University and American National Bureau of Stan headed by Wu, confirmed that the 'law' of parity is indeed violated in the case of weak interactions. For this prediction, Lee and Yang were awarded the 1957 Nobel Prize for Physics, and the Einstein Commemorative Award from Yeshiva University in the same year. Lee's other research interests have been in the areas of statistical mechanics, theory and turbulence.

LEEUWENHOEK, Anton van
(1632–1723)

Dutch cloth merchant and amateur scientist, born in Delft. Apprenticed in Amsterdam and educated as a businessman, he became skilled in grinding and polishing lenses to inspect cloth fibres. He returned to Delft around 1652 and became a draper and haberdasher. Through his second wife, Cornelia Walmius, he joined Delft's more intellectual circles. His civic posts included chamberlain to the sheriffs of Delft, alderman and winegauger. With his microscopes, each made for a specific investigation, he discovered microscopic animalicules (protozoa) in water everywhere (1674), bacteria in the tartar of teeth (1676), that all males produce spermatozoa – even fleas, lice and mites – and he described their copulation and life cycles. Independently, he discovered blood corpuscles (1674), blood capillaries (1683), striations in skeletal muscle (1682), the structure of nerves (1717) and plant microstructures among endless other observations. After his death, his daughter Maria auctioned his collection of 248 microscopes and 178 separate lenses. Extant lenses magnify from 30 × to 266 × . His discoveries, in more than 110 letters (published 1684), were described with the assistance of Delft's anatomist, Cornelis-Gravesande, and introduced to the Royal Society by Graaf. Leeuwenhoek was elected FRS in 1680.

LEHMANN, Inge
(1888–1993)

Danish geophysicist, born in Copenhagen. She was educated at a school founded by Niels Bohr's aunt, where girls were encouraged to study the same subjects as boys, and entered Copenhagen University in 1907 to read mathematics. After a year at Newnham College, Cambridge (1910–1911), she pursued a career in insurance until 1918, although she maintained an interest and contacts in science. Her studies were resumed in 1918 and she received her degree in 1920. She was awarded her PhD at the University of Copenhagen in 1928. From 1928 until her retirement in 1953, she was chief of the seismological department of the newly founded Danish Geodetic Institute, taking responsibility for the seismological stations in Greenland. Lehman's research involved the interpretation of seismic events detected by European stations; in this work she discovered that the presence of a distinct inner core of the Earth was required to explain the data received from large epicentral distances. In collaboration with Gutenberg she also endeavoured to resolve a velocity structure compatible with Harold Jeffreys revised travel time tables, and found a low-velocity layer at 200 kilometres depth. This fitted well with European seismic data and became accepted generally.

LEIBNIZ, Gottfried Wilhelm
(1646–1716)

German mathematician and philosopher, remarkable also for his encyclopaedic knowledge and diverse accomplishments outside these fields. He was born in Leipzig, the son of a professor of moral philosophy, studied there and at Altdorf, showing great precocity of learning, and in 1667 obtained a position at the court of the Elector of Mainz on the strength of an essay on legal education. There he codified laws, drafted schemes for the unification of the churches, and was variously required to act as courtier, civil servant and international lawyer, while at the same time he absorbed the philosophy, science and mathematics of the day, especially the work of Descartes, Isaac Newton, Pascal and Boyle. In 1672 he was sent on a diplomatic mission to Paris, where he met Nicolas Malebranche and Huygens, and he went on in 1676 to London where his discussions with mathematicians of Newton's circle led later to an unseemly controversy as to whether he or Newton was the inventor of the infinitesimal calculus. Leibniz had published his system of differential calculus in *Nova Methodus pro Maximis et Minimis* ('New Method for the Greatest and the Least') in 1684; Newton published his in 1687, though he could relate this to earlier work. The Royal Society formally declared for Newton in 1711, but the controversy was never really settled; it is Leibniz's notation that appears in modern calculus. In 1676 he visited Baruch Spinoza in the Hague on his way to take up a new, and his last, post as librarian to the Duke of Brunswick at Hanover. Here he continued to elaborate his mathematical and philosophical theories (although he did not publish them), and maintained a huge learned correspondence. He also travelled in Austria and Italy in the years 1687–1690 to gather materials for a large-scale history of the House of Brunswick, and went in 1700 to persuade Frederick I of Prussia to found the Prussian Academy of Sciences in Berlin, of which he became the first president. He died in Hanover without real recognition and with almost all his work unpublished. Leibniz was perhaps the last universal genius, spanning the whole of contemporary knowledge. He made original contributions to optics, mechanics, statistics, logic and probability theory; he built calculating machines, and contemplated a universal language; he wrote on history, law and political theory; and his philosophy was the foundation of eighteenth-century rationalism. His best-known philosophical doctrine was that the world is composed of an infinity of simple, indivisible, immaterial, mutually isolated 'monads' which form a hierarchy the highest of which is God; the monads do not interact causally but constitute a synchronized harmony with material phenomena.

LEISHMAN, Sir William Boog
(1865–1926)

Scottish bacteriologist, born in Glasgow. He obtained his MD from the University of Glasgow in 1886, and later became Professor of Pathology at the Army Medical College and Director-General of the Army Medical Service (1923). In 1900 he discovered the protozoan parasite (*Leishmania*) responsible for the disease known variously as kala-azar and dumdum fever. He went on to develop the widely used 'Leishman's stain' for the detection of parasites in the blood. Leishman also made major contributions to the development of various vaccines, particularly those used against typhoid, and it was as a result of his work that mass vaccination was introduced in 1914 for the British army. He was knighted in 1909, and elected FRS in 1910.

LELOIR, Luis Frederico
(1906–)

French-born Argentinian biochemist, born in Paris. Educated in Buenos Aires and at Cambridge, he worked mainly in Argentina where he set up his own Research Institute in 1947. Following unproductive work on the adrenals and diabetes, he recognized the proteolytic action of renin from the kidneys on a precursor yielding the hormone angiotensin, an important hormone which increases blood pressure (1940). He then turned to glucose metabolism and discovered the enzyme glucose 1-phosphate kinase (1949), and that the product of the reaction, glucose 1,6-bisphosphate, is a coenzyme of the glycolysis pathway enzyme, phosphoglucomutase. Extending these studies he identified galactokinase, and showed that the product, galactose 1-phosphate, is converted into glucose 1-phosphate. In the 1950s Leloir linked these reactions to the formation of the energy storage polysaccharide glycogen, and showed that the glucose is added by a stepwise transfer process in which the reactive intermediate is a nucleotide derivative, uridine diphosphateglucose (UDPG), and further, that galactose is converted to glucose by a similar mechanism. For this work, of medical significance and complimenting that of Carl Cori, he was awarded the Nobel Prize for Chemistry in 1970, becoming the Argentinian to be so honoured.

LEMAITRE, Georges Henri
(1894–1966)

Belgian astrophysicist and cosmologist, born in Charleroi. After studies in engineering at the University of Louvain and voluntary service in World War I in which he was decorated with the Belgian *Croix de Guerre*, he turned to mathematical and physical sciences and obtained his doctorate in 1920. Three years later he was ordained as a Catholic priest and in the same year he obtained a travelling scholarship from the Belgian government. This took him to Cambridge, Harvard and MIT. In Cambridge he came under the strong influence of Eddington. In 1927 he published his first major paper on the model of an expanding universe and its relation to the observed redshifts in the spectra of galaxies. In the 1930s he developed his ideas on cosmology, and from 1945 onwards he put forward the notion of the 'primeval atom' which is unstable and explodes, starting what is now called the Big Bang, the beginning of the expanding universe. Lemaitre received many honours from all over the world. He was the first recipient of the Eddington Medal of the Royal Astronomical Society (1953) and in 1960 became the President of the Pontifical Academy of Sciences.

LENZ, Heinrich Friedrich Emil
(1804–1865)

Russian born German physicist and geophysicist, born in Dorpat (now Tartu). He studied chemistry and physics at Dorpat University. His first post was as a geophysical observer during a scientific voyage around the world (1823–1826) on the sloop *Predpriatie*. He was appointed Professor of Physics at St Petersburg (1836) and a member of the Russian Academy of Sciences. In 1834 he formulated 'Lenz's law' governing induced current, according to which the emf induced in a circuit is such as to oppose the flux change giving rise to it, and is credited with discovering the dependence of electrical resistance on temperature (Joule's law). It soon became clear that these physical laws were special cases of the law of conservation of energy.

LEUCKART, Kari George Friedrich Rudolf
(1822–1898)

German zoologist, born in Helmstedt. He studied at Göttingen, and became Professor of Zoology at Giessen (1850) and Leipzig (1869). A pioneer of parasitology, he described the complex life cycles of many parasites such as tapeworms and liver flukes. He was able to demonstrate that the disease trichiniasis in man was due to infection by a roundworm. He also showed that the radiata did not comprise a natural group and that the radial symmetry found both in the coelentrates and echinoderins (starfish) did not imply close phylogenetic relationship. Between 1863 and 1876 he wrote his great parasitological treatise *Parasites of Man* (translated 1886).

LEVI-MONTALCINI, Rita
(1909–)

Italian neuroscientist, born and educated in Turin, where she graduated in medicine in 1936. She began studying the mechanisms of how nerves grow, but form 1939 onwards was prevented, as a Jew, from holding an academic position, and worked from a home laboratory. During the latter part of World War II she served as a volunteer physician. In 1947 she was invited by Viktor Hamburger to Washington University in St Louis, where she remained until 1981 (Research Associate 1947–1951; Associate Professor 1951–1958; Professor 1958–1981), when she moved to Rome. Her work has primarily been on chemical factors that control the growth and development of cells, and she isolated, originally from mouse salivary glands, a substance now called nerve growth factor that promoted the development of sympathetic nerves. Her work continued on locating further sources of the factor, on determining its chemical nature, and examining its biological activity in isolated tissues and whole neonatal and adult animals. She revealed that there were many diverse sources of the factor, such as mouse cancer cells and snake venom glands, that it was chemically a protein, and that cells are most responsive to its effects during the early stages of differentiation. This work has provided powerful new insights into processes of some neurological diseases and possible repair therapies, into tissue regeneration, and into cancer mechanisms. In 1986 she shared the Nobel Prize for Physiology or Medicine with Stanley Cohen.

LEWIS, Sir Thomas
(1881–1945)

Welsh cardiologist and clinical scientist, born in Cardiff. He received his pre-clinical training at University College, Cardiff. In 1902 he went to University College Hospital, in London, where he remained as student, teacher and consultant until his death. Starling stimulated his interest in cardiac physiology and the physician Mackenzie awakened his curiosity about diseases of the heart. He was the first to master completely the use of the electrocardiogram, and he and his students established the basic parameters which still govern the interpretation of electrocardiograms in health and disease. Through animal experiments he was able to correlate the various electrical waves recorded by an electrocardiograph with the sequence of events during a contraction of the heart. This enabled him to use the instrument as a diagnostic aid when the heart had disturbances of its rhythm, damage to its valves or changes due to high blood pressure, arteriosclerosis and other conditions. During his later years he turned his attention to the physiology of cutaneous blood vessels and the mechanisms of pain. He conducted experiments on himself in an attempt to elucidate the distribution of pain fibres in the nervous system and to understand patterns of referred pain. He fought for full-time clinical research posts to investigate what he called 'clinical science'. This broadening of his interests was signalled when he changed the name of the journal he had founded in 1909, from *Heart* to *Clinical Science* (1933). His textbooks of cardiology went through multiple editions and translations. A compulsive worker, he spent a month's holiday each year fishing and photographing birds. He was knighted in 1921.

LEWONTIN, Richard Charles
(1929–   )

American geneticist, born in New-York City, one of the major figures in the development of population genetics in the second half of the twentieth century. He received his postgraduate training (1951–1954) under Dobzhansky at Columbia University, and subsequently held positions at North Carolina State College, the University of Rochester, the University of Chicago and Harvard, where he was appointed Alexander Agassiz Professor of Zoology in 1973. Lewontin is equally at home in the theoretical and experimental dimensions of his field, and has contributed substantially to both areas. The central theme to his research is the problem articulated by Dobzhansky: how to examine the distribution of variation within and among natural populations. His theoretical contributions have primarily been in the area of multi-locus theory, while his empirical work, which involves detailed studies of the genetics of natural populations of *Drosophila* (fruit fly) species, has been dominated by what he has termed 'the struggle to measure variation'. He synthesized much of the work in these fields by himself and others in his influential 1974 book, *The Genetic Basis of Evolutionary Change*. In 1966, in two seminal papers with University of Chicago biochemist J. L. Hubby, Lewontin introduced gel electrophoresis as a means of assaying variation in protein sequences. The widespread application of this technique resulted in the discovery of large amounts of variation, and indirectly led to Motoo Kimura's development of the neutral theory of molecular level – that at the molecular level, most evolutionary changes result from mutations with no selective advantages. He has continued to apply technological developments biochemistry to population genetic problems; the first study of nucleotide sequence polymorphism was carried out in his laboratory in 1983. Lewontin has also published widely on human genetics, especially with reference to the problems of a sociobiological analysis of human behaviour (e.g. *Not in Our Genes: Biology, Ideology & Human Nature*, 1984) and is a politically active Marxist.

LIE, (Marius) Sophus
(1842–1899)

Norwegian mathematician, born in Nordfjordeide. He studied at Christiania (now Oslo) University, then supported himself by giving private lessons. After visiting Klein in Berlin, and studying with Klein under Camille Jordan in Paris, a Chair of Mathematics was created for him in Christiania. In 1886 he succeeded Klein as Professor of Mathematics at Leipzig, but he returned once more to Christiania in 1898. His study of contact transformations arising from partial differential equations led him to develop an extensive theory of continuous families of transformations, now known as Lie groups. A natural geometer, he thought it preferable to write up his work in the language of analysis; his ideas were not always clear, and it was not until the later efforts of Cartan and Weyl that modern theories of Lie groups and algebras began to develop. These theories have become a central part of twentieth-century mathematics, and have important applications in quantum mechanics and the theory of elementary particles.

LIND, James
(1716–1794)

Scottish physician, born in Edinburgh. He first served in the navy as a surgeon's mate, then after qualifying in medicine at Edinburgh, became physician at the Royal Naval Hospital at Haslar. In 1747 he conducted a classic therapeutic trial, dividing 12 patients suffering from scurvy into six groups of two, treating each group with a different remedy. The two sailors given two oranges and a lemon each day responded most dramatically. His work on the cure and prevention of scurvy helped induce the Admiralty in 1795 at last to issue the order that the navy should be supplied with lemon juice, and during the Napoleonic Wars the British navy suffered far less scurvy than the French. Lind also stressed cleanliness in the prevention of fevers, and wrote major treatises on scurvy, fevers and the diseases encountered by Europeans in tropical climates. His writings are full of sensible practical advice, couched in a broad environmental framework.

LINDERSTRØM-LANG, Kaj Ulrik
(1896–1959)

Danish biochemist, born in Copenhagen and trained at the city technical college. He worked for a few months at the National Research Institute for Animal Husbandry in Copenhagen before joining the staff of the Carlsberg Laboratory as assistant to Sørenson. In 1938 Sørensen retired and Linderstrøm-Lang, became professor and head of the chemistry department. In 1926, with Sørensen, he first titrated the dissociable protons from proteins using the hydrogen electrode to measure pH, and in 1927 he reported the use of organic solvents to lower the dissociation constant (pK) of an ionizing group whose value was too high to measure by normal methods. His work on proteolytic (digestive) enzymes led him into a major study of protein structure. By using deuterium labelling, he found that the rate at which the protons of a polypeptide exchanged with the medium was either slow (associated with hydrogen-bonded structures within the molecule) or fast, if more superficially located, and variable, reflecting protein inherent structural instability. From his own and other observations he classified globular protein structure into three divisions (1951): primary (amino acid sequence plus disulphide bridges), secondary (helices and pleated sheets based on hydrogen bonding) and tertiary (folding to form the active protein) which, together with quaternary structure (subunit association), has become the basis of all modem teaching on this subject. In 1935 Linderstrøm-Lang was elected to the Danish Royal Society as its youngest member. He became a Foreign member of the Royal Society in 1956.

LINNAEUS, Carolus (Carl von Linné)
(1707–1778)

Swedish naturalist and physician, the founder of modern scientific nomenclature for plants and animals. Born in Rashult, the son of the parish pastor, he studied medicine briefly at Lund and then botany at Uppsala, where he was appointed lecturer in 1730. He explored Swedish Lapland (1732) and published the results in *Flora Lapponica* (1737), then travelled in Dalecarlia in Sweden and went to Holland for his MD (1735). In Holland he published his system of botanical nomenclature in *Systema Naturae* (1735), followed by *Fundamenta Botanica* (1736), *Genera Plantarum* (1737) and *Critica Botanica* (1737), in which he used his so-called 'sexual system' of classification based on the number of flower parts, for long the dominant system. His major contribution was the introduction of binomial nomenclature of generic and specific names for animals and plants, which permitted the hierarchical organization later known as systematics. He returned to Sweden in 1738 and practised as a physician in Stockholm, and in 1741 became Professor of Medicine and Botany at Uppsala. In 1749 he introduced binomial nomenclature, giving each plant a latin generic name with a specific adjective. His other important publications included *Flora Suecica* and *Fauna Suecica* (1745), *Philosophia Botanica* (1750), and *Species Plantarum* (1753). His manuscripts and collections are kept at the Linnaean Society in London, founded in his honour in 1788. In his time he had a uniquely influential position in natural history.

LIPPIVIANN, Gabriel Jonas
(1845–1921)

French physicist, born in Hollerich, Luxembourg, to French parents who settled in Paris. He was appointed Professor of Mathematical Physics at the Faculty of Sciences in Paris (1883), Professor of Experimental Physics at the Sorbonne (1886), and the Director of the Laboratory of Physical Research. While on a scientific mission in Germany, he began research in electrocapillarity in the laboratory of Kirchhoff, which led on his return to Paris to his invention of a very sensitive mercury capillary electrometer. He made many contributions to instrument design, including developing an astatic galvanometer, the coelostat which made it possible to photograph (or observe) a region of the sky for an extended period without apparent movement, and a new form of seismograph. For his technique of colour photography based on the interference phenomenon, subsequently also used by Lord Rayleigh (1842–1919), he was awarded the 1908 Nobel Prize of Physics. He was elected FRS the same year.

LIPSCOMB, William Nunn
(1919–)

American inorganic chemist, born in Cleveland, Ohio. He studied at Kentucky and Caltech, and was appointed Professor of Chemistry at Harvard in 1959. He deduced the molecular structures of a group of boron hydride compounds (boranes) by X-ray crystal diffraction analysis in the 1950s, and went on to develop novel theories to explain their chemical bonds. His ingenious experimental and theoretical methods were later applied by him and others to a variety of related chemical problems. He was awarded the Nobel Prize for Chemistry in 1976.

LISTER, Joseph, Lord
(1827–1912)

English surgeon, the 'father of antiseptic surgery', born in Upton, Essex. He was the son of the microscopist Joseph Jackson Lister (1786–1869). After graduating from London University in arts (1847) and medicine (1852), he became house surgeon at Edinburgh Royal Infirmary to the surgeon James Syme, whose daughter he married in 1856. He was successively a lecturer on surgery, Edinburgh; Regius Professor of Surgery, Glasgow (from 1859); Professor of Clinical Surgery, Edinburgh (from 1869) and King's College Hospital, London (1877–1893); and President of the Royal Society (1895–1900). In addition to important observations on the coagulation of the blood and the microscopical investigation of inflammation, his great work was the introduction of his antiseptic system (1867), which revolutionized modern surgery. His system was inspired by Pasteur's work on the role of micro-organisms in fermentation, putrefaction and other biological phenomena. Lister began soaking his instruments and surgical gauzes in carbolic acid, a well-known disinfectant. His early antiseptic work was primarily concerned with the operative reduction of compound fractures and the excision of tuberculous joints. Both conditions would previously have been treated with amputation, since the operation mortality for conservative surgical treatment would have been very high. Antiseptic surgery led to aseptic techniques, whereby the organisms causing wound infections were, whenever possible, excluded from the surgical field altogether. Listerian procedures made it possible for surgeons to open the adnominal, thoracic and cranial cavities without fatal infections supervening. Lister himself rarely ventured into the bodily cavities, confining himself mostly to the limbs and superficial parts of the body. He was, however, a technically gifted surgeon who popularized the use of sutures made of catgut. He worked later in his life on the aetiology of wound infection and was an ardent advocate of the value of experimental science for medical and surgical practice. He was much revered in life and was the first medical man to be elevated to the peerage.

LOCKYER, Sir (Joseph) Norman
(1836–1920)

English astronomer, born in Rugby. On leaving school he became a clerk at the war office to which he remained technically attached (1857–1875), devoting as much time as he could spare to science. In 1868 he designed a spectroscope for observing solar prominences outside of a total eclipse and succeeded in doing this independently of Janssen who had used the same principle a few months earlier. Lockyer shared with Janssen the credit for this discovery, commemorated by a medallion issued by the French government. In the same year he postulated the existence of an unknown element which he named helium (the 'Sun element'), an element not found on Earth until 1895 by Ramsay. He also discovered and named the solar chromosphere. In 1875 he became a member of the staff of the Science Museum in South Kensington, London. His researches gave rise to unconventional ideas such as his theory of dissociation, whereby atoms were believed to be capable of further subdivision, and his meteoritic hypothesis which postulated the formation of stars out of meteoric material. Among other activities, Lockyer took part in eclipse expeditions and made surveys of ancient temples for the purpose of dating them by astronomical methods. He was the founder (1869) and first editor of the scientific periodical *Nature*, and was knighted in 1897. His solar physics observatory at South Kensington was transferred to Cambridge University in 1911, but Lockyer remained active in a private observatory which he set up in Sidmouth, Devon, until his death.

LOEB, Jacques
(1859–1924)

German-born American biologist, born in Mayen. Educated in philosophy at Berlin, and in medicine at Strassburg, Loeb obtained his MD in 1884, and in 1886 was appointed to an assistantship at Würzburg. An interest in the philosophy of the will led to research which attempted to show, in the animal world, phenomena analogous to plant tropisms. He began to publish in this area in 1888, showing that certain caterpillars move towards light even when their food is in the opposite direction. He emigrated to the USA in 1891, and held various university appointments before becoming head of the general physiology division at the Rockefeller Institute for Medical Research (1910–1924). He conducted pioneering work on artificial parthenogenesis. He demonstrated that unfertilized sea-urchin eggs, made to start segmentation by osmotic changes, could subsequently grow into larvae (1899). A champion of materialism in philosophy, of mechanistic explanations in science, and a socialist in politics, Loch became well known to the American public. His writings included *Dynamics of Living Matter* (1906) and *Artificial Parthenogenesis and Fertilisation* (1913).

LOEWI, Otto
(1873–1961)

German pharmacologist, born in Frankfurt. Educated at Strassburg and Munich, he was appointed Professor of Pharmacology at Graz (1909–1938). Forced to leave Nazi Germany in 1938, he became research professor at New York University College of Medicine from 1940. From 1901 he worked for a time alongside Dale in the laboratories of Starling at University College London, and confirmed current theories that a chemical substance released by the nerve terminal transmitted the stimulus invoking contraction of an isolated frog heart. In 1921 he showed that the substance released by stimulating one isolated heart (Wagusstoff) caused contraction in another. He subsequently identified several possible transmitter substances and distinguished Wagusstoff as acetylcholine, rather than adrenaline or a similar substance, because (as predicted by Dale) its action was prolonged by inhibition of the enzyme that caused its breakdown (acetylcholinesterase) with escrine. Anticholinesterase drugs are now used to increase muscle strength in patients with myasthenia gravis, an inherited muscular dystrophy. Loewi shared with Dale the 1936 Nobel Prize for Physiology or Medicinc for investigations on the chemical transmission of nerve impulses. He was elected a Foreign Member of the Royal Society in 1954.

LONGUET-HIGGINS, (Hugh) Christopher
(1923–)

English theoretical chemist, born in Lenham, Kent. He studied chemistry at Balliol College, Oxford, where he was a research fellow from 1946 to 1948. From 1948 to 1952 he was lecturer and then reader in theoretical chemistry at the University of Manchester before briefly holding the Chair of Theoretical Physics at King's College, London (succeeding Coulson), and then in 1954 becoming Plummer Professor of Theoretical Chemistry at Cambridge (succeeding Lennard-Jones). From 1968 to 1974 he was Royal Society Research Professor in the Department of Artificial Intelligence at the University of Edinburgh and from 1974 to his retirement in 1988 he held a similar position at the University of Sussex. For some 20 years he made fundamental contributions to the molecular orbital theory of organic and inorganic chemistry, successfully predicting the failure of resonance to explain convincingly the reactivity of biphenylene compounds and the formation of complex molecules containing a metal and cyclobutadiene. He used symmetry arguments to predict the course of various electrocyclic reactions, thus paralleling the work of Robert Woodward and Hoffmann. Following his move to Edinburgh in 1968, he embarked on a second phase of research in which he has worked on problems of the mind, including language acquisition, music perception and speech analysis. He was elected FRS in 1958 and a foreign associate of the US National Academy of Sciences in 1968, and gave the Gifford Lectures in 1972.

LORENZ, Konrad Zacharias
(1903–1989)

Austrian zoologist and ethologist, born in Vienna, the son of a surgeon. He studied in Vienna, and along with Tinbergen, advocated the observation of animal behaviour under natural conditions rather than in the laboratory, thus founding in the late 1930s the school of animal behaviour called ethology. Lorenz and colleagues mainly studied the behaviour of birds, fish and some insects whose behaviours contain a relatively high proportion of stereotyped elements or fixed action patterns. These could be used in the same way as morphological characters as the basis for comparisons between species. Thus Lorenz was able to compare the courtship of different species of duck and suggest how the different patterns had arisen and subsequently evolved. He was the first to describe imprinting and release mechanisms. In the former, newly hatched birds, such as goslings and jackdaws, become imprinted on the first moving object they see and treat the object thereafter as a member of their own species and even as a potential mate. Releaser mechanisms are innate behaviours which are elicited by very specific stimuli, for example the red breast of the robin releasing aggression. Although Lorenz was criticized for the rigidity of these and other ethological concepts, the ethological approach had a profound influence on subsequent animal behaviour studies. In *On Aggression* (1963), he argued that aggressive behaviour in man is an inherited drive and can be channelled into non-destructive activities such as sport. The validity of this view of human aggression is still disputed. In addition to his scientific writings he wrote popular accounts of his work with animals such as *King Solomon's Ring* (1949) and *Man Meets Dog* (1950). He shared the 1973 Nobel Prize for Physiology or Medicine with Tinbergen and Karl von Frisch.

LORENZ, Ludwig Valentin
(1829–1891)

Danish physicist, born in Elsinore. He trained as a civil engineer at the Technical University of Denmark in Copenhagen, going on to teach at the Danish Military Academy where he became professor in 1866. His researches concerned many areas of optics, heat and the electrical conductivity of metals. In his early work on the nature of light itself, he adopted a model based on elasticity to describe wave propagation. He eventually abandoned this and concentrated instead on finding a phenomenological description of the way light passes through matter. Pursuing this he advanced a mathematical description for light waves (1863), showing that under certain conditions double refraction will occur. The publication of his work on relating refraction and specific densities of media (1869) preceded that of Lorentz by one year; the result became known as the Lorentz-Lorenz formula. From observations of the scattering of sunlight in the atmosphere, Lorenz made the first fairly accurate estimate of Avogadro's number (1890). Two years after the publication of Maxwell's famous paper on electromagnetic theory, without knowledge of Maxwell's results, Lorenz published his own theory of electromagnetism.

LOWRY, (Thomas) Martin
(1874–1936)

English chemist, born in Bradford, West Yorkshire. He studied chemistry under Henry Armstrong at the City and Guilds Institute, South Kensington, and from 1896 was for some years Armstrong's assistant. In 1906 he became lecturer in chemistry at Westminster Training College, and in 1912 he moved to Guy's Hospital Medical School. From 1913 he was head of the chemistry department at Guy's, and he became the first Professor of Chemistry in any London medical school. During World War I he was engaged in the development of explosives. In 1920 Lowry became the first to hold a chair of physical chemistry at Cambridge, where he remained until his death. His earliest researches were on camphor derivatives and in particular the changes in optical rotation (mutarotation) which occur when these are treated with acids or bases as catalysts in solution. This ultimately led to his redefinition of the terms acid and base (1923) in the same way as that advocated independently by Brønsted in the same year. In his later days he worked extensively on optical rotatory dispersion, foreshadowing its importance many years later as a structural tool in organic chemistry. He was elected FRS in 1914, appointed CBE in 1920, and gave the Bakerian Lecture in 1921. From 1928 to 1930 he was President of the Faraday Society.

LUBBOCK, Sir John, 1st Baron Avebury
(1834–1913)

English politician and biologist, born in London, the son of the astronomer Sir J. W. Lubbock (1803–1865). From Eton he went into his father's banking house at the age of 14, becoming a partner in 1856. He served on several educational and currency commissions, and in 1870 was returned for Maidstone as a Liberal MP, and then in 1880 for London University – from 1886 to 1900 as a Liberal Unionist. He succeeded in passing more than a dozen important measures, including the Bank Holidays Act (1871), the Bills of Exchange Act, the Ancient Monuments Act (1882) and the Shop Hours Act (1889). He was Vice-Chancellor of London University (1872–1880), President of the British Association (1881), Vice-President of the Royal Society, President of the London Chamber of Commerce, Chairman of the London County Council (1890–1892), and much else. Scientifically, he is best known for his researches on primitive man and on the habits of bees and ants; he published *Prehistoric Times* (1865, revised 1913), *Origin of Civilisation* (1870) and many books on natural history. He was a neighbour of Charles Darwin, and a friend and counsellor over many years. He was a member of the 'X' Club (together with T. H. Huxley, Joseph Dalton Hooker and others) which conspired to replace the ecclesiastical establishment with a scientific one.

LUDWIG, Karl Friedrich Wilhelm
(1816–1895)

German physiologist. Born in Witzenhausen, Ludwig became a medical student in Marburg in 1834. His was a stormy student career: duelling left him with a heavily scarred lip, and friction within the university authorities drove him to study elsewhere. In 1840 he returned to Marburg and was teaching there by 1846. Later he taught in Zürich, Vienna and Leipzig. In 1865 he helped establish the famous Institute of Physiology in Leipzig. Ludwig's work proved fundamental to modern physiology. He denied there was any role for vital force, seeking explanations of living processes in the paradigms of physics and chemistry. In formulating such views, he was much influenced by his friend, the chemist Bunsen. Ludwig proved fruitful in devising medical instruments, especially diagnostic technology. In 1846 he developed the kymograph and used it to garner much information about the circulation and respiration. In 1859 he designed the mercurial blood-pump, in 1867 the stream gauge, and in 1865 perfusion, a method of maintaining circulation in an isolated organ. His blood-pump allowed examination of blood gases and respiratory exchange. His research focused on the operation of the heart and kidneys, on the lymphatic system, and on salivary secretion. The problem of the operation of secretion was an enduring preoccupation. The circulation of the blood also attracted his attention. He investigated the relations of blood pressure to heart activity and probed the role of muscles in the fluidity of the blood. He proved an immensely energetic and influential teacher.

LURIA, Salvador Edward
(1912–1991)

American biologist, born in Turin, Italy. He graduated in medicine at Turin University in 1935, and went on to the Radium Institute in Paris to study medical physics, radiation and techniques of working with phage, the bacterial virus. When Italy entered World War II, Luria emigrated to the USA, where he taught at Indiana University; James Watson was among his students there. In 1943, with Delbrück, Luria showed that bacteria undergo mutations. He went on to demonstrate in 1951 that phage genes can also mutate, and that different strains of phage can exchange and recombine genes. With Hershey and Delbrück Luria founded the Phage Group, committed to using phage to investigate genetics. During the repressive McCarthy era, he was refused a visa to travel to a scientific conference in Oxford (1952). In 1961 he was awarded the Nobel Prize for Physiology or Medicine, jointly with Delbrück and Hershey, for discoveries related to the role of DNA in bacterial viruses. Critical of the cost of the US defence and space budgets, he gave part of his prize money to pacifist groups. He wrote *General Virology* (1953), which became a standard textbook and published his autobiography *A Slot Machine, A Broken Test-Tube, An Autobiography* in 1985.

LYAPUNOV, Aleksandr Mikhailovich
(1857–1918)

Russian mathematician, born in Yaroslavl, the son of an astronomer. He studied at St Petersburg where he came under the influence of Chebyshev, and then taught at Kharkov University. He returned to St Petersburg as professor in 1901. Lyapunov is principally associated with important mathematical methods in the theory of the stability of dynamical systems, related to Poincaré's work although largely discovered independently of it. He committed suicide after his wife's death from tuberculosis.

LYNEN, Feodor Felix Konrad
(1911–1979)

German biochemist, born and educated in Munich. He studied under Wieland (1935–1937) before joining the Department of Chemistry at the University of Munich (1942). He later became professor there (from 1947) and Director of the Max Planck Institute for Cell Chemistry and Biochemistry (1954–1979). In 1951 he isolated coenzyme A from baker's yeast and showed that it formed acetyl-SCoA, an important intermediate in lipid metabolism, via a thioester bond, thereby adding a new dimension to the metabolic energy concepts of Lipmann. He also showed that the conversion of acetyl-S-CoA to form malonyl-S-CoA, a precursor of fatty acid biosynthesis, involved the uptake of carbon dioxide by the vitamin biotin (1958). With Ochoa, he showed in 1953 that acyl thioesters of ethanolamine would substitute for the full coenzyme A molecule in enzyme reactions and he used these to study the pathway of fatty acid degradation and the formation of acetoacetate (ketone bodies). At the same time as Konrad Bloch, Lynen independently isolated isopentanyl pyrophosphate and contributed evidence towards elucidating the biosynthesis of cholesterol. He also worked on the biosynthesis of cysteine, terpenes, rubber and fatty acids. He gave the first of the newly instituted Otto Warburg Medal Lectures of the German Chemical Society in 1963, and was awarded the Nobel Prize for Physiology or Medicine in 1964 jointly with Bloch.

LYON, Mary Frances
(1925–   )

English biologist, born in Norwich. She obtained a scholarship to Girton College, Cambridge University, where she graduated 1946, received a PhD in 1950 and an ScD in 1968. In 1950 she joined the UK Medical Research Council's staff, and has worked since 1955 at their Radiobiology Unit, Harwell/Chilton. From 1962 she headed its Genetics Division, and she was Deputy Director from 1986 until 1990, when she officially retired. She is a Foreign Associate of the US

National Academy of Sciences and Foreign Honoraria Member of the Genetics Society of Japan. She chaired the Committee on Standardized Genetic Nomenclature for Mice (1975–1990), as well as the Mouse Genome Committee of the Human Genome Organisation (HUGO). Lyon has published on many aspects of mammalian genetics and mettagenesis. Her name is particularly associated with 'Lyon hypothesis' of random inactivation of the mammalian X chromosome, which she propounded in 1961. She suggested that one of the two X chromosomes in female animals is inactivated in early development becoming the 'sex chromatin' found in the nucleus of normal female, but not male cells, that females are in effect mosaics of the different genetic cell lines (characterized by which of the X chromosomes is switched off). This idea has been widely confirmed and has proved to be of great value in studies on clinical genetics and imprinting. She has extended knowledge of the mammalian X, especially with respect to human-mouse homologies. Her long-term research on the t-complex region of mouse chromosome 17 has elucidated many puzzling features and made it the most thoroughly studied part of the mouse genome, and her studies of the genetic effects of low radiation doses and female germ-cell exposures have strengthened genetic risk assessment. Through her leadership of mouse committees and her work on mouse genetic compilations, she demonstrated the immense value of mouse genetics in helping us to understand the mammalian genome and to tackle the problems of hereditary disease. She was elected a Fellow of the Royal Society in 1973, and was awarded its Royal Medal in 1984.

M

MACH, Ernst
(1838–1916)

Austrian physicist and philosopher, born in Chirlitz-Turas, Moravia. Until his teens he was educated almost entirely by his father. After a few years at the local gymnasium he entered the University of Vienna. By 1860 his doctorate was complete, and he taught mathematics and physics before being appointed Professor of Mathematics at Graz in 1864. In 1867 he moved to Prague, where he was Professor of Experimental Physics for 28 years. Finally, in 1895, he was elected to the Chair of History and Theory of the Inductive Sciences at Vienna. A crippling stroke in 1897 forced his resignation from active research in 1901, but he continued to write and served as a member of the Austrian parliament. Between 1873 and 1893 Mach experimented on the rapid flow of air over projectiles and wave propagation. New optical measuring techniques allowed him to obtain remarkable photographs of gas jets and projectiles in flight with accompanying shock waves. Its name is associated with the ratio of speed of a body, or of the flow of a fluid, to the speed of sound in the same medium (Mach number). From the early 1860s Mach stressed the interdependence of physical concepts with the physiological and psychological processes of sensory perception: 'psychophysics'. Within his positivist philosophy, all knowledge was a conceptual organization of sensory experience. Scientific statements must be empirically verifiable and should merely summarize our experience economically; theories dependent on metaphysical constraints or referring to entities not reducible to sensory elements (e.g. absolute space, absolute time, atoms) were to be rigorously rejected. Mach undertook a searching historical critique of Newtonian mechanics within these terms. The 'Mach principle', so named by Einstein in 1918, asserted that the inertia of a body depended on its relationship with the matter of the entire universe: it was meaningless to assign inertia to an isolated body. Einstein admitted the crucial role played by Mach's work in the formation of relativity theory which, ironically, Mach opposed.

MACNAMARA, Dame (Annie) Jean
(1890–1968)

Australian physician, born in Beechworth, Victoria. Educated at Melbourne University, she worked in local hospitals where she developed a special interest in 'infantile paralysis'. During the poliomyelitis epidemic of 1925, she tested the use of immune serum, and convinced of its efficacy, she visited England, the USA and Canada with the aid of Rockefeller scholarship. With Burnet, she found that there was more than one strain of polio virus, a discovery which led to the development of the Salk vaccine. She also supported the experimental treatment developed by Elizabeth Kenny, and introduced the first artificial respirator (iron lung) into Australia. She was created a DBE in 1935, and later became involved in the controversial introduction of the disease myxomatosis as a means of controlling the rabbit population of Australia. In the early 1950s it was estimated that as a result of her efforts the wool industry had saved over £30 million.

MAGNUS, Heinrich Gustav
(1802–1870)

German physicist, born in Berlin, discoverer of the 'Magnus effect'. He entered the University of Berlin in 1822, was awarded a doctorate five years later for a dissertation on tellurium, and then moved to Sweden to work under Berzelius who became his lifelong friend. He returned to Berlin in 1928 and continued his researches in the field of chemistry for a while, then seemed to change direction towards physics, and in 1845 he became Professor of Technology and Physics at the University of Berlin. He worked on a very wide range of topics including thermoelectricity, the boiling of liquids, electrolysis (Magnus's rule relates to the electrolytic deposition of metals from solutions of their salts), optics, mechanics, magnetism, fluid mechanics and aerodynamics. It was in this last field that he studied the flow of air over rotating cylinders and in 1853 discovered and evaluated the Magnus effect – the sideways force experienced by a spinning ball, which is responsible for the swerving of golf or tennis balls when hit with a slice. The phenomenon was exploited many years later in the Flettner rotor ship, which used large vertical rotating cylinders as its motive power, and successfully crossed the Atlantic in 1926.

MALPIGHI, Marcello
(1628–1694)

Italian anatomist and microscopist, born near Bologna, where he studied philosophy and medicine. He later became professor at Pisa, Messina and Bologna (1666), and from 1691 served as Chief Physician to Pope Innocent II. Malpighi was an early pioneer of histology, plant and animal, conducting a remarkable series of microscopic studies of the structure of the liver, lungs, skin, spleen and brains, many of which were published in the *Philosophical Transactions of the Royal Society*. Studying the lung, Malpighi showed it was linked to the venous system on one side and to the arterial system on the other, thereby corroborating the insights of William Harvey (1578–1657). He also further extended Harvey's work by his discovery of the capillaries. He wrote a treatise on the silkworm, giving the first full account of an insect, the silkworm moth, and investigated muscular cells. In the 1670s he became involved in plant anatomy, discovering stomata in leaves and delineating the formation of the plant embryo.

MALTHUS, Thomas Robert
(1766–1834)

English economist and clergyman, born at Rookery, near Dorking. He was Ninth Wrangler at Cambridge in 1788, was elected fellow of his college (Jesus) in 1793, and in 1797 became curate at Albury, Surrey. In 1798 he published anonymously his *Essay on the Principle of Population*, and having travelled to several countries to collect more data, the next edition appeared in 1803, bearing his name. Malthus argued that population increases faster than food supplies, and that population growth should be kept in check. Charles Darwin read Malthus in 1838, after his return from his travels on HMS *Beagle*, and was greatly influenced by him, seeing in the struggle for existence a mechanism for producing new species natural selection. The work was also a strong influence on Wallace, who developed a theory of evolution around the same time. In 1805 Malthus became Britain's first Professor of Political Economy, at the newly established East India College in Haileybury. His other works included *An Inquiry into the Nature and Progress of Rent* (1815) and *Principles of Political Economy* (1820).

MANSON, Sir Patrick
(1844–1922)

Scottish physician, born in Old Meldrum, Aberdeen, known as 'Mosquito Manson' from his pioneer work with Sir Ronald Ross in malaria research. He studied medicine in Aberdeen and then practised in the East in China (from 1871) and Hong Kong (from 1883), where he helped start and was the first Dean of a school of medicine that became the University of Hong Kong. In China, he studied a chronic disease called elephantiasis and showed that it is caused by a parasite spread through mosquito bites. This was the first disease to be shown to be transmitted by an insect vector. In 1890 Manson set up practice in London, where he became the leading consultant on tropical diseases and was appointed medical adviser to the Colonial Office; in 1899 he helped to found the London School of Tropical Medicine. He was the first to argue that the mosquito is host to the malaria parasite (1877), and encouraged Ross in his own researches, acting as his chief London agent. Manson's *Tropical Diseases* (1898) helped define the new specialism, emphasizing the importance of entomology, helminthology and parasitology in understanding the diseases peculiar to tropical climates.

MARCONI, Gugiielmo, Marchese
(1874–1937)

Italian physicist and inventor, born in Bologna of wealthy parents, an Italian father and Irish mother. Educated for a short time at the Technical Institute of Livorno, but mainly by private tutors, he became fascinated by the discovery of electromagnetic waves by Heinrich Hertz and started experimenting with a device to convert them into electricity. His first successful experiments in wireless telegraphy were made at Bologna in 1895, and in 1898 he transmitted signals across the English Channel. In 1899 he erected a wireless station at La Spezia, but because of the continuing indifference of the Italian government to his work, he decided to establish the Marconi Telegraph Co in London. In 1901 he succeeded in sending signals in Morse code across the Atlantic. He patented in 1902 the magnetic detector, and in 1905 the horizontal directional aerial. He shared the 1909 Nobel Prize for Physics with Ferdinand Braun. He later developed short-wave radio equipment, and established a worldwide radio telegraph network for the British government. From 1921 he lived on his yacht, the *Elettra* and in the 1930s he was a strong supporter of the Italian Fascist leader Mussolini

MARIOTTE, Edmé
(c.1620–1684)

French experimental physicist and plant physiologist, probably born in Chazeuil, Burgundy. Mariotte was prior of Saint-Martin-de-Beaumont-sur-Vingeanne and moved from Dijon to Paris in the early 1670s. Following his election to the Paris Academy of Sciences (1666), Mariotte's skilled experimental work and prolific literary output were tied to the society's activities. He attracted attention as a physiologist, comparing plant sap to the blood circulating in animals. In 1668 he announced his controversial discovery of the eye's blind spot. He studied pendulums and falling bodies (1667–1668), and published an exposition of the laws of elastic and inelastic collisions (1673). In the comprehensive review *de la nature de l'air* (1679), Mariotte restated the law bearing his name in France (elsewhere attributed to Boyle) and used it to estimate the height of the atmosphere. In the same work he discussed the connection between variations in barometric pressure, the winds and the weather. His hydrodynamical *Traité du mouvement des eaux* (1686) was widely read. Mariotte also discussed scientific methodology, hydrology, optics (including the rainbow), astronomy and the strength of materials.

MARSH, James
(1789–1846)

English chemist, born in London. He worked at the Royal Arsenal in Woolwich and was subsequently assistant to Faraday at the Royal Military Academy, where his salary was only 30 shillings a week. His most significant work was on poisons; he is best known for devising a sensitive test for arsenic (the Marsh test) which was published in 1836 and quickly translated into French and German. Any arsenic in a suspected substance is converted to the arsine; it then passes through a heated glass tube where it decomposes, leaving a brown arsenic deposit. He also wrote on electromagnetism and invented a percussion tube for ship's cannon made from quills. He died in poverty in London.

MARSH, Othniel Charles
(1831–1899)

American palaeontologist, born in Lockport, New York and wealthy by inheritance. He studied at Yale, at New Haven, and in Germany, and became first Professor of Palaeontology at Yale in 1866, without a salary or classes to teach. From 1870 to 1873 he led a series of expeditions through the western territories making spectacular discoveries of vertebrate fossils; this led him into bitter clashes with Cope who organized rival dinosaur collecting expeditions. They frequently disagreed about the description and significance of the newly discovered vertebrate faunas, and this ultimately led to the sensational airing of the bitter dispute on the front pages of the *New York Herald* in January 1890, with the result that federal financial support for palaeontology was withdrawn. Marsh discovered (mainly in the Rocky Mountains) over a thousand species of extinct American vertebrates, including dinosaurs and the mammals uintatheres and brontotheres. By 1874 he was able to establish an evolutionary lineage for horses using the fossil remains which he had assembled. He also contributed to the documentation of evolutionary changes with his discovery of Cretaceous birds with teeth. His major contribution to stratigraphical palaeontology was an early discussion of the Miocene–Pliocene boundary. Marsh was the first vertebrate palaeontologist of the US Geological Survey (1882–1892). He published over 300 papers, including *Odontornites: A Monograph on the Extinct Toothed Birds of North America* (1880) and *Dinosaurs of North America* (1896).

MASON, Sir (Basil) John
(1923–)

English physicist and meteorologist, born in Docking, Norfolk. He graduated from Nottingham University in 1948, and moved to Imperial College, London, where he set up his own section on cloud physics under Sir David Brunt. He worked on the study of rain-making processes, electrification in thunderstorms, ice nucleation and other aspects of cloud physics. His book, *The Physics of Clouds* (1957), is a classic. He was awarded the DSc in 1956 and became the first and only Professor of Cloud Physics (1961–1965) at Imperial College. Elected FRS in 1965, he became Director-General of the Meteorological Office (1965–1983), where his inspired leadership, scientific ability and administrative skills had full rein: the first operational numerical weather predictions were made, the office carried out research on the effect of Concorde on stratospheric gases and much research was carried out using meteorological radar. From 1983 to 1989 he organized, through the Royal Society and the Swedish and Norwegian academies, a thorough interdisciplinary study of acid rain which finally established the facts. He was very active in international circles, being Chairman of the Global Atlantic Tropical Experiment 1978 and Chairman of the Joint Organising Committee of the World Climate Research Programme. Knighted in 1979, he has received many awards including the Royal Medal of the Royal Society (1991).

MATUYAMA, Motonori
(1884–1958)

Japanese geophysicist, born in Oita Prefecture, the son of a Zen abbot. He was educated at Hiroshima Normal College where he studied physics and mathematics, and graduated in physics from Kyoto University (1911), where he subsequently became an instructor and then assistant professor (1916). In geophysical work with Chamberlin at Chicago (1919), he studied the physics of ice movement and returned to Japan as Professor of Theoretical Geology at the Imperial University (1921). From 1926 he worked on the remanent magnetism of basalts from Japan and Manchuria, and made the first successful link of magnetic reversals with the geological timescale. Matuyama was responsible for the extension of a national gravity survey into Korea and Manchuria (1927–1932), and in 1934 commenced a marine gravity survey of the Japan trench. He published numerous papers on the physics of the lithosphere and interior of the Earth, seismology and magnetism, and conducted research into physical methods of locating underground resources.

MAXWELL, James Clerk
(1831–1879)

Scottish physicist, born in Edinburgh, the son of a lawyer. One of the greatest theoretical physicists the world has known, he was nicknamed 'Dafty' at school (the Edinburgh Academy) because of his gangling appearance. At the age of 15 he devised a method for drawing certain oval curves, which was published by the Royal Society of Edinburgh. He studied mathematics, physics and moral philosophy at Edinburgh University, where he published another paper, on rolling curves, and later graduated from

Cambridge University as Second Wrangler. He was appointed Professor of Natural Philosophy at Marischal College, Aberdeen (1856), and King's College, London (1860), but resigned in 1865 to pursue his researches at home in Scotland. In 1871 he was appointed the first Cavendish Professor of Experimental Physics at Cambridge, where he organized the Cavendish laboratory. During his brilliant career he published papers on the kinetic theory of gases, linking the properties of single molecules and bulk matter, and established theoretically the nature of Saturn's rings (1857), later to be confirmed by Keeler; he also investigated colour perception and demonstrated colour photography with a picture of tartan ribbon (1861). However, his most important work was on the theory of electromagnetic radiation, with the publication of his great *Treatise on Electricity and Magnetism* in 1873, which treated mathematically Faraday's theory of electrical and magnetic forces considered as action in a medium rather than action at a distance. He showed that oscillating electric charges could generate propagating electromagnetic waves, with a theoretical wavespeed almost exactly the same as the experimentally determined speed of light; this was the first conclusive evidence that light consisted of electromagnetic waves. 'Maxwell's equations', describing mathematically the wave propagation, were later developed by Heaviside and independently by Heinrich Hertz. Maxwell envisaged electromagnetic radiation carried by the 'ether', the hypothetical medium which pervaded all space, allowing light to travel from the distant stars; but his equations survived unchanged when the revelations of the Michelson-Morley experiment finally laid to rest the notion of the ether and Einstein's revolution revised most of classical physics. Maxwell also predicted the possible existence of radiation beyond ultraviolet and infrared, and suggested that electromagnetic waves could be generated in a laboratory – as Hertz was to demonstrate in 1887.

MAYALL, Nicholas Ulrich
(1906–)

American astrophysicist, born in Moline, Illinois, and educated at the University of California, where he graduated in 1928 and received his PhD in 1934. He served on the staff of Mount Wilson Observatory (1929–1931) and Lick Observatory (1933–1942), and subsequently worked at MIT Radiation Laboratory (1942–1943) and Caltech (1943–1945), before returning to Lick Observatory in 1945. After 15 years at Lick, he was appointed Director of Kitt Peak Observatory, Arizona, where he remained until his retirement in 1971. Mayall's principal researches were in the field of optical observations of galaxies. At Mount Wilson in the early 1930s, he collaborated with Hubble in making surveys over the sky of faint galaxies, to study their distribution in space. His work at Lick Observatory included optical radial velocity measurements of the Andromeda galaxy, demonstrating its rotation (1950), and (with W. W. Morgan) the development of a system of classifying galaxies from their composite spectra to indicate their stellar content (1957). In 1961 Mayall was made chairman of a committee to plan and oversee the building of a 4-metre telescope on Kitt Peak. This famous and successful instrument, named the Mayall telescope, was officially inaugurated in 1973 on the occasion of the fifth centenary of the birth of Copernicus.

MAYOW, John
(1640–1679)

English chemist and physiologist, born in Bray, Cornwall. He studied medicine at Oxford, and thereafter practised as a physician, probably dividing his time between London and Bath. He noted the similarities between combustion and respiration, in particular that both use up only a small proportion of the available air. He also noted that the remainder of the air does not support life, extinguishes a lighted candle, and is insoluble in water. Mayow suggested that the function of respiration is to convey life-giving nitrous particles from the air to the blood. He also suggested that respiration is the source of animal heat, and pointed out that the foetus breathes through the placenta. Mayow's originality and the extent of his influence, if any, on Lavoisier, are the subject of debate. He died in London.

McCLINTOCK, Barbara
(1902–1992)

American geneticist, born in Hartford, Connecticut. She received a PhD in botany in 1927 from Cornell, where she worked from 1927 to 1935. Later she held posts at the University of Missouri (1936–1941) and Cold Spring Harbor (1941–1992). In 1927, with Harriet Creighton, she showed that changes in the chromosomes of maize resulted in physical changes in the colour of the corn kernels; this ultimate proof of the chromosome theory of heredity was published in 1931. In the 1940s she showed how genes in maize are activated and deactivated by 'controlling elements' – genes that control other genes, and which can be copied from chromosome to chromosome. She presented her work in 1951 at a Cold Spring Harbor symposium, but its significance was lost on the attendees who mainly worked with bacteria. It was not until the 1970s, after the work of Jacob and Monod, that her work began to be appreciated. At a 1976 symposium, McClintock's research was acknowledged with the introduction of the term 'transposon' to describe her 'controlling elements'. Finally in 1983, she was awarded the Nobel Prize for Physiology or Medicine. She continued to work on maize genetics at Cold Spring Harbor until her death in 1992.

McCREA, Sir William Hunter
(1904–   )

Irish theoretical astrophysicist and mathematician, born in Dublin. After being educated at Chesterfield Grammar School in Derbyshire and Trinity College, Cambridge, McCrea lectured in mathematics at Edinburgh University and Imperial College, London, later becoming Professor of Mathematics at Queen's University in Belfast (1936). After World War II he moved to Royal Holloway College, London, and then to the University of Sussex in 1966. In 1934, with Edward Milne, McCrea was the founder of modern Newtonian cosmology, applying classical theories of physics with considerable success to the primordial gas cloud that condensed to form the galaxies. He also extended Mach's viewpoint, suggesting that the Heisenberg's uncertainty principle applies to light as it travels, and that our knowledge of the universe therefore deteriorates as we extrapolate both to great distances and back in time. McCrea was the first to emphasize the effect of turbulence in condensing gas clouds and applied this theory to the formation of planets, stars and globular clusters, and specifically to the angular momentum of these condensing systems. He made important contributions to Dirac's 'large number hypothesis' in which the number 1039 figures prominently, and to discussions on relativity and the low-probability transitions of electrons between energy states in atoms. In 1975 McCrea proposed that comets may have been formed in the high-density interstellar clouds that are found in galactic spiral arms, and that these comets were then picked up periodically by the solar system. He also suggested that the Earth, Moon and Mars were formed from a single body that differentiated and then split up. He was elected FRS in 1952, and knighted in 1985.

McLAREN, Anne Laura
(1927–)

British geneticist. She was educated at the University of Oxford, and after postdoctoral research at University College, London (1952–1955), and the Royal Veterinary College, London (1955–1959), she joined the staff of the Agricultural Research Council Unit of Animal Genetics in Edinburgh in 1959. She has been Director of the Medical Research Council's Mammalian Development Unit since 1974. Elected FRS in 1975, she received the Scientific Medal of the Zoological Society of London in 1967. McLaren has published prolifically in the fields of reproductive biology, embryology, genetics and immunology, and is best known for her discovery and isolation of the embryonal carcinoma cell line. This cell type was isolated from the testes of mouse embryos and has the capability of differentiating into many different cell types in culture. It therefore provides the opportunity to study the environmental and genetic requirements which cause embryonic cells to differentiate along particular pathways of development. Because embryonic cells also have similarity to the malignant state, this cell type is also of great value in studying the nature of carcinogenesis.

McMILLAN, Edwin Mattison
(1907–1991)

American atomic scientist, born in Redondo Beach, California. He was educated at Caltech and Princeton University, where he took his doctorate in 1932. He joined the staff of the University of California at Berkeley, moving to the Lawrence Radiation Laboratory when it was founded within the university in 1934. In 1940, following up the work of Fermi who had split the uranium atom by bombarding it with low-velocity neutrons, McMillan and Abelson synthesized an element heavier than uranium (the heaviest of the naturally occurring elements, with an atomic number of 92) by bombarding uranium with neutrons, in the Berkeley cyclotron. They called this new silvery metal 'neptunium' after the most distant of the planets then known. The synthesis of neptunium (atomic number 93), the first 'transuranic' element, marked the beginning of a new epoch in science and in world affairs. The following year Seaborg, also working at Berkeley, synthesized plutonium, leading on to the development of the atomic bomb. McMillan spent the rest of World War II working on radar and sonar, and on the atomic bomb at Los Alamos. In 1944 he established the principle of phase stability in accelerated particles, and as a result was able to modify the Berkeley cyclotron into a 'synchrocyclotron' which could accelerate particles whose speed could be travelling at unknown speeds as well as particles whose speed could be calculated. This improvement presaged the development of the present-day nuclear accelerator, used in the search for the fundamental particles. McMillan was appointed to the chair at Berkeley in 1946 and to the directorship of the Lawrence Radiation Laboratory in 1958, retiring in 1973. From 1968 to 1971 he was Chairman of the US National Academy of Sciences. He was awarded the Nobel Prize for Chemistry, jointly with Seaborg, in 1951.

MEDAWAR, Sir Peter Brian
(1915–1987)

British zoologist and pioneering immunologist, born in Rio de Janeiro of an English mother and a Lebanese father. He was educated at Marlborough and in zoology at Magdalen College, Oxford. During World War II he investigated methods of skin grafting for burn victims. From this work arose the realization that homograft rejection occurred by the same immunological mechanism as the response to foreign bodies. He was appointed Professor of Zoology at Birmingham University (1947–1951), Jodrell Professor of Comparative Anatomy at University College, London (1951–1962), and Director of the National Institute for Medical Research at Mill Hill from 1962. His research was concerned with the problems of tissue rejection following transplant operations. In 1960 he shared the Nobel Prize for Physiology or Medicine with Burnet, for researches into immunological tolerance. They showed that prenatal injection of tissues from one individual to another resulted in the subsequent suppression of rejection in the recipient to the donor's tissues. This mimicked the situation existing between identical twins. He considered it important to explain science to the lay public and gave the brilliant Reith Lectures on 'The Future of Man' in 1959. His writings included *The Uniqueness of the Individual* (1957), *The Art of the Soluble* (1967), *Pluto's Republic* (1982) and his autobiography *Memoirs of a Thinking Radish* (1986). He was elected FRS in 1949 and knighted in 1965.

MEITNER, Lise
(1878–1968)

Austrian physicist, born in Vienna. Educated at the University of Vienna, she became professor at Berlin (1926–1938) and member of the Kaiser Wilhelm Institute for Chemistry (1907–1938), where together with Hahn she set up a laboratory for studying nuclear physics. In 1917 she shared with Hahn the discovery of the radioactive element protactinium. In 1938 she fled to Sweden to escape persecution by the Nazis. Shortly afterwards, Hahn observed radioactive barium in the products of uranium bombarded by neutrons. He wrote to Meitner telling her of the discovery, and with her nephew Otto Frisch, she proposed that the production of barium was the result of the uranium nucleus being spit in two by nuclear fission. Frisch was able to verify the hypothesis within a few days. Meitner worked in Sweden until retiring to England in 1960. Recently nuclear physicists named the element of atomic number 108 after her.

MELVILL, Thomas
(1726–1753)

Scottish scientist noted for his research on optics, probably born in Glasgow. He studied theology at Glasgow University where he was a friend of Alexander Wilson, later the first Professor of Astronomy. He used a prism to examine the colour of flames, detecting the yellow line of sodium when salt, ammoniac and other substances were introduced into burning alcohol. This early attempt at spectroscopy made no impact on the scientific community and no other serious investigations were carried out until Wollaston discovered the dark lines in the Sun's spectrum in 1802. Melvill also attempted to explain why different colours of light bend by different amounts when passing from one medium to another, suggesting that they travel at different speeds. His early death may explain his lack of influence; he died in Geneva at the age of 27.

MENDEL, Gregor Johann
(1822–1884)

Austrian botanist, born near Udrau. Entering an Augustinian cloister in Brünn in 1843, he was ordained as a priest in 1847. After studying science at Vienna (1851–1853), he returned and later became Abbot in 1868. In the experimental garden of the monastery, Mendel bred peas, and grew almost 30,000 plants between 1856 and 1863. He artificially fertilized plants with specific characteristics for example crossing species that produced tall plants with those that produced short plants, and counting the numbers of tall and short plants which appeared in the subsequent generations. All the plants of the first generation produced were tall; the next generation consisted of some tall and some short, in proportions of 3:1. Mendel suggested that each plant received one character from each of its parents, tallness being 'dominant' and shortness being 'recessive' or hidden, appearing only in later generations. His experiments led to the formulation of 'Mendel's law of segregation' and his 'law of independent assortment'. These results were published in 1865; although Mendel wrote regularly to Nägeli, his ideas were not taken up by others, and he gradually discontinued his work on peas. Later analysis of his results revealed that although he had arrived at the correct conclusions, Mendel's data did not show enough statistical variation to be plausible experimental values, and must have been adjusted in some way to agree with the expected results. In 1877 he was instrumental in introducing regular weather reports for the local farmers. His concepts have become the basis of modern genetics.

MENDELEYEV, Dmitri Ivanovich
(1834–1907)

Russian chemist, famous for drawing up the periodic table which explains the relationship between the elements. He was born in Tobolsk, Siberia, the 14th child of liberal middle class parents. He studied at St Petersburg and Heidelberg, Germany, where he collaborated briefly with Bunsen and investigated the behaviour of gases, formulating the idea of critical temperature – the highest temperature at which a gas can be liquefied by pressure alone. Mendeleyev was appointed professor at St Petersburg Technical Institute in 1863 and at the University of St Petersburg in 1866. Russian society then being in a state of upheaval after the abolition of serfdom, he began to study chemical problems relating to agriculture and petroleum, and never lost interest in these practical concerns. In 1869, in the course of writing a textbook, he tabulated the elements in ascending order of their atomic weight and found that chemically similar elements tended to fall into the same columns. Several attempts had already been made to group the elements by their chemical properties and to relate chemical behaviour to atomic weight (e.g. by Döbereiner and Newlands). Mendeleyev's great achievement was to realize that certain elements still had to be discovered and to leave gaps in the table where he predicted they would fall. In the second version of his table, published in 1871, the 63 known elements exhibited a striking pattern (or 'periodicity' as Mendeleyev termed it), with families of elements all having atomic weights which varied by integer multiples of eight times the atomic weight of hydrogen. At first the periodic table was largely rejccted by the scientific world, but as each new element that was subsequently discovered fitted into it perfectly, scepticism turned to enthusiasm. It also soon became obvious that the table made it possible to predict the properties of the still undiscovered elements. Thus it added enormously to scientific understanding and provided a framework for further research. However, the underlying reason for the periodicity of the elements remained unexplained until the structure of the atom – in particular the arrangement of the electrons around the nucleus – came to be understood. Mendeleyev continued to refine the table for the next 10 years, meanwhile continuing his work on gases, studying solutions, and taking up aeronautical research (he made a solo ascent in a balloon in 1887). In 1890 he was forced to resign his position at the university because the authorities feared the effect of his liberal views on students, but he was subsequently put in charge of a project to produce smokeless fuel and made Director of the Central Board of Weights and Measures. He died in St Petersburg. The transuranic element mendelevium (atomic number 101) is named in his honour.

MERRIFIELD, (Robert) Bruce
(1921–)

American chemist, born in Fort Worth, Texas. He obtained both his BA (1943) and PhD (1949) from the University of California at Los Angeles. He joined the Rockefeller Institute for Medical Research in 1949, and appointed assistant professor at Rockefeller University in New York City in 1957 and full professor in 1966. His research has been centred upon the laboratory synthesis of proteins. Before Merrifield developed his method of synthesis, the process of linking amino acids was painfully slow, even for the separation of quite a small protein. He simplified the process by devising a procedure whereby the amino acids were linked the correct order on the surface of a microscopic bead of polystyrene. This process has now been automated and computer controlled, allowing the ready synthesis of small quantities of quite large proteins. Merrifield was awarded the Nobel Prize for Chemistry for this work in 1984.

METCHNIKOFF, Elle, originally Ilya Ilyich
(1845–1916)

Russian embryologist and immunologist, born in Ivanovka, the Ukraine. After graduating from the University of Kharkov in 1864, he studied invertebrate and fish embryology at several European centres. He received a doctorate from the University of St Petersburg in 1867, and after spells of teaching and research at St Petersburg and Odessa, he took a research post at Messina, Italy. It was here that he began his immunological studies. He observed how mobile cells in starfish larvae surround, engulf and destroy foreign bodies. He called these cells phagocytes, and hypothesized that the role of phagocytes in vertebrate blood is to fight invasion by bacteria. In 1886 Metchnikoff returned to Odessa to become director of the new Bacteriological Institute. He studied the action of phagocytes during infections in dogs, rabbits and monkeys. Harrassed by journalists and physicians for lacking medical qualifications, Metchnikoff left Russia in 1887. Pasteur offered him the directorship of a laboratory at the Pasteur Institute in Paris. As the role of phagocytes became accepted, Metchnikoff turned to other problems. After investigating aging and death, he advocated eating large quantities of yoghurt to promote good health. Metchnikoff was awarded the 1908 Nobel Prize for Physiology or Medicine jointly with Ehrlich.

MICHAELSON, Sidney
(1925–1991)

English computer scientist, born in London and educated at Imperial College, London. From 1949 to 1963 he lectured there, pioneering the design of digital computers incorporating the principle of microprogramming. In 1963 he moved to Edinburgh University, where he founded the Department and held the Chair of Computer Science (1966–1991). There he initiated work in the UK on system software to provide shared interactive multiple access to computers and led a diverse range of research activities, from computational theory to VLSI (very large scale integration) design. He was a leader in field of stylometry, the science of computer-based analysis of literary texts for authorship and chronology, and fought to promote the professional recognition of computer scientists.

MICHEL, Hartmut
(1948–)

German biochemist, born in Ludwigsburg, West Germany. Following his early career at the universities of Tübingen, Würzburg, where he became research group leader (1977–1979). Munich and the Max Planck Institute of Biochemistry, he was appointed Director of the Max Planck Institute of Biophysics in Frankfurt in 1987. In 1981 he devised a method of producing a large, well-ordered crystal of the membrane-bound photosynthetic reaction centre of the purple bacterium *Rhodopseudomonas viridis*. Michel Huber and Deisenhofer then collaborated to determine its structure by X-ray crystallography. By 1985 they were able to report the complete structure, which confirmed and elaborated predictions about how the energy transfer process in photosynthesis operated. The exact locations of the individual amino acids have still to be determined, but the overall structure appears as a cylinder of seven alpha helices with a central waterchannel. For this discovery of the structure of a membrane protein. Michel shared the 1988 Nobel prize for Chemistry with Huber and Deisenhoer. He published *Crystallization of Membrane Proteins* in 1990.

MICHELSON, Albert Abraham
(1852–1931)

German-born American physicist, born in Strelno (now Strzelno, Poland). His family emigrated to the USA when he was aged four. He graduated from the US Naval Academy in 1873, and after teaching at the Academy, the Case School of Applied Science in Cleveland and Clarke University, he was appointed Professor of Physics at Chicago in 1892. His lifelong passion was precision measurement in experimental physics using the latest technology. He made a number of determinations of the speed of light, the most precise in 1924–1926, when over a 22-mile course between mountains in southern California, he determined by optical means the velocity to be 299,796 $\pm$ 4 kilometres per second. He is chiefly remembered for the

Michelson-Morley experiment to determine ether drift, the negative result of which set Einstein on the road to the theory of relativity. The interferometer which he invented for this experiment was developed subsequently for spectroscopic studies which revealed the hyperfine structure of spectral lines, and provided measurements of the breadths of spectral lines. The latter investigations showed the appropriateness of Babinet's choice of the red cadmium spectral line as a wavelength standard, and in 1894 Michelson measured the metre and showed that it contained 1,553,163.5 wavelengths of this radiation. He also developed a stellar interferometer for measuring the sizes and separations of celestial bodies which cannot be resolved even with the largest telescopes. His first measurements with this instrument, published in 1898, were of the separation of the double star Capella; later he measured the angular sizes of some of the satellites of Jupiter and the star Alpha Orionis. In 1898 he invented the echelon grating, an ultra-high resolution device for the study and measurement of hyperfine spectra. Michelson became the first American scientist to win a Nobel Prize when he was awarded the Nobel Prize for Physics in 1907. A member (1888) and President (1923–1927) of the US National Academy of Sciences, his many honours included foreign membership of the Royal Society (1902), and the award of the society's Copley Medal (1907). Towards the end of his life he successfully investigated the origin of the irridescent colours exhibited by some birds and insects. Advanced forms of his spectral interferometer and stellar interferometer have recently enabled important advances in physics and astronomy.

MICHIE, Donald
(1923–)

British specialist in artificial intelligence, born in Rangoon, Burma. Educated at Rugby School and Balliol College, Oxford, he served in World War II with the Foreign Office at Bletchley Park (1942–1945), where work on the Colossus code-breaking project acquainted him with computer pioneers such as Turing, Newman and T. H. Flowers. With Turing he discussed the mechanization of thought processes and chess-playing machines. After a biological career in experimental genetics, he developed the study of machine intelligence at Edinburgh University as Director of Experimental Programming (1963–1966) and Professor of Machine Intelligence (1967–1984). He is Editor-in-Chief of the *Machine Intelligence* series, of which the first 12 volumes span the period 1967–1990. Since 1986 he has been chief scientist at the Turing Institute which he founded in Glasgow in 1984. In publications such as *The Creative Computer* (1984) and *On Machine Intelligence* (1974), he has argued that computer systems are able to generate new knowledge. His research contributions have primarily been in the field of machine learning.

MIDGLEY, Thomas Jr
(1889–1944)

American engineer and inventor, born in Beaver Falls, Pennsylvania. He graduated in mechanical engineering at Cornell University in 1911. During World War I, on the staff of the Dayton (Ohio) Engineering Laboratories (1916–1923), he worked on the problem of 'knocking' in petrol engines, and by 1921 found tetraethyl lead to be effective as an additive to petrol, used with 1,2-dibromoethane to reduce lead oxide deposits in the engine. He also devised the octane number method of rating petrol quality. Since 1980 there has been rising concern that the lead emitted in vehicle exhausts constitutes a health hazard. As president of the Ethyl Corporation from 1923, he also introduced Freon 12 as a non-toxic non-inflammable agent for domestic refrigerators. Again there is now concern that chlorofluorocarbons (CFCs), such as Freon, cause destruction of the ozone layer in the upper atmosphere with damaging climatic and other effects as a result of the increased passage of ultra-violet radiation. He died tragically by accidental strangulation through the failure of a harness he used to help him rise in the morning, needed because he was a polio victim.

MILANKOVITCH, Milutin
(1879–1958)

Yugoslav geophysicist, born in Dalj. He was educated at the Institute of Technology in Vienna where he received a PhD in 1904. He then moved to the University of Belgrade where he remained for the rest of his career. During 1914–1918 he was held prisoner of war, but was allowed to continue his studies in the library of the Hungarian Academy of Sciences in Budapest. Studying the possible astronomical cycles which produce climatic variations, he attempted to reconstruct the palaeoclimates of the Earth. He realized that the key to past climates was the amount of solar radiation received by the Earth, and that this varies according to latitude and depends upon the Earth's orbit, a 21,000 year precession which determines which hemisphere receives greater radiation, and the tilt of the Earth's rotational axis, which changes over a 40,000 year period from 21.8 to 24.4 degrees. Using these parameters he reconstructed the classic theoretical radiation curves for the past 650,000 years for comparison with observed climatic cycles. These are known as the Milankovitch cycles and are still in use.

MILLER, Stanley Lloyd
(1930–)

American chemist, born in Oakland, California. He studied at California University and taught there from 1960. His best-known work was carried out in Chicago in 1953, and concerned the possible origins of life on Earth. Inspired by the theories of Oparin and J. B. S. Haldane, with Urey, he passed electric discharges (simulating thunderstorms) through mixtures containing reducing gases (hydrogen, methane, ammonia and water) which Haldane had suggested were likely to have formed the early planetary atmosphere. After some days, analysis showed the presence of some typical organic substances, including aldehydes, carboxylic acids, amino acids and urea. In later work, Miller used other mixtures containing carbon dioxide and carbon monoxide, and the developmental patterns of different compounds with time were analysed, hydrogen cyanide and cyanogen appeared first as the ammonia concentration declined, while amino acids were formed more slowly. Formation of the Oparin-Haldane 'primeval soup' is now accepted as the most plausible theory for the generation of complex organic molecules on Earth, although the probable subsequent path from these chemicals to a living system is still hotly debated.

MILLIKAN, Robert Andrews
(1868–1953)

American physicist, born in Illinois. He studied at Oberlin College and Columbia University where he received his PhD in 1895. After working at Berlin and Göttingen he became Michelson's assistant at the University of Chicago, where he was appointed professor in 1910. In 1921 he moved to Caltech where he established the experimental physics laboratory. At Chicago he refined the oil drop technique for measuring the electron's charge that had been developed by J. J. Thomson. He observed the motion of electrically charged drops of oil as they floated between two parallel plates with a voltage applied across them. By measuring the speed of fall of the droplets for different voltages, he was able to show that the charge on each was always a multiple of the same basic unit – the charge on the electron which he measured very precisely. In studies of the photoelectric effect he confirmed Einstein's theoretical equations and gave an accurate value for Planck's constant. For all these achievements he was awarded the 1923 Nobel Prize for Physics. He also investigated cosmic rays, a term that he coined in 1925.

MILNE, John
(1859–1913)

English seismologist, born in Liverpool. He was educated at King's College and the Royal School of Mines, London. He began his career as a mining engineer in the UK and Germany, then spent two years in Newfoundland and travelled in Egypt, Arabia and Siberia. Always a keen traveller, he joined an expedition in 1874 to locate Mount Sinai and on being appointed Professor of Geology in Tokyo (1875–1894), travelled there by camel across Mongolia. In Japan he took up an interest in earthquakes, becoming a supreme authority, for which he was awarded the Order of the Rising Sun. He pioneered modern seismology and introduced precise physical measurements. In 1892, with colleagues, he developed a seismometer to record horizontal components of ground motion which became used on a worldwide basis. From this time seismology advanced rapidly and began to be applied to the study of the internal structure of the Earth. Milne devised methods of locating distant earthquakes and early traveltime curves for seismic wave arrivals, initiated experiments using explosives, and compiled *A Catalogue of Destructive (Japanese) Earthquakes AD 7 AD 1899* (1912). On his retirement to the Isle of Wight (1895) he ran a private seismological observatory, regularly issuing a bulletin summarizing data from a worldwide network of seismological stations which he set up with his own instruments. A forerunner of the International Seismological Summary he was a prolific writer, publishing *Earthquakes and other Earth Movements* (1886) and *Seismology* (1898).

MILSTEIN, Cesar
(1927–)

Argentinian-born British molecular biologist and immunologist, born in Bahía Blanca. He graduated from Buenos Aires University with a BSc in chemistry in 1945, and worked on the active sites of enzymes in Sanger's laboratory in Cambridge (1958–1961), where he obtained a PhD in 1960. In 1961 he returned to Argentina to become Head of the Division of Molecular Biology at the National Institute of Microbiology. When many members of the institute were dismissed following the military coup, Milstein resigned in protest and returned to Cambridge, where he has been on the staff of the Medical Research Council at the Laboratory of Molecular Biology since 1963. He has conducted important research into antibodies, proteins that are produced by the immune system cells in response to foreign molecules called antigens. By fusing cells from tumours of the immune system (myelomas) with cells producing one antibody, to form a 'hybridoma', it became possible to maintain production of the antibody. This technique of 'monoclonal' antibodies developed in 1975 with Köhler, has become widespread in the commercial development of new drugs and diagnostic tests. In 1984, Milstein was awarded the Nobel Prize for Physiology or Medicine with Köhler and Jerne.

MINKOWSKI, Hermann
(1864–1909)

Russian-born German mathematician, born near Kovno. He was educated at the University of Königsberg, where he received his PhD in 1885. He was later appointed professor at Königsberg (1895), Zürich (1896), where he taught Einstein, and at Göttingen (1902), where he was a close friend and colleague of Hilbert. Minkowski won a prize for work in the theory of numbers from the Parisian Academy of Sciences when aged only 18, and went on to discover a new branch of number theory, the geometry of numbers. In his most important work he gave a precise mathematical description of spare-time as it appears in Einstein's relativity theory; the four-dimensional 'Minkowski space' was described in *Space and Time* (1907). He died of unexpected complications arising from appendicitis.

MINOT, George Richards
(1885–1950)

American physician, born into a prominent Boston family and educated at Harvard College and Medical School with which, except for three postdoctoral years at Johns Hopkins, he was associated all his working life. Using special staining techniques on blood smears, he began to study anaemia, and was able to demonstrate that some anaemias are caused by the failure of the bone marrow to make enough red blood cells, whereas others are induced when blood cells are destroyed too quickly. From 1925, working with Murphy, he examined clinically George Whipple's observation that dogs made anaemic through repeated bleedings improved significantly when fed liver. Minot and Murphy gave large amounts of raw liver to patients diagnosed as suffering from pernicious anaemia, at that time a fatal disease. Their patients rapidly improved, although the 'intrinsic factor' contained in liver was not identified for another 20 years. The 'intrinsic factor' was necessary to the absorption and utilization of folic acid, used by the body in the production of red blood cells. Minot, a diabetic, was one of the earliest patients to benefit from insulin therapy. With Murphy and Whipple, he shared the 1934 Nobel Prize for Physiology or Medicine.

MITCHELL, Maria
(1818–1889)

American astronomer, born in Nantucket into a serious-minded Quaker family. Her father's activities included regulating chronometers for whaling ships in which his daughter took part from an early age. In 1836 the US Coast Survey equipped an observatory at their home as a local station with her father in charge, where

Maria, a librarian in the local Athenaeum, had an opportunity to practise astronomy. Her discovery of a comet in 1847 brought her to the public notice and earned for her (1848) the King of Denmark's Gold Medal for first discoverers of telescopic comets, and election to the American Academy of Arts and Sciences, as its first woman member. Her first professional commission was the computing of tables of the planet Venus for the American Ephemerides and Nautical Almanac, a duty she performed for 20 years (1849–1968). In 1865 she was appointed Professor of Astronomy at the newly founded Vassar College for women at Poughkeepsie where she was an inspiring teacher and a doughty campaigner in the women's rights and anti-slavery movements. In failing health she retired to her native Nantucket in 1988.

MITCHELL, Peter Dennis
(1920–1992)

English biochemist, born in Mitcham, Surrey. He graduated from Cambridge and taught there (1943–1955) and at Edinburgh University (1955–1963) before founding his own research institute, the Glynn Research Laboratories at Bodmin in Cornwall (1964), to extend his studies of the way in which energy is generated inside cells at the molecular level. Scientists knew that the energy comes from the adenosine triphosphate (ATP) molecule, and that somehow ATP is produced from adenosine diphosphate (ADP) by a process known as oxidative phosphorylation, possibly regulated by undiscovered enzymes. In the 1960s Mitchell rejected the idea that oxidative phosphorylation occurs via an active chemical intermediate, and proposed an entirely novel theory in which electron transport from reducing coenzymes to form water causes the formation of a proton gradient across the inner mitochondrial membrane. It was suggested that this 'chemi-osmotic gradient' directly drives the synthesis of ATP from ADP and inorganic phosphate in special membrane structures where protons re-enter the mitochondrion. The mechanism is unknown but possibly involves structural changes in energy-dependent proteins operating in an essentially water-free environment. Although at first greeted with scepticism, his views became widely accepted and are now supported by a considerable body of evidence. Mitchell's theory was formally acknowledged by his award of the unshared Nobel Prize for Chemistry in 1978.

MÖBIUS, August Ferdinand
(1790–1868)

German mathematician, born in Schulpforta. As professor at Leipzig, he worked on analytical geometry, statics, topology and theoretical astronomy. He introduced barycentric coordinates and the 'Möbius net', thus extending Cartesian coordinate methods to projective geometry, and showed how they could be used to express a duality between points and lines (discovered earlier by Joseph Diaz Gergonne). He gave a straightforward algebraic account of statics, following the work of the French mathematician Louis Poinsot, and in this way used vectorial quantities before vectors as such entered mathematics. He also discovered a novel type of duality in three-dimensional space in this connection. In topology he investigated which surfaces can exist, and became one of the discoverers of the 'Möbius strip' (a one-sided surface formed by giving a rectangular strip a half-twist and then joining the ends together). This shows that a sense of orientation is not necessarily part of the intrinsic geometry of a surface. He also examined in detail the possible types of three-dimensional spaces which can be created by similar gluing constructions.

MOHL, Hugo von
(1805–1872)

German botanist, born in Stuttgart. Although he had a classical education, from childhood it was clear that his vocation would combine botany and optics. After studying medicine at Tübingen he was Professor of Physiology at Bern (1832–1835) and Professor of Botany at Tübingen (1835–1872). Early research included studies of climbing plants, and pioneering work on stomatal movement. He constructed a microscope and published a manual on microscopy, *Mikrographie, oder Anleitung zur Kenntnis und zum Gebrauche des Mikroskops* (1846). His most lasting researches, published as two fundamental papers and in *Die vegetabilischen Zellen* (1846), lay in the field of plant cell structure and physiology, where his meticulous observations were the first attempts at cytochemistry. He differentiated the cell membrane, nucleus, cellular fluid, utricle, and a substance he called protoplasm. This term had earlier been used by the Czech physiologist Purkinje to denote the embryonic material of eggs; Mohl was the first person (1846) to use the term protoplasm in plant cell biology. For Mohl, protoplasm was a preliminary substance in cell generation, a quite different sense to the modern usage which dates from Schultze (1861). He was also the first to clearly explain osmosis, and discovered that the secondary walls of plant cells are fibrous.

MOISSAN, (Ferdinand Frédéric) Henri
(1852–1907)

French chemist, born in Paris. He studied chemistry and pharmacy in Paris and qualified as a pharmacist in 1879. He taught at the School of Pharmacy in Paris, becoming Professor of Toxicology and Inorganic Chemistry in 1886 and moving to the Chair of Inorganic Chemistry at the University of Paris in 1900. Moissan was noted for his teaching and experimental work. He was the first to isolate fluorine (1886). In 1892 he invented the electric arc furnace, which in its simplest form consisted of two blocks of lime with a space in the middle for a crucible and grooves for carbon electrodes. With the high temperatures that could be reached for the first time, he reduced the oxides of uranium and tungsten; prepared carbides, borides and hydrides; and synthesized rubies. His claims to have synthesized diamonds, however, were met with scepticism. Moissan is regarded as the founder of high-temperature chemistry, and both his furnace and his discoveries were soon shown to have many industrial applications. For his work on fluorine he was awarded the Nobel Prize for Chemistry in 1906. He died in Paris.

MONOD, Jacques Lucien
(1910–1976)

French biochemist, born in Paris. After graduating from the University of Paris (1931), a Rockefeller Fellowship enabled him to work in Thomas Hunt Morgan's laboratory at Columbia University. He returned to France and received his PhD from the Sorbonne for his thesis on bacterial growth. He joined the French Resistance during World War II, and after the war began work at the Pasteur Institute in Paris, becoming Head of the

Cellular Biochemistry Department in 1954 and Director in 1971. From 1967 he was also Professor of Molecular Biology at the College de France. Monod worked closely with Jacob on the genetic control mechanisms of the bacterium *Escherichia coli*. Together they developed the theory of the operon system, whereby a regular gene binds to a specific section of the DNA strand (the operator), and deactivates the structural genes which code for the protein. They suggested the term 'messenger RNA' (mRNA), which is transcribed from the DNA, molecule and is carried to the cytoplasm for protein production. In 1965 Monod and Jacob shared the Nobel Prize for Physiology or Medicine with Lwoff. In 1968 in Paris, Monod supported the students in their battles with the university establishment. He published *Chance and Necessity* in 1970, a biologically based philosophy of life.

MONRO, Alexander
(1697–1767)

'Monro Primus', Scottish anatomist, born in London. Himself a surgeon's son, Monro founded a three-generation dynasty of anatomy professors that dominated anatomy teaching at Edinburgh for 126 years. He studied in London (under William Cheselden), in Paris, and in Leiden (under Boerhaave). From 1719 he lectured at Edinburgh on anatomy and surgery, serving as professor from 1725 to 1759. He played a key part in founding the Edinburgh Royal Infirmary and in promoting medicine within the university. His energetic and well organized teaching deploying a wide range of effective preparations but no great classroom use of dissection. He expedited the rise of Edinburgh as a popular centre for medical training. He wrote *Osteology* (1726), *Essay on Comparative Anatomy* (1744), *Observations Anatomical and Physiological* (1758) and *Account of the Success of Inoculation of SmallPox in Scotland* (1765) – works that reveal a skilled communicator of medical knowledge rather than a profound researcher. It was partly through the orientation of his interests that the Edinburgh ideal of the Practitioner integrated the physician and the surgeon.

MONTAGNIER, Luc
(1932–)

French molecular biologist. He was educated at the Collège de Chatellerault, the University of Poitiers and the University of Paris, where he was subsequently appointed to a number of research posts. He became Laboratory Head of the Radium Institute (1965–1971) in Paris, and since 1972 has worked at the Pasteur Institute, as Head of the Viral Oncology Unit, professor (since 1985) and Head of the Department of AIDS and Retroviruses (since 1990). Since 1974 he has also been Director of Research at the National Centre for Scientific Research. Montagnier has published widely in molecular biology and virology, and is now credited with the discovery of the HIV virus. He and his team first isolated the HIV virus from a Frenchman with AIDS in 1983. However, an American team led by Robert Gallo claimed to have independently discovered the virus and Montagnier's virus was discredited; the journal *Nature* declined to publish his findings. Around 10 years after the discovery, it was shown that Gallo's original findings were erroneous and that the virus he eventually used to develop and patent an HIV blood test actually came from Montagnier's laboratory.

MORGAGNI, Giovanni Battista
(1682–1771)

Italian physician, born in Forli. Graduating in Bologna, Morgagni taught anatomy there and later in Padua. Active throughout his life in anatomical research, his great work *De Sedibus et Causis Morborum per Anatomen Indagatis* (1761) was not published until he was 80. It was grounded on over 600 post-mortems and was written in the form of 70 letters to an anonymous medical confrèré. Case by case, Morgagni described the clinical aspects of illness during the patient's lifetime, before proceeding to detail the post-mortem findings. His object was to relate the illness to the lesions established at autopsy. Morgagni did not use a microscope and he regarded each organ of the body as a complex of minute mechanisms. His book may be seen as a crucial stimulus to the rise of morbid anatomy, especially when physicians also made use of the techniques of percussion, developed by Auenbrugger, and ausculation, pioneered by Laénnec. Morgagni himself made significant discoveries. He was the first to delineate syphilitic tumours of the brain and tuberculosis of the kidney. He grasped that where only one side of the body is stricken with paralysis, the lesion lies on the opposite side of the brain. His explorations of the female genitals, of the glands of the trachea, and of the male urethra also broke new ground. He is judged the father of the science of pathological anatomy.

MORGAN, Thomas Hunt
(1866–1945)

American geneticist and biologist, born in Lexington, Kentucky. He graduated in zoology from Kentucky State College in 1886, and received his PhD from Johns Hopkins University in 1890. He became Professor of Experimental Zoology at Columbia University (1904–1928) and then at Caltech (1928–1945). Morgan started out with many objections to Mendel's theory of heredity, and in 1908, he began work on *Drosophila*, the fruit fly, to test out Mendel's ideas. From his breeding programme, he found that certain traits are linked (for example, only male flies have the 'white eye' characteristic), but that the traits are not always inherited together. From his work, he suggested that certain traits are carried on the X chromosome, that traits can cross-over to other chromosomes and that the rate of crossing-over could be used as a measure of distances along the chromosome. His laboratory became known as the 'fly room', and attracted many researchers. Morgan's co-workers included Hermann Müller, Sturtevant and C. B. Bridges, with whom he wrote *The Mechanism of Mendelian Heredity* (1915), which established the chromosome theory of inheritance in confirmation of Mendel's work. Morgan was awarded the 1933 Nobel Prize for Physiology or Medicine. His many other books included *Evolution and Adaptation* (1911), *The Theory of the Gene* (1926) and *Embryology and Genetics* (1933).

MORISON, Robert
(1620–1683)

Scottish botanist, born in Aberdeen and educated locally, graduating from Aberdeen University in 1638. His subjects included Hebrew, as his parents wished him to enter the ministry. Times during the reign of Charles I were turbulent, however, and Morison became a Royalist. After recovery from a head wound received in battle, he escaped to France and took his MD at Angers (1648). From around 1949/50 to 1660 he managed the gardens of the Duke of Orleans and travelled throughout France collecting plants. At Blois, Charles II invited Morison to accompany him to England and appointed him as senior physician, King's Botanist and superintendent of the royal gardens. In 1669 he became Professor of Botany at Oxford, shortly after publication of his *Praeludia Botanica* which contained the basis of his classification system. In 1672 he published *Plantarum Umbelliferarum Distribution Nova*. A pioneering work, this was the earliest botanical monograph. The rest of Morison's life was spent compiling *Plantarum Historiae Universalis Oxoniensis* (1680–1699). Five of the projected 15 parts on herbaceous plants were published in his lifetime; the remaining 10 parts were posthumously edited by Jacob Bobart (1641–1719). Morison died suddenly when a coach pole struck him in a London street.

MOSELEY, Harry (Henry Gwyn Jeffroys)
(1887–1915)

English physicist, born in Weymouth and educated at Oxford University where he graduated in 1910. He then joined Rutherford in Manchester before returning to the University of Oxford in 1913. Using a crystal diffraction technique, he measured the frequencies of characteristic X-ray lines in the spectra of over 30 different metals, and showed that they varied regularly from element to element in order of their position in the periodic table. He suggested that these regular changes were related to the nuclear charge and would allow the atomic numbers of elements to be calculated. Discontinuities in the spectral series made it clear that a number of elements were missing from the periodic table and allowed prediction of their properties; these elements were sought and soon discovered. Moseley's work was an important step in advancing knowledge of the nature of the atom, firmly establishing that the properties of the elements are determined by atomic number rather than atomic weight. He was killed in action of Gallipoli.

MOTT, Sir Nevill Francis
(1905–)

English physicist, born in Leeds. He studied mathematics at Cambridge and became a lecturer and Fellow there working with Rutherford. At the age of 28 he became Professor of Theoretical Physics at Bristol (1933–1954), where his group developed the theory of dislocations, defects and the strengths of crystals. It was here that he studied the electronic behaviour of 'Mott transitions' – transitions between metals and insulators. He returned to Cambridge in 1954 to become Cavendish Professor of Physics, decisively shaping the Cavendish Laboratory's research activities. In 1965 he officially retired and returned to full-time research to work on the new area of non-crystalline semiconductors, contributing considerably to the understanding of the fundamental physical and electrical properties of these materials. For his work on the electronic properties of disordered materials he won the 1977 Nobel Prize for Physics jointly with Philip Anderson and van Vleck. Mott has been one of the major theoretical physicists of this century, opening new and complex areas of the solid-state physics and materials science. He was knighted in 1962.

MÜOLLER, (Karl) Alex(ander)
(1927–)

Swiss physicist, born in Basle. He was educated at the Swiss Federal Institute of

Technology, Zürich, where he received his PhD in 1958. After five years at the Battelle Institute in Geneva (1958–1963) he joined the IBM Zürich Research Laboratory working in the area of solid-state physics. For the discovery of new low-temperature superconductors, Müller shared the 1987 Nobel Prize for Physics with Bednorz.

MÜLLER, Johannes Peter
(1801–1858)

German physiologist, born in Coblenz. Son of a successful shoemaker, Müller entered the newly founded Bonn University in 1819 where he excelled. In 1826 he was appointed to the Chair of Physiology at Bonn, and in 1833 moved to Berlin University. His first work covered problems in animal locomotion, but he won fame for his precocious researches in embryology, in 1820 engaging in the Bonn prize question: does the foetus breathe in the womb? His experimentation on a sheep revealed changes in the blood colour entering and leaving the foetus; these indicated that it did indeed respire. He also showed early interest in the eye and vision, investigating the eye's capacity to respond not just to external but also to internal stimuli (the play of imagination as well as organic malfunctions). His later work was wide-ranging; he studied electrophysiology, the glandular system (particularly the relations between the kidneys and the genitals), the human embryo and the nervous system. He was convinced of the need for pathology to make full use of microscopy. He worked on zoological classification, dealing especially with marine creatures. In 1840 he proposed the law of specific nerve energies, that is, the claim that each sensory system will respond in the same way to a stimulus whether this is mechanical, chemical, thermal or electrical. However stimulated, the eye always responds with a sensation of light, the ear with a sound sensation. In Müller's view, man did not immediately perceive the external world, but only indirectly, through the effects on his sensory systems; hence introspection was integral to human biology. Müller took a profound interest in the more philosophical problems of life. He upheld the view that vitality was animated by some kind of life force that defied reduction to the purely physical; he was convinced there was a soul separable from the body. Such idealist beliefs were rejected by the next generation of more materialist German physiologists like Ludwig and the physicist Helmholtz. Müller's *Handbuch der Physiologie des Menschen* (1833–1840) was extremely influential, and he himself was probably the most significant life scientist and medical theorist in Germany in the first half of the nineteenth century.

MÜLLER, Paul Hermann
(1899–1965)

Swiss chemist, inventor of the insecticide DDT. He was born in Olten and educated at the University of Basle. From 1925 onwards he worked at the experimental laboratory of Johann Rudolf Geigy, where he later became deputy head of pest control. He is known for his work on insecticides, particularly for discovering and developing DDT (dichlorodiphenyltrichloroethane) which was first marketed in 1942. DDT is extremely toxic to a wide variety of insects, attacking their nervous system and effectively eradicating the carriers of diseases like malaria, yellow fever and typhus, and plant pests such as the Colorado beetle. It was used in tropical areas during World War II and after the war in many parts of the world. In the 1960s it became clear that many species quickly became resistant to it, and that it is such a stable compound that its cumulative effect in the food chain are very destructive; its use was therefore discontinued. For the discovery of DDT, Müller was awarded the 1948 Nobel Prize for Physiology or Medicine. He died in Basle.

MULLIS, Kary Banks
(1944–)

American biochemist, born in Lenoir, North Carolina. He was educated at Georgia Institute of Technology and the University of California. In the early 1980s, while working for Cetus Corporation in California, he discovered a technique known as the 'polymerase chain reaction' (PCR), which allows tiny quantities of DNA to be copied millions of times to make analysis practical. It is now used in a multitude of applications, including tests for the HIV virus and the bacteria which cause tuberculosis, forensic science and evolutionary studies of the genetic material in fossils. For this work Mullis was awarded the 1993 Nobel Prize for Chemistry (jointly with Michael Smith). Since 1988 he has worked as an independent consultant for various laboratories.

MUNK, Walter Heinrich
(1917–)

Austrian-American physical oceanographer and geophysicist, born in Vienna. He travelled to the USA in 1932 and was educated at Caltech (BS 1939, MS 1940) and the University of California, where he received a PhD in oceanography in 1947. At Scripps Institution of Oceanography he was appointed assistant professor of geophysics (1947–1949), associate professor (1954–1959), and professor from 1954. He was also associate director of their Institute of Geophysics and Planetary Physics (1959–1982). During World War II his predictions for days of non-high sea swell and surf were used in the allied landings in North Africa and in the Pacific, and probably saved many lives. Using more accurate clocks than Chandler, he determined that a July day is two milliseconds shorter than a January day (1961). They attributed variations in daily length to season shifts in terrestrial air masses, ocean tides, the distribution of glaciers and changes within the Earth's Core. With colleagues, Munk developed a new method for predicting tides, was an initiator of the 'Mohole deep drill project' and developed ocean acoustic tomography (three-dimensional modelling of the ocean temperature field). The latter method is currently being applied in tests for global warming.

MURRAY, Joseph Edward
(1919–)

American surgeon, born in Milford, Massachusetts. Murray attended Harvard Medical School and graduated MD in 1943. He spent 1944–1947 in military service at the Valley Forge General Hospital in Pennsylvania. After discharge from the army he joined the staff of the Peter Bent Brigham Hospital in Boston, where he became Chief Plastic Surgeon (1951–1986). From the 1940s a team at the Brigham was attempting to overcome the reactions of the immune system to transplanted kidneys. In 1954 Murray and his colleagues first successfully transplanted a kidney between identical twins, which greatly stimulated the clinical and laboratory studies. Animals were subjected to X-rays and drugs in attempts to suppress the immunological reactions. A successful transplant was performed between non-identical twin brothers using the X-ray treatment. In 1961 the first transplant using an unrelated kidney and the immunosuppressive drug azathioprine took place. By the following year this technique was shown to be successful. Soon kidney transplants became common, and systems were established for finding donors. Murray was awarded the 1990 Nobel Prize for Physiology or Medicine with Donnall Thomas.

N

NÄGELI, Karl Wilhelm von
(1817–1891)

Swiss botanist, born in Kilchberg, near Zürich. Professor at Munich (from 1858), he was one of the early writers on evolution. He investigated the growth of cells and originated the micellar theory relating to the structure of starch grains, cell walls and other cell organelles. According to this theory, the molecules making up these substances lie together in a regular arrangement when dry. In contact with water, molecules become surrounded by an aqueous envelope. He originated the concept of cell organelles, describing the membrane surrounding chloroplasts, and their increase by division. He distinguished between two types of cell formation – vegetative and reproductive – and observed the cell nucleus dividing into two parts before the cell itself divided. His description of 'transitory cytoblasts' was almost certainly the first observation of chromosomes. He emphasized that the cell contents secrete the cell walls at their surfaces, so that the wall is a secondary structure. He distinguished between meristem and permanent tissue and between primary and secondary meristems. His most significant advance was probably the recognition of phloem as a fundamental tissue, containing sieve-tubes, and forming vascular bundles.

NAPIER, John
(1550–1617)

Scottish mathematician, the inventor of logarithms, born at Merchiston Castle, Edinburgh. He matriculated at St Andrews University in 1563, travelled on the Continent, and settled down to a life of literary and scientific study. A strict Presbyterian, he published religious works and believed in astrology and divination, and for defence against Philip II of Spain, he devised warlike machines (including primitive tanks). He described his famous invention of logarithms in *Mirifici Logarithmorum Canonis Descriptio* ('Description of the Marvellous Canon of Logarithms', 1614); formulated to simplify computation, his system used the natural logarithm base c, but was modified soon after by Briggs to use the base 10. Napier also devised a calculating machine, using a set of rods, called 'Napier's bones', which he described in his *Rabdologiae* (1617).

NATHANS, Daniel
(1928–   )

American microbiologist, born in Wilmington, Delaware, of Russian-Jewish immigrants. He graduated from the University of Delaware in 1950 before pursuing post-graduate studies at Washington University School of

Medicine, where he obtained a medical degree in 1954. From 1955 to 1957 he worked at the National Cancer Institute of the National Institutes of Health. As a guest at the Rockefeller University, New York (1959–1962), he researched protein biosynthesis. From 1962 he was Professor at Johns Hopkins University and worked on SV40 virus. Nathans pioneered the use of restriction enzymes (which had recently been isolated by Hamilton Smith) to fragment DNA molecules, enabling him to make the first genetic map (of the circular SV40 DNA) and to identify the location of specific genes on the DNA. For this work he shared the 1978 Nobel Prize for Physiology or Medicine with Smith and Werner Arber.

NATTA, Giulio
(1903–1979)

Italian chemist, born in Imperia, near Genoa. He first studied mathematics at the University of Genoa but switched to chemical engineering as a student at Milan Polytechnic Institute, where he later became assistant lecturer in chemistry. In 1933 he was appointed Professor of Chemistry at the University of Pavia, and two years later he became Director of the Institute of Physical Chemistry in Rome. He was appointed Professor of Chemistry in Turin in 1937 and from 1938 held the post of Professor and Director of the Milan Institute of Industrial Chemistry until his retirement in 1973. His early work concerned heterogeneous catalysts used in a number of industrial processes, and in 1938 he initiated a programme for the production of artificial rubber. His most important work, used the organomettalic catalysts developed by Ziegler for the polymerization of propene to give polypropylene containing uniformly oriented methyl groups. Such polymers have a high melting point and great strength, and have become very important in industrial and commercial fields. In 1963 he shared the Nobel Prize for Chemistry with Ziegler.

NAVIER, Claude Louis Marle Henri
(1785–1836)

French civil engineer, born in Dijon. His father died when he was 14 and he came under the influence of his mother's uncle, the eminent civil engineer Emiland-Maric Gauthey. Educated at the École Polytechnique and the École des Ponts et Chaussées, for much of his life he taught at one or the other of these schools, being principally occupied in developing the theoretical basis of structural mechanics and the strength of materials, as well as the work done by machines. When Gauthey died leaving the unfinished manuscript of a treatise on bridges he completed the work, publishing it with editorial notes in three volumes between 1809 and 1816. He was responsible for the construction of a number of elegant bridges over the river Seine, but one of his most ambitious designs, a suspension bridge in Paris, encountered both engineering and political problems, to such an extent that it was dismantled just before completion. He published a number of important treatises on various aspects of structural mechanics, emphasizing the importance of being able to predict the limits of elastic behaviour in structural materials; his formulae represent some of the greatest advances in structural analysis ever made.

NEEDHAM, Joseph
(1900–)

English biochemist and historian of Chinese science, born in London. He was educated at Cambridge, and remained there as a university demonstrator in biochemistry (1928–1933), reader (1933–1966), and Master of Gonville and Caius College (1966–1976). Inspired by the earlier work of H. Spelman, Julian Huxley and de Beer, he experimented extensively on the nature of the organizers of the so-called 'morphogenic field' in amphibian development. The presence of a hormone (possibly a sterol) was inferred from the observation that dead tissue from the head end of an embryo would organize a second head if transplanted to the head end of a live embryo, but not if transplanted to the tail end. A first grade organizer induced the neural axis, and a second grade organizer produced the eye-cup and lens. Although no firm findings emerged. Needham's work provided the foundation for further studies. His early publications such as *Man a Machine* (1927), *The Sceptical Biologist* (1929) and *Chemical Embryology* (1931) were scientifically oriented, but his historical preoccupations emerged in *A History of Embryology* (1934) and *History is on Our Side* (1945). During World War II he headed the British Scientific Mission in China and became Scientific Counsellor at the British Embassy there. From 1946 to 1948 he was Director of the Department of Natural Sciences, UNESCO. He also published *Chinese Science* (1946) and *Science and Civilisation in China* (12 vols, 1954–1984), an exposition of the foremost significance in the history of science and Chinese historical achievement. In addition he published on the history of acupuncture, Korean astronomy and clocks, among a vast body of other work.

NÉEL, Louis Eugène Félix
(1904–)

French physicist, born in Lyons. A graduate the École Normale Supérieure, he was later Professor of Physics at Strasbourg University (1937–1940). In 1940 he moved to Grenoble and became the driving force in making it one of the most important scientific centres in France, becoming Director of the Centre for Nuclear Studies there in 1956. His research has been concerned with magnesium solids. At a time when three states of magnetism had been identified and explained (dia-, para- and ferromagnetism), he postulated a fourth, antiferromagnetism (1936). He argued for a crystal model in which two lattices with their magnetic fields acting in opposite directions are interlaced. Their opposing magnetic fields would cancel, leaving the crystal with little observable magnetic field. His predictions were verified by experiment in 1938, and fully confirmed by neutron diffraction techniques in 1949. He later explained the strong magnetism found in ferrite materials such as magnetite (1948), demonstrating that if the magnetic field of one of the two lattices (mentioned above) were stronger than the other there would be an observable magnetic field. His work on ferrimagnetic materials saw great application in the coating of magnetic tape, the permanent magnets of motors and the magnetic storage media used by computers. He was awarded the 1970 Nobel Prize for Physics jointly with Alfvén. He has also studied the past history of the Earth's magnetic field.

NEHER, Erwin
(1944–)

German biophysicist, born in Landsberg. He was educated in physics at the Technical University of Munich and the University of Wisconsin gaining his PhD from Munich in 1970. He joined the Max Planck Institute for Psychiatry in Munich (1970–1972) and then moved as a research associate to the Max Planck Institute for Biophysical Chemistry in Göttingen in 1972, becoming Director of the Membrane Biophysics Department in 1983. In 1976, after a sabbatical year at Yale University he succeeded, with Sakmann, in recording the electric currents through single ion channels in biological membranes, the existence of which were suggested by the work of Sir Alan Hodgkin and Sir Andrew Huxley. They developed the 'patch-clamp' technique, touching a cell membrane with the tip of a glass pipette filled with a saline solution, and by applying suction through the pipette, creating a seal which isolated a small section of the membrane. This permitted precise biophysical measurements to be made over a discrete area, a method that has revolutionized cell physiology. In 1991 Neher and Sakmann shared the Nobel Prize for Physiology or Medicine for this work.

NEWLANDS, John Alexander Reina
(1837–1898)

English chemist, born in London. He spent a year at the Royal College of Chemistry and fought with Garibaldi in 1860. From 1868 to 1888 he was Chief Chemist to a sugar refinery at Victoria Docks, London and later set up as an independent analyst and consultant. By 1963 Newlands had begun to build on earlier observations by Döbereiner and others that there was a relationship between the chemical properties of elements and their atomic weight. In 1865 he drew up a table of 62 elements arranged in eight groups in ascending order of atomic weight in order to illustrate what he described as the 'law of octaves'. Each element in the table was numbered, with hydrogen (the lightest element) as 1, osmium (the heaviest element then known) as 51, and elements which apparently had the same atomic weight sharing the same number. Newland's hypothesis was given a rough reception, because under his scheme, not all chemically similar elements fell into appropriate groups. It was to be Mendeleyev who made the critical leap forward when he realized that spaces should be left for undiscovered elements, thus allowing the known elements to fall into groups which demonstrated a true periodicity. After Mendeleyev published his periodic table, Newlands claimed priority and was eventually awarded the Davy Medal of the Royal Society in 1887. He died in London.

NEWTON, Alfred
(1829–1907)

English zoologist, born in Geneva. In 1866 he was appointed the first Professor of Zoology and Comparative Anatomy at Cambridge. He made visits to Lapland, Spitzbergen, the West Indies and North America on ornithological expeditions, and was instrumental in having the first Acts of Parliament passed for the protection of birds. His ornithological writings included *A Dictionary of Birds* (1893–1896) and he was editor of *Ibis* (1870–1872).

NEWTON, Sir Isaac
(1642–1727)

English scientist and mathematician, born in Woolsthorpe, Lincolnshire. Educated at Grantham Grammar School and Trinity College, Cambridge, in 1665 he committed to writing his first discovery on fluxions (an early form of differential calculus); and in 1665 or 1666 the fall of an apple in his garden suggested the train of thought that led to the law of gravitation. He turned to study the nature of light and the construction of telescopes. By a variety of experiments upon sunlight refracted through a prism, he concluded that rays of light which differ in colour differ also in refrangibility – a discovery which suggested that the indistinctness of the image formed by the object-glass of telescopes was due to the different coloured rays of light being brought to a focus at different distances. He concluded (correctly for an object-glass consisting of a single lens) that it was impossible to produce a distinct image, and was led to the construction of reflecting telescopes; and the form devised by him is that which reached such perfection in the hands of William Herschel and Rosse. Newton became a Fellow of Trinity College, Cambridge (1667), and Lucasian Professor of Mathematics in 1669. By 1684 he had demonstrated the whole gravitation theory, which he expounded first in *De Motu Corporum* (1684). Newton showed that the force of gravity between two bodies, such as the Sun and the Earth, is directly proportional to the product of the masses of the bodies and inversely proportional to the square of the distance between them. He described this more completely in *Philosophiae Naturalis Principia Mathematica* (1687); this great work was edited and financed by Halley. In the *Principia* he stated his three laws of motion: (1) that a body in a state of rest or uniform motion will remain in that state until a force acts on it; (2) that an applied force is directly proportional to the acceleration it induces, the constant of proportionality being the body's mass ($F = ma$); and (3) that for every 'action' force which one body exerts on another, there is an equal and opposite 'reaction' force exerted by the second body on the first. He also wrote *Opticks* (1703). The part he took in defending the rights of the university against the illegal encroachments of James II procured him a seat in the Convention parliament (1689–1690). In 1696 he was appointed Warden of the Mint, and was Master of the Mint from 1699. He again sat in parliament in 1701 for his university. He solved two celebrated problems proposed in June 1696 by Jean Bernoulli, as a challenge to the mathematicians of Europe; and performed a similar feat in 1716, by solving a problem proposed by Leibniz. He superintended the publication of Flamsteed's *Greenwich Observations*, which he required for the working out of his lunar theory – not without much argument between himself and Flamsteed. He was also involved in several priority disputes with Hooke in the latter's capacity as Secretary of the Royal Society. In the controversy between Newton and Leibniz as to priority of discovery of the differential calculus and the method of fluxions, Newton acted secretly through his friends. The verdict of science is that the methods were invented independently, and that although Newton was the first inventor, a great debt is owing to Leibniz for the superior facility and completeness of his method. In 1705 Newton was knighted by Queen Anne; he lies buried in Westminster Abbey. Newton also devoted much time to the study of alchemy and theology, and among a mass of more or less worthless material on these subjects he left some substantial discourses on transmutation, a remarkable manuscript on the prophecies of Daniel and on the Apocalypse, a history of creation, and a large number of miscellaneous tracts.

NICHOLSON, (Edward) Max
(1904–)

English bureaucrat, ornithologist, and conservation pioneer. He was educated at the university of Oxford, and was one of the founders of the Oxford Ornithological Society (1921); this led to the establishment of the Oxford Bird Census and hence to the British Trust for Ornithology (1932) and the Edward Grey Institute for Field Ornithology (1938) Nicholson was Director-General of the Nature Conservancy (1952–1965): in this role he stimulated and established conservation work in the UK and throughout the world. He was the author of *Birds in England* (1926), *How Birds Live* (1927), *Birds and Men* (1951), *The New Environmental Revolution* (1970) and *The New Environmental Age* (1987).

NICOL, William
(1768–1851)

Scottish geologist and physicist, born in Edinburgh, where he lectured in natural philosophy at the university. In 1828 he invented the Nicol prism, which utilizes the doubly refracting property of Iceland Spar, and which proved invaluable in the investigation of polarized light. It was also of fundamental importance in studies of minerals under the microscope. He also devised a new method of preparing thin sections of rocks for the microscope, by cementing the specimen to the glass slide and then grinding it until it was possible to view it by transmitted light, thus revealing the mineral's properties and internal structure. The technique was initially developed to examine the minute details of fossil and recent wood, and Nicol himself prepared a large number of thin sections to this end. Many of these sections were described by Henry Witham in his *Observations on Fossil Vegetables* (1831). Nicol's reluctance to publish delayed the widespread use of thin sections for some 40 years until Sorby and others introduced them into petrology.

NICOLLE, Charles Jules Henri
(1866–1936)

French physician and microbiologist, born in Rouen into a medical family, and educated in Rouen and Paris. His aptitude for research was stimulated at the Pasteur Institute by Émile Roux and Metchnikoff. From 1892 to 1902 he tried unsuccessfully to establish a research laboratory in Rouen. He then became Director of the Pasteur Institute in Tunis, which he and his colleagues turned into a leading research centre, working on the mode of spread, prevention and treatment of a number of diseases, including leishmaniasis, toxoplasmosis, Malta fever and typhus. His discovery that typhus is spread by lice (1909) had important implications during World War I and led to his award, in 1928, of the Nobel Prize for Physiology or Medicine. From 1932 he lectured each year at the College de France in Paris, but maintained his base in Tunis. He was a man of wide erudition who wrote novels, short stories and philosophical works.

NIEPCE, Joseph Nicéphore
(1765–1833)

French chemist and pioneer of photography, born in Chalon-sur-Saône. He served under Napoleon and in 1795 became administrator Nice. With enough inherited wealth to support himself, he was able to devote himself to research from 1801 onwards. He experimented with the new technique of lithography, using a camera obscura to project an image onto a wall, then tracing round the image in the time-honoured fashion. Being a poor draughtsman, he decided to look for ways of fixing the image automatically. In 1822, using silver chloride paper and a camera, he achieved a temporary image of the view outside his workroom window, but could not fix it. In 1826 he succeeded in making a permanent image using a pewter plate coated with bitumen of Judea, an asphalt which hardens on exposure to light. This historic negative, which Niepce termed a 'heliograph', is now in the Gernsheim Collection at the University of Texas at Austin. From 1829 Niepce collaborated with Daguerre in the search for materials which would reduce the exposure time but he died, in Saint-Loup-de-Varennes, before any progress was made. Although Niepce is known principally for his photographic work, he was active in other fields; he invented a method to extract sugar from pumpkin and beetroot, and with his brother Charles, built the Pyreolophore motor.

NIEUWLAND, Julius Arthur
(1878–1936)

Belgian-American chemist whose researches led to the synthesis of neoprene, the first commercially successful synthetic rubber. He was born in Hansbeke, Belgium, and after his parents emigrated to the USA he was educated at the University of Notre Dame, South Bend, Indiana. He then studied for the priesthood at the Catholic University of America, Washington DC. In 1904 he gained his doctorate with a study of acetylene which was to be fundamental to his subsequent research. He spent his teaching career at Notre Dame, as Professor of Botany from 1904 to 1918 and Professor of Organic Chemistry from 1918 to 1936. While studying the reaction between acetylene and arsenic trichloride in the course of his doctoral work, Nieuwland made a highly toxic gas and discontinued his research because of its deadly nature. It was subsequently developed into a poison gas (lewisite) and used in World War I. In 1920 Nieuwland discovered that, in the presence of a catalyst, acetylene could be polymerized – that is, its molecules could be condensed into much larger molecules to form quite different compounds. The product of the polymerization was a mixture of substances including divinylacetylene. In 1925 Nieuwland began to develop this reaction in collaboration with the chemists of Du Pont de Nemours, synthesizing neoprene (at first known as Duprene) in 1929. Synthetic rubber was put on the market in 1932. Nieuwland died in Washington DC.

NIRENBERG, Marshall Warren
(1927–)

American biochemist, born in New York City. He was educated at the universities of Florida and Michigan, and worked from 1957 at the National Institutes of Health in Bethesda, Maryland. Following the demonstration of the model of DNA by James Watson and Crick in 1953, it had been proposed that there are different combinations of three nucleotide bases (triplets or 'codons') in nucleic acid chains in DNA and RNA, each coded for a different amino acid in the biological synthesis of proteins, and that this was the fundamental process in the chemical transfer of inherited characteristics. However, the precise nature of the code remained unknown; there were 64 possible combinations of bases, and only 20 amino acids to be coded. Nirenberg attacked the problem of the 'code dictionary' by synthesizing a nucleic acid with a known base sequence, and then finding which amino acid it converted to protein. With his success, Khorana and others soon completed the task of deciphering the full code, which Nirenberg showed was 'universal' by examining *Escherichia coli*, a toad and the guinea pig as representative living organisms

(1967). Nirenberg, Khorana and Holley shared the Nobel Prize for Physiology or Medicine in 1968 for this work.

NOBEL, Alfred
(1833–1896)

Swedish chemist and industrialist, the inventor of dynamite and the founder of the Nobel prizes. Descended from several generations of scientists and inventors, Nobel was born in Bernhard, but spent much of his childhood in St Petersburg, Russia, where his father had an explosives factory and was experimenting with an underwater mine. Alfred was educated by tutors and became fluent in several languages. He studied chemistry in Paris, worked in the USA with Swedish-born John Ericsson, and returned to Sweden in 1859. In 1863 he and his father began to investigate nitroglycerine, an explosive oil that had been prepared by Sobrero in 1847. In 1863 Nobel invented the Nobel patent detonator, a metal cap containing an explosive compound, mercury fulminate. When the mercury was detonated, the shock detonated the main charge. The principle of detonating explosives by an initial smaller shock, rather than by heat, transformed blasting techniques. In 1865 Nobel opened the first factory for the manufacture of nitroglycerine but it shortly blew up, killing five people including his younger brother. Forbidden by the government to reopen it, he was reduced at one point to continuing his experiments on a barge. He discovered by chance that kieselguhr, a porous silaceous earth used as a packing material, would absorb large quantities of nitroglycerine and make it safer to handle. The result, which Nobel named dynamite, was first patented in 1867. Nobel later found that nitroglycerine formed a colloidal solution with nitrocellulose in the form of guncotton and that this was more powerful than dynamite, less subject to shock, and resistant to water and corrosion. It became known as gelignite. Nobel's other important achievement, around 1879, was to produce a practically smokeless Powder (Ballistite) by adding 10 percent of camphor to dynamite. He patented many other inventions – he made synthetic gutta-percha and mild steel for armour-plating and he also had large holdings in oilfields in Russia. He amassed a huge fortune and spent much of his life travelling from country to country monitoring his interests. Towards the end of his life, he investigated problems in chemistry and biology at his laboratory in San Remo. He hoped that his explosives would decrease, rather than increase, the chances of war throughout the world and in many respects remained a pacifist. He never married and left his total wealth to fund the rich prizes which bear his name. Since 1901 these have been awarded annually for physics, chemistry, physiology or medicine, literature and contributions to peace (the prize for economics was not established until 1968 and was funded by the Bank of Sweden). Nobel died in San Remo. The transuranic element synthesized in 1958 with an atomic number of 102 was named 'nobelium' in his honour.

NOETHER, (Amalie) Emmy
(1882–1935)

German mathematician, born in Erlangen. The daughter of the mathematician Max Noether, she studied at Erlangen and Göttingen. Though invited to Göttingen in 1915 by Hilbert, as a woman she could not hold a full academic post at that time, but worked there in a semi-honorary capacity until, expelled by the Nazis as a Jew, she emigrated to the USA in 1933 to Bryn

Mawr College and Princeton. She was one of the leading figures in the development of abstract algebra, working in ring theory and the theory of ideals. The theory of Noetherian rings has been an important subject of later research. She developed it to provide a neutral setting for problems in algebraic geometry and number theory with a view to enabling their essential features to stand out from the technicalities.

NOGUCHI, Hideyo
(1876–1928)

Japanese-born American bacteriologist, born in Inawashiro. He graduated from Tokyo Medical College and worked in the USA from 1900. At the Rockefeller Institute, in New York, he successfully cultured the spirochaete bacterium *Treponema pallidum* which causes syphilis. This enabled him to devise a diagnostic skin test for the disease using an emulsion of his culture. As a result of this, he was awarded the Order of the Rising Sun in his home country in 1915. He then went on to show that Oroya fever is caused by the bacterium *Bertonella bacilliformis*, which is transmitted to humans by sandflies. In 1927 he went to West Africa to obtain confirmatory evidence that yellow fever is a viral disease. He succeeded in proving this, but, just before his departure to return to New York, he contracted the disease, from which he died shortly afterwards.

NOLLET, Jean Antoine
(1700–1770)

French abbé and physicist, born in Pimprez near Noyan. While following an ecclesiastical, course of study in Paris he became interested in science, and devoted the rest of his life to research, writing and lecturing, and making scientific instruments. From 1730 he collaborated for a time in electrical researches with Dufay, Réaumur and others, taking a leading part in the popularization of experimental science in France. In 1748 he became the first Professor of Physics at the Collège de Navarre, in Paris, and in the same year he discovered and gave a clear explanation of the phenomenon of osmosis. He invented an early form of electroscope, and improved the Leyden jar (an early form of capacitor) invented by Musschenbroek.

NOROYI, Ryoji
(1937–)

Japanese chemist. Educated at Kyoto University, where he received his BSc (1961) and PhD (1967), he carried out postdoctoral work at Harvard. In 1966 he discovered transition-metal-catalysed asymmetric reactions in homogeneous phases. This discovery has led to his continued interest in the synthesis of chiral molecules using organometallic reagents. One of his major findings is the widely applicable homogeneous asymmetric hydrogenation process, in which different optical isomers may be specifically synthesized. Such syntheses are of particular value in the pharmaceutical industry. His discoveries in this area are important in that the chirality (the optical isomerism) of the catalyst is passed on to the reaction product – the chemical multiplication of chirality. Thus stereoselective organic synthesis may be performed; this is of immense importance in making physiologically active compounds such as pharmaceuticals. Noroyi's newly created tools have opened up efficient routes to such diverse compounds as terpenes, alkaloids, prostaglandins, nucleosides and nucleosides. He won the Chemical Society of Japan award in 1984.

NORRISH, Ronald George Wreyford
(1897–1978)

English physical chemist, born in Cambridge. He won a scholarship to Emmanuel College at Cambridge in 1915, but did not commence his studies until 1919. In the meantime he saw active service in the Royal Field Artillery and spent six months as a prisoner of war. After taking a first class in both parts of the Natural Sciences Tripos, he carried out research under Rideal. Following various junior appointments at Cambridge, he became H. O. Jones Lecturer in Physical Chemistry in 1928 and in 1937 he was promoted to Professor of Physical Chemistry and head of department, a post he held until his retirement in 1965. He was one of the founders of modem photochemistry, and also made advances in the area of chain reactions. His most important innovation (1945), in association with George Porter, was flash photolysis. In this technique a very brief flash of intense light causes photochemical change and immediately afterwards the unstable chemical intermediates produced can be studied by means of their absorption spectra. For this work Norrish was awarded the Nobel Prize for Chemistry jointly with Porter and Eigen in 1967. He was elected a Fellow of the Royal Society in 1936 and received its Davy Medal in 1958. He received the Faraday Medal of the Chemical Society in 1965 and its Longstaff Medal in 1969, and served as President of the Faraday Society from 1953 to 1955.

NORTHROP, John Howard
(1891–1987)

American biochemist, born in Yonkers, New York. Educated at Columbia University, he became Professor of Bacteriology at the University of California at Berkeley (1949–1962). In 1930, using an alcohol water mixture, he crystallized pepsin, the protein digesting enzyme of the stomach, showed it to be a protein and estimated its molecular weight. This was followed by the purification of other macromolecules, the gut protease, chymotrypsin, and its precursor, chymotrypsinogen (1935), a trypsin inhibitor from the pancreas (1937), and diphtheria toxin in crystalline form. He isolated the first bacterial virus, and was the first to equate the biological function of an enzyme with its chemical properties. He published *Crystalline Enzymes* (1939) which describes his important discovery and purifying proteins by 'salting out'; he also discovered the fermentation process used in the manufacture of acetone. For their studies of methods of producing purified enzymes and virus products, Northrop, Wendell Stanley and Sumner shared the 1946 Nobel Prize for Chemistry. From 1947 Northrop studied the autolysis (self-digestion) of pepsin and trypsin, and the effects of plant and animal proteases on living organisms.

NOSSAL, Sir Gustav Joseph Victor
(1931–   )

Australian immunologist, born in Bad Ischl, Austria. He arrived in Australia in 1939 and was educated at the universities of Sydney and Melbourne. He was appointed Research Fellow at the Walter and Eliza Hall Institute of Medical Research (1957–1959), and worked as Assistant Professor of Genetics at Stanford University (1959–1961) before returning to Melbourne as Deputy Director (immunology) in 1961. He proceeded to the directorship in 1965, when he also became Professor of Medical Biology at Melbourne University. His research work has been on immunity, which has strong experimental evidence in support of the clonal selection theory of Burnet, and his discovery of the 'one cell–one antibody' rule is crucial to modern work in immunology. He has written several popular books on immunology, and also on the progress of medical science, including *Antibodies and Immunity* (1971), *Medical Science and Human Goals* (1975) and *Reshaping Life: Key Issues in Genetic Engineering* (1984). He was knighted in 1977, and elected FRS in 1982.

NOYCE, Robert Norton
(1927–1990)

American physicist and electronics engineer, born in Burlington, Iowa. As a physics major at Grinell College, Iowa, he learned of the invention of the transistor from Grant Gale, the college's physics professor and a friend of Bardeen. He went on to MIT, where he received his PhD in 1953, and two years later he joined Shockley's semiconductor laboratory. In 1957, Noyce, a Swiss-born physicist named Jean Hoern and six others left Shockley and founded Fairchild Semiconductor in Silicon Valley. Here, using Hoerni's planar process, Noyce developed and perfected the planar integrated circuit, which led directly to the invention of a commercially feasible integrated circuit. With Kilby who worked on the microchip independently, Noyce is regarded as the co-inventor of the integrated circuit. In 1961 Fairchild introduced its first chips and Noyce's company prospered: between 1957 and 1967 revenues rose from a few thousand dollars to $130 million, and the number of employees grew from the original eight to 12,000. Noyce also co-founded Intel, the chip manufacturer.

NUTTALL, Thomas
(1786–1859)

English-born American naturalist, born in Settle, Yorkshire. A printer by trade, in 1808 he emigrated to Philadelphia, Pennsylvania, where he took up botany, accompanied several scientific expeditions between 1811 and 1834, and discovered many new American plants. He wrote *Genera of North American Plants* (1818), and became Curator of the Botanical Garden at Harvard (1822–1832). His *Introduction to Systematics and Physiological Botany* was published in 1827. While at Harvard he also turned his attention to ornithology, and published *A Manual of the Ornithology of the United States and Canada* (1832). His two-volume work *North American Silva* was published in 1842. In the same year he returned to England to fulfil the conditions of an inheritance, and remained there until his death.

O

OATLEY, Sir Charles
(1904–   )

English electronic engineer and inventor, born in Frome, Somerset. He graduated in physics from St John's College, Cambridge (1925), and shortly afterwards, joined the staff of King's College, London. During World War II he was a member of the Radar Research and Development Establishment, and in 1945 returned to Cambridge where in 1960 he became Professor of Electrical Engineering. Efforts by Zworykin and others, in the early 1940s to construct a practical electron microscope had met with only limited success, but Oatley realized in 1948 that newly developed circuits and components might overcome at least some of the problems. One of his research students, D. McMullan, produced a prototype instrument, which embodied many of the features of modern electron microscopes. Further development at Cambridge resulted in a scanning electron microscope being manufactured commercially in 1960, capable of producing three-dimensional images at magnifications of 100,000 or more. Oatley was elected FRS in 1969 and knighted in 1974.

OCHOA, Severo
(1905–   )

American geneticist, born in Luarca, Spain. He obtained his MD in Madrid (1929) and worked at several research centres, including Heidelberg (with Meyerhof on muscle physiology) and Oxford (with Peters on thiamine), before emigrating to the USA, where he accepted a post at the Washington University School of Medicine in St Louis. He later settled at the New York University School of Medicine, where he became full professor in 1946. Ochoa isolated two of the enzymes which catalyse part of the Hans Krebs cycle, and this led him to study the energetics of carbon dioxide fixation in photosynthesis, from 1948. He went on to isolate the enzyme polynucleotide phosphorylase (1955), later used for the first synthesis of artificial RNA, and established its properties and wide distribution in plants and animals. In 1961 he adopted Nirenberg's approach to solving the amino acid genetic code and determined a number of base triplets (codons) based on uridine. He also studied the direction of Protein synthesis along the DNA (1965), and the initiation factors associated with binding N-formylmethionine, the first amino acid in a bacterial peptide sequence (1967). For his contributions to the elucidation of the genetic code he was awarded the 1959 Nobel Prize for Physiology or Medicine, jointly with Arthur Kornberg.

ODUM, Eugene Pleasants
(1913–)

American ecologist, born in Newport, New Hampshire, and educated at the universities of North Carolina and Illinois. He was Callaway Professor Emeritus of Ecology and the Alumni Foundation Distinguished Professor Emeritus of Zoology at the University of Georgia at Athens. His research interests have included ecological energetics, estuarine and wetland ecology, and physiological and population ecology. He has stressed the view that ecosystem theory provides a common denominator for man and nature and that neither can be considered in isolation. These views are expounded in his *Ecology and Our Endangered Life-support Systems* (1989). He is also the author of three widely used textbooks, *Fundamentals of Ecology* (1953), *Ecology* (1975) and *Basic Ecology* (1982). His honours include the Institut de la Vie Prize (1975) awarded by the French government, and the Crafoord Prize of the Royal Swedish Academy of Sciences (1987).

OHM, Georg Simon
(1787–1854)

German physicist born in Erlangen, Bavaria, the son of a locksmith. He withdrew from the University of Erlangen because his overindulgence in dancing, billiards and ice-skating had incurred his father's displeasure. After a spell in Switzerland teaching mathematics, he completed his studies at Erlangen. He later became professor at Nuremberg (1833–1849) and Munich (1849–1854). His 'Ohm's law', relating voltage, current and resistance in an electrical circuit, was published in 1827, and was followed by a long struggle before its importance was recognized. Wheatstone was an early adherent. His work on the recognition of sinusoidal sound waves by the human ear as pure tones (1843) received similar treatment until it was rediscovered by Helmholtz. He was awarded the Royal Society's Copley Medal (1841) and was elected a Foreign Member of the society in 1842. The SI unit of electrical resistance is named after him.

OKAZAKI, Reiji
(1930–1975)

Japanese biochemist, born in Hiroshima, and aged 14 when the world's first atomic bomb was dropped on his home town. Having graduated in science at Nagoya University, he remained there as a lecturer and was appointed professor in 1967. Working with bacteria and bacteri-ophage, Okazaki was the first to identify, by buoyant density measurements, the DNA-RNA fragments named after him (1967). These units of DNA replication, of length about 1000–2000 nucleotides in prokaryotes and 100–200 nucleotides in eukaryotes, resolved the dilemma of how DNA was simultaneously synthesized in opposite directions, but always corresponding to the opposing polarity (the 5′ to 3′ direction) of the two DNA strands. Continuous synthesis occurred on one strand while 'Okazaki fragments' subsequently joined together, built up on the other. With Arthur Kornberg he was the first to recognize the 'primer' function of the short RNA sequence to which the DNA is attached, the RNA being subsequently excised and replaced by DNA by the 'Kornberg enzyme'. He also studied the RNA free, so called pseudo-Okazaki fragments produced by certain bacterial mutants or derived by degradation of normal DNA. Okazaki was awarded the Asashi Prize in 1970. He developed leukaemia, and his health deteriorated until he died of heart failure.

OLDHAM, Richard Dixon
(1858–1936)

Irish geologist and seismologist, born in Dublin, discoverer of the Earth's core. Educated at Rugby and the Royal School of Mines, he was a member of the Geological Survey of India (1878–1903) and for some of this time, Director of the Indian Museum in Calcutta. In 1903 he resigned his post as superintendent of the survey, partly because of ill-health and returned to England. His important report on the Assam Earthquake of June 1897 distinguished for the first time between primary and secondary seismic waves and was able to characterize many other phenomena of earthquake activity. He proved the generality of his notions about the different types of seismic waves with reference to six other earthquakes in *On the Propagation of Earthquake Motion to Great Distances* (1900), and laid the foundations of what is now one of the principal branches of geophysics. In 1906 he established from seismographical records the existence of the Earth's core. He was the author of *Bibliography of Indian Geology* (1888), *Catalogue of Indian Earthquakes* (1883) and many other works on Indian geology.

OLIPHANT, Sir Mark (Marcus Laurence Elwin)
(1901–   )

Australian nuclear physicist, born in Adelaide. He studied there and at Trinity College and the University of Cambridge, where he received his PhD in 1929. He then worked at the Cavendish Laboratory in Cambridge where in 1934, with Rutherford and Harteck, he discovered the tritium isotope of hydrogen by bombarding deuterium with deuterons. In 1937 he became professor at Birmingham University, where he designed and built a 60-inch cyclotron particle accelerator, completed after World War II. He worked on the Manhattan project at Los Alamos (1943–1945) to develop the nuclear bomb, but at the end of hostilities strongly argued against the American monopoly of atomic secrets. In 1946 he became Australian representative of the UN Atomic Energy Commission. He was later appointed research professor at Canberra University (1950–1963) and designed a proton synchrotron accelerator for the Australian government. From 1971 to 1976 he served as Governor of South Australia. He was elected FRS in 1937, and knighted in 1959.

OLSEN, Kenneth Harry
(1926–   )

American computer engineer and entrepreneur, born in Bridgeport, Connecticut, into an evangelical Scandinavian Protestant family. He studied electrical engineering at MIT, where he obtained a master's degree in 1952 and joined Forrester's pioneering computer group in 1950. Soon afterwards Olsen became an on-site engineer at IBM, but in 1956 he left to establish his own computer company, the Digital Equipment Corporation (DEC) in Maynard, Massachusetts. DEC defined and then exploited a new niche in the growing computer industry – the market for minicomputers, or 'interactive' machines, that were less expensive and easier to use than mainframes. Aided by brilliant engineers, such as Gordon BM from MIT, Olsen launched the PDP-8 – the first successful minicomputer – in the early 1960s. By 1986 DEC was the second largest US computer company behind IBM, and Olsen was described by *Fortune* as 'America's most successful entrepreneur'. In 1992, however, after DEC had suffered heavy losses and stagnating sales in the minicomputer market, Olsen was forced to resign as the chief executive.

ONSAGER, Lars
(1903–1976)

Norwegian-American chemical physicist, born in Christiania (now Oslo). He was trained at the Technical University of Norway as a chemical engineer, but pursued further studies in mathematics in preparation for working on difficult problems in theoretical physics and chemistry. He worked in Zürich with Debye from 1926 to 1928 and then went to the USA, where he spent the rest of his life. After periods at Johns Hopkins University and Brown University, Rhode Island, he settled at Yale, where he advanced from assistant professor to associate professor between 1934 and 1945, when he became Gibbs Professor of Theoretical Chemistry. He held this position until 1972, when he moved to the University of Miami as Distinguished University Professor. Onsager's work with Debye was on strong electrolytes, for which he developed an extension of the Debye-Hückel theory. However, he is best known for his pioneering work on the thermodynamics of irreversible processes, which he put on a sound basis. The fundamental equations in this field are called the 'reciprocal relations' and are commonly known by his name. For this work he was awarded the Nobel Prize for Chemistry in 1968. The theory of irreversible thermodynamics was developed further by Prigogine.

OORT, Jan Hendrik
(1900–1992)

Dutch astronomer, born in Franeker. He studied under Kapteyn at the University of Gröningen, and worked mainly at the Leiden Observatory in Holland (1924–1970), becoming director there in 1945. He proved (1927) by observation that our galaxy is rotating, and calculated the distance of the Sun from the centre of the galaxy, initially locating it 300,000 light years away. He found that the Sun has a period of revolution around the galactic centre of just over 200 million years. This enabled him to make the first calculation of the mass of galactic material interior to the Sun's orbit, this being some 1,011 solar masses. In 1932 he made the first measurement that indicated that there is dark matter in the galaxy, concluding that the visible stars near the sun could account for only around half the mass implied by the velocity of these stars perpendicular to the galactic plane. Beginning in 1944, Oort traced the structure of the galactic disc by using radio telescopes to detect the 21 centimetre wavelength radiation that is emitted by atomic hydrogen. In 1946 he realized that the filamentary nebulae called the Cygnus Loop is a supernova remnant. In 1950 he extended Õpik's suggestion concerning the huge circular reservoir of comets surrounding the solar system. These have maximum distances from the Sun of some 100,000 astronomical units and are thus susceptible to being perturbed by passing stars. This 'Oort cloud' was the suggested source of long-period comets. In 1956, with Theodore Walraven, he discovered the polarization of the radiation from the Crab nebula indicating that it was produced by synchrotron radiation from electrons moving at high speeds along magnetic field lines.

OPPENHEIMER, (Julius) Robert
(1904–1967)

American nuclear physicist, born in New York City. He studied at Harvard where he graduated in 1925, the University of Cambridge and under Born at Göttingen, where he received his doctorate in 1927. He returned to the USA and established schools of theoretical physics at Berkeley and Caltech. His work included studies of electron-positron pairs, cosmic-ray theory and deuteron reactions. During World War II, he led a team which pioneered theoretical studies on building the atomic bomb, and he was later selected as leader of the atomic bomb project. He set up the Los Alamos laboratory and brought together a formidable group of scientists. After the war he became Director of the Institute for Advanced Studies at Princeton and continued to play an important role in US atomic energy policy from 1947. He used his political influence to promote peaceful uses of atomic energy and was bitterly opposed to the development of the hydrogen bomb. In 1953 when the US government turned against him and declared him a security risk, he was forced to retire from political activities; later, however, he received the Enrico Fermi Award of the Atomic Energy Commission (1963). He delivered the BBC Reith Lectures in 1953.

OSBORN, Henry Fairfield
(1857–1935)

American palaeontologist and zoologist, born in Fairfield, Connecticut. He studied at Princeton, and became Professor of Zoology at Columbia University and concurrently Curator of Vertebrate Palaeontology at the American Museum of Natural History (1891–1910). Retaining a research professorship at Columbia, he was President of the American Museum of Natural History from 1908 to 1933. Although known as an autocratic leader, he revolutionized museum display with innovative instructional techniques and the acquisition of spectacular specimens, especially dinosaurs. He popularized palaeontology, mounting skeletons in realistic poses with imaginative backdrops. His many publications include *The Age of Mammals* (1910), *Man of the Old Stone Age* (1915) and *The Origin and Evolution of Life* (1917). His major scientific contribution was a vast monograph on *Proboscidea* published posthumously in two volumes (1935–1942).

OSLER, Sir William
(1849–1919)

Canadian-American-British physician, born in Bond Head, Ontario, and educated at Toronto and McGill universities. After graduating in medicine, he toured Britain and Germany for scientific training and in 1874 became Professor of Medicine at McGill. Chairs at the University of Pennsylvania (1884–1889) and Johns Hopkins (1889–1904) followed, and he was subsequently appointed to the Regius Chair of Medicine at Oxford. His years at Johns Hopkins were his best; during this time he produced many clinical papers, monographs on cerebral palsy and chorea, and his textbook *The Principles and Practice of Medicine* (1892). This codified the scientific clinical practice of his time and was frequently revised and translated. An advocate of full-time clinical training and research, Osler made a number of bedside and pathological observations of permanent value. He made an early study of the platelets, described hereditary haemorrhagic telangiectasis (Osler-Rendu-Weber disease), polythaemia vera (Vaguez-Osler disease) and infection of the heart valves (endocarditis). He was instrumental in founding the Association of Physicians of Great Britain and Ireland, a society devoted to encouraging clinical research. He was also an elegant stylist, an ardent bibliophile and bibliographer, and advocate of humane values in a world of science. He was made a baronet in 1911. By the time of his death, he was revered throughout the English-speaking world as a kind of patron saint of patient-oriented scientific medicine; the reverence continues unabated through Osler societies, lectures and prizes in many places, including Japan. His last years were clouded by the death of his only son during World War I.

OSTWALD, (Friedrich) Wilhelm
(1853–1932)

German physical chemist, born in Riga, Latvia. He studied chemistry at the University of Dorpat (now Tartu), taking the *Candidat* examinations in 1875. After holding various posts as an assistant at Dorpat, he was appointed Professor of Chemistry at the Riga Polytechnic in 1881. In 1887 he moved to Leipzig as Professor of Physical Chemistry, taking early retirement in 1906. He spent the rest of his life on his country estate in Saxony, devoting himself to various scientific, literary and other intellectual activities. With van't Hoff and Arrhenius, Ostwald is regarded as one of the founders of physical chemistry. At Dorpat he worked on the measurement of chemical affinity by observing changes in the physical properties of solutions as a result of chemical reactions. During his Riga period he used rates of reaction to study chemical affinity and he measured the 'affinity coefficients' of many acids, particularly organic acids, through studies of their catalytic behaviour. His results were greatly illuminated by the electrolytic dissociation theory of Arrhenius, which Ostwald did much to promote. In Leipzig he built up a great school of physical chemistry, which attracted students from all over the world. His studies of electrolytic conductivity (resulting in Ostwald's dilution law) and of the electromotive force of cells were carried out in Leipzig. He founded the journals *Zeitschrift fur physikalische Chemie* in 1887 and *Annalen der Naturphilosophie* in 1901. His various books were very influential, notably his *Lehrbuch der Chemie* (2 vols, 1883–1887). In his long retirement he worked on the theory of colour perception, an interest which arose from his skill as a landscape painter. He was also an able musician and was interested in various systems of philosophy. For his work on catalysis, Ostwald was awarded the Nobel Prize for Chemistry in 1909. He became an Honorary Fellow of the Chemical Society in 1898 and received its Faraday medal in 1904.

OTTO, Nikolaus August
(1832–1891)

German engineer, born in Holzhausen, Nassau. The son of a farmer, he left school at 16 to work in a merchant's office, but soon moved to Cologne where he became interested in the gas engines of Etienne Lenoir. By 1861 he had built a small experimental gas engine, and three years later he joined forces with Eugen Langen (1833–1895), an industrialist who had studied at Karlsruhe Polytechnic, to form a company for the manufacture of such engines. Dissatisfied with the limitations of Lenoir's low compression low speed gas engines, Otto devised the four-stroke cycle which bears his name and obtained a patent for it in 1877. His engines were so successful that other manufacturers sought ways in which to evade the restrictions of the patent, and eventually it was discovered that the principle of the four-stroke cycle had been outlined by Beau de Rocha, in a patent dated 1862, though he had not built even a prototype engine at the time. Finally in 1886 Otto's patent was cancelled, but by that time over 3,000 of his so-called 'silent otto' engines had been sold, and the Otto cycle remains today the operating principle of the great majority of the world's internal combustion engines.

OWEN, Sir Richard
(1804–1892)

English zoologist and palaeontologist, born in Lancaster. He studied medicine at Edinburgh and at St Bartholomew's Hospital in London, and became curator at the Royal College of Surgeons. In 1856 he was appointed superintendent of the natural history department of the British Museum, and was instrumental in the establishment of the separate British Museum (Natural History) (now the Natural History Museum), becoming its first director in 1881. He was the zoologist of Victorian England, and published 400 scientific papers as well as a number of important books, including *British Fossil Mammals and Birds* (1846), *A* History of British Fossil Reptiles (1849–1884), and an influential essay on *Parthenogenesis*. He named and reconstructed numerous celebrated fossils, including the giant moa bird *Dinornis*, the dinosaur *Iguanodon* and the earliest bird, *Archaeopteryx*. He coined the term 'dinosaur' – ('terrible lizard') Owen studied in detail the homologies between apparently dissimilar structures in organisms, and drew the crucial distinction between homologous and analogous organs. However, he remained implacably opposed to evolution; for him, homologies were variants on a divine plan or 'archetype', not evidence of common descent. He was a virulent and outspoken opponent of Charles Darwin and T. H. Huxley; Darwin stated that 'his power of hatred was unsurpassed', and put it down to jealousy at the success of the *Origin of Species*. Owen accepted a knighthood in 1884, having previously declined the honour in 1842.

OXBURGH, Sir (Ernest) Ronald
(1934–)

English geologist. Educated at the University of Oxford and Princeton University, he became a lecturer at Oxford (1960–1978) and Professor of Mineralogy and Petrology at Cambridge (1978–1991). He was also chief scientific advisor to the Ministry of Defence (1988–1993). Oxburgh has undertaken important petrological and geochemical research, particularly with his studies of the origin and distribution of radiogenic helium in the Earth's crust. He was elected FRS in 1978, and knighted in 1992.

P

PALADE, George Emil
(1912–)

Romanian cell biologist, born in Iasi. Trained as a doctor in Bucharest, he became Professor of Anatomy there until he emigrated to the USA in 1946. He worked at the Rockefeller Institute, New York (1946–1972) and from 1972 headed cell biology at Yale Medical School. Since 1990 he has been Professor of Cellular and Molecular Biology at the University of California, San Diego. In the 1950s Palade developed a method of separating components of the cell known as 'cell fractionation'. Using the newly introduced technique of electron microscopy, he described the components of the cell; the mitochondria, the endoplasmic reticulum, the Golgi apparatus and ribosomes. He showed that protein synthesis occurs on strands of RNA in the ribosomes which are attached to the membranous endoplasmic reticulum. The proteins are then carried through the cell in sacs, called vacuoles, before being released into the extra-cellular fluid. For his work in cell biology he shared the 1974 Nobel Prize for Physiology or Medicine with Albert Claude and de Duve.

PANDER, Christian Heinrich
(1794–1865)

Russian-born German anatomist and crucial figure in modern embryology. Born in Riga, Pander studied at Dorpat, and subsequently at Berlin, Göttingen and Wurzburg, where he was befriended by Baer. He took his MD in 1817. He undertook valuable research on chick development in the egg, in particular demonstrating the embryonic layers named after him, and coining the term blastoderm for the trilaminar structure (from the Greek *blastos* meaning 'germ', and *derma*, 'skin'). Pander never followed up his early findings, leaving the field of embryological research to others, including Baer. He spent much time travelling in 1820 acting as a naturalist on a Russian mission to Bokhara. In 1826 he was elected a member of the St Petersburg Academy of Sciences, and a year later he retired to his estates around Riga.

PAPANICOLAOU, George Nicholas
(1883–1962)

Greek-born physiologist and microscopic anatomist, born in Kimi, the son of a physician. He received his MD from Athens University (1904) and a PhD from Munich University (1910). He moved to the USA in 1913, becoming assistant in pathology at the New York Hospital and, in 1914, assistant in anatomy at Cornell Medical College. All his research was conducted at these two institutions until 1961, when he was appointed Director of the Miami Cancer Institute, although he died three months later. He became Professor of Clinical Anatomy at Cornell in 1924 and was Emeritus Professor from 1949. Papanicolaou's research on reproductive physiology led him to the discovery that the cells lining the wall of the guinea pig vagina change with the oestrus cycle. Similar changes take place in women, but more importantly, Papanicolaou noticed that he could identify cancer cells from scrapings from the cervixes of women with cervical cancer. He subsequently pioneered the techniques, now familiar as the 'pap smear', of microscopical examination of exfoliated cells for the early detection of cervical and other forms of cancer.

PAPPUS OF ALEXANDRIA
(fourth century)

Greek mathematician. He wrote a mathematical *Collection* covering a wide range of geometrical problems, some of which inspired Descartes and contributed to the development of projective geometry in modern times. It is of great importance for the historical understanding of Greek mathematics. In it he described the economical work of bees, and discussed the isoperimetric problem (the claim that of all curves of a given length the circle encloses the greatest area). He generalized Pythagoras's theorem to triangles that are not right-angled, wrote on the trisection of the angle and devices to square the circle, and offered commentaries on Euclid's *Elements* and Ptolemy's *Almagest*.

PARACELSUS, (a name coined for himself by Theophrastus Bombastus von Hohenhelm)
(1493–1541)

German alchemist, physician and self-styled seer. The name referred to the celebrated Roman physician Celsius and meant 'beyond' or 'better than' Celsus. Paracelsus was born in Einsieden, Switzerland, but moved when young to Villach, a mining area in southern Austria. His father was a physician who taught chemistry at the borough school and here the boys were trained to assay local ores, giving Paracelsus an early grounding in metallurgy and chemistry. When he was 14 he began to wander from one university to another in search of inspired teachers, and he is said to have graduated in Vienna and taken his doctorate in Ferrara. Finding universities fossilized in their attitudes and insulated from real life, he then spent many years exploring Europe, including England and Scotland. He served as an army surgeon in the Netherlands and Italy, and was captured by the Mongols during a visit to Russia; finally he travelled through Egypt, Arabia and the Holy Land to Constantinople. In all these lands he studied contemporary medical practice and the medical lore of the common people, always looking for ways to encourage 'the latent forces of Nature' in all healing processes. In 1526 he was appointed town physician in Basle and lecturer in chemistry at the university. Already famous for his erudition and revolutionary views on medicine and religion, he raged against medical malpractices and the fashion for patent medicines. He also criticized the Catholic Church and the new Lutheran doctrines. He taught in German, not Latin, another innovatory move which displeased the authorities who preferred learning to be kept from the populace. In 1528, having antagonized all the vested interests in the town, he had to flee for his life, and spent most of the rest of his life as an itinerant preacher and physician conducting a chemical research wherever he could find a laboratory, and writing mystical tracts and works on medicine, chemistry and alchemy. By the end of the 1730s he was to some extent re-established and numbered the rich and influential among his patients, although he continued to treat the poor without charge. He died in Salzburg where his grave became a place of pilgrimage for the sick. Despite his erratic and violent nature, echoed in his writings, Paracelsus had enormous influence, particularly through the emphasis he laid on observation and experiment and need to assist – rather than hinder – natural processes. He believed that wounds should be allowed to heal naturally without interference except for cleaning. He stated that diseases had external causes and that every disease had its own characteristics, thus reversing the traditional view that disease was generated within the patient and followed an unpredictable course. He made careful studies of tuberculosis and silicosis (both miners diseases); recognized that there was a connection between goitre and the minerals in drinking water; researched the role of acids in digestion; was the first to recognise congenital syphilis; studied hysteria; and advocated that lunatics should be treated kindly. Paracelsus was also an able chemist who discovered many techniques which became standard laboratory practice, such as concentrating alcohol by freezing it out of its solution. He also prepared drugs with due regard to their purity and advocated carefully measured doses, both important steps forward in medicine.

PARDEE, Arthur Beck
(1921–)

American biochemist, born in Chicago. He trained and worked at the University of California, Berkeley (1942–1961), spending some time at Caltech (1943), the University of Wisconsin (1947–1949) and the Pasteur Institute, Paris (1957–1958), before becoming professor at Princeton (1961–1967). Since 1975 he has been professor of Pharmacology at Harvard. His early work was on biological oxidation processes with Van Potter (of homogenizer fame), notably on the inhibition of succinate dehydrogenase by malonate, and with Pauling on tumour metabolism and quantification of the strength of antibody reactions. Pardee became interested in how cells control their own synthetic processes, and in 1958 worked with Monod on the lac operon of *Escherichia coli.* He also discovered feedback control of amino acid synthesis (threonine synthetase). In order to study DNA repair and the regulation of DNA synthesis, he developed a gentle method of synchronizing the cellcyle (cyclical growth and division) of initially bacteria and then eurocaryotic cells in culture. In this connection he exploited a range of specific inhibitors (bromouracil, chlorammphenicol) in relation to such processes as the regulation of bacterial division by protein X (recA gene control) and the requirement for branched chain fatty acids (1971), the regulation of ribonucleotide reductase (1978), the formation of Okazaki fragments (1977) and histone synthesis (1978). From around 1980 Pardee investigated the effects of serum peptides, including insulin-like growth factor, and the ubiquitin system (1984).

PARKES, Alexander
(1813–1890)

English chemist, the inventor of celluloid, born in Birmingham. While working for a brass founder in Birmingham he devised a way of electroplating fragile natural objects, such as flowers, and a spider's web which he had electroplated was presented to Prince Albert. Parkes was the first person to show that phosphorus added to metal alloys increased their strength. In 1841 he discovered that rubber could be vulcanized without heat if treated with a solution of carbon bisulphide. This process was used to waterproof fabric and the patent was sold to Macintosh. While superintending the construction of a coppersmelting works in south Wales in the early 1850s, Parkes discovered a way to extract silver from lead. In this method zinc and lead are melted together; the zinc then reacts with the silver and any other metals in the lead, and the newly formed compounds – being lighter than silver – float to the surface and can be skimmed off. The Parkes process was soon used worldwide. Around 1855 Parkes was searching for a substitute for ivory and produced a synthetic material from a mixture of chloroform and castor oil. Although it was flexible and durable, it didn't achieve commercial success in Britain. The process was refined in the USA and the product renamed 'celluloid'. Parkes died in West Dulwich, now part of London.

PASCAL, Blaise
(1623–1662)

French mathematician, physicist and theologian, born in Clermont-Ferrand, the son of the local President of the Court of Exchequer. His mother having died, the family moved to Paris (1630), where the father, a considerable mathematician, personally undertook his children's education. Blaise was not allowed to begin a subject until his father thought he could easily master it. Consequently it was discovered that the 11 year old boy had worked out for himself in secret the first 23 propositions of Euclid, calling straight lines 'bars' and circles 'rounds'. Inspired by the work of Desargues, at 16 he published an essay on conics which Descartes refused to believe was the handiwork of a youth. It contains his famous theorem on a hexagram inscribed in a conic. Father and son collaborated in experiments to confirm Torricelli's theory, unpalatable to the schoolmen, that nature does not, after all, abhor a vacuum. These experiments consisted in carrying up the Puy de Dôme two glass tubes containing mercury, inverted in a bath of mercury, and noting the fall of the mercury columns with increased altitude. This led on to the invention of the barometer, the hydraulic press and syringe. In 1647 he patented a calculating machine, later simplified by Leibniz, which Blaise had built to assist his father in his accounts. His correspondence with Fermat in 1654 laid the foundations of probability theory. Following his experience of a religious revelation in 1654, he joined his sister at the Jansenist retreat at Port-Royal (1655), and concentrated on theological works. Around that time he gave up mathematics almost completely, although in one last important contribution he solved the long-standing problem of the area of the cycloid; his publication of this work heralded the invention of the integral calculus.

PASTEUR, Louis
(1822–1895)

French chemist, the father of modern bacteriology, born in Dôle. He studied at Besançon and at the École Normale

Supérieure, and held academic posts at Strasbourg, Lille and Paris, where in 1867 he became Professor of Chemistry at the Sorbonne. His work was at first chemical, notably on tartrate crystals; he noticed that the crystals appeared in two mirror-image forms, one which would rotate plane polarized light to the left, and the other which would rotate light to the right. He discovered a living ferment, a micro-organism comparable in its powers to the yeast plant, which would, in a solution of paratartrate of ammonia, select for food the 'right-handed' tartrates alone, leaving the 'left-handed'. He went on to show that other fermentations – lactic, butyric, acetic – are essentially due to organisms, not spontaneous generation. He greatly extended Schwann's researches on putrefaction, and gave valuable rules for making vinegar and preventing wine disease, introducing in this work the technique of 'Pasteurization', a mild and short heat treatment to destroy pathogenic bacteria. After 1865 he tackled silkworm disease, his research leading to the revival of the silk industry in southern France following a devastating parasitic disease which was killing the silkworms; he also investigated injurious growths in beer, splenic fever, and fowl cholera. Pasteur's 'germ theory of disease' represented one of the greatest discoveries of the nineteenth century – he began to realize that disease was communicable through the spread of micro-organisms. He showed that it was possible to attenuate the virulence of injurious micro-organisms by exposure to air, by variety of culture, or by transmission through various animals. He thus demonstrated that sheep and cows vaccinated with the attenuated bacilli or anthrax were protected from the evil of subsequent inoculation with virulent virus; by the culture of antitoxic reagents, the prophylactic treatment of diphtheria, tubercular disease cholera, yellow fever and plague was also found effective. He introduced a similar treatment for hydrophobia (rabies) in 1885. The Pasteur Institute of which he became first director, was founded in 1888 for his research.

PAUL, Wolfgang
(1913–)

German physicist, born in Lorenzkirch. He studied in Munich and at the Technical University in Berlin where he received his doctorate in 1939. He later joined the staff of the University of Göttingen (1944), becoming professor there in 1950. He simultaneously held a teaching post at Bonn University from 1952. He developed the 'Paul trap' to constrain electrons and ions within a small space for study; this allowed important advances in the accuracy with which atomic properties could be measured, and has been used in important tests of modern atomic theory. For this work he shared one-half of the 1989 Nobel Prize for Physics with Dehmelt, who had developed a similar technique (the other half of the prize was awarded to Norman Ramsey).

PAULING, Linus Carl
(1901–1997)

American chemist, born in Portland, Oregon. He was educated at Oregon State College and Caltech, receiving his PhD in 1925. After postdoctoral work in Munich, Zürich, and Copenhagen, he was on the chemistry faculty at Caltech from 1927 to 1963, as full professor from 1931. Pauling's early work on crystal structures (1928) led to their rationalization in terms of ionic radii and greatly illuminated mineral chemistry. He then turned to the quantum-mechanical treatment of the chemical bond and made many important contributions, including the concept of the 'hybridization of orbitals', central to understanding the shapes of molecules. This period of his work generated two influential books – *Introduction to Quantum Mechanics* (1935, with E. Bright Wilson) and *The Nature of the Chemical Bond* (1939). His interest in complex molecular structure led him into work in biology and medicine; he studied the structures of proteins and antibodies, and investigated the nature of serological reactions and the chemical basis of hereditary disease. During the last 20 years of his life he advocated the use of vitamin C in combating a wide range of diseases and infections, and his views have generated controversy. He was also a controversial figure for his work in the peace movement and his criticism of nuclear deterrence policy. Pauling was awarded the Nobel Prize for Chemistry in 1954 and the Nobel Peace Prize in 1962. He became a Foreign Member of the Royal Society in 1948 and was awarded its Davy medal in 1947. He was elected an Honorary Fellow of the Chemical Society in 1943.

PAVLOV, Ivan Petrovich
(1849–1936)

Russian physiologist, born near Ryazan, the son of a priest. Pavlov studied natural sciences, graduating from St Petersburg in 1875, and medicine, receiving his doctorate in 1879. From 1886 he worked at the Military Medical Academy in St Petersburg. He became Professor of Pharmacology (1890), Professor of Physiology (1895), and Director of the Institute of Experimental Medicine (1913). Pavlov's work was concerned with three main areas of physiology: the circulatory system (1874–1888), the digestive system (1879–1897) and higher nervous activity including the brain (1902–1936). He developed a series of surgical preparations for the study of digestion in dogs, and his investigations, which included studies of the nervous control of salivation and the role of enzymes in digestion, provided insights of great value for the clinical pathology of the gut. For his work on digestion, Pavlov was awarded the Nobel Prize for Physiology or Medicine in 1904. Pavlov is most famous for his work which showed that if a bell is sounded whenever food is presented to a dog, it will eventually begin to salivate when the bell is sounded without food being presented. This he termed a 'conditioned' or acquired reflex, established by the involvement of the cortex of the brain in modifying innate reflexes. He regarded this phenomenon as of fundamental importance, and it was the starting point for subsequent studies of experimental psychoses, human psychic disorders, and his theories of animal and human behaviour.

PEANO, Giuseppe
(1858–1932)

Italian mathematician, born in Cuneo. He was educated at the University of Turin, where he later taught and became Extraordinary Professor of Infinitesimal Calculus (1890) and full professor in 1895. He did important work on differential equations and discovered continuous curves passing through every point of a square. This deepened his distrust of intuitive mathematics, and he moved to mathematical logic. He advocated writing mathematics in an entirely formal language, and the symbolism he invented became the basis of that used by Bertrand Russell and Whitehead in their *Principia Mathematica*. He also promoted

Interlingua, a universal language based on uninflected Latin.

PECQUET, Jean
(1622–1674)

French anatomist, born in Dieppe. Pecquet studied in Paris, proceeding to Montpellier where he received his MD in 1652. He was to serve as physician to many Paris notables. He built upon, but corrected, Aselli's researches on the lacteals. As Pecquet described in his *Experimenta nova anatomica* (1651), while performing a vivisection experiment on a digesting frog, he was the first to see clearly the thoracic duct, through which the lacteal vessels discharge chyle into the veins. His discovery was warmly welcomed by Caspar Bartholin, who clearly distinguished the chylous vessels from the lymphatics; but debate regarding the lymphatic and lacteal system remained intense throughout the rest of the century.

PEDERSEN, Charles
(1904–1990)

American chemist, born in Pusan, Korea. Although his mother was Japanese and his father a Norwegian mining engineer, he studied chemical engineering in the USA (University of Dayton, Ohio) and took a master's degree in organic chemistry at MIT. Throughout his life he worked for Du Pont de Nemours as a research chemist, and the work for which he is best known was not published until 1967, almost at the end of his professional life. By accident he prepared a cyclic polyether, a molecule shaped rather like a crown and to which the name 'crown ether' was given. He found that many compounds of this type bind alkali metal ions (sodium, potassium) very strongly, making alkali metal salts soluble in organic solvents. His discovery initiated the study of guest-host chemistry and gave insight into the means by which metal ions are transported across membranes in living organisms. He retired in 1969 and shared the Nobel Prize for Chemistry with Cram and Lehn in 1987.

PEIRCE, Charles Sanders
(1839–1914)

American philosopher, logician and mathematician, born in Cambridge, Massachusetts, the son of the mathematician Benjamin Peirce. He graduated from Harvard in 1859 and began his career as a scientist, working for the US Coast and Geodetic Survey from 1861. In 1879 he became a lecturer in logic at Johns Hopkins University, but he left in 1894 to devote the rest of his life in seclusion to the private study of logic and philosophy. In his scientific work, he developed the theory of gravity measurement using pendulums, and conducted gravity experiments in Europe and North America. He also made an early determination of the metre in terms of a wavelength of light. In philosophy, he was a pioneer in the development of modern formal logic and the logic of relations, but is best known as the founder of pragmatism, which he later named 'pragmaticism' to distinguish it from the work of the philosopher William James. His theory of meaning helped establish the new field of semiotics, which has become central in linguistics as well as philosophy. His enormous output of papers was collected and published posthumously.

PELL, John
(1610–1685)

English mathematician and clergyman, born in Southwick, Sussex. A brilliant student at Cambridge, he was appointed Professor of Mathematics at Amsterdam in 1643 and lecturer at the New College, Breda, in 1646. Employed by Oliver Cromwell, first as a mathematician and later in 1654 as his agent, he went to Switzerland in an attempt to persuade Swiss Protestants to join a Continental Protestant league led by England. In 1661 he became Rector at Fobbing in Essex and in 1663 Vicar of Laindon. In mathematics, he is remembered chiefly for the equation $x2 = Ay2 + 1$ for integers x, y and A, mistakenly named after him by Leonhard Euler, and for introducing the division sign into England.

PENFIELD, Wilder Graves
(1891–1976)

American-born Canadian neurosurgeon, born in Spokane, Washington. After undergraduate studies in Princeton, he went to Oxford University as a Rhodes Scholar in 1914, and studied physiology under Sherrington, but the outbreak of World War I interrupted his studies. Wounded in the war he returned to the USA, where he finished his medical education at Johns Hopkins University. Further scientific study in Oxford and Spain prepared him for his experimental neurosurgical work, which he developed in conjunction with surgical practice in New York at the Presbyterian Hospital and the College of Physicians and Surgeons at Columbia University. In 1928 he moved to a neurosurgical appointment at McGill University where he was instrumental in founding the world-famous Montreal Neurological (1934–1960). An outstanding practical neurosurgeon, his experimental work on animals and on the exposed brains of conscious human beings helped in understanding the higher functions of the brain, and the causes of symptoms of brain disease such as epilepsy, and the mechanisms involved in speech. He became a Canadian citizen in 1934. Following his retirement in 1960, he began a successful second career as a novelist and biographer.

PENROSE, Lionel Sharples
(1898–1972)

English geneticist, born in London. He studied at St John's College, Cambridge, receiving his BA in 1921. Following postgraduate research in psychology at Cambridge, he worked in psychiatry in Vienna, developing a strong interest in mental illness. Returning to the UK in 1925, he studied medicine and wrote his MD thesis on schizophrenia. During the 1930s he carried out a major survey into the causes of mental illness, the Colchester Survey. As a Quaker and pacifist, he spent World War II in Canada, as Director of Psychiatric Research for Ontario. In 1945 he became Galton professor of Eugenics at London. Penrose objected strongly to the attempts of 'eugenics' to improve the characteristics of human populations through the application of genetics, and changed his title to Professor of Human Genetics, having changed the name of the journal of the Galton Laboratory from *Annals of Eugenics* to *Annals of Human Genetics* in 1954. Under his direction (1945–1965), the Galton Laboratory became an international centre for human genetics, conducting varied research projects, for example on Down's syndrome, the mapping of genes on chromosomes, and palm and finger prints, using statistical and cytogenetic techniques. He wrote *The Biology of Mental Defect* (1949) and *An Outline of Human Genetics* (1960).

PENZIAS, Arno Allan
(1933–)

American astrophysicist, born in Munich, Germany. A refugee with his family from Nazi Germany, he was educated at Columbia University, New York, and joined the Bell Telephone Laboratories in 1961, finally becoming Executive Director of Research and Communication Science there. In 1963 Penzias and his colleague Robert Wilson were assigned the task of tracing the radio noise that was interfering with Earth-satellite-Earth Communications. Using a 20-foot horn reflector they discovered a 7.3-centimetre wavelength extraterrestrial signal that seemed to show no directional variation and did not vary with the positions of the Sun and stars. They also measured the corresponding 'temperature' (the temperature of a black body which would characteristically emit radiation of this wavelength) of this background radiation, first obtaining a value of 3.5 kelvin but subsequently revising this to 3.1 kelvin. At first it was assumed that the horn reflector was responsible, but in fact Penzias and Wilson had discovered the residual relic of the intense beat that was associated with the birth of the universe following the hot Big Bang. This was the cosmic microwave background radiation predicted to exist by Gamow and Alpherin 1948. In 1970, with Wilson and K. B. Jefferts, Penzias discovered the radio spectral line of carbon monoxide at 2.6 millimetres; this has since been used as a tracer of galactic gas clouds. Penzias and Wilson were awarded the Nobel Prize for Physics in 1978, along with Kapitza.

PERKIN, William Henry Jr
(1860–1929)

English chemist, born in Sudbury, the eldest son of William Henry Perkin Sr. After study at the Royal College of Science in London, he went to Wurzburg (1880–1882) to work with Wislicenus and to Munich (1882–1886) to work with Baeyer. He returned to Britain to take up a position at Heriot-Watt College in Edinburgh, and in 1892 became professor at Owens College in Manchester. His final position was as Waynflete Professor of Chemistry at Oxford (1912–1929). As a young man in Germany he worked on small ring compounds, and his results were used by Baeyer in his ring strain theory. Subsequent work was entirely concerned with the elucidation of the structures of natural products by degradation studies. He had particular success with a number of alkaloids including berberine, harmine, cryptopine, strychnine and brucine. Although a fine chemist and a skilled practical worker, he did not have the innovative talent of his father or indeed, of his brothers-in-law Kipping and Arthur Lapworth. He was elected FRS in 1890.

PERKIN, Sir William Henry Sr
(1838–1907)

English chemist, born in Shadwell, London. He became fascinated by chemistry while at school and in 1853 enrolled at the Royal College of Chemistry to study under Hofmann. At the age of 17 he became Hofmann's personal assistant, but his first major discovery was made in a private laboratory he fitted out in his home. Here, in 1856, Perkin attempted to synthesize quinine, much in demand for the treatment of malaria, by oxidizing allyltoluidine with potassium dichromate. He was not successful, but after conducting the same experiment with toluidine, he was able to extract a brilliant purple dye, subsequently named mauveine. With the encouragement of the dyeing firm J. Pullar and Son, Perth,

Perkin at the age of 18 built a factory near Sudbury to manufacture mauveine. A craze for things purple swept Victorian society; Perkin's venture was a great commercial success and initiated the modern synthetic dyestuffs industry, which introduced a new range of colours into human life. He also devised synthetic procedures for the production of the natural dye alizarin. In 1874 Perkin sold his dye works in order to devote himself exclusively to his chemical researches, which reached an important point in 1881 when he observed the magnetic rotatory power of a number of the organic compounds he had made. He continued many investigations into the relationship between molecular constitution and the physical properties of organic compounds. He held no academic post and conducted his research in his private laboratory. Such is his standing among organic chemists that the section of the Royal Society of Chemistry concerned with organic chemistry is called the Perkin Division. He received many honours both in Britain and in the USA, including election to the Royal Society in 1866 and a knighthood in 1906. His three sons all became distinguished chemists.

PERRIN, Jean Baptiste
(1870–1942)

French physicist, born in Lille. He was educated at the École Normale Superieure in Paris and from 1898 to 1940 was on the physical chemistry staff of the University of Paris, as full professor from 1910. In 1940 he escaped to the USA following the invasion of France; he died in New York in 1942. Perrin's earliest work was on cathode rays and helped to establish their nature as negatively charged particles. However, he is most remembered for his studies of the Brownian movement. Through elegant experimental work he demonstrated that the suspended particles which show Brownian motion in colloidal solutions essentially obey the gas laws and he used such systems to determine a fairly accurate value for the Avogadro number. His book *Les Atomes* (1913), which described this work, became a classic. He was awarded the Nobel Prize for Physics in 1926. Elected a Foreign Member of the Royal Society in 1918, he served as President of the French Academy of Sciences in 1938.

PERSOON, Christiaan Hendrik
(1761–1836)

South African botanist, born at the Cape of Good Hope. He studied theology at Halle, medicine at Leiden, and medicine and natural science at Göttingen. In 1799 he received a PhD at Erlangen, and he moved to Paris in 1802. He corresponded widely and influenced many other botanists. His *Synopsis Fungorum* (1801) is considered to be the basis of modern mycology. Persoon's work is the starting point for the nomenclature of the Gasteromycetes, Uredinales and Ustilaginales. He sought to describe briefly all phanerogams then known in *Synopsis plantarum* (1805–1807), and he also produced the incomplete *Mycologia Europaea*. In 1818 he expressed the incorrect view that some fungi grew from spores while others were formed by spontaneous generation. He gave his botanical collections to the Rijksherbarium in Leiden in 1828, in exchange for a pension. Leiden publishes the mycological journal *Persoonia* named after him.

PERT, Candace, née Beebe
(1915–   )

American pharmacologist born in Manhattan, New York. Educated at Bryn Mawr College and Johns Hopkins University Hospital Medical School, she received her PhD in pharmacology in 1974. She stayed at Johns Hopkins as Research Fellow (1974–1978) and Research Pharmacologist (1978–1982) until her appointment as chief of the section of brain chemistry of the National Institute of Mental Health. Her doctoral research, under the supervision of Snyder, as stimulated by the realization that the highly specific effects of synthetic opiates at very small doses indicated that they must bind to highly selective target receptor sites. She began a search for these sites, using radioactively labelled compounds. In 1973 she first reported the presence of such receptors in specialized areas of the mammalian brain. From this arose the suggestion that there might be natural opiate-like substances in the brain that used these sites, as later discovered by Kosterlitz and Hughes. Her research continues on the chemical characteristics of brain tissue and the relationships of chemicals to neural functioning.

PERUTZ, Max Ferdinand
(1914–   )

Austrian-born British biochemist, born in Vienna. After graduating at Vienna, he emigrated to the UK (1936) where, apart from a brief alliance with chymotrypsin, he has worked single-mindedly on the structure of haemoglobin at Cambridge ever since. He became Director of the Medical Research Council (MRC) Unit for Molecular Biology. His determination of haemoglobin structure reflects the development of X-ray crystallography. By 1939, Perutz had determined the first crystal parameters indicating four haem groups in one plane, and observed effects shown later (1953) by Kendrew to reflect structural changes in the molecule. The presence of the alpha helix, discovered by Pauling in 1951, was predicted in the same year to occur in haemoglobin by Perutz, Crick and Sir Lawrence Bragg. By the time Kendrew had published the structure of myoglobin at 2 angstroms resolution, Perutz had determined the haemoglobin structure to 5.5 angstroms. Amino acid sequence studies by others, facilitating comparison between myoglobin and haemoglobin, allowed Perutz to predict the detailed distribution of amino acids in haemoglobin (1964); confirmation at 2.8 angstroms was completed in 1969. Perutz and Kendrew were awarded the 1962 Nobel Prize for Chemistry. Since then, Perutz has established the nature of the subunit interactions in relation to oxygen binding and its control, the effects of genetic variants, the evolutionary development and numerous other aspects of haemoglobin. His publications include *Mechanisms of Cooperativity and Allosteric Regulation in Proteins* (1990). He was elected FRS in 1954.

PFEFFER, Wilhelm Friedrich Philipp
(1845–1920)

German botanist, born near Cassel. Trained as a pharmacist, he became a specialist in plant physiology, was appointed professor successively at Bonn, Basle, Tübingen and Leipzig, and was noted particularly for his researches on osmotic pressure. He experimented with aniline dyes on plant cells, and was able to establish the nature and structure of vacuoles. He concluded that living protoplasts could modify their permeability, allowing non-osmosing substances to pass. He hinted at the role of plasmodesmata in solute transport between cells. He also suggested that streaming movements within cells might be a factor in solute transport, and alluded to the importance of transpiration in plant growth and development. Pfeffer's work on chloroplasts showed the effects of light intensity and other physical and chemical factors on the process of photosynthesis. He also studied nitrogen metabolism in leguminous plants. His *Handbuch der Pfanzenphysiologie* (1881) was a standard work on plant physiology for many years.

PFEIFFER, Richard Friedrich Johannes
(1858–1945)

German bacteriologist, born near Posen (now Poznan, Poland). He studied under Koch, and became professor at Berlin (1894) Konigsberg (1899) and Breslau (1901). It was during the influenza epidemic of 1889–1892 that he discovered the bacillus *Haemophilus influenzae* in patients' throats (1892) and suggested that it was the cause of the disease. However, it was later shown that influenza is caused by a viral infection and that the bacillus is responsible for many of the complications. Pfeffer's most significant discovery was his observation, for the first time, of a complex immune reaction (1894). He injected live cholera vibrios into guinea pigs which had already been immunized, then extracted some of the germs. Examining the extract under a microscope, he observed the germs becoming motionless then swelling and finally disintegrating, in a process which he named 'bacteriolysis'. He showed that the same process occurred *in vitro*, and that the reaction would cease when heated to over 60 degrees Celsius. This work led Bordet to study the immune system, resulting in the discovery of complement. Pfeiffer published books on hygiene and microbiology. He was presumed dead in 1945.

PIAZZI, Giuseppe
(1746–1826)

Italian astronomer, born in Ponte, northern Italy. He became a Theatine monk (1764), was appointed Professor of Mathematics at Palermo in Sicily in 1780, and founded on behalf of the Bourbon government two observatories, at Naples and at Palermo. After 22 years of observing at Palermo with a vertical circle made by the English optician Jesse Ramsden, he published a monumental catalogue of 7,646 stars (1813). In the course of his observation he discovered, on the night of 1 January 1801, the very first minor planet (or asteroid) which he named Ceres after the tutelary deity of Sicily.

PICCARD, Auguste Antoine
(1884–1962)

Swiss physicist, born in Basle. He became Professor at Brussels in 1922 and held posts at Lausanne, Chicago and Minnesota universities. With his brother Jean Felix, he ascended 16–18 kilometres by balloon (1931–1932) into the stratosphere. In 1948 he explored the ocean depths off west Africa in a bathyscaphe constructed from his own design. His son Jacques, together with an American naval officer, Donald Walsh, established a world record by diving more than seven miles in the US bathyscaphe *Trieste* into the Marianas Trench of the Pacific Ocean in January 1960.

PICKERING, Sir George White
(1904–1980)

English medical scientist, born in Whalton, Northumberland. He graduated in natural sciences from Pembroke College, Cambridge (1926), and trained in medicine at St Thomas's Hospital, qualifying in 1928. He was appointed to the staff of the Medical Research Council (1931) and spent eight years working with Sir Thomas Lewis at University College Hospital in London. In 1932 he presented results of experiments on headaches and went on to work on hypertension. He studied the causes of increased peripheral resistance, hereditary factors, and aspects of atheromatous disease. He also conducted important experimental work on the mechanism of pain in peptic ulcers. Pickering became lecturer in cardiovascular pathology in 1936, Professor of Medicine at St Mary's Hospital Medical School (1939–1956) and Regius Professor of Medicine at Oxford (1956–1968). He was a key figure in medical education in Britain from the 1950s and wrote widely on historical and cultural issues. He was knighted in 1957.

PICTET, Marc-Auguate
(1752–1825)

Swiss physicist, born in Geneva. Pictet qualified in law (1774), but his true interests lay in the natural sciences. He was appointed to the Chair of Philosophy at the Geneva Academy in 1786 on the resignation of his mentor, Horace Benedict de Saussure. Pictet was prominent in the public life of Geneva, respected by Napoleon, and active in the scientific circles of post-revolutionary France. His studies in geology, meteorology and astronomy were characteristically wide-ranging. Best known throughout Europe was the *Essai sur le Feu* (1790) recording researches on heat and hygrometry. There he described how radiant heat could be reflected, like light. Pictet's journal, the *Biblioteque Britannique*, sustained intellectual contact between Britain and the Continent during the Napoleonic Wars.

PICTET, Raoul Pierre
(1846–1929)

Swiss physicist, born in Geneva. Between 1868 and 1870 he studied physics and chemistry in Geneva and in Paris, returning to Geneva to devote himself to the study of low temperatures in connection with developing refrigeration techniques. He became Professor of Industrial Physics at the University of Geneva in 1879. In 1886 he moved to Berlin to establish an industrial research laboratory and to market his refrigeration inventions. Later he returned to Paris. Working in Geneva in 1877, he liquefied oxygen in bulk by 'cascade' cooling and compression. Cailletet in Paris liquefied oxygen independently at around the same time, but in much smaller quantities. Pictet later made an erroneous claim to have liquefied hydrogen.

PINCUS, Gregory Goodwin
1903–1967)

American physiologist who developed the oral contraceptive pill, born in Woodbine, New Jersey. He graduated in science from Cornell University in 1924, undertook postgraduate study at Harvard (1924–1927) and then visited Europe, working at Cambridge and Berlin universities (1927–1930). He was a member of the biology faculty at Harvard (1930–1938) and of the experimental zoology department at Clark University (1938–1945). In 1944 he established, with Hudson Hoagland, the Worcester Foundation for Experimental Research, which became internationally renowned for work on steroid hormones and mammalian reproduction. In 1951, influenced by the birth control campaigner Margaret Sanger, he began work on developing a contraceptive pill. With John Rock and Min Chueh Chang he studied the antifertility effect of those steroid hormones which inhibit ovulation in mammals; in this way refertilization is prevented during pregnancy. Synthetic hormones became available in the 1950s and Pincus organized field trials of their antifertility effects in Haiti and Puerto Rico in 1954. The results were successful, and oral contraceptives ('the pill') have since been widely used, despite concern over some side effects. Their success is a pharmaceutical rarity; synthetic chemical agents do not usually show nearly 100 percent effectiveness in a specific physiological action, or have such remarkable social effects.

PLANCK, Max Karl Ernst Ludwig
(1858–1947)

German physicist, born in Kiel. Planck's father was Professor of Civil Law at the University of Kiel. When the family moved to Munich in 1867, Planck entered the Maximilian Gymnasium. Although musically gifted and fascinated, even at an early age, by philology, he decided to study science at the University of Munich from 1874. In Berlin (1877–1878) he attended lectures given by Kirchhoff and Relmboltz, and read Clausius's papers. Returning to Munich he completed his doctorate on the second law of thermodynamics (1879) and then lectured in the university until he became Professor of Theoretical Physics at Kiel (1885). Late in 1888 he succeeded Kirchhoff as Professor of Theoretical Physics at Berlin, where he remained until his retirement (1926). From 1880 Planck had studied the foundations of thermodynamics and also chemical equilibrium. Later, stimulated by experiments in progress at the Physikalisch-Technische Reichanstalt, he combined thermodynamics with electrodynamics in studies of black-body radiation. Where classical theories failed to match experimental results, Planck's formulation (1900) modelled them with great accuracy. He had relied on Boltzmann's statistical interpretation of the second law of thermodynamics; and he had also assumed the 'discontinuous' emission of small discrete packets of energy: 'quanta'. This revolutionary work earned Planck the Nobel Prize for Physics in 1918, by which time conceptions of subatomic processes had changed completely; on such microscopic scales quantum theory took over. Einstein soon developed the quantization idea to explain the photoelectric effect (1905), and Niels Bohr applied quantum theory to problems of atomic structure (1913). Planck continued to hope for a synthesis of deterministic classical physics and quantum methods, and opposed the extreme indeterminism of Heisenherg. In 1930 Planck was elected President of the Kaiser Wilhelm Institute, but he resigned in 1937 in protest against the actions of the Nazi regime. He remained in Germany, however, and was eventually reappointed President of the renamed Max Planck Institute. His life was overshadowed with personal tragedy; his son Erwin was executed for involvement in the 1944 plot against Hitler's life.

PLAYFAIR, Lyon, 1st Baron Playfair
(1819–1898)

Scottish chemist and statesman, born in Chunar, India. He studied medicine at Glasgow and Edinburgh, and under Liebig at Giessen. He worked as an industrial chemist in Manchester in association with Mercer, and later developed Mercer's ideas on catalysis. He was appointed Professor of Chemistry at the School of Mines in London, and ended his academic career as Professor of Chemistry at Edinburgh. In 1868 he was elected to Parliament, and from 1880 to 1883 he served as Deputy Speaker in the House of Commons. Playfair described the nitroprussides, worked on vapour densities, and translated the works of Liebig, but his principal importance was as an administrator. He was a prominent member of many government committees, for example on public health, the Irish potato famine, and the reform of the civil service. For a period he was inspector of the schools of science in London. He was a member of the committee which organized the Great Exhibition of 1851, and in 1853 he was appointed secretary to the newly formed Department of Science and Art, subsequently helping to establish the Royal College of Science and the South Kensington Museum (which later became the Victoria and Albert Museum and the Science Museum). One of the first scientists to hold important public positions, he worked throughout his life to promote scientific and technical education, and to encourage industry to make use of scientific advances. He was elected to FRS in 1848, knighted in 1883 and created a baron in 1892. He died in London.

PLINY, Galus Plinius Secundus, known as 'the Elder'
(AD 23–79)

Roman writer on natural history. He came from a wealthy north Italian family owning estates at Novum Comum (Como), where he was born. He was educated in Rome, and when about 23 entered the army and served in Germany. He became colonel of a cavalry regiment and a comrade of the future emperor Titus, wrote a treatise on the throwing of missiles from horseback, and compiled a history of the Germanic wars. He also made a series of scientific tours in the region between the Ems, Elbe and Weser, and the sources of the Danube. Returning to Rome in AD 52, he studied for the bar, but withdrew to Como and devoted himself to reading and authorship. Apparently for the guidance of his nephew, he wrote his *Studiosus*, a treatise defining the culture necessary for the orator, and the grammatical work, *Dubius Sermo*. By Nero, he was appointed procurator in Spain, and following his brother-in-law's death (AD 71) became guardian of his sister's son, Pliny the Younger, whom he adopted. Vespasian, whose son Titus he had known in Germany, was now emperor, and became a close friend; but court favour did not wean him from study, and he brought down to his own time the history of Rome by Atifidius Bassus. A model student, amid metropolitan distraction he worked assiduously, and by lifelong application filled the 160 volumes of manuscript which, after using them for his universal encyclopedia in 37 volumes, *Historia Naturalis* (AD 77), he bequeathed to his nephew. In AD 79 he was in command of the Roman fleet stationed off Misenum when the great eruption of Vesuvius was at its height. Eager to witness the phenomenon as closely as possible, he landed at Stabiae (Castellamare), but had not gone far before he succumbed to the stifling vapours rolling down the hill. His *Historia Naturalis* alone of his many writings survives. Under that title the ancients classified everything of natural or non-artificial origin. Pliny adds digressions on human inventions and institutions, devoting two books to a history of fine art, and dedicates the whole to Titus. His observations, made at second-hand, show no discrimination between the true and the false, between the probable and the marvellous, and his style is inartistic, sometimes obscure. But he supplies information on an immense variety of subjects about which, but for him, we should have remained in the dark.

POINCARÉ, Jules Henri
(1854–1912)

French mathematician, born in Nancy. He studied at the École Polytechnique, and as an engineer at the École des Mines became Professor of mathematics in Paris in 1881. Following the work of Immanuel Fuchs and Klein he created the theory of automorphic functions, which blended new ideas in the theories of groups, non-Euclidean geometry and complex functions, and showed the importance of topological considerations in differential equations. Many of the basic ideas in modern topology – such as triangulation, homology, the Euler-Poincaré formula and the fundamental group – are due to him. In a paper on the three-body problem (1889) he opened up new directions in celestial mechanics, and began the study of dynamical systems in the modern sense. He tried to keep the French tradition of applied mathematics alive through his influential lecture courses on such topics as thermodynamics, optics, magnetism and electricity in the sense of Maxwell and Heinrich Hertz. In 1905 he came close to anticipating Einstein's theory of special relativity, showing that the Lorentz transformations form a group. In his last years he published several articles (later collected as books) on the philosophy of science and scientific method, including *Science et methode* (1909). He opposed the move towards the axiomatic foundations of mathematics and logic, and advocated a form of intuitionism.

POISSON, Simeon Denis
(1781–1840)

French mathematical physicist, born in Pithiviers. He was educated at the École Polytechnique under Laplace and Lagrange and became the first Professor of Mechanics at the Sorbonne, achieving a leading position in the French scientific establishment. He published extensively on mathematical physics, and although his work was criticized for lack of originality by many of his contemporaries, his contributions to potential theory and the transformation of equations in mechanics by means of Poisson brackets have proved of lasting worth. He is also remembered for his discovery of the 'Poisson distribution', a special case of the binomial distribution in statistics.

POLANYI, Michael
(1891–1976)

Hungarian-British physical chemist, social scientist and philosopher, born in Budapest. He qualified in medicine at the University of Budapest in 1913, but he was already seriously interested in chemistry, and had written on thermodynamics and topics in medical biochemistry. He studied physical chemistry with Georg Bredig at the Technische Hochschule, Karlsruhe, on several occasions. During World War I he was a medical officer in the Austrian army. From 1920 to 1923 he worked at the Kaiser Wilhelm Institute for Fibre Chemistry in Berlin, and then moved to the Institute of Physical Chemistry under Haber. In 1933, finding the political climate of Germany increasingly unpleasant, Polanyi accepted the Chair of Physical Chemistry at Manchester University. He built up an excellent school of physical chemistry, but his interests were already moving to wider cultural and philosophical matters. In 1949 he gave up the Chair of Physical Chemistry and was given a personal Chair in Social Studies. In 1958 he moved to Oxford as a Senior Research Fellow of Merton College. He was elected FRS in 1944. In Berlin Polanyi worked on X-ray diffraction by fibres and then began his studies of chemical kinetics which continued for some 25 years. He did important experimental work, particularly on reactions of free atoms with small molecules in the gas phase, in the expectation that such reactions would prove amenable to theoretical treatment. He was himself much involved in the development of transition state theory, and this was central to his chemical interests in Manchester. In this theory it is supposed that all reactions proceed via a highly energetic molecular entity, the activated complex, and this idea is fundamental to the understanding of rates and mechanisms of chemical change. His social and philosophical interests are best indicated by the titles of some of his books: *The Contempt of Freedom* (1940), *Full Employment and Free Trade* (1945), *Science, Faith and Society* (1946), *Personal Knowledge* (1958) and *Knowing and Being* (1969). His writings often met with suspicion and criticism in philosophical circles.

PONCELET, Jean Victor
(1788–1867)

French engineer and geometrician, born in Metz. A military engineer during Napoleon's Russian campaign, he was taken prisoner by the Russians on the retreat from Moscow. During this time he thought over the geometry he had learned from Monge and his book on this subject *Traité des propriétés projectives des figures* (1822) made his name and revived interest in the development of projective geometry. He became Professor of Mechanics at Metz (1825–1835) and Paris (1838–1848). Poncelet sought to found geometry on basic principles as general as those of algebra, so that for example, one argument concerning a line and a conic would suffice for the cases where the line cuts, touches, or does not meet the conic. Disillusioned by hostile contemporary criticism directed at his approach, and despite the many fertile methods his book contained, he turned to the theory of machines and applications of mathematics to technology.

POPE, Sir William Jackson
(1870–1939)

English chemist, born in London. After school in London he studied at Finsbury Technical College and the Central Technical College at South Kensington, where he worked with Henry Armstrong. In 1897 he became head of the chemistry department of the Goldsmiths' Institute and in 1901 he moved to the Municipal School of Technology in Manchester. His final appointment in 1908, at the early age of 38, was to the Chair of Organic Chemistry at Cambridge. He worked on a number of topics, including mustard gas, camphor, organometallic compounds and photographic sensitizers, but he is best known for his work on optical activity. In 1899 he reported the synthesis and resolution of enantiomeric nitrogen compounds, and at Cambridge repeated this with sulphur and selenium. For his scientific work he was awarded the freedom and livery at the Goldsmiths Company in 1919, and served as prime warden (1928–1929). He as knighted in 1919.

POPOV, Aleksandr Stepanovich
(1859–1905)

Russian physicist, born in Bogoslavsky, the son of an Orthodox priest. After graduating from the seminary in Perm (now Molotov), he studied physics at St Petersburg University, and while still a student he worked at the Elektrotekhnik artel (1881), which ran the first Russian small generating plants and arc light installations. He was appointed as an instructor at the Russian Navy's Torpedo School in Kronstadt (1881), and later professor at the St Petersburg Institute of Electrical Engineering (1905). Independently of Marconi, he is acclaimed in Russia as the inventor of wireless telegraphy (1895). He was the first to use a suspended wire as an aerial.

PORTA, Giovanni Battista delta
(1535–1615)

Italian natural philosopher, born in Naples. Of a noble family, he was probably self-educated, although few details of his life are known apart from his numerous writings. He was keenly interested in every aspect of Renaissance natural philosophy, including the occult sciences. He wrote on such varied subjects as physiognomy, fortification, palmistry and natural magic, metallurgical technology, crystallography and the classification of plants, besides producing several comedies. Most of his original work was in optics and the application of steam. He was one of the first to make a serious study of the camera obscura, and found that the sharpness of the image could be increased by placing a lens at the pin-hole, thus turning it into a remote ancestor of the box camera. He discovered that the condensation of steam in a closed vessel leaves an empty space (the concept of a vacuum was then unknown) into which more liquid may be drawn, and he designed a rudimentary steam-pump to supply water to a fountain – in this case, the ancestor of Thomas Savery's atmospheric steam-pump. Porta founded a number of scientific academies, including one in Naples for the study of the secrets of nature. This may have been the cause of his interrogation by the Inquisition as a result of which his works were banned from 1592 to 1598. He was admitted in 1610 as a member of the Accademia dei Lincei in Rome.

PORTER, George, Baron Porter of Luddenham
(1920–)

English physical chemist, born in Stainforth, Yorkshire. After taking a BSc in chemistry at Leeds University, he became a radar officer in the Royal Naval Volunteer Reserve from 1941 to 1945. He then entered Emmanuel College Cambridge, for his PhD. From 1949 to 1954 he held junior posts at Cambridge and after a brief stay at the British Rayon Research Association, he became Professor of Physical Chemistry at Sheffield University in 1955. In 1963 he transferred to the Firth Chair of Chemistry. He left Sheffield to become Resident Professor and Director of the Royal Institution, where he remained until 1985. His researches have mainly been concerned with extremely rapid reactions in the gas phase, especially photochemical reactions and with photochemistry generally. In the late 1940s, Norrish and Porter developed the technique of flash photolysis, which became important in the study of very rapid gas reactions; with Eigen, they were awarded the 1967 Nobel Prize for Chemistry for this work. In flash photolysis a very brief flash of intense light causes photochemical change and immediately afterwards the unstable chemical intermediates produced can be studied by means of their absorption spectra. When the technique was first introduced, the species had to survive for a few milliseconds for study to be possible, but through improvements over the past 40 years the timescale involved has moved from milliseconds to seconds. Porter has been prominent in the refinements of techniques which have made this possible and in exploring the ever-widening range of fast processes which may be studied. In later years he became prominent as a spokesman for science in the UK. While at the Royal Institution he was very successful as a popularizer of science, particularly in the Christmas Lectures shown on television. He was knighted in 1972, admitted to the Order of Merit in 1989, and made a life peer in 1990. He was elected FRS in 1960, received Royal Society's Davy (1971) and Rumford (1978) medals, and served as its president from 1985 to 1990. Porter was President of the Chemical Society from 1970 to 1972, and received the society's Faraday medal in 1980.

PORTER, Rodney Robert
(1917–1985)

English biochemist, born in Newton le Willows, Lancashire. He studied there and with Sanger in Cambridge (1946–1949), where they developed the 2,4-dinitrofluorobenzene method for determining the N-terminus of a protein, and then worked at the National Institute for Medical Research (1949–1960) and St Mary's Hospital Medical School in London (1960–1967) before becoming Professor of Biochemistry at Oxford in 1967. His discovery that the plant protease papain cleaves the Y-shaped IgG (immunoglobulin) into the stem and two separate arms (1959), plus the finding that these fragments were solubilized by salt solutions without activity loss, opened the way for a detailed study of immunoglobulin structure, including the location of the antigen binding sites on the ends of the arms and the function of the stem region. He was the first to propose the bilaterally symmetrical four-chain structure which is the basis of all immunoglobulins. Edelman carried out complementary structural studies on immunoglobulins in the USA, and for this work they were jointly awarded the Nobel Prize for Physiology or Medicine in 1972. Porter followed this work with studies of the proteases of the related complement system, particularly the components called C1-, C2- and C3-convertase (C4b,2a). He was elected a Fellow of the Royal Society in 1964, and was awarded its Royal (1973) and Copley (1983) medals. In 1985 he was run over and killed crossing the road.

POWELL, John Wesley
(1834–1902)

American geologist, born in Mount Morris, New York, of English parents. Almost wholly self-educated, he served in the American Civil War rising to the rank of major, but lost his right arm during the battle of Pittsburgh Landing. He led daring boat expeditions down the Colorado river through the Grand Canyon (1869), demonstrating that the canyon resulted from the river erosion of rock strata which were being progressively uplifted; the river preserved its level but the mountains were being raised up. From 1874 to 1880 he directed the US Geological and Geographical Survey of Territories, organizing the 'Powell Surveys' to explore and map various parts of the USA. He subsequently became second director of the combined US Geological Survey (1880–1894), and after retirement he continued to work on behalf of the Bureau of Ethnology, which he co-founded in 1984. Powell undertook extensive work on water supplies in arid regions, and also wrote on crustal movements and human evolution.

PRAIN, Sir David
(1857–1944)

Scottish botanist, born in Fettercairn, Aberdeenshire, educated at Fettercairn Parish School and Aberdeen Grammar School. His early ambition was to enter banking, but his former Fettercairn headmaster intervened and sent him to Aberdeen University. Although his studies were broken by three spells as a schoolmaster, he finally graduated in medicine from Edinburgh University, and in 1884 entered the Indian Medical Service. Unexpectedly, due to the sickness of Lewis Brace, then Curator of the Calcutta Herbarium, Prain was sent to Calcutta Royal Botanic Garden in 1885. When Brace returned, Prain resumed military duties. When Brace again fell ill, Prain became Curator of the Calcutta Herbarium (1887–1898) and eventually Superintendent of the Royal Botanic Garden and Cinchona Department (1898–1906). His many publications include *Noviciae Indicae* (1905, based on papers published between 1889 and 1904), *The Species of* Pedicularis *of the Indian Empire* (1890), *Bengal Plants* (1903), papers on Labiatae, Leguminosae and Papaveraceaem, and reports on Indian hemp, wheats of Bengal, mustards, yams and indigo. From 1906 to 1922 he was Director of the Royal Botanic Gardens, Kew, where he had to turn his attentions to African botany. Knighted in 1912, he served on the committees of innumerable bodies, including the John Innes Horticultural Institute and the Imperial College of Tropical Agriculture, Trinidad.

PREGL, Fritz
(1869–1930)

Austrian chemist, born in Laibach (now Ljubljana, Slovenia). He studied medicine at Graz University and spent most of his working life there, becoming Professor of Medical Chemistry in 1913. Finding that traditional methods of analysis were useless when applied to the minute quantities of biochemical materials that he wished to investigate, he devised new techniques for microanalysis, including a balance which could weigh within an accuracy of 0.001 milligrams. His innovations were fundamental to the development of biochemistry, and brought him the Nobel Prize for Chemistry in 1923. He died in Graz.

PRELOG, Vladimir
(1906–1998)

Swiss chemist, born in Sarajevo. After the assassination of Archduke Ferdinand in Sarajevo, Prelog's family moved to Zagreb, where he attended the local gymnasium. From 1924 until 1929 he studied chemistry at the Institute of Technology in Prague. Unable to fund an academic position, he worked for the chemical company G. J. Driza (1929–1935). He then joined the staff of the University of Zagreb, where he was successful in synthesizing adamantane, a fascinating molecule related to diamond and previously found only in Czech petroleum. When the Nazis invaded Yugoslavia he went to Zürich as a guest of Leopold Ruzícka who was able to find him a position at the Federal Institute of Technology. He was appointed associate professor in 1947 and full professor five years later. Prelog spent his professional life studying the shapes of molecules. This study included the phenomenon of chirality (the characteristics of certain molecules that they can exist in left- or right-handed forms, i.e. optical isomers), particularly important among naturally occurring molecules. He used X-ray crystallography to determine molecular shapes and linked reactivity with stereochemistry. With R. S. Calin and Christopher Ingold he devised a system of naming stereoisomers based on a set of sequence rules. As a result, he became interested in other problems of stereochemical description, such as group theory, graph theory and chemical topology. He also studied the synthesis and chemistry of a number of natural products, including the enzyme fatty acid synthetase. In 1975 he shared the Nobel Prize for Chemistry with Cornforth for his contribution to our understanding of the stereochemistry of enzyme catalysis.

PRESTWICH, Sir Joseph
(1812–1896)

English geologist, hydrologist and prehistorian, born in Pensbury, Clapham. He was educated in Paris, Reading and at University College, London. Until the age of 60, he was a wine merchant in the family business, his business interest aiding rather than restricting his geological studies with frequent trips to France and Belgium as well as around Britain. He undertook early studies of the stratigraphy of Coalbrookedale, Shropshire, publishing *The Geology of Coalbrookedale* (1836), and on the correlation of the English Eocene with that of France. His principal work was on the stratigraphical position of flint implements and human remains in England and France, helping to confirm the antiquity of early man. His work on *The Water-bearing Strata of the Country around London* (1851) was a standard authority. In 1874 Prestwich became Professor of Geology at Oxford; it was there that he produced his two-volume work on *Geology, Chemical and Physical, Stratigraphical and Palaeontological* (1886–1887). He was knighted in 1896, shortly before his death.

PRIESTLEY, Joseph
(1733–1804)

English clergyman and chemist, renowned for his work on gases, particularly oxygen. He was born in Fieldhead, Leeds, to a dissenting family at a time when membership of the Church of England was a prerequisite for entrance to English universities and most positions of any importance. Intending to become a preacher, he was educated at the dissenting academy at Daventry. There he studied books on philosophy and science as well as theology, but it was to be his religious and political views, rather than his scientific achievements, which shaped his career. After preaching at Downham Market, Norfolk, and Nantwich, Cheshire, where his leanings towards Arianism (denying the divinity of Christ) troubled his congregation, he became a very successful teacher at Warrington Academy replacing the traditional classical syllabus with history, science and English literature. He also began the studies of electricity which earned him a fellowship of the Royal Society in 1766 and the friendship of Benjamin Franklin whom he met in London. The following year Priestley was appointed minister at Mill Hill Chapel, Leeds, and soon afterwards began to study gases and to write pamphlets criticizing the government's treatment of the American Colonies. Continuing his search towards religious truth, he now leaned towards the Unitarians, a sect founded in 1774 who disregarded the doctrine of the Trinity and placed great faith in reason. From 1772 to 1780 Priestley was librarian and travelling companion to Lord Shelbourne, based at Cable, Wiltshire, and subsequently junior minister at the New Meeting House, Birmingham. He was soon part of the Lunar Society, an informal but influential body of local savants, and included members such as Matthew Boulton, Watt and Erasmus Darwin among his friends. His revolutionary sympathies continued to attract attention, particularly once the French Revolution began, and in 1791 a mob of townsfolk sacked his house and fired his chapel. Priestley and his family fled to London and three years later emigrated to the USA where he settled in Northumberland, Pennsylvania, where he died 10 years later. Priestley's work on gases made him the most respected pneumatic chemist in Europe. When he began his experiments, only three gases were known: air (not then recognized as a mixture of gases), carbon dioxide and hydrogen. He added another 10, including hydrogen chloride, sulphur dioxide, ammonia and nitrous oxide (laughing gas). His discoveries were made by heating common laboratory substances such as 'oil of vitriol; (sulphuric acid) or 'spirits of salt' (a solution of hydrochloric acid) in a retort, with or without another reagent. He then collected any gas that was evolved over a pneumatic bath containing mercury or water. His most momentous discovery took place in 1774 when he heated mercury in air to form a 'crust'. On further heating this yielded a colourless odourless gas which made a lighted candle burn more brightly and did not kill his laboratory mice. Priestly recognized that this gas was new to science but he did not confirm that it was an element, nor did he identify it with the new gas prepared by Scheele. Because he accepted the traditional view of combustion – that inflammable substances contain invisible fiery essence called phlogiston which they lose on burning – he called this gas 'dephlogisticated air'. When he was in Paris with Lord Shelbourne in autumn 1774 he repeated this experiment for Lavoisier who perceived its significance and in a long series of experiments showed its central role in combustion and named it 'oxygene'. Priestley never accepted Lavoisler's discovery that burning is a chemical reaction in which a substance combines with oxygen, and continued to defend the phlogiston theory until his death. Among Priestley's many religious writings are *Letters to a Philosophical Unbeliever* (1774), *Disquisition relating to Matter and Spirit* (1777) and *History of Early Opinions Concerning Jesus Christ* (1786).

PRINGLE, Sir John
(1707–1782)

Scottish physician, born in Roxburgh, the youngest son of a baronet. He studied at St Andrews and Edinburgh universities before spending two years in Leiden, where he received his MD in 1730. Returning to Edinburgh, he practised medicine and held a Chair of Pneumatical and Ethical Philosophy (1734–1745). His family connections helped him obtain a commission as physician to the commander of the British military forces on the Continent, and his performance shortly advanced him to the post of Physician General to British forces in the Low Countries. A subsequent career in London was spectacularly successful, bringing a succession of royal appointments, his own baronetcy and election to the presidency of the Royal Society of London (1772). Throughout his life, Pringle was an active physician despite the many other demands on his time. His many medical observations, still not published today, are kept at the Royal College of Physicians in Edinburgh. On the other hand, his *Observations on Diseases of the Army* (1752) remains a classic of military hygiene, went through many editions and translations, and established many principles for preventing typhus, dysentery and other common diseases of soldiers and others who live in crowded conditions.

PROKHOROV, Alexander Mikhailovich
(1916–)

Russian physicist, born in Atherton, Australia, of Russian emigré parents. After the Russian Revolution his family returned to the USSR (1917). He graduated from Leningrad (now St Petersburg) University in 1939, and after serving with the Red Army during World War II, he took a junior post at the Lebedev Physical Institute, rising to become Deputy Director in 1968. In 1952, with his colleague Basov, he postulated the possibility of constructing a 'molecular generator' for the amplification of electrochemical radiation. He later achieved this using a beam of molecular ammonia. He went on to describe a new way in which atomic systems could be employed to produce amplification of microwaves. This led to the development of the maser and eventually the laser (terms that stand for Microwave/Light Amplification by Stimulated Emission of Radiation). For this work he won the 1964 Nobel Prize for Physics jointly with Basov and Townes.

PROUST, Joseph Louis
(1754–1826)

French analytical chemist, born in Angers. After studying pharmacy and chemistry in Paris he spent most of his working life in Spain. In the early 1780s he conducted aerostatic experiments with Pilatre de Rozier and Charles, and as one of the first people to make an ascent in a balloon, which he did in 1784 with the King and Queen of France among the onlookers. He was appointed Professor of Chemistry at the Royal Artillery College at Segovia and Director of the Royal Laboratories at Madrid (1789–1808), after which he returned to France. Proust made two significant advances in analytical chemistry: he developed the use of hydrogen sulphide as a reagent (important because the differing solubilities of sulphide makes it possible to separate them) and he gave the results of his analyses in terms of percentage weights. By means of the percentages he realized that the proportions of the constituents in any chemical compound are always the same, regardless of what method is used to prepare it. He announced this discovery, known as the 'law of definite proportions', in 1794. Not all his contemporaries accepted his findings, his principal adversary in a renowned controversy being Berthollet. Although Proust was correct in his observations, the reason why reagents behave in this way did not become clear until Dalton formulated his atomic theory in 1803. Proust died in Angers.

PTASHNE, Mark Steven
1940–)

American molecular biologist, born in Chicago. He was educated at Reed College where he obtained his BA in 1961, and at Harvard University where he received his PhD in 1968. He remained at Harvard as a lecturer in biochemistry (1968–1971), and since 1971 has been professor there. The concept that genes can be specifically activated or de-activated, which has been essential to our understanding of the normal and abnormal functioning of cells, originated from genetic studies of *Escherichia coli.*

These studies implied the existence of a repressor protein that binds to a transcription unit called the 'lac operon' to 'turn off' the b-galactosidase gene, when lactose is absent from the cell. The purification of the lactose repressor allowed the mechanisms to be explored by the use of mutant repressor molecules. It was shown by Ptashne and others that the lac repressor inhibits transcription of the lac operon by binding to a specific nucleotide sequence, 21 base pairs long, called the operator. Since the operator overlaps the promoter, binding of the repressor to the operator prevents RNA V polymerase binding to the promoter and thus transcription of the adjacent gene sequences. When lactose is present in the cell, a breakdown product dislodges repressor from the operator and allows RNA polymerase attachment to the promoter. Thus the b-galactosidase gene is de-repressed only in the presence of lactose. Similar mechanisms are found in higher organisms, and the lactose system has been of crucial importance as a model.

PTOLEMY (Claudius Ptolemaeus)
(c. AD 90–168)

Egyptian astronomer and geographer, who flourished in Alexandria. His 'great compendium of astronomy' seems to have been denominated by the Greeks *megisti*, 'the greatest', from which the Arab name *Almagest* by which it is generally known was derived. With his *Tetrabiblos Syntaxis* is combined another work called *Karpos* or *Centiloquium* because it contains a hundred aphorisms – both treat astrological subjects, so have been held by some to be of doubtful genuineness. Then there is a treatise on the fixed stars or a species of almanac, the *Geographia*, and other works dealing with mapmaking, the musical scale and chronology. Ptolemy, as astronomer and geographer, held supreme sway over the minds of scientific men down to the sixteenth and seventeenth centuries; but he seems to have been not so much an independent investigator as a corrector and improver of the work of his predecessors. In astronomy he depended almost entirely on Hipparchos. However, as his works form the only remaining authority on ancient astronomy, the system they expound is called the *Ptolemaic System*; like the system of Plato and Aristotle, this was an attempt to reduce to scientific form the common notions of the motions of the heavenly bodies. The Ptolemaic astronomy, handed on by Byzantines and Arabs, assumed that the Earth is the centre of the universe, and that the heavenly bodies revolve round it. Beyond and in the ether surrounding the Earth's atmosphere were eight concentric spherical shells, to seven of which one heavenly body was attached, the fixed stars occupying the eighth. The apparent irregularity of their motions was explained by a complicated theory of epicycles. As a geographer, Ptolemy is the corrector of a predecessor, Marinus of Tyre. His *Geography* contains a catalogue of places, with latitude and longitude; general descriptions; and details regarding his mode of noting the position of places – by latitude and longitude, with the calculation of the size of the Earth. He constructed a map of the world and other maps.

PURKINJE, or PURKYNE, Jan Evangelista
(1787–1869)

Czech physiologist. Born in Libochowitz and educated by monks, Purkinje trained for the priesthood. He then studied philosophy, and finally graduated in medicine, with a dissertation on vision that gained him Johann Wolfgang von Goethe's friendship. He rose to become professor at Breslau and later in Prague. In 1825 he began to use a compound microscope, making important new observations. After 1830, much of Purkinje's work centred on cell observations, deploying compound microscope to delineate nerve fibres. In 1837 he outlined the key features of the cell theory, to be more fully propounded in 1839 by Schwann. Around the same time he described nerve cells with their dendrites and nuclei and the flash-like cells ('Purkinje cells') in the cerebellar cortex. In 1838 he observed cell division, and in the following year promoted the word 'protoplasm' in the modern sense. He made improvements in histology, including early utilization of a mechanical microtome in place of a razor to procure thin tissue slices. Purkinje also experimented on the physiological basis of subjective feelings. He took up the study of vertigo and was interested in the peculiarities of the eyes, performing fascinating self-experimentation on the visual effects of pressure applied to the eyeball. The effect of being able to see in one's own eye the shadows of the retinal blood vessels is now know as 'Purkinje's figure'.

PYTHAGORAS
(sixth century BC)

Greek philosopher, sage and mathematician born in Samos. Around 530 BC he settled in a Greek colony in southern Italy, where he established a religious community of some kind. He may later have moved to Megapontum, after persecution. He wrote nothing, and his whole life is shrouded in myth and legend. Pythagoreanism was first a way of life rather than a philosophy, emphasising moral asceticism and purification, and associated with doctrines of the transmigration of souls, the kinship of all living things and various ritual rules of abstinence (most famously, 'do not eat beans'). He is also associated with mathematical discoveries involving the chief musical intervals, the relations of numbers, the proof of the theorem on triangles which bears his name, and with more fundamental beliefs about the understanding and representation of the world of nature through numbers. The equilateral triangle of ten dots, the tetracys of the decad, itself became an object of religious veneration, referred to in the Pythagoran oath 'Nay, by him that gave us the *tetracys* which contains the fount and root of ever-flowing nature'. The Pythagoreans were the first to recognize that the square root of 2 is an irrational number, i.e. cannot be expressed as a fraction of two whole numbers, and they have said to have murdered one of the members who disclosed this knowledge to others. It is impossible to disentangle Pythagoras's own views from the later accretions of mysticism and neoplatonism, but he had a profound influence on Plato and later philosophers, astronomers and mathematicians.

Q

QUASTEL, Juda Hirsch
(1899–1987)

British biochemist, born in Sheffield. He read chemistry at Imperial College, London, and in 1921 moved to Cambridge to work for his PhD on bacterial metabolism with Frederick Gowland Hopkins. In 1930 he became staff biochemist at Cardiff City Mental Hospital, where he began pioneer studies on biochemical aspects of mental disease. He proceeded on three main, complementary fronts: examining the normal and abnormal metabolism of brain tissue; coordinating his laboratory research with clinical investigations such as the effects of barbiturates on brain metabolism, and developing improved live function tests for schizophrenia; and investigating the neurochemistry of neuroactive chemicals, especially the effects of amphetamine, and the synthesis of acetylcholine, a chemical then receiving considerable interest because of the contemporary work of Dale and Loewi. During World War II he worked at Rothamsted Experimental Station (1940–1947) on soil biochemistry in a programme to improve soil fertility. Utilizing techniques from his studies on the brain, he examined the structure and metabolic profiles of different types of soil, developed artificial chemical conditioners to improve quality, and produced a particularly powerful selective herbicide. Despite his interest in this work and his willingness to continue it after the war, in 1947 he accepted an invitation to Montreal to return completely to brain chemistry. As Professor of Biochemistry at McGill University and as Deputy Director of a biochemical research unit of Montreal General Hospital, he worked on a wide range of neurochemical problems, as well as on intestinal absorption. On retirement from McGill in 1964 he moved to the Kinsman Laboratories at Vancouver, working most notably on glutamic acid metabolism in the brain.

R

RAINWATER, (Leo) James
(1917–   )

American physicist, born in Council, Idaho. He was educated at Caltech and Columbia University, and during World War II he contributed to the Manhattan Project to develop the atomic bomb. He became Professor of Physics at Columbia University in 1952 and was Director of the Nevis Cyclotron Laboratory there (1951–1953, 1956–1961). Rainwater's work led to the unification of two theoretical models of the atomic nucleus. Measurements such as the electric quadropole moment of the nucleus had indicated that some nuclei were deformed, a phenomenon which remained unexplained by either of the two existing nucleus models (the liquid drop model and the shell model), which both assumed that nuclei are symmetric. Rainwater proposed that distortions of the nucleus could be due to some of the nucleons distorting the central symmetric potential. His colleague Aage Bohr later developed this idea into the collective model of nuclei with Mottelson. Rainwater also worked with Fitch on the studies of muonic X-rays which gave the first indications of the nuclear radius. He also developed an improved theory of how high-energy particles scatter as they pass through material to explain aspects of the behaviour of cosmic-ray muons. He shared the 1975 Nobel Prize for Physics with Bohr and Mottelson for the development of the collective model of the nucleus.

RAMAN, Sir Chandrasekhara Venkata
(1888–1970)

Indian physicist, born in Trichinopoly. He was educated at Madras University, then spent 10 years working in the Indian Finance Department before becoming Professor of Physics at Calcutta (1917–1933). In 1930 he was awarded the Nobel Prize for Physics for important discoveries in connection with the scattering of light by transparent materials. He showed that the interaction of vibrating molecules with photons passing through altered the spectrum of the scattered light, either increasing or decreasing it by a fixed amount. This 'Raman effect' enabled the probing of molecular energy levels and became an important spectroscopic technique used throughout the world. His other research interests included the vibration of musical instruments and the physiology of vision. He was the first Indian Director of the Indian Institute of Science (1933–1948), where he instigated work on light scattering by colloids, Brillouin scattering and the diffraction of light by ultrasonic waves. He was a founder of the Indian Academy of Science (1934) and in 1947 he founded the Raman Institute in Mysore where he remained as Director until his death. His outstanding contributions to Indian Science made him a national hero, and he was knighted by the British government in 1929.

RAMANUJAN, Srinivasa
(1887–1920)

Indian mathematician, born in Eroda, Madras, one of the most remarkable self-taught prodigies in the history of mathematics. The child of poor parents, he taught himself mathematics from an elementary English textbook. Although he attended college, he did not graduate. While working as a clerk, he was persuaded to send over 100 remarkable theorems that he had discovered to Godfrey Hardy at Cambridge, including results on elliptic integrals, partitions and analytic number theory. Hardy was so impressed that he arranged for him to come to Cambridge in 1914. There Ramanujan published many papers, some jointly with Hardy. The most remarkable was an exact formula for the number of ways an integer can be written as a sum of positive integers. Having no formal training in mathematics, he arrived at his results by an almost miraculous intuition, often having no idea of how they could be proved or even what the form of an orthodox proof might be. He was elected FRS and a Fellow of Trinity in 1918, but soon returned to India suffering from poor health; he died shortly after.

RAMON Y CAJAL, Santiago
(1852–1934)

Spanish physician and histologist, born in Petilla de Aragon in the Spanish Pyrenees. Apparently considered indolent as a child he was apprenticed to both a barber and a cobbler, but was unsuccessful at both occupations and in some desperation his physician father tried to interest him in medicine. He responded positively, and graduated from Zaragoza University in 1873. He joined the Army Medical Service and served in Cuba where he contracted malaria and was soon discharged through ill-health. He returned to Zaragoza for further anatomical training and in 1883 began his academic career, being appointed Professor of Anatomy at Valencia until 1886, then Professor of Histology at Barcelona (1886–1892) and finally Professor of Histology and Pathological Anatomy at Madrid (1892–1922). His major work was on the microstructure of the nervous system and he utilized the specialized histological staining techniques of the Italian Golgi. The two men disagreed about their interpretations of neural structure. Ramon y Cajal maintaining that nerve cells were discrete, and that there was no physical continuity between one cell and another, although they shared the 1906 Nobel Prize for Physiology or Medicine. Ramon y Cajal provided detailed histological descriptions of many regions of the brain, including the spinal cord, the cerebellum and the retina and demonstrated several distinct patterns in different parts of the cerebral cortex. He also proposed that the dendrites of a nerve cell receive information which is then transmitted through the axon; and he instigated significant work on the processes of degeneration and regeneration of nerves. He wrote many articles and books in Spanish, some of which have been translated into English, including an autobiographical account: *Recollections of my Life* (1937).

RAMSAY, Sir William
(1852–1916)

Scottish chemist who discovered the rare gases of the atmosphere. He was born in Glasgow and studied classics at Glasgow University and chemistry at Tübingen. He then became an assistant at Anderson's College, Glasgow, Professor of Chemistry at University College, Bristol (1880–1887), and Principal from 1881 to 1887. He subsequently became professor at University College, London (1887–1913). At Glasgow he studied the alkaloids, and at Bristol he worked on the vapour pressure of liquids. In 1892 he set out to discover why nitrogen derived from the air always had a higher atomic weight than nitrogen prepared in the laboratory, a problem which had been highlighted by Lord Rayleigh (1842–1919). Ramsay believed that atmospheric nitrogen must contain a small percentage of a heavier gas and proved this to be the case in 1894. Taking a sample of air, he removed the oxygen by sparking and the nitrogen by combining it with heated magnesium, and found a residual inert gas which occupied 1 percent of the original volume. This was later named 'argon'. In 1895 Ramsay isolated a light inert gas resembling argon by boiling a mineral called cleivite. Spectroscopic analysis showed that this gas was helium, which Lockyer and Edward Frankland had discovered in the spectrum of the Sun nearly 30 years earlier. Subsequently helium was shown to exist in the atmosphere, and in 1903 Ramsay and Soddy demonstrated that it was formed by the decay of thorium and uranium minerals, a discovery which was to prove critical to the understanding of nuclear reactions. Argon and helium fitted neatly into Mendeleyev's periodic table, but spaces further down the same group suggested that heavier inert gases remained to be discovered in the atmosphere. Working with Travers, Ramsay liquefied air, collected the last fraction to evaporate, removed the oxygen and nitrogen, and examined the small residue spectroscopically. In 1898 they found the green and yellow lines of krypton, the crimson of neon and the blue lines of xenon. Further research confirmed the inert nature of these gases and their atomic weights. In 1907, to exclude the possibility that other undiscovered elements were present, Ramsay investigated the residue of 100 tons of liquid air but found nothing. The following year, returning to his earlier studies of radioactivity, he obtained radon – discovered by Dorn in 1900 – in sufficient quantities to show that it belonged to the same family as helium and the other inert gases. Ramsey was elected FRS in 1888, knighted in 1904 and awarded the Nobel Prize for Chemistry in the same year. He died in Hazlemere, Buckinghamshire.

RAMSEY, Frank Plumpton
(1903–1930

English philosopher and mathematician, born in Cambridge. He read mathematics at Trinity College, Cambridge, and went on to be elected a Fellow of King's College when he was only 21. In his tragically short life (he died after an operation) he made outstanding contributions to philosophy, logic, mathematics and economics, to an extent which was only properly recognized many years after his death. To him is due the ingenious idea that large subsets of a set with some structure must also carry at least some of that structure. This has important application in number theory, and especially in combinatorics and graph theory. Typically, results in Ramsey theory assert that complete disorder is impossible. He was much stimulated by his Cambridge contemporaries Bertrand Russell, whose programme of reducing mathematics to logic he ingeniously defended and developed, and Ludwig Wittgenstein, whose *Tractatus* he was among the first both to appreciate and to criticize, rejecting the idea of ineffable metaphysical truths beyond the limits of language with the famous remark, 'What we can't say we can't say, and we can't whistle it either'. The best of his work is collected in *Philosophical Papers* (1990).

RAMSEY, Norman
(1915–)

American physicist, born in Washington DC. He was educated at Columbia University and at Cambridge, and after various teaching posts in the USA became Associate Professor of Physics at Harvard in 1947. Since 1966 he has been Higgins Professor of Physics there. Ramsey was awarded the 1989 Nobel Prize for Physics jointly with Paul and Dehmelt for his development of the 'separated field' method. In this an electromagnetic field applied to a beam of atoms or molecules induces transitions between specific energy states. This allowed the energy of atomic transitions to be measured with great accuracy and led to the development of the caesium clock, the atomic clock which now provides international time standards. Ramsey has also contributed to the development of the hydrogen maser.

RANKINE, William John Macquorn
(1820–1872)

Scottish natural philosopher and engineer, born in Edinburgh. His father retired from the army to become a railway engineer. Macquorn studied at Edinburgh University (1836–1838), where he achieved distinction in the natural philosophy class of James Forbes. In the summer of 1838 he began a civil engineering apprenticeship in Ireland. During the 1840s railway mania he worked in Scotland. In 1849 he began to publish extensively on elasticity and the new mechanical theory of heat, basing his writings on an idiosyncratic 'hypothesis of molecular vortices'. With Kelvin and Clausius he shaped the new thermodynamics, particularly in its practice dimension (he patented an elaborate air engine with his friend James Robert Napier of the famous Clyde shipbuilding family). Rankine introduced the term 'actual' (kinetic) and 'potential' energy. Later he proposed an abstract 'science of energetics' which aimed to unify physics. From 1855, as Regius Professor of Civil Engineering and Mechanics at Glasgow, Rankine espoused a fertile harmony between the theoretical and practical worlds which he straddled. His textbooks (on prime movers, machinery, shipbuilding and applied mechanics), calling on first-hand experience of Glasgow industry, were enormously popular and effective in the establishment of an 'engineering science'. Rankine was a keen musician and a prolific writer. His dryly humorous *Songs and Fables* were published posthumously in 1874.

RAOULT, François Marie
(1830–1901)

French physical chemist, born in Fournes, near Lille. He studied in Paris, but financial problems forced him to abandon the course, and from 1853 to 1867 he taught science in various schools, completing his doctorate in 1863. In 1867 he began to teach chemistry at the University of Grenoble, and in 1870 he was promoted to the Chair of Chemistry, which he occupied until his death. His doctoral research concerned heat changes in chemical cells and their relation to electromotive force. Raoult is remembered most, however, for his work on the freezing points and vapour pressures of solutions (1878–1892). His findings provided the basis for methods of determining molecular weights. The generalization that the vapour pressure of solvent above a solution is proportional to the mole fraction of solvent in the solution is known as Raoult's law. Raoult also found that electrolyte solutions showed anomalies, which proved later to be important in connection with the ionic dissociation theory of Arrhenius. He became an Honorary Fellow of the Chemical Society in 1898.

RAPHAEL, Ralph Alexander
(1921–)

English chemist, born in Dublin. Having studied chemistry at Imperial College, London as an undergraduate he worked with Sir Ian Heilbron and Sir Ewart Jones for his PhD. After a brief period with May and Baker he returned in 1946 to Imperial College as an ICI Fellow. This was followed by a lectureship in Glasgow, where he developed the synthesis of a number of natural products from acetylenic precursors. Before promotion to the chair in Glasgow (1957) he had a short stay at Queen's University, Belfast. In 1973 he was invited to the Chair of Organic Chemistry at Cambridge. His prowess as a synthetic organic chemist is surpassed only by his skill as an after-dinner speaker. He was elected FRS in 1962 and awarded a CBE for services to chemistry in 1982.

RAY, John
(1627–1705)

English naturalist, born in Black Notley, near Braintree, Essex. He was educated at Catharine Hall and Trinity College, Cambridge, and became a Fellow of Trinity in 1649. He was appointed as a lecturer in Greek (1651), mathematics (1653) and humanity (1655). In 1658 he made his first botanical tour, through the Midlands and north Wales, and in 1660 he published *Catalogus Plantarum circa Cantabrigiam Nascentium*, the first catalogue of the plants of a particular district to be published in England. In 1662 he resigned his fellowship at Cambridge, rather than take the oath of Act of Uniformity after the Restoration. Accompanied and subsidized by a wealthy fellow naturalist Francis Willoughby, he toured extensively in Europe (1663–1666), with the aim of jointly preparing a systematic description of the organic world, Ray dealing with plants, Willoughby covering animals. In 1682 Ray published *Methodus Plantarum Nolla* in which he first divided flowering plants into monocotyledons and dicotyledons and demonstrated the true nature of buds. Earlier, in his *Catalogus Plantarum Angliae* (1670), he was the first to separate cryptogams, from higher plants. His *Synopsis Methodica Stirpium Britannicorum* (1690) was the first systematic English Flora. After Morison's death (1683), Ray began his major work, *Historia Generalis Alantarum* (3 vols 1686–1704), which described about 690 plants. From around 1690 he worked mainly on insects; his zoological work, in which he developed the most natural pre-Linnaean classification of the animal kingdom has been considered of even greater importance than his botanical achievements. He was elected FRS in 1667.

RAYLEIGH, John William Strutt, 3rd Baron
(1842–1919)

English physicist, born near Maldon, Essex. In 1865 he graduated from Trinity College, Cambridge, as Senior Wrangler and Smith's Prizeman and was elected a Fellow (1866). He inherited his father's title of third baron 1873 and continued to work in his laboratory in the family mansion, Terling Place, Essex. He later succeeded Maxwell as Professor of Experimental Physics at Cambridge (1879–1884). He was Professor of Natural Philosophy at the Royal Institution (1888–1905). President of the Royal Society (1904–1908) and became Chancellor of Cambridge University in 1908. In 1871 he married Evelyn Balfour, sister of the Scottish statesman Lord Balfour. Rayleigh's work included valuable studies and research on vibratory motion, in both optics and acoustics. With Ramsay he was the discoverer of argon (1894), and for this and his work on gas densities, he was awarded the Nobel Prize for Physics in 1904. He also studied radiation, his research leading to Rayleigh-Jeans formula which accurately predicts the long-wavelength radiation emitted by hot bodies. The problems produced by the failure of this theory for short-wavelength radiation were solved by Plank's suggestion that energy is emitted in discrete quanta – this idea, which Rayleigh never found satisfactory, was to dramatically alter the course of physics. His books included *The Theory of Sound* (1877–1878) and *Scientific Papers* (1899–1900).

RÉAUMUR, René Antoine Ferchault de
(1683–1757)

French natural philosopher, born in La Rochelle. The earlier years of his life are obscure, but in 1703 he moved to Paris and five years later became a member of the Academy of Sciences. He was put in charge of a government-sponsored project to assemble information on all the arts, industries and professions in France, and in gathering the data required for the monumental *Description des arts et métiers* he acquired a very wide knowledge of contemporary science and technology. He developed improved methods for producing iron and steel, being among the first to recognize the importance of carbon as a constituent of steel; the cupola furnace for melting grey iron was developed by him in 1720. He became one of the greatest naturalists of his age, publishing the six volume *Memoires pour servir d l'histoire des insectes* (1734–1742), the first serious and comprehensive entomological work. In 1740 he produced an opaque form of porcelain which is still known as Reaumur porcelain. His thermometer of 1731 used a mixture of alcohol and water instead of mercury, with 80 degrees between the freezing and boiling points of water – the Reaumur temperature scale. Through his research into digestion in birds and animals he established that it was a chemical rather than a mechanical (grinding) process, and by 1752 he was able to isolate samples of gastric juices and demonstrate the process in the laboratory.

RECORDE, Robert
(c.1510–1558)

English mathematician, born in Tenby. He studied at Oxford, and in 1545 took his MD at Cambridge. He practised medicine in London, and was in charge of mines in Ireland, but died in prison after losing a lawsuit brought against him by the Duke of Pembroke. He wrote the first English text book, an elementary arithmetic and algebra, which became the standard works in Elizabethan England, including *The Ground of Artes* (1543) and *The Whetstone of Witte* (1557). The books are presented as dialogues, and introduced the equals sign to mathematics.

REES, Sir Martin John
(1942–)

British astrophysicist, born in York. He was educated at Trinity College, Cambridge, and became a staff member of the Institute of Theoretical Astronomy there. After holding fellowships in the USA and a short period as professor at Sussex University (1972–1973), he returned to Cambridge as Plumian Professor of Astronomy and Experimental Philosophy (1973–1991) and Director of the Institute of Astronomy (1977–1982, 1987–1991). Since 1992 he has been royal Society Research Professor. Rees has made important contributions to the study of stellar systems, galaxies and the nature of the invisible 'dark matter' known to exist in the universe, but his best-known work is in the study of active galactic nuclei. He demonstrated that the relatively rapid brightness variations observed in quasers and active galaxies could be best understood if the nuclei contained gas which is outflowing at almost the speed of light and predicted that when projected onto the plane of the sky, such motions would be observed as apparently faster than the speed of light, or 'superluminal' motion. Observational evidence for this appeared in the 1970s. Rees also showed that the strong radio-emitting regions which lie far from the visible boundaries of some active galaxies could be produced by highly collimated beams of particles moving outwards from the nuclei at almost the speed of light; this has formed part of the current theory of the mechanisms of these mysterious objects. He was elected FRS in 1979 and became President of the Royal Astronomical Society in 1992, the year in which he was knighted.

REGIOMONTANUS, (originally Johannes Muller)
(1436–1476)

German mathematician and astronomer, who took his name from his Franconian birthplace, Konigsberg (*Mons Regius*). He studied at Vienna, and in 1461 accompanied Cardinal Bessarion to Italy to learn Greek. In 1471 he settled in Nuremberg, where he was supported by the patrician Bernhard Walther. The two laboured at the *Alphonsine Tables*, and published *Ephemerides 1475–1506* (1473), a work used extensively by Christopher Columbus. Regiomontanus established the study of algebra and trigonometry in Germany and wrote on waterworks, burning-glass, weights and measures, and the quadrature of the circle. He was summoned to Rome in 1474 by Pope Sixtus VI to help to reform the calendar, and died there.

REICHENBACH, Karl, Baron von
(1788–1869)

German natural philosopher and industrialist, born in Stuttgart. He was educated at the University of Tübingen, and had interests in metallurgical factories and steel works. He designed a new oven for making charcoal with which it was possible to collect the volatile products. From these he isolated creosote and paraffin in the 1830s. Around 1844, after studying animal magnetism, he believed that he had discovered a new force which he called 'Od', intermediate between electricity, magnetism, heat and light, and recognizable only by the nerves of sensitive persons, usually women. He expended much energy on publications and demonstrations, trying to convert the scientific world to his views, neglecting his businesses and living as a recluse. He also wrote on the geology of Moravia and on meteorites, of which he had a large collection (now in the University of Tübingen). He died in Leipzig.

REMAK, Robert
(1815–1865)

German physician and pioneer in electrotherapy, born in Poznan. Remak studied pathology and embryology as a student of Johannes Muller in Berlin. He remained there to develop a general practice and to pursue a university career, although he was prevented from obtaining a senior teaching post because he was a Jew. In his early twenties he did significant work on the microscopy of the nerves. In 1838 he located the lyelin sheath of the main nerves, and also showed that the axis-cylinder (axon) arises in the spinal cord and runs continuously. In this and in further work, he discerned that nerves are not merely – as they had been viewed for centuries – structureless hollow tubes; rather they possessed a flattened solid structure. A pioneer embryologist, Remak was one of the first fully to depict cell division and to hold that all animal cells came from pre-existing cells. He discovered the 'fibres of Remak' (1830), and the nerve cells in the heart known as Remak's ganglia (1844).

REYNOLDS, Osborne
(1842–1912)

English engineer, born in Belfast of a Suffolk family. After spending two years with a local mechanical engineer to gain some practical experience, he went to the University of Cambridge where he graduated Seventh Wrangler in 1867 and was elected to a fellowship at Queen's College. The following year he was appointed the first Professor of Engineering at Manchester, where he remained for 37 years. He possessed a combination of outstanding abilities in intuitive mechanical design and mathematical analysis, and investigated a wide range of engineering and physical problems. Some of his best work was in the field of hydrodynamics; in 1875, for example, he took out a patent which proposed the use of fixed guide vanes in centrifuged pumps, a feature found in many such pumps today. He also designed the first multi-stage centrifugal pump. Other areas in which he carried out significant research work included cavitation in hydraulic machines, dynamic stresses, thermodynamics and heat transfer, and lubrication. The 'Reynolds number', a dimensionless ratio characterizing the dynamic state of a fluid, takes its name from him. He was elected a Fellow of the Royal Society in 1877 and was awarded its Royal Medal in 1888.

RICHARDS, Dickinson Woodruff
(1895–1973)

American physician, born in Orange, New Jersey. Educated at Yale, he specialized in cardiology, which he taught at Columbia University (1928–1961), becoming Professor of Medicine there from 1947. With Cournand, Richards developed Forssman's technique of cardiac catheterization into an important procedure for studying blood pressure, oxygen tension and a variety of other physiological variables in health and disease. Their work led to better understanding and treatment of shock, and provided the basis for much of modern cardiology, including the non-invasive surgical treatment of a number of conditions. The three men shared the Nobel Prize for Physiology or Medicine in 1956. Richards was a man with wide cultural attainments and an eloquent advocate for improved standards of medical care for the disadvantaged and elderly.

RICHARDS, Theodore William
(1868–1928)

American chemist, born in Germantown, Pennsylvania. He was educated at Haverford College and at Harvard, where he was awarded a PhD in 1888. After a year of visiting laboratories in Europe, he became an assistant in analytical chemistry at Harvard, thereafter climbing the academic ladder to full professor by 1901. He held the Erving Chair of Chemistry from 1912 until his death. In his PhD research under Josiah Cooke, Richards redetermined very accurately the mass ratio for the combination of hydrogen. This introduction to the methods of atomic weight determination was of lasting influence, and such work was dominant throughout his career. Over the course of 40 years the atomic weights of 25 elements were determined by Richards and his students. His investigation of the variation of the atomic weight of lead with source proved the existence of isotopes (1914). Richards also carried out much work on various topics in physical chemistry, notably thermochemistry (e.g. heats of reaction, combustion, dilution), electrochemistry (investigating the electromotive force of chemical cells) and physical properties of elements and compounds, particularly measurements of compressibility. He was awarded the Nobel Prize for Chemistry in 1914 for his work on atomic weights. Richards also received the Davy Medal of the Royal Society in 1910 and was made a Foreign Member in 1919. He became an Honorary Fellow of the Chemical Society in 1908 and was awarded its Faraday medal in 1911.

RICHET, Charles Robert
(1850–1935)

French physiologist, born in Paris, the son of a surgeon. Richet obtained a medical degree at the University of Paris in 1877 and was appointed to the Chair of Physiology a decade later. During his career he conducted research in many fields, including hypnosis, pain, digestion, muscle contraction and animal heat. In pioneering work on serum therapy, he conceived the idea that the blood of animals that are resistant to a harmful bacterium might contain a substance which could be used to confer immunity on a non-resistant animal. Having demonstrated that the principle was sound using a staphylococcus infection in dogs and fowl, Richet and his colleague Jules Hericourt attempted to prepare a serum for the treatment of tuberculosis, but found they could only delay the course of the disease. Some years later, when conducting experiments which attempted to define a toxic dose of an extract of sea-anemone tentacles, Richet and Paul Portier found that dogs that had survived experiments in which they had been injected with the poison were subsequently much more sensitive to it. This was a rather surprising phenomenon, the opposite of what might have been expected from his previous work, and he called this ‘anaphylaxis’. He suggested that the mechanism, however, might be similar to that of immunity, resulting from the production of substances in the blood. He confirmed that this is the case by making normal animals anaphylactic by injecting them with serum from anaphylactic animals. It soon became apparent that anaphylactic reactions were not so uncommon, and would have to be taken account of by physicians in serum therapy. Richet was awarded the 1913 Nobel Prize for Physiology or Medicine for this work.

RICHTER, Charles Francis
(1900–1985)

American seismologist, born near Hamilton, Ohio. Educated at Stanford University (AB 1920), he began a PhD at Caltech on atomic theory, but before completing it he was offered a position at the Carnegie Institute of Washington. There he finished his PhD (1928), became fascinated with seismology and met Gutenberg with whom in 1932 he was co-creator of the famous Richter scale, the original instrumental scale for determining the energy released by an earth quake. The magnitude scale ranged from 1 to 9, and a magnitude increase of one unit corresponded to the release of about 30 times the seismic energy. Richter's name has been popularly ascribed to all the later magnitude scales devised and that in use today. In 1937 he returned to Caltech where he spent the rest of his career, as research assistant and Professor of Seismology from 1952. He played a key role in establishing the southern California seismic array and published *Seismicity of the Earth* (1954, with Gutenberg) and *Elementary Seismology* (1958).

RICHTER, Jeremias Beniamin
(1762–1807)

German scientist who established stoichiometry as a branch of chemistry. Born in Hirschberg, he spent seven years in the engineering corps of the Prussian army, mathematics and philosophy at the University of Konigsberg, and after some years as an independent chemical consultant was appointed chemist to the Royal Porcelain Works in Berlin, Richter believed that all chemical reactions are guided by mathematical laws. He analysed a vast number of compounds to determine the proportions of their reagents by weight and described this new mathematical approach as 'stoichiometry'. Although his results demonstrated that agents do indeed combine in fixed proportions, he never formulated this observation to a law and it was Proust, working independently in France and Spain who first proposed the 'law of definite proportions' in 1794. Richter, in the course of his research on proportions noticed that when double decomposition takes place between two neutral salts, the compounds formed in the reaction are also neutral. This not only supported his theory of fixed ratios but also helped later chemists in their study of valence. Richter's work was little noticed in his lifetime but its importance became apparent after the atomic theory of Dalton provided the explanatory framework for his hypothesis in 1803. He died in Berlin.

RIEMANN, (Georg Friedrich) Bernhard
(1826–1866)

German mathematician, born in Breselenz. He studied in Göttingen under Gauss and in Berlin under Dirichlet, whom he succeeded as Professor of Mathematics at Göttingen in 1859. He was forced to retire by illness in 1862 and died or tuberculosis in Italy. His first publication (1851) was on the foundations of the theory of functions of a complex variable, including the result now known as the Riemann mapping theorem. This asserts that any simply connected region (roughly speaking a single region with no holes) can be mapped by a complex function onto a disc so that the boundaries correspond. In this general it is false, but under restriction it is true, and the theory has been a profound contribution to the subject. In a later paper on Abelian functions (1857), he introduced the idea of 'Riemann surface' to deal with 'multi-valued' algebraic functions; this was to become a key concept in the development of analysis. His famous lecture in 1854, 'On the hypotheses that underlie geometry', given in the presence of the aged Gauss, first presented his notion of a 'manifold', an n-dimensional curved space, greatly extending the non-Euclidean geometry of Bolyai and Lobachevski. These ideas were essential in the formulation of Einstein's theory of general relativity, and have led to the modern theory of differentiable manifolds which now plays a vital role in theoretical physics. Riemann's name is also associated with the zeta function, which is central to the study of the distribution of prime numbers; the 'Riemann hypothesis' is a famous unsolved problem concerning this function.

RIESZ, Frigyes (Frederic)
(1880–1956)

Hungarian mathematician, born in Gyor. He studied at Zürich, Budapest and Göttingen, then returned to Hungary as a lecturer at the University of Szeged. In 1945 he was appointed Professor of Mathematics at the University of Budapest. He worked in functional analysis, integral equations and subharmonic functions, and developed a new approach to the Lebesgue integral. His work was instrumental in allowing the matrix and wave mechanics methods developed simultaneously during the emergence of quantum theory to be identified as equivalent. Riesz's textbook on functional analysis, *Leçons d'analyse fonctionnelle* ('Lessons of Functional Analysis', 1952) written with Bela Szokefalvi-Nagy (1887–1953), is a classic.

RIGHI, Augusto
(1850–1920)

Italian physicist, born in Bologna, famous for his work on electromagnetic waves. Educated in his home town, he taught physics at Bologna Technical College (1873–1880), leaving to take up the newly established Chair of Physics at the University of Palermo. He was Professor of Physics at Padua (1885–1889) and later returned to a professorship at Bologna, where he taught until his death. He invented an induction electrometer (1872) capable of detecting and amplifying small electrostatic charges; the principal of the design was very similar to that of Van de Graaff's accelerator (1933). He formulated mathematical descriptions of vibrational motion, a problem addressed by Lissajous only a few months earlier. Righi also discovered magnetic hysteresis (1880), and whilst working at Palermo he discovered that bismuth exhibits a Hall effect thousands of times stronger than that of gold. He introduced the term 'photoelectric effect' to describe the increase in the conductivity of air gaps between electrodes when the air is exposed to ultraviolet light. Embarking on a series of experiments that he hoped would demonstrate that Hertzian waves obeyed the laws of classical optics, such as those of interference and diffraction, he generated Hertzian waves with wavelengths of just 26 millimetres. In doing so he was the first person to generate 'microwaves', thereby opening a whole new area of the electromagnetic spectrum to research and subsequent application. His experiments with microwaves did indeed confirm that Hertzian waves behaved very similarly to visible light and showed comparable interference, diffraction, refraction and absorption effects. His *L'onica delle oscillazioni electriche* (1897) which summarized these results is considered a classic of experimental electromagnetism. By the turn of the century he had turned to work on X-rays and the Zeeman effect. In 1903 he wrote (with B. Dessau) the first paper on wireless telegraphy, *La telegrafia senza fila.* He also worked on the conduction of gases under various conditions of pressure and ionization, and on improvements to the Michelson-Morley experiment from 1918.

RITTER, Johann Wilhelm
(1776–1810)

German physicist, born in Samitz, Silesia (now in Poland), the son of a Protestant pastor. He was trained as an apothecary, before studying medicine at the University of Jena. He taught at Jena and at Gotha before becoming a full member of the Bavarian Academy of Sciences in Munich (1804). While working in Jena he discovered the ultraviolet rays in the spectrum by means of their darkening effect on silver chloride (1801), but his chief contributions were in electrochemistry and electrophysiology. He demonstrated in 1800 that galvanic electricity was a manifestation of electricity, like static electricity, by means of the electrolysis of water and collected the constituent gases of hydrogen and oxygen (this experiment was first performed with static electricity); he made the first dry cell (1802); and a secondary cell, or accumulator (1803) and was the first to propose an electrochemical series. He saw electrical action in terms of chemical behaviour. Although his research was empirical, his conceptual framework was coloured by the German Romantic movement and *Naturphilosophie*, and this adversely affected the acceptance of his work abroad, but he did strongly influence Humphry Davy.

ROBBINS, Frederick Chapman
(1916–   )

American physiologist and paediatrician, born in Auburn, Alabama, the son of a plant physiologist. Robbins trained at Missouri University and Harvard University Medical School, where he graduated in 1940. He served as intern at the Children's Hospital Medical Center, Boston (1941–1942), before spending four years with the Army Medical Corps. Returning to Boston he completed his paediatric training and then joined Enders and Weller at the Infectious Diseases Research Laboratory of the Children's Hospital. Robbins, Enders and Weller had all worked on improving techniques for cultivating viruses, and they decided to apply their improvements to a new attempt to cultivate the poliomyelitis virus. Their success, which relied upon the use of antibiotics, led to quicker and cheaper means of diagnosis and was also an important step in the development of a polio vaccine. For this work the three scientists were awarded the 1954 Nobel Prize for Physiology or Medicine. Robbins was Professor of Paediatrics at Case Western Reserve University, Cleveland, from 1952 until 1980.

ROBERTS, Richard
(1943–   )

British molecular biologist, born in Derby. Educated at the University of Sheffield, he moved to the USA in 1969 and worked at Cold Spring Harbor

Laboratory in New York from 1972. Since 1992 he has been research director at New England Biolabs in Beverly, Massachusetts. In 1977, he announced his intriguing discovery that genes contain sections of DNA now known as 'introns' which carry no genetic information. He shared the 1993 Nobel Prize for Physiology or Medicine jointly with Sharp, who had independently come to the same conclusion around the same time.

ROBERTSON, Howard Percy
(1903–1961)

American mathematician and cosmologist, born in Hoquiam, Washington. He was educated at the University of Washington, where he received his first degree in 1922 and at Caltech, where he graduated PhD in 1925. Following a spell as Research Fellow at Göttingen, Munich and Princeton, he became Assistant Professor of Mathematics at Caltech (1927–1929). He joined the staff of Princeton University in 1928, becoming Professor of Mathematical Physics in 1938, and in 1947 returned to Caltech as Professor of Mathematical Physics, a post which he held until his death. Robertson's researches were in the fields of differential geometry, relativity and theoretical cosmology. The Poynting-Robertson effect (1937) was the result of his rigorous application of relativistic theory to the effect first noted by Poynting (1903) of the Sun's radiation pressure on micrometeorites. In cosmology his name is also given to the expression developed by him for four-dimensional relativistic line-element applicable to a homogeneous isotropic expanding universe (1933).

ROBIN, Gordon de Quetteville
(1921–)

Australian geophysicist and glaciologist, born and educated in Melbourne where he received a BSc in physics (1940) and an MSc (1942). He joined the Royal Australian Naval Volunteer Reserve and served on submarines (1942–1945), and later moved to the UK where he became a research student and lecturer at Birmingham University. He moved to the Australian National University as Senior Fellow (1957–1958), returning to the UK in 1958 to become Director of the Scott Polar Research Institute, Cambridge (1958–1982). He was Secretary (1958–1970) and President (1970–1974) of the International Scientific Committee on Antarctic Research. During meteorological and glaciological work in Antarctic, Robin obtained the first seismic traverse from the coast to the Antarctic Plateau and developed a theory of flow temperature distribution in ice sheets. His research group revolutionized the study of sheets by using radio echo sounding techniques and studied the penetration of ocean waves into fields of pack ice. He made first proposal for studying polar ice sheets by satellite altimetry (1963), and in publications on ice sheets and glaciers investigated their interaction with processes which produce global warming.

ROBINS, Benjamin
(1707–1751)

English mathematician and father of the art of gunnery, born into a Quaker family in Bath. He became a teacher of mathematics in London, published several treatises, commenced his experiments on the resisting force of the air to projectiles, studied fortification, and invented the ballistic pendulum to measure the energy of objects in flight. In

1735 he refuted George Berkeley's objections to Isaac Newton's calculus, in a treatise entitled *Newton's Methods of Fluxions*. His *New Principles of Gunnery* appeared in 1742, and inspired Leonhard Euler and Lambert to investigate the subject. Engineer to the East India Company (1749), he died in Madras. His works were collected in 1761.

ROBINSON, Sir Robert
(1886–1975)

English chemist, born in Chesterfield. The son of a manufacturer of surgical dressings, he was sent to Manchester University in 1902 to study chemistry as it was thought it would be useful to the family firm. He came under the influence of William Henry Perkin Jr, and on graduation entered Perkins' research laboratory to study the dyewood colouring matter brazilian. An interest in the chemistry of coloured natural products remained with him for the rest of his life. He also started work on alkaloids while with Perkin. During his time in Manchester, Robinson established friendships with Arthur Lapworth, Weizmann and Norman Haworth. In 1912 he was invited to a chair of organic chemistry in Sydney, where he continued to study topics started in Manchester and also undertook work on material from eucalyptus oils. After only two years he returned to Britain as professor at the University of Liverpool, and entered upon a new and very significant phase of his researches on alkaloids, including his famous synthesis of tropinone. Rather surprisingly he left Liverpool in 1919 to become Director of Research for the British Dyestuffs Corporation based in Huddersfield. However, the move was not a success and in 1920 he resigned to become professor at the University of St Andrews, although he maintained close contact with the dyestuffs industry for many years. He continued research on natural products with significant results and also developed many of his ideas on mechanistic organic chemistry. At this time he proposed a description of aromaticity in terms of a stable sextet of electrons. This was a happy time for Robinson but when the chair at Manchester fell vacant in 1922 he moved there. By this time his fame was great enough to attract many overseas students. After a brief spell at University College London (1928–1930) he became Waynflete Professor of Chemistry at Oxford, where he remained until 1955. It was the culmination of his scientific career, but he became increasingly irascible and entered into a long altercation with Christopher Ingold on mechanistic organic chemistry. During World War II he played an important role in the development of penicillin. Even after retirement he maintained an interest in chemistry. He was a close friend of Robert Maxwell and founded a number of learned journals. He served on many government committees, was knighted in 1939, awarded the Order of Merit in 1949, and became President of the Royal Society in 1945. He received the Nobel Prize for Chemistry in 1947 for his work on the chemistry of natural products.

ROEMER, Olaus
(1644–1710)

Danish astronomer, born in Aarhuus. He worked at the Paris Observatory (1672–1681) where he discovered that the intervals between successive eclipses of a satellite in Jupiter's shadow were less when Jupiter and the Earth were approaching than when they were receding (1675). He concluded that this was due to the finite velocity of light – a new discovery and from the observed intervals and the known rate of motion of Jupiter and the Earth obtained the first estimate of the velocity of light. Roemer also invented the transit instrument, a telescope movable only in the meridian, which greatly increased the accuracy attainable in the determination of both time and right ascension. In 1681 he returned as royal mathematician and University Professor of Astronomy to Copenhagen, where he remained for the rest of his life.

ROHRER, Heinrich
(1933–)

Swiss physicist, born in Buchs, co-recipient of the 1986 Nobel Prize for Physics with Binnig for their development of the tunnelling electron microscope. He was educated at the Swiss Institute of Technology in Zurich, completing his first degree in Physics in 1955 and his PhD in 1960. He moved to the USA to take up a research post at Rutgers University in New Jersey (1961–1963). Returning to Zurich in 1963, he joined the IBM Research Laboratory, where he ultimately became manager in the physics department. His work with Binnig on the scanning tunnelling microscope began in 1978; within two years they had constructed their first instrument, and by the end of the third had completed the instrument that won them the Nobel Prize (jointly with Ruska). The operation of the microscope is based on a quantum-mechanical effect, namely the tunnelling of electrons across a narrow gap between two surfaces. The microscope uses a needle with a tip of just one atom positioned close above the surface of a sample. With a potential difference applied between the needle and the sample a tunnelling current flows, and the needle can be scanned over the surface of the sample to build up a picture of the surface topography. The microscope is able to resolve features horizontally only 6 angstroms apart, and in the vertical direction can detect distances of 0.1 angstrom, one-thirtieth the size of an average atom. The scanning tunnelling microscope is now found increasingly frequently in laboratories around the world and sees particular application in the development of small solid-state electronic devices. In 1984 Rohrer was co-recipient with Binnig of two international awards, the Hewlett Packard Europhysics Prize and the King Faisal International Prize.

ROME DE L'ISLE, Jean-Baptise Louis
(1736–1790)

French crystallographer, born in Gray. In 1756 he entered the Royal Corps of Artillery and Engineering and saw active service in the French Indies, where he was captured by the English in 1761 and transported to China. He returned to France in 1764, published his first work on freshwater polyps (1766) and then devoted himself to the study of mineralogy and chemistry. In his *Essai de cristallographie* (1772), he identified 110 crystal forms. In 1779 he became involved in controversy with Buffon concerning the theory of a central terrestrial fire and the eventual cooling of the Earth. His major work, *Cristallographie* (1783), described more than 450 crystal forms. In the course of making terracotta crystal models with his assistant, he devised a contact goniometer and together they discovered the law of the constancy of interfacial angles, whereby the angle between similar faces in a particular mineral is always the same. 1784 saw the publication of his book on the external characteristics of minerals, stating that form, density and hardness were sufficient criteria to permit the identification of any mineral species. He also published a work comparing 'the weights and measures of antiquity with modern counterparts' (1784). He had a strained relationship with Hauy, whose ideas became more widely accepted.

ROMER, Alfred Sherwood
(1894–1973)

American palaeontologist, born in White Plains, New York. He studied at Columbia University and became Professor of Vertebrate Palaeontology at Chicago (1923–1934), and then Professor of Zoology and Director of the Museum of Comparative Zoology at Harvard (1934–1965). He was interested in the evolution of early vertebrates, in particular the transition from water to land, and wrote papers on the importance of the development of the amniotic egg and other preadaptations for terrestrial life. This formed the basis of his *The Vertebratė Story* (1959). One of the standard texts in its area was *Vertebrate Palaeontology* (1933), and *The Vertebrate Body* (1949) was a comprehensive text on comparative anatomy widely used in universities throughout the English speaking world.

RONALDS, Sir Francis
(1788–1873)

English inventor, born in London, the son of a merchant. Interested in the application of electricity, he invented an electric clock whose motive power was the recently introduced high-intensity dry battery, and in the same year (1816, published in 1823) an electrostatic telegraph which he constructed in his garden at Hammersmith. His offer of the telegraph to the Admiralty was refused; the device was rather impractical and practical versions did not appear until the later development of electromagnetic devices. He also invented a system of automatic photographic registration for meteorological instruments (1845). In 1843 Ronalds was made Honorary Director and Superintendent of the Kew Meteorological Observatory, set up by the British Association for the Advancement of Science. He retired in 1852 and lived for many years in Italy. He was elected FRS in 1844, and was knighted in 1870 for his pioneering work on the electric telegraph.

RØNTGEN, Wilhelm Konrad von
(1845–1923)

German physicist, born in Lennep, Prussia. He studied mechanical engineering at Zürich. After teaching at Strassburg, he was appointed Professor of Physics successively at Giessen (1879), Wurzburg (1888), where he succeeded Kohlrausch, and Munich (1899–1919). At Wurzburg in 1895, while investigating the properties of cathode rays with a 'Crookes's tube', he discovered the electromagnetic rays which he called X-rays (known also as Rontgen rays), because of their unknown properties, and for his work on them he was awarded in 1896, jointly with Lenard, the Rumford Medal, and in 1901 the first Nobel Prize for Physics. He also did important work on the heat conductivity of crystals, the specific heat of gases, and the magnetic effects produced on dielectrics, predicted by Heaviside as a consequence of Maxwell's theory.

ROSCOE, Sir Henry Enfield
(1833–1915)

English chemist, born in London, grandson of the English historian William Roscoe. He was educated at University College London, and the University of Heidelberg where, with Bunsen, he carried out research on quantitative photochemistry. In 1857 he was appointed Professor of Chemistry at Owens College, Manchester. His students and staff were encouraged to follow their own interests, with the result that

Manchester led many new areas of research, and became famous at home and abroad. He had the first Chair of Organic Chemistry in Britain and also taught crystallography, thermal chemistry and gas analysis. Roscoe encouraged links with industry and was energetic in promoting lectures on science for the general public. He was Liberal MP for South Manchester from 1885 to 1895 and Vice-Chancellor of London University from 1896 to 1902. Roscoe's own research was considerable. With Bunsen he showed that when chlorine and hydrogen combine in the presence of light, the extent of the photochemical action varies inversely with the distance from the light source and is directly proportional to its intensity. In 1865 he isolated vanadium from copper ores in the Cheshire mines. Previously vanadium, which had been discovered by Sefstrom, had only been found in very small quantities and its properties were imperfectly known. Roscoe showed that it belongs to the same family as phosphorus and arsenic. He was also the author of influential textbooks. He was elected FRS in 1863, knighted in 1884, President of the British Association for the Advancement of Science 1887, and a privy councillor from 1909. He died in Leatherhead, Surrey.

ROSE, William Cumming
(1887–1984)

American biochemist, born in Greenville, South Carolina. He studied at Yale and Freiburg, Germany, before returning to the USA where he worked at the University of Texas from 1913 to 1922. He was subsequently appointed professor at Illinois University, where he remained until his retirement in 1955. Following the demonstration by Frederick Gowland Hopkins that some amino acids appeared essential for growth, Rose began accumulating purified amino acids for experiments to determine the dietary importance of each of the 20 amino acids that occur in proteins (1924). By 1938 he could reliably state their nutritional significance in the diet, and he identified 10 indispensable amino acids. This led to the currently used division into essential and non-essential amino acids (1948) and his first list of human minimum and recommended daily requirements for each (1949). As a member of numerous nutritional committees he amended these values in 1957. During this work he isolated a new amino acid, threonine (1934), showed that methionine (1937) and valine (1939) are essential in the diet, and demonstrated that valine can form glucose. Rose also revealed that creatinine is formed from creatine, and studied the detoxification of benzoic acid by hippurate formation.

ROSKY, Herbert
(1935–   )

German chemist, born in a small village in East Prussia, who moved to West Germany after World War II and studied at the University of Göttingen. He received his doctorate in 1963 for work on metal fluorides. He then spent a year at the Du Pont Experimental Station in Wilmington, Delaware, where he collaborated with Muetterties on phosphorus anion chemistry. Returning to Germany he became Professor of Chemistry at Frankfurt University, and he is Professor at Göttingen. He is a synthetic inorganic chemist whose principal interest remains the preparation of fluorides of both main group and transition metals. He won the Wohler Prize in 1960, the Leibniz Award in 1987 and the Alfred Stock Memorial Award, in 1990.

ROSS, Sir Ronald
(1857–1932)

British physician, born in Almora, India, the son of an army officer. He studied medicine at St Bartholomew's Hospital in London and entered the Indian Medical Service in 1881. He learned bacteriological and microscopical techniques on various furloughs in England, during one of which he met Manson (1894). Manson told Ross of his belief that malaria is transmitted through mosquito bites and Ross returned to India determined to investigate this possibility. He discovered the malaria parasite in the stomachs of mosquitoes which had bitten patients suffering from the disease, and by 1898 had worked out the life cycle of the malaria parasite for birds. He returned to England in 1899 to lecture at the newly founded Liverpool School of Tropical medicine and to lead several expeditions in Africa concerned with mosquito eradication. Elected FRS in 1901, he was knighted in 1911. He moved to London in 1912 where, from 1926, he directed the Ross Institute. He was a gifted if eccentric mathematician who also wrote poetry and romances. His award of the 1902 Nobel Prize for Physiology or Medicine was contested by Giovanni Grassi (1854–1925), an Italian parasitologist who had independently and almost simultaneously worked out the life cycle of the human malaria parasite.

ROSSE, William Parsons, 3rd Earl of
(1800–1867)

Irish astronomer and landowner, born in York, England, and educated at Trinity College, Dublin, and Oxford University where he graduated in mathematics (1822). With the help only of the workers on his feudal estate of Birr Castle at Birr (then Parsonstown) he constructed on the model of William Herschel's instruments a gigantic metal-mirror telescope 6 feet in diameter and 52 feet in focal length, the 'leviathan of Parsonstown'. With this, the largest telescope ever built until the 100-inch reflector in California (1917), he discovered the spiral structure of the nebula Messier 51 near the tail of the Great Bear (1845), the first ever observation of a spiral galaxy. Drawings of this and of many other nebulae including spirals were published in Rosse's catalogue of nebulae (1850). The giant mirror is preserved in the Science Museum, London.

ROSSI, Bruno
(1905–)

Italian-American physicist, born in Venice. He was educated at the universities of Padua (1923–1925) and Bologna (1925–1927) and subsequently took up an assistantship in the physics department at the University of Florence (1928–1932). In 1932 he was appointed Professor of Physics at the University of Padua, where he carried out early research on the properties of cosmic rays. He showed that these energetic particles reaching the Earth from space have the capacity to traverse great thicknesses of matter. He also demonstrated that the incident primary radiation from space may collide with atoms such as oxygen and nitrogen in the atmosphere causing nuclear reactions and generating cascades of secondary particles, now called 'showers'. This primary radiation he found to be positively charged (in fact cosmic rays consist mainly of protons), and he was able to show that not all of the particles making up primary cosmic radiation could produce particle showers. In 1939 he moved first to the University of Manchester, and then to the USA, where he spent time continuing his work on cosmic rays at the University of Chicago before becoming Professor of Physics at Cornell University in 1940. Five years later he moved to MIT where he spent the rest of his working life. His work has included investigations of meson decay, and among his many publications were *Cosmic Rays* (1964, with S. Olbert) and *Introduction to the Physics of Space* (1970).

ROTHLISBERGER, Hans
(1923–)

Swiss glaciologist, born in Langnau, near Berne. He entered a teacher training college then changed to study petrology at the Swiss Federal Institute of Technology (ETH) in Zurich, where he received a diploma in 1947. In his doctoral thesis on seismic velocity in sedimentary rocks (1954), he described a new method for grain size determinations. With an interlude as contract scientist to the US army Snow, Ice and Permafrost Research Establishment (1957–1961) he held various appointments at ETH, finally becoming head of the glaciology section of the Laboratory of Hydraulics, Hydrology and Glaciology (1980–1987). During expeditions to Baffin Island in 1950 and 1953, he undertook seismic soundings across glaciers and with colleagues made the first ascent of Mount Asgard. His main contributions lie in glacial hazard assessment, and in englacial and subglacial drainage of meltwater. Rothlisberger made the important identification of the equivalence of the rate of melting caused by frictional heat of the running water and the rate of conduit closure caused by ice overburden pressure. A meltwater conduit incised in ice at the glacier bed is referred to as a Rothlisberger channel. He published *Seismic Exploration in Cold Regions* (1972), and was President of the International Glaciological Society (1984–1987).

ROUS, (Francis) Peyton
(1879–1970)

American pathologist, born in Baltimore. Educated at Johns Hopkins University and Medical School, he became assistant (1909–1910), associate (1910–1912), associate member (1912–1920) and member (1920–1945) of the Rockefeller Institute for Medical Research in New York City. From 1909 he began studying a sarcoma in chickens, which he demonstrated was caused by a virus. He also showed that the sarcoma could be successfully transplanted only in closely related chickens, thus encouraging others to use genetically similar strains of laboratory animals. He returned to this line of work in the 1930s, when a rabbit tumour was also shown to be caused by a virus; Rous found that coal tar and the virus could potentiate each other in making the tumour malignant. The discovery of many other oncogenic viruses from the 1950s made his early work more widely appreciated, and he shared with Charles Huggins the 1966 Nobel Prize for Physiology or Medicine. For more than 50 years, Rous was an editor of the *Journal of Experimental Medicine*, published at the Rockefeller Institute.

ROUX, (Pierre Paul) Emile
(1853–1933)

French bacteriologist, born in Confolens, Charente. He studied at the universities of Clermont-Ferrand and Paris, where he became assistant to Pasteur, and in 1904 succeeded him as Director of the Pasteur Institute. Roux worked with Pasteur on most of the latter's major medical discoveries. He tested the anthrax vaccine, and did much of the early work on the rabies vaccine. With Yersin he showed that the symptoms of diphtheria are caused by a lethal toxin produced by the diphtheria bacillus, rather than by the bacteria themselves. Behring and Kitasato had shown in 1890 that infected guinea pigs produce an antitoxin in the blood. Using the same principle, Roux obtained large quantities of blood serum containing the antitoxin from horses, and tested it on patients in 1894. In four months the mortality rate had fallen from 51 to 24 percent. He also made important contributions to research into syphilis.

ROUX, Wilhelm
(1850–1924)

German anatomist and physiologist, born in Jena. Roux studied medicine at Jena University, before spending 10 years (1879–1889) at the anatomical institute in Breslau. In 1895 he was appointed head of anatomy at Halle, where he remained until 1921. Roux was prolific both as a researcher and as a propagandist for a particular, highly mechanistic, vision of science. He accomplished extensive practical and theoretical work on experimental embryology within the paradigm of *Entwicklungsmechanik*, or development mechanics. A follower of Haeckel and Charles Darwin and a convinced evolutionist. Roux sought to understand evolutionary processes at the cellular and molecular levels, focusing in particular upon experimental embryology, and applying the findings of physics and chemistry. On the basis of extensive and often violent embryological experimentation, Roux endorsed Weismann's idea that the material of the germ plasm is passed on intact from parent to offspring; he thereby pointed to a physical basis for heredity.

ROXBURGH, William
(1751–1815)

Scottish botanist, born in Craigie, Ayrshire. He was educated at Edinburgh University, where he studied under John Hope. By Hope's influence he became a surgeon's mate with the East India Company and ultimately assistant surgeon in Madras (1780–1793). Stationed at Samulcotta, he cultivated cinnamon, annatto, nutmeg, coffee, peppers and indigo, studied sugar-growing and silkworm-rearing and made large collections of plants, and was appointed the East India Company's 'Botanist in the Carnatic'. Many of his early collections were lost in a flood (1787). He prepared a series of illustrations of local plants, using an Indian artist, published as *Plants of the Coast of Coromandel* (3 vols, 1795–1819). He became Superintendent of Calcutta Botanic Garden (1793–1813), but while there was forced twice to return to Britain because of illness, before his health completely broke down in 1813, when he finally left India. He left behind the manuscript of his *Flora Indica*, which was eventually edited by Wallich and partially published (2 vols, 1820–1824) by William Carey, with a later complete edition (3 vols, 1832). Roxburgh's book remains valuable for its notes on Indian economic botany and its accurate descriptions.

RUBBIA, Carlo
(1934–   )

Italian-born American physicist, born in Gorizia and educated at Pisa, Rome and Columbia universities. From 1960 he worked at CERN (the European nuclear research centre) in Geneva. In 1971 he accepted a Chair in Physics at Harvard University, simultaneously continuing his research at CERN. Rubbia was head of the team that discovered the W and Z bosons which mediate the weak nuclear force. Although indirect evidence for the existence of these particles had been available for some time, this was the first direct proof and put the unified theory of electromagnetic and weak forces (the 'electroweak' theory) on a firm experimental footing. Rubbia pointed out that because the proton and anti-proton have exactly the same mass and opposite charges within experimental limits, they will describe exactly the same orbit but in opposite directions when placed in a magnetic field. This allowed the existing Super Protons Synchrotron (SPS) accelerator to be upgraded to become a collider of protons and anti-protons, achieving enough energy to produce W and Z bosons for the first time. For this work he shared the 1984 Nobel Prize for Physics with van der Meer. As Director-General of CERN (1989–1993), he was the driving force behind the LEP (an electron-positron collider) and LHC (Large Hadron Collider) projects.

RUBIN, Gerald Mayer
(1950–)

American molecular biologist, born in Boston. He was educated at MIT where he obtained a BS in 1971 and at the University of Cambridge where he received his PhD in 1974. Subsequently he became Helen Hay Whitney Foundation Fellow at Stanford University School of Medicine in California (1974–1976) and Assistant Professor of Biological Chemistry at the Sidney Farber Cancer Institute, Harvard Medical School (1977–1980). He was then appointed as a staff member at the Carnegie Institute of Washington, Baltimore, from 1980 until 1983, when he moved to his current position as John D. McArthur Professor of Biological Chemistry at the University of California at Berkeley. He has simultaneously been Investigator at the Howard Hughes Medical Institute since 1987. Rubin was responsible for the first production of transgenic fruit flies. This involves the introduction of specific cloned genes into the germ line of the fruit fly. When used in conjunction with the rapid production of genetic mutations in fruit flies, this technique provides a powerful tool for the study of the regulation of specific gene control. For example, artificially introduced genes can be tracked throughout the development of the fruit fly to determine in which tissue and at what developmental stage specific genes are active. In addition the introduced genes can be mutated in specific ways so that the function of precise DNA sequences in the correct expression of their accompanying gene can be determined.

RUBNER, Max
(1854–1932)

A notoriously taciturn German physician who specialized in physiology and hygiene. Rubner attended the University of Munch, later holding chairs at Marburg and Munich and rising to become Knoch's successor in the Chair of Hygiene at Berlin. He worked on many aspects of metabolism, definitively establishing the applicability of the principle of the conservation of energy to living organisms. Impressed by energy physics, Rubner attempted to asses the life-careers of mammals with regard to the energy consumption of their protoplasm and its transformation into growth. Against the background of food shortages during World War I, he increasingly directed his researches into problems of nutrition.

RUDBECK, Olof
(1630–1702)

Swedish physiologist and intellectual, born in Vasteras. A bishop's son, Rudbeck studied medicine at the University of Uppsala and early embarked on research into animal economy. He quickly accepted the views of William Harvey (1574–1657) and familiarized himself with recent work on the lymphatic system. Independently of Pecquet, he discovered the thoracic duet, through which the lacteal vessels discharge chyle into the veins. There followed a violent priority dispute involving Pecquet, who had published first and Thomas Bartholin, who had separately been working in Copenhagen on the lymphatic system. Further studies at Leiden were followed by appointment to a medical chair in Uppsala. Rudbeck devoted massive energies to the improvement of the buildings and intellectual standing of the university. In later years Rudbeck pursued botanical studies, developing a tradition of which Linnaeus was the later beneficiary.

RUMFORD, Benjamin Thompson, Count
(1753–1814)

English-American administrator and scientist, born in Woburn, Massachusetts. He was an assistant in a store and a school teacher, until in 1772 he married a wealthy widow. He became a major in the 2nd New Hampshire regiment, but left his wife and baby daughter and fled to England in 1776, possibly due to political suspicion. He gave valuable information to the government as to the state of America during the Revolution, and received an appointment in the Colonial Office. In England he began the experiments with gun-powder which continued throughout the rest of his life. He was elected FRS in 1779, and by 1782 he was back in America, with a lieutenant-colonel's commission. After the peace he was knighted, and in 1784 entered the service of Bavaria. In this new sphere he reformed the army, drained the marshes round Marmheim, established a military academy, planned a poor-law system, introduced the cultivation of the potato, disseminated knowledge of nutrition and domestic economy, improved the breeds of horses and cattle, and laid out the English Garden in Munich. For these services he was made head of the Bavarian war department and count of the Holy Roman Empire. In the former capacity he was responsible for the arsenal at Munich, where he observed the immense amount of heat generated by the boring of cannon. He was able to deduce experimentally that there was a relationship between the work done and the heat generated, and this led to his championing of the vibrato (rather than the caloric) theory of heat. He reported his findings to the Royal Society in 1798 in one of the classic papers of experimental science, *An Experimental Inquiry concerning the Source of Heat Excited by Friction*. He also invented the Rumford shadow photometer, designed the so-called Rumford oil lamp, and introduced the concept of the standard candle which became the international standard of luminosity until the middle of the present century. During a visit to England (1795–1796) he endowed the two Rumford medals of the Royal Society, and also two of the American Academy, for researches in light and heat. In 1799 he left the Bavarian service, returned to London, and founded the Royal Institution; in 1802 he moved to Paris, married Lavoisier's widow (1804) and lived in her villa at Auteuil, where he died.

RUNGE, Friedlieb Ferdinand
(1795–1867)

German dye chemist and pioneer of chromatography, born near Hamburg. He studied medicine in Berlin, Göttingen and Jena, and then spent three years visiting factories and laboratories throughout Europe. He was appointed to the Chair of Technical Chemistry at Breslau in 1828; from 1831 to 1852 he managed a chemical factory at Oranienburg, and thereafter worked independently as a chemical consultant. At Oranienburg he distilled coal tar and investigated the products which boiled off at different temperatures. In this way he isolated phenol (carbolic acid) aniline (cyanol) and other compounds. In 1834 he patented the process for obtaining the dye 'aniline black' from cyanol. He conducted experiments with a process that can now be identified as paper chromatography, an important technique for separating and identifying the components of a mixture of soluble substances based on the fact that every compound has a different capillarity. He placed a solution of the mixture on absorbent paper and demonstrated that the components separated out into coloured rings of different radii. This technique was further developed by Schonbein and Tswett. Runge died in Oranienburg.

RUSKA, Ernst August Friedrich
(1906–1988)

German electrical engineer, born in Heidelberg. He was educated at the Technical University of Munich (1925–1927) and moved to the Technical University of Berlin in 1931. In the same year he developed the world's first electron microscope. Just four years earlier, Davisson and Germer of Bell Telephone Laboratories in the USA had demonstrated experimentally that electron beams behave with both particle- and wave-like properties – one of the basic principles of quantum mechanics. The wavelengths of high-energy electrons can be less than 1 angstrom, which is about a third of the size of an average atom and about a thousand times smaller than the wavelength of visible light. The resolving power of a microscope is limited by the wavelength of the illuminating beam, and so by using electron beams, Ruska's electron microscope was able to resolve details on a scale of a few angstroms. This allowed a magnification of one million times, substantially better than the maximum magnification of around 2,000 times possible with a light microscope. To focus the electron beams, Ruska used electromagnetic lenses made of current carrying coils. Throughout World War II he continued his development work on the electron microscope in a converted bakery, supported by the German engineering firm Siemens. Most of his career was dedicated to further improving the capabilities of the electron microscope, his areas of work including many aspects of electromagnetic and permanent magnetic lens design and improvements to the transmission and reflection electro microscopes. It was not until 1976 that he received a half share of the Nobel Prize for Physics along with the inventors of the tunnelling electron microscope, Binnig and Rohrer.

RUSSEL, Bertrand Arthur William, 3rd Earl Russell
(1872–1970)

Welsh philosopher, mathematician, prolific author and controversial public figure throughout his long and extraordinary active life. He was born in Trelleck, Gwent; his parents died when he was very young and he was brought up by his grandmother, the widow of Lord John Russell, the Liberal prime minister and 1st Earl. He was educated privately, and in mathematics and philosophy at Trinity College, Cambridge. He graduated in 1894, was briefly British Embassy attache in Paris, and became a Fellow of Trinity in 1895. His most original contributions to mathematical logic and philosophy are generally agreed to belong to the period before World War I, as expounded for example in *The Principles of Mathematics* (1903), which argues that the whole of mathematics could be derived from logic, and the monumental *Principia Mathematica* (1910–1913, with Whitehead), which worked out this programme in a fully developed formal system and stands as a landmark in the history of logic and mathematics. Russell's famous 'theory of types', and his 'theory of descriptions' belong to this same period. Politics became his dominant concern during World War I and his active pacifism caused the loss of his Trinity fellowship in 1916 and his imprisonment in 1918, during the course of which he wrote his *Introduction to Mathematical Philosophy* (1919). Later he wrote on politics and education, and founded a progressive school. The rise of Fascism led him to renounce his pacifism in 1939; his fellowship at Trinity was restored in 1944, and he returned to England after the war to be honoured with the Order of Merit, and to give the first BBC Reith Lectures in 1949. He was awarded the Nobel Prize for Literature in 1950. After 1949 he became increasingly preoccupied with the cause of nuclear disarmament, taking a leading role in the Campaign for Nuclear Disarmament and later the remarkable correspondence with various world leaders. In 1961 he was again imprisoned for his part in a sit-down demonstration in Whitehall. The last major publications were his three volumes of *Autobiography* (1967–1969).

RUSSELL, Sir Frederick Stratten (1807–1984)

English marine biologist, born in Doncaster and educated at Gonville and Caius College, Cambridge. He served with distinction in the Royal Naval Air Service during World War I and then joined the scientific staff of the Plymouth Laboratory of the Marine Biological Association, of which he was a director between 1946 and 1965. Russell's early research included some pioneering work on the phenomenon of diurnal vertical migration by zooplankters, but it is for his work on medusae (*The medusae of the British Isles*, 2 vols. 1953, 1970), larval fish (*Eggs and planktonic stages of British marine fish,* 1976) and the long-term dynamics of zooplankton communities that he is best known. Hydoid and scyphozoan cnidarians typically have two markedly different phases in their life-cycle – a free-swimming medusoid (jelly-fish) phase and a benthic polyploid phase. By painstaking laboratory rearing, Russell and co-workers succeeded in solving major taxonomic problems to link the medusa to the polyp phase for many species. He initiated long-term sampling and analysis of macrozooplankton off the Eddystone Lighthouse (1924–1934, 1946–1980), and noted that certain large and readily identifiable planktons could be used as 'indicators' of the source of the water mass with which they were associated. His Plymouth studies led to the elucidation of the natural variations in local plankton communities, and the environmental changes these variations produce; the dynamics of this phenomenon are now known as the 'Russell cycle'. The value of Russell's long-term work lies in its potential use in the assessment of large-scale changes in water mass circulation, and possible climate change, in addition to predictions for commercially important fish populations. He was elected FRS in 1938 and knighted in 1965.

RUSSELL, Henry Norris
(1877–1957)

American astronomer, born in Pyster Bay, New York. After studies at Princeton and Cambridge, UK, he was appointed Professor of Astronomy in Princeton in 1905 and six years later director of the university observatory there. A theoretician of the highest calibre, he developed with Shapley methods for the calculation of the orbits and dimensions of eclipsing binary stars and for the determination of the distances of double stars. He worked on the theory of stellar atmospheres from the analysis of stellar spectra. His most famous achievement was the formulation of the Hertsprung-Russel diagram (1913) correlating the spectral types of stars with their luminosity which became of fundamental importance for the theory of stellar evolution.

RUSSELL, John Scott
(1802–1882)

Scottish engineer, born in what was then the village of Parkhead near Glasgow. He graduated MA from the University of Glasgow at the age of 17 and moved to Edinburgh where he taught mathematics and natural philosophy. In 1932 on the death of Leslie he provisionally accepted an invitation of the University of Edinburgh to occupy the Chair of Natural Philosophy, and it was only after some hesitation that he decided not to stand as a candidate for the professorship. Instead he went ahead with his scheme to build and operate steam coaches on the roads of Scotland, and in 1834 a service between Glasgow and Paisley was inaugurated. After only a few months of successful operation, however, an accident involving several fatalities put an end to the project. He had been engaged at the same time in testing ships' hulls of many different shapes and sizes on the Union Canal near Edinburgh, and from the results he developed his wave-line principle of naval architecture, although it was subsequently shown to have been unduly influenced by the narrowness of the waterway. He was, however, one of the first to advocate and adopt a scientific approach to the design of ships, and he took a leading part in the building of Brunel's *Great Eastern*, launched in 1858 from his yard on the Thames. Russell was one of the founders of the Institution of Naval Architects (1860) and served as Vice-President for many years.

RUTHERFORD, Ernest, 1st Baron Rutherford of Nelson
(1871–1937)

New Zealand-born British physicist, one of the greatest pioneers of subatomic physics, born in Spring Grove (later Brightwater) near Nelson, the fourth of 12 children of a wheelwright and flax-miller. Winning scholarships to Nelson College and Canterbury College, Christchurch, his first research projects were on magnetization of iron by high-frequency discharges (1894) and magnetic viscosity (1896). In 1895 he was admitted to the Cavendish Laboratory and Trinity College Cambridge on a scholarship. There he made the first successful wireless transmissions over two miles. Under the brilliant direction of J. J. Thomson, Rutherford discovered the three types of uranium radiation; alpha, beta and gamma rays. In 1898 he became Professor of Physics at McGill University, Montreal, where with Soddy, he formulated the theory of atomic disintegration to account for the tremendous heat energy radiated by uranium. In 1907 he became professor at Manchester. While there he asked his students Geiger and Ernest Marsden to investigate the scattering of alpha particles from gold foil; they observed that a tiny fraction of those striking the gold foil were deflected back from it. Rutherford compared this astonishing result to firing a pistol at a piece of paper, and finding that the bullet bounced back. He derived a formula to describe the scattering of alpha particles from the gold foil based on a model of the atom where electrons orbit a compact nucleus, and its predictions were verified by Geiger and Marsden. However, the theory ignored the problem that classically the electrons should radiate away all their energy and fall into the nucleus (this was solved by Niels Bohr, then at Manchester, using Planck's idea of energy quanta; thus the quantum theory of atoms was born). The new model revealed the importance of the atomic number – the charge on the nucleus. During World War I, Rutherford did research on submarine detection for the admiralty. In 1919, in a series of experiments, he discovered that alpha-ray bombardments induced atomic transformation in atmospheric nitrogen, liberating hydrogen nuclei. The same year he succeeded J. J. Thomson to the Cavendish professorship at Cambridge. In 1920 he predicted the existence of the neutron, later discovered by his colleague Chadwick. Like Niels Bohr, Rutherford's contribution to science was not only his own research but also his influence on the work of his contemporaries. He was awarded the Nobel Prize for Chemistry in 1908. He published nearly 150 original papers, and his books included *Radioactivity* (1904), *Radioactive Transformations* (1906) and *Radioactive Substances* (1930).

RUZICKA, Leopold
(1887–1976)

Swiss chemist, born in Vukovar, Croatia. He trained at the Technische Hochschule Karisrube under Staudinger on the chemistry of the pyrethrins and other insecticides in the chrysanthemum (1912), and continued this work at the Polytechnic at Zürich (1916), where he was appointed to an unsalaried professorship in 1923. From 1921 he collaborated with a perfume factory on the synthesis of aromatic terpenes (such as civetone, muscone and jasmone). After a short stay in Utrecht (1926–1929), he returned as Professor of Organic and Inorganic Chemistry to Zürich, where he remained for the rest of his career. Rutka's work on perfumes introduced him to the multimembered ring structures (terpenes) which he studied in detail. In the 1930s he discovered their structural relationship to the steroids, which led him to synthesize lanosterol and enunciate in 1953 the biogenic isoprene rule by which these five carbon compounds combine to form steroids. In 1935 he was able to announce the synthesis of the still undiscovered male hormones, testosterone and methyltestosterone. With Butenandt, he was awarded the 1939 Nobel Prize for Chemistry; he used the substantial royalties from his discoveries to found an art gallery in Zürich for paintings by Dutch and Flemish seventeenth-century masters.

RYDBERG, Johannes Robert
(1854–1919)

Swedish physicist, born in Halmstad. He was educated at the University of Lund where he remained throughout his career, becoming professor in 1901. He worked on the classification of optical spectra. From his studies he developed en empirical formula relating the frequencies of spectral lines, incorporating the constant known by his name, which was later derived by Niels Bohr using his quantum theory of the atom.

RYLE, Sir Martin
(1918–1984)

English physicist and radio astronomer, born in Brighton, son of John Ryle, a distinguished physician and Professor of Medicine. He was educated at Bradfield College, Berkshire, and Christ Church, Oxford, graduating in 1939 at the outbreak of World War II. For the duration of the war he was involved in important research in the field of radar. At the end of the war he joined the Cavendish laboratory in Cambridge where he investigated the emission of radio waves from the Sun and improved on the low resolving power of radio telescopes by the introduction of the radio analogue to Michelson's optical interferometer. Using two instead of one radio antenna he was able to pinpoint radio sources such as Sunspots with considerable accuracy (1946). He then turned to studies of radio waves from the universe and identified localized radio sources with distant galaxies many millions of light years away. He found that the numbers of radio sources increased as their intensities decreased (1955), indicating that there were more galaxies per unit volume the further one looked into space and back in time, a result which pointed to an evolving universe starting with a Big Bang. He mapped radio sources by his ingenious method of 'aperture synthesis' (developed from 1960 onwards) in which a number of similar small radio telescopes are moved successively to different separations and their output processed by computer, made to simulate the performance of a huge single telescope. Ryle was one of the most outstanding scientists of his generation. He was awarded a knighthood in 1966, was appointed to the first Chair of Radio Astronomy in Cambridge in 1969, and in 1972 became the first Astronomer Royal to come from the field of radio astronomy. In 1974 he received the Nobel Prize for Physics with his colleague Hewish.

S

SABATIER, Paul
(1854–1941)

French chemist, born in Carcassonne. After secondary education at the École Normale Superieure he taught briefly at a lycée in Nimes and then moved to Paris to work with Berthelot. He received his doctorate in 1880 and moved to Bordeaux for a year before taking an established post at Toulouse. Although he was offered posts elsewhere, he chose to remain in Toulouse for the rest of his life. In 1913 he became one of the first scientists to be elected to one of six chairs newly created by the Academy of Sciences for provincial members. He made a number of interesting discoveries in inorganic chemistry, but is best known for his discovery of catalysed hydrogenation of unsaturated organic compounds, such as the conversion of ethene to ethane over reduced nickel. He showed that this technique was applicable to many classes of organic compounds and as a synthetic procedure it has been widely used. For example, catalytic hydrogenation is used to produce margarine from vegetable oils. He proposed a theory (which has stood the test of time) to explain heterogeneous catalysis, involving the formation of unstable intermediates on the surface of the catalysts. He took little interest in the commercial application of his studies but received the Nobel Prize for Chemistry in 1912.

SABIN, Albert Bruce
(1906–1993)

Polish–American microbiologist, born in Bialystok, Russia (now in Poland). He was educated at New York University, where he received his MD in 1931, and in 1946 he was appointed Research Professor of Pediatrics at the University of Cincinnati. After working on developing vaccines against dengue fever and Japanese B encephalitis, he became interested in the polio vaccine and attempted to develop a live attenuated vaccine (as opposed to Salk's killed vaccine). He succeeded in persuading the Russians to help with the testing of his live virus, and in 1959 he was able to produce the results of 4.5 million vaccinations. His vaccine was found to be completely safe, possessing a number of advantages over that of Salk: it gave a stronger, longer-lasting immunity and could be administered orally. Consequently there was a widespread international adoption of the Sabin vaccine in the early 1960s. Some years later, he reported a major advance in cancer research, claiming to have evidence in support of the viral origin of human cancer. Later, however, he rejected his own experimental results.

SABINE, Sir Edward
(1788–1883)

Irish solder, physicist, astronomer and explorer, born in Dublin. Educated at Marlow and the Royal Military Academy at Woolwich, he was commissioned in the Artillery and served in Gibraltar and Canada. He accompanied his lifelong friend Sir James Clark Ross as astronomer on John Ross's expedition to find the North-West Passage in 1818 and on William Parry's Arctic expedition of 1819–1920. He conducted valuable pendulum experiments to determine the shape of the Earth at Spitzliergen and in tropical Africa (1821–1823), and devoted the rest of his life to work on terrestrial magnetism. By means of magnetic observatories established in British colonies, he made the important discovery that there is a correlation between variations in the Earth's magnetism and solar activity, which followed a 10–11 year cycle. He retired from the army in 1877 as a Major-General. Sabine's gull is named after him. He was elected FRS (1818), and served for many years as the Royal Society's president (1861–1871). He was knighted in 1869. An adroit politician, he was involved with the reforms of the Royal Society in the 1840s, and was a leading figure in the British Association for the Advancement of Science. His brother, Joseph Sabine (1770–1837), Inspector-General of Taxes, was a noted botanist.

SABINE, Wallace Clement Ware
(1868–1919)

American physicist, born in Richwood, Ohio. After graduating from Ohio State University in 1886, he entered Harvard, receiving the AM in 1888. Apart from service during World War I (he was effectively the chief scientist and authority on instruments for the US air force), Sabine was attached to Harvard throughout his life, from 1908 as Dean of the Graduate School of Applied Science which he had initiated. When Harvard opened its Fogg Art Museum (1895), the auditorium was found to be 'monumental in its acoustic badness'. Set to remedy this situation, Sabine introduced a quantitative element into the previously empirical study of 'architectural acoustics'. He analysed the problem in terms of the size, shape and materials of a room affecting the reverberation time: his prescription of 22 hair-felt blankets made the theatre functional. By 1898 he had devised the Sabine formula (linking reverberation time, total absorptivity and room volume) and with its aid advised on the new Boston Symphony Hall (1898–1900). The unit of sound absorbing power (the sabin) was named after him.

SACHS, Julius von
(1832–1897)

German botanist, born in Breslau (now Wroclaw, Poland), one of nine children of a poor engraver. He was befriended by Purkinje and enabled to study at the University of Prague, then became botany lecturer at an agricultural college near Boon; from 1868 he held the post of Professor of Botany at Wurzburg. There he carried out important experiments, especially on the influence of light and heat upon plants, and the organic activities of vegetable growth. Sachs is regarded as the founder of modern plant physiology. He took up water culture to establish the mineral requirements of plants. He observed the conversion of sugar into starch in chloroplasts, and suggested that enzymes are involved in the conversion of oil into starch, starch into sugar, and proteins into soluble nitrogen compounds. He contributed the volume *Experimental-Physiologie* to Hofmeister's *Handbuch der physiologischen Botanik* in 1865. His *Lehrbuch der Botanik* (1868) and its English translation *Textbook of Botany* (1875) exerted widespread influence.

SAGAN, Carl Edward
(1934–)

American astronomer, born in New York City. After studying at Chicago and Berkeley, he worked at Harvard then moved to Cornell, becoming Professor of Astronomy and Space Science in 1970. Interested in most aspects of the solar system, Sagan has done work on the physics and chemistry of planetary atmospheres and surfaces. He has also investi gated the origin of life on Earth and the possibility of extraterrestrial life. He was an active member of the imaging team associated with the Voyager mission to the outer planets, and since 1983 he has given considerable thought to the concept of the nuclear winter, a deep winter triggered by some catastrophic event in the Earth's atmosphere. In the 1960s he worked on the theoretical calculation of the Venus greenhouse effect. Sagan and James Pollack were the first to advocate that temporal changes on Mars were non-biological and were in fact due to wind-blown dust distributed by seasonally changing circulation patterns. Through books and a television programme, *Cosmos*, Sagan has done much to interest the general public in this aspect of science. His *Cosmic Connection* (1973) dealt with advances in planetary science; *The Dragons of Eden* (1977) and *Broca's Brain* (1979) helped to popularise recent advances in evolutionary theory and neurophysiology. Sagan is President of the Planetary Society and is a strong proponent of SETI – the search for extraterrestrial intelligence.

SAHA, Meghnad
(1894–1956)

Indian astrophysicist, born in Dacca (now in Bangladesh) the son of a small shopkeeper. He was educated at Presidency College, Calcutta, and subsequently visited Europe on a travelling scholarship. He taught at Allahabad University, and in 1938 was appointed Professor of Physics at Calcutta. He worked on the thermal ionization that occurs in the extremely hot atmospheres of stars and in 1920 demonstrated that elements in stars are ionized in proportion to their temperature ('Saha's equation'). This thermal ionization theory created a point of departure for the physical interpretation and classification of the spectra of stars. Saha later moved to nuclear physics, and played a prominent role in the creation in India of an institute for its study, which was named after him.

SAINTE-CLAIRE DEVILLE, Henri Etienne
(1818–1881)

French chemist, born on St Thomas, West Indies (now part of the Virgin Islands). During medical studies in Paris, he became attracted to chemistry. From 1845 to 1851 he was Professor of Chemistry at the University of Besançon. In 1851 he returned to Paris as successor to Balard at the École Normale Supérieure, and from 1853 he also gave lectures in chemistry at the Sorbonne. He was made an Honorary Fellow of the Chemical Society in 1860. His early work was on turpentine and other natural products, but later his interests moved to inorganic chemistry. In 1849 he isolated nitrogen pentoxide. He devised a process for the large-scale production of aluminium, which involved the reduction of the chloride by sodium (1855). He also developed related processes for the large-scale production of boron, silicon and titanium. Sainte-Claire Deville is also remembered for his extensive measurements of the vapour densities of substances at high temperatures, which led to the recognition of reversible thermal dissociation, of great significance for the theory of chemical equilibrium. His latter days were clouded by ill-health and he took his own life.

SAKHAROV, Andrei Dimitrievich
(1921–1989)

Soviet physicist and dissident born in Moscow, the son of a scientist. He graduated in physics from Moscow State University in 1942 and was awarded his doctorate for work on cosmic rays. He worked under Tamm at the Lebedev Institute in Leningrad. There he principally worked on nuclear fusion and proposed use of a 'magnetic bottle' to contain the plasma; this has been developed into the Tokamak design for a fusion reactor. He was mainly responsible for the development of the Soviet hydrogen bomb and in 1953 became the youngest-ever entrant to the Soviet Academy of Sciences. He also studied cosmology and proposed that the reason for the dominance of matter over antimatter in the universe is connected to CP violation, a complicated and unexpected phenomenon observed in certain particle interactions. During the early 1960s he became increasingly estranged from the Soviet authorities because of his campaigning for a nuclear test-ban treaty, peaceful international coexistence and improved civil rights within the USSR. In 1975 he was awarded the Nobel Peace Prize, but in 1980, during a Cold War crackdown against dissidents, he was sent into internal exile in the 'closed city' of Gorky. There he undertook a series of hunger strikes in an effort to secure permission for his wife, Yelena Bonner, to receive medical treatment overseas. Under the personal orders of Mikhail Gorbachev, he was eventually released in December 1986. He continued to campaign for improved civil rights and in 1989 was elected to the Congress of the USSR People's Deputies.

SAKMANN, Bert
(1942–)

German electrophysiologist, born in Stuttgart and educated in medicine at Munich. Currently Director of the Max Planck Institute for Medical Research, Heidelberg, his work with Neher revolutionized cell physiology with the invention of the 'patch-clamp' recording technique. This made it possible to record the electrical activity of very small areas of membrane, by eliminating the membrane's electrical noise, which improved the sensitivity of previously available methods by a factor of a million. The patch-clamp method of studying minute changes in electrical current caused by the movement of ions has been widely used to study electrical activity in whole cells. Sakmann and colleagues have studied the relationship between the protein structure of ion channels and their function, and the patch-clamp technique has facilitated their studies of neural transmission in the central nervous system, similar to those achieved quite easily in the more accessible peripheral nervous system. Sakmann and Neher's method has promoted a new approach to research in many fields, including studies of nerve impulse propagation along axons; the process of egg fertilization; the regulation of the heartbeat; and investigations into the cellular mechanisms of disease. In 1991 they shared the Nobel Prize for Physiology or Medicine.

SALAM, Abdus
(1926–)

Pakistani theoretical physicist, born in Jhang. Educated at Punjab University and Cambridge, he became Professor of Mathematics at the Government College of Lahore and at Punjab University (1951–1954). He lectured at Cambridge (1954–1956) and in 1957 became Professor of Theoretical Physics at Imperial College of Science and Technology, London. His concern for his subject in developing countries led to his setting up the International Centre of Theoretical Physics in Trieste in 1964. In 1979 he was awarded the Nobel Prize for Physics, with Steven Weinberg and Glashow. Independently each had produced a single unifying theory of both the weak and electromagnetic interactions between elementary particles. The predictions of the 'electroweak' theory were confirmed experimentally in the 1970s and 1980s.

SALISBURY, Edward James
(1886–1978)

English botanist and ecologist, born in Harpenden, Hertfordshire, and educated at University College London. He was a senior lecturer in botany at East London (now Queen Mary) College (1914–1918) before joining the staff of University College (1918–1943; Professor of Botany, 1929–1943). There, he and Felix Eugen Fritsch (1879–1954) jointly prepared many botanical textbooks, including *Plant Form and Function* (1938). He was particularly interested in the quantity of seeds produced by plants, laboriously counting and weighing the seeds of more than 240 species. The results were published in *The Reproductive Capacity of Plants* (1942). From 1943 to 1956 he was Director of the Royal Botanic Gardens, Kew, and initiated a postwar development phase which included the construction of the Australian House (1952). He continued to publish on plant ecology, including two of his best-known books, *Downs and Dunes* (1952) and *Weeds and Aliens* (1961). A hallmark of all his research was his special gift of synthesizing data from many different fields into a unified whole. His ability to popularize his subject shone in his textbooks and in his synthesis of biology, ecology and horticulture. *The Living Garden* (1935). He founded the British Ecological Society in 1917.

SALISBURY, originally MARKHAM, Richard Anthony
(1761–1819)

English botanist and horticulturist, born in Leeds, the son of a cloth manufacturer. Educated at Edinburgh University, he assumed his surname to fulfil the conditions of a bequest. From 1805 to 1816 he was Secretary of the Horticultural Society of London. His illustrated work *Icones Stirpium Rariorum Descriptionibus* was published in 1791. He wrote extensively producing several works on English botany and classification, including *Prodromus Stirpium* (1796), *Generic Characters in the English Botany* (1806), and *Paradisus Londinesis* (1805–1808), but earned opprobrium for unethical professional behaviour and the unwarranted changing of botanical names. His *Genera Plantarum* was edited and published posthumously (1866).

SALK, Jonas Edward
(1914–)

American virologist, born in New York City, the son of a garment worker. He was educated in medicine at the New York University College of Medicine where he obtained his MD in 1939, and subsequently taught there and at several other schools of medicine or public health. In 1963 he became Director of the Salk Institute in San Diego, California, previously known as the Institute for Biological Studies. Some of his early research had been on the influenza virus, but by the time he moved to California, he had attracted worldwide attention for his work on the 'Salk vaccine' against poliomyelitis. In 1949 it became known that there were three types of polio virus, and it was not until 1954 that Salk developed his killed virus vaccine. To be accepted, this had to overcome opposition which had continued from 1935 when killed and attenuated vaccines given to over 10,000 children had proved ineffective and unsafe. However, the evaluation of the 1954 trial showed that Salk's vaccination was 80–90 percent effective and by the end of 1955, over 7 million doses had been administered. Later the vaccine was overtaken by Sabin vaccine, which used a live attenuated strain and could be given orally instead of by injection, which Salk's vaccine required.

SAMUELSSON, Benqt Ingemar
(1934–)

Swedish biochemist, born in Halmstad. He entered the medical school of the University of Lund where he worked in Bergstrom's laboratory. In 1958 he moved with Bergstrom to the Karolinska Institute in Stockholm. There Samuelsson graduated in medicine (1961) and was appointed Assistant Professor of Medical Chemistry. After a short period of research at Harvard University he returned to the Karolinska Institute. Bergstrom had been working on prostaglandins – substances which act like hormones as chemical messengers, but which have a more localized action. Samuelsson studied the biosynthesis of prostaglandins showing that they are produced from an unsaturated fatty acid arachidonic acid, found in certain foodstuffs. He elucidated the details of the relevant biochemical pathways. Samuelsson and Bergstrom showed that one group of prostaglandins, known as the E series, relaxes blood vessel walls and lowers blood pressure, while another, the F series, has the opposite effect. The E series may be used in treating some circulatory diseases, while the F series has been used to induce abortion. Samuelsson became Professor of Medical Chemistry at the Royal Veterinary College in Stockholm (1967–1972) and then Professor of Chemistry at the Karolinska. He shared the 1982 Nobel Prize for Physiology or Medicine with Bergstrom and Vane.

SANDAGE, Allan Rex
(1926–)

American astronomer, born in Iowa City. He studied at Illinois University and Caltech before joining the Hale Observatories, initially as an assistant to Hubble. In 1960 he made the first optical identification of a quasar. With Thomas Matthews, a junior Colleague, he found a faint optical object at the same location as the compact radio source 3C 48 and found it to have a very unusual spectrum. This was soon shown by Maarten Schmidt to be the result of a huge Doppler redshift, implying that the object, now known as a quasar, is receding from the Earth at enormous speed. This is now interpreted to suggest that it is extremely remote and as luminous as hundreds of galaxies. Sandage went on to identify many more quasars via this peculiarity of their spectra, and showed that most quasars are not radio emitters.

SANDERS, Howard Lawrence
(1921–   )

American marine biologist, born in Newark, New Jersey. He was educated at the universities of British Columbia and Rhode Island, and received his PhD at Yale in 1955. He worked throughout his career at Woods Hole Oceanographic Institution in Massachusetts, where he was appointed as a research associate (1955) and senior scientist (1965). He remained there as Scientist Emeritus following his retirement in 1986. From 1969 to 1980 he simultaneously held posts as Adjunct Professor and research affiliate at the State University of New York, and as Associate Professor at Harvard. His early research concerned the benthic fauna of shallow water invertebrates; Sanders discovered what proved to be the first species of a hitherto undescribed and primitive class of crustacea, the Cephalocarida. Working from the research vessels *Atlantis* and *Chain*, he then focused his attention on obtaining quantitative samples of deep-sea benthos from the north-west Atlantic continental shelf down to abyssal depths. From the earliest samples it became clear that deep-sea communities were characterized by large numbers of very small species which had been overlooked in previous surveys; the patterns of species richness and diversity were extraordinarily high, equalled only by tropical shallow-water ecosystems. These quantitative data, and the development of sampling devices to facilitate the project, revolutionized deep-sea biology. During the mid 1960s, Sanders and colleagues began to formulate the 'stability-time' hypothesis to account for the exceptional levels of species diversity of deep-sea benthic communities. The hypothesis has become a pivotal (if not controversial) concept in marine ecology and was a catalyst in the development of other more general concepts of the structuring of ecological communities. During the 1970s he also became a leading expert on oil spills and their environmental consequences.

SANGER, Frederick
(1918–   )

English biochemist, born in Rendcombe, Gloucestershire. He was educated at the University of Cambridge, where he received his BA in 1939 and a PhD in 1943. As a Quaker, he was exempted from military service during World War II. Since 1951 he has been on the staff of the Medical Research Council in Cambridge. In the 1940s Sanger devised methods to deduce the sequence of amino acids in the chains of the protein hormone insulin. By the 1950s he had deduced the sequence of the 51 amino acids in its two-chain molecule, and found the small differences in this sequence in insulin from pigs, sheep, horses and whales. For this he was awarded the Nobel Prize for Chemistry in 1958. He then turned to the structure of nucleic acids, working on RNA and DNA. Using a highly ingenious combination of radioactive labelling, gel electrophoresis and selective enzymes, his group was able, by 1977, to deduce the full sequence of bases in the DNA of the virus phi X 174, with over 5,400 bases. Such methods led to the full base sequence of the Epstein-Barr virus by 1984. For his nucleic acid work Sanger shared (with Walter Gilbert and Berg) the 1980 Nobel Prize for Chemistry, and became the first scientist to win two Nobel prizes in this field.

SARS, Michael
(1805–1869)

Norwegian marine biologist, born in Bergen, son of a sea captain. He studied theology and between 1828 and 1854 was vicar at several Norwegian seaside communities. It was during this period that he carried out much of his research into marine biology. In 1854 he was made Extraordinary Professor of Zoology at Christiania (now Oslo) University. He was one of the founders of marine zoology and made numerous expeditions to collect specimens. He made several trips to the northern seas and to the Mediterranean, to the Adriatic (1851) and to Naples and Messina (1852–1853). His particular interests were in the migration of marine organisms and the life cycles of marine invertebrates. Because of the dissimilarity of larval and adult forms, the relationship between the two had scarcely been realized before the work of Sars. He demonstrated for the first time the phenomenon of alternation of generations in the coelenterates, i.e. the existence of sessile and free-living forms succeeding one another in the same species. He was first to describe the veliger larva of marine molluscs (1837) and the bipinnaria larva of starfish (1844). Using a deep-sea dredge he collected many specimens from depths of up to 450 fathoms with his son Georg, and revealed a hitherto unsuspected fauna. The most spectacular discovery was made from 300 fathoms near Lofoten; the first specimen of a living crinoid, a form previously believed to have become extinct during the Mesozoic era. With his son George, who himself became a well-known marine biologist, he published *On Some Remarkable Forms of Animal Life from the Great Deeps off the Norwegian Coast* (1868).

SAUSSURE, Horace Benedict de
(1740–1799)

Swiss physicist and geologist, born in Conches, near Geneva. At the age of 22, following a distinguished college career, he became Professor of Physics and Philosophy at Geneva (1762–1788), although his first love was botany. He travelled in Germany, Italy and England; he crossed the Alps by several routes and ascended Mont Blanc (1787). *Voyages dans les Alpes*, describing his observations on minerology, geology, botany and meteorology, was published in several volumes between 1779 and 1796. Saussure explained the Alpine topography by erosion and called attention to the evidence of great horizontal disturbance in the strata, but wrongly asserted that a diluvial current rather than glacial action had distributed the boulders around the Alps. The mineral saussurite is named after him.

SAVART, Felix
(1791–1841)

French physician and physicist, born in Mezieres in the Ardennes, the son of a military engineer. Even during his medical studies at the military hospital at Metz and at the University of Strasbourg, he showed an early interest in the physics of sound, in particular of that of a violin, and he presented several papers on the subject. He succeeded Fresnel as a member of the Paris Academy (1827), and was appointed Professor of Experimental Physics at the College de France in Paris in 1828. He invented 'Savart's wheel' for measuring tonal vibrations, and the 'Savart quartz plate' for studying the polarization of light. With Biot he discovered the law (named after them) defining the intensity of magnetic field produced at a given point near a long straight current-carrying conductor. His brother, Nicolas, an officer in the engineering corps, also studied vibration.

SAYERS, James
(1912–)

English physicist, born in Ballymena, County Antrim, Northern Ireland. He was a member of the British team associated with the Manhattan project at Los Alamos in World War II, to develop the nuclear bomb. He became Professor of Electron Physics at Birmingham (1946) and in 1949 received a government award for his work on the cavity magnetron valve, which was of great importance in the development of radar.

SCHALLY, Andrew Victor
(1926–)

Polish-born American biochemist, born in Wilno (now Vilnius, Lithuania). He fled from Poland at the time of the German invasion in 1939 and studied at the National Institute for Medical Research in London and McGill University in Montreal. He later worked at the Baylor Medical School (1957–1962) and Tulane University (from 1962). Following a suggestion by Geoffrey Harris in Cambridge that special 'releasing factors' from the brain hypothalamus stimulate the release of hormones from the pituitary gland, Schally, and independently Guillemin, discovered an assay system for this in vitro (1955). Thus began their well-documented, highly competitive race to isolate corticotrophin-releasing hormone (CRH), one of several such factors. Schally used a million pig pituitaries and reported the isolation and structure of other factors, thyrotropin-releasing hormone (TRH) in 1969, luteinizing-hormone-releasing hormone (KH-RH) in 1971, and somatostatin, which inhibits the release of growth hormone (1976). CRH was eventually isolated by W. Vale in 1981. Schally's success depended heavily on the participation of Folkers and a talented Japanese chemist, Matsuo. He shared the 1977 Nobel Prize for Physiology or Medicine with Guillemin and Yalow. Schally also studied the distribution and function of adrenocortical trophic hormone (ACTH), somatostatin and other factors.

SCHAUDINN, Fritz Richard
(1871–1906)

German zoologist and microbiologist, born in Roseningken, East Prussia. He studied philology at the University of Berlin, but later turned to zoology, and after research work in Berlin he became Director of the Department of Protozoological Research at the Institute for Tropical Diseases in Hamburg (1904). Schaudinn demonstrated the amoebic nature of tropical dysentery and distinguished the organism which causes the disease, *Entamoeba histolytica*, from its beneficial relative *Escherichia coli* which inhabits the human intestinal lining. With the dermatologist Erich Hoffmann, he discovered the spirochaete which causes syphilis (1905), *Spirochaeta pallida*, now known as *Treponema pallidum*. A treatment for the disease, first used on patients in 1911, was subsequently developed by Ehrlich. In a wide range of investigations, Schaudinn demonstrated that human hookworm infection is contracted through the skin of the feet, researched malaria and made important contributions to zoology.

SCHAWLOW, Arthur Leonard
(1921–)

American physicist, and co-inventor of the laser, born in Mount Vernon, New York. He studied at Toronto and Columbia (1949–51) where he worked with Townes and married his sister. He moved to Bell Telephone Laboratories (1951–1961) and subsequently became Professor of Physics at Stanford University in 1961. Townes and Schawlow collaborated to extend the maser principle (Microwave Amplification by Stimulated Emission of Radiation) to light, thereby establishing the feasibility of the laser (Light Amplification by Stimulated Emission of Radiation). Although they played a central role in laying down the theoretical framework of the laser and had started to construct one, it was Maiman who constructed the first working ruby laser in 1960 at Hughes Research Laboratories in California. From the early 1970s Schawlow worked on the development of laser spectroscopy. Emitting light of a very narrow range of frequencies and in an intense beam, lasers were an attractive light source for investigating the energy levels of electrons in atoms and molecules. The full potential of laser spectroscopy was not, however, realized until a new type of laser using organic dyes had been developed. The wavelengths of the light these dye lasers produced could be tuned over a significant range. Working with German-born physicist Theodor Hansch, Schawlow made precise measurements of the energy levels the electron can occupy in the hydrogen atom, allowing the value of the Rydberg constant to be determined with unprecedented accuracy. For his work on laser spectroscopy Schawlow shared the 1981 Nobel Prize for Physics with Bloemergen and Kai Siegbahn.

SCHEELE, Carl Wilhelm
(1742–1786)

Swedish chemist who discovered chlorine, preceded Priestley in the preparation of oxygen, and identified many important, chemical compounds. He was born in Stralsund (now in Germany) and was apprenticed to an apothecary, later working at Malmo, Stockholm, Goteborg and Uppsala, where he arrived in 1770. There he was granted the use of a laboratory and met Gahn. In 1775, the year that he was elected to the Stockholm Royal Academy of Sciences, he moved to Koping, where he became the town pharmacist. In the 1760s Scheele began to investigate air and fire, and soon came to doubt the received view, first propounded by George Stahl, that substances contain a vital essence (referred to as phlogiston) which they lose when they burn. By 1772, according to his laboratory notes he had realized that air was a mixture of two compounds, one of which encouraged burning and another which prevented it. He prepared a gas identical to the inflammable component of air from a number of compounds including mercuric oxide and saltpetre (potassium nitrate). His discovery was made known to the scientific world by Berginan some months before Priestley announced a similar discovery in August 1774. In a letter written in September the same year, Scheele himself passed on information about his experiments to Lavoisier, who subsequently discovered the true nature of combustion and named the new flammable gas 'oxygine'. Even without oxygen Scheele's discoveries would have placed him in the first rank of chemists. In 1770 he encouraged Gahn to look for phosphorus (which had been recently discovered by Brand) in animal bones and subsequently developed a method of extracting it, a very useful discovery as previously the only known source had been urine. In 1771 Scheele's studies of fluorine led to the discovery of hydrofluoric acid. In 1773 he discovered chlorine in the course of experiments with pyrolusite (a manganese ore) which he treated with hydrochloric acid. He noticed that the new gas dissolved gold and also acted as a bleach. In 1775 his extensive experiments on arsenic resulted in the discovery of copper arsenide, a green pigment which was known as 'Scheele's green'. In his only book, *Abhandlung von der Luft und der Feuer* (1777, translated as *Chemical Observations and Experiments on Heat, Light and Fire* 1780), he gave the first accurate description of hydrogen sulphide, which he had prepared by heating sulphur in hydrogen. In 1781 Scheele distinguished between two very similar minerals, plumbago (graphite) and molybdena, discovering the metal molybdenum in the process. His investigations of plant and animal material were fundamental to the development of organic chemistry. He also devised new methods of analysing fragile organic material, avoiding strong heat and often precipitating the new substance from solution. In this way he identified many important organic acids including citric, oxalic, lactic and tartaric acids. He isolated prussic (hydrocyanic) acid in 1783. He died in Koping.

SCHEUTZ, Pehr George
(1785–1873)

Swedish lawyer, publisher, and builder of a calculating machine. Born in Jonkoping, he was educated in law at Lund. In the 1830s he read about Babbage's 'difference engine' and then designed his own machine, later completed by his son, Edvard Scheutz.

SCHIAPARELLI, Giovanni Virginio
(1835–1910)

Italian astronomer, born in Savigliano, Piedt. He graduated at Turin in 1854 and later worked under Wilhelm Struve at Pulkova before becoming head of the Brera observatory, Milan, from 1860. Schiaparelli worked on the relationship between meteors and comets, and was the first to realize that the daily variation of the meteor count rate was only to be expected if the causative dust particles were moving on orbits similar to those of the short-period comets rather than the planets. He also provided the first identification of a specific meteoroid stream with a specific comet, the pair being the Perseids and comet Swift-Tuttle (1862). In 1877, having re-equipped the observatory, Schiaparelli began observations of Mars in order to produce a detailed map. Not only did he detect linear markings on the surface that he termed *canali* (i.e. channels), but he also noticed that they changed as a function of the martian season, sometimes splitting into two and sometimes disappearing altogether. From his 1882–1889 observations of Mercury he concluded that this planet orbited the Sun in such a way that one of its hemispheres always pointed sunwards; he found the same effect occurring for Venus.

SCHIKARD, Wilhelm
(1592–1635)

German polymath, born in Herrenberg. Educated at the University of Tubingen, he was the pioneer in the construction of calculating machines. His 'calculating clock' was designed and built around 1623, but lay forgotten until the discovery of Schikard's papers allowed its reconstruction in 1960.

SCHIMPER, Andreas Franz Wilhelm
(1856–1901)

German botanist, son of Wilhelm Philipp Schimper, born in Strassburg (now Strasbourg, France). He was educated there under de Bary and later accepted a post of the botanic gardens in Lyons, but soon moved to work with Sachs at the University of Wurzburg. He visited Florida and the West Indies, where his interest in plant geography was first awakened. He also studied epiphytic and carnivorous plants. In 1882 he went to work with Strasburger in Bonn. He continued to travel widely, visiting Brazil, Ceylon and Java, then joining the Valdivia expedition to study the planktonic vegetation of the Canaries, Kerguelen, Seychelles, the Congo, Sumatra and Cape Province. Professor at Basle (1898–1901), he was noted as a plant geographer, and divided the continents into floral regions. He also proved, in 1880, that starch is a source of stored energy for plants. He published *Pflanzengeographie auf physiologischer Grundlage* ('Plant Geography on Physiological Principles') in 1898.

SCHIMPER, Karl (or Carl) Friedrich
(1803–1867)

German naturalist and poet, born in Mannheim, cousin of Wilhelm Philipp Schimper. He studied theology at Heidelberg, and collected plants in the south of France, where he studied medicine. While teaching at Munich he explored the Alps under the service of the King of Bavaria. He is remembered as a pioneer of modern plant morphology. He was notable for his work on phyllotaxis, and in geology for his theory of prehistorical alternating hot and cold periods. Despite his talents, he failed to secure any academic post. Many of his scientific ideas were published as poems; several hundred are known. He experimented with dioecious mosses, and discussed the arrangement of leaves on plant stems according to geometrical principles. He started many projects, but completed almost none. He contributed to F. W. L. Succow's *Flora Manhemiensis*, and to F. C. L. Spenner's *Flora Friburgensis*. The modern concepts of ice ages and climatic cycles were probably originated by Schimper; he also speculated that mountain building was due to the contraction of the Earth.

SCHIMPER, Wilhelm Philipp
(1808–1880)

German botanist, born in Dosenhain, cousin of Karl Friedrich Schimper and father of Andreas Schimper. He studied philosophy, philology, theology and mathematics at the University of Strasbourg, where he became Director of the Natural History Museum in 1835. He was an authority on mosses and co-authored *Bryologia Europaea* (6 vols, 1836–1855). He studied the Triassic flora of the Vosges region, and contributed the volume *Paleoophytologie* to Zittel's *Handbuch der Paleoontologie* in 1879. He was a superbly competent observer, making exceptionally detailed descriptions, but he was not an original thinker.

SCHMIDT, Bernhard Voldemar
(1879–1935)

Estonian optical instrument-maker, born on the island of Naissaur near Tallin. Having worked, on leaving school, as a telegraph operator and photographer, he studied at the Chalmers Institute in Goteborg, Sweden, and joined the engineering school at Mittweida in Germany (1901) where he established a reputation as an optician and where in 1926 he installed a small observatory. In the same year he joined the staff of the Hamburg Observatory in Bergedorf, where he developed his idea of a coma-free mirror and completed his first 0.5 metre Schmidt telescope in 1932. His optical system overcame the aberrations produced by spherical mirrors by introducing a specially shaped correcting plate at their centre of curvature. His invention was of great importance to optical astronomy, as it provided extremely fine image definition over a field of several degrees. The best-known Schmidt telescope is that on Mount Palomar (1949), with an aperture of 1.2 metres and focal length 3 metres used for the photographic survey of the northern sky. The UK Schmidt telescope of the same dimensions sited in New South Wales, Australia (1973), extended the sky survey to the southern hemisphere.

SCHMIDT, Jonannes
(1877–1933)

Danish fisheries biologist, born in Jagerspris. Schmidt's work focused on problems relating to the early life history of marine fish. In 1904 he captured a *Leptocephyalus* (a small leaf-shaped fish which later undergoes metamorphosis to become an elver) off the coasts of the Faroes whilst surveying for cod eggs. That observation started him on a 20-year investigation throughout the North Atlantic on the research vessel *Dana*, which culminated in his discovery of the true breeding area of the *Anguilla anguilla* eel. Previously, Italian scientists had erroneously suggested that the spawning grounds of the European eel lay in the vicinity of Messina. Schmidt revealed that breeding actually occurs in the Atlantic Ocean in the vicinity of the Sargasso Sea, at approximately 400 metres depth and at a temperature of 17 degrees Celsius. It is now known that spawning occurs there in February each year; larvae rise to the surface where they are transported by the Gulf Stream towards the coats of Europe and North Africa. After drifting for approximately 22 months the *Leptocephali* arrive over the European continental shelf in November, and there they metamorphose to elvers and start to seek fresh water. The elvers enter European rivers in the early spring of the following year when they are about 28 months old. No mature eels ever re-enter fresh water and therefore we know that, although spawning adults have never been found, they must die after reproduction at sea.

SCHMIDT, Maarten
(1929–)

Dutch-born American astronomer, born in Gröningen. He was educated there and at Leiden, and moved to the USA in 1959 to join the staff of the Hale Observatories. He became Professor of Astronomy in 1964 and Director of the Hale Observatories in 1978. Schmidt made important contributions in the study of galactic structure and dynamics by analysing the distribution of matter in our galaxy, but is best known for his astounding results in the study of quasars. He studied the spectrum of an optically identified quasar and discovered that the peculiarities of its spectrum were caused by a massive Doppler redshift: it appeared to be receding from the Earth at nearly 16 percent of the speed of light. Such high velocities are now interpreted as implying that quasars are very distant objects, which must therefore be as luminous as hundreds of galaxies to be visible on Earth. He also found that the number of quasars increases with distance from Earth, providing evidence for the 'Big Bang' theory for the origin of the universe.

SCHOENHEIEMR, Rudolf
(1898–1941)

German-born American biochemist, born in Berlin. He studied there and taught in Germany for 10 years before moving to the USA in 1933. There, working at Columbia University with Urey, he used two new isotopes discovered by Urey (deuterium, to replace stable hydrogen atoms in a molecule, and heavy nitrogen) to trace biochemical pathways (1935). His work showed that many materials of the human body (e.g. depot fats, proteins and bone) previously regarded as static, are steadily degraded and replaced (1935–1937). New dietary lipid was found to be partly stored and partly utilized along with lipid released from the body's fat depots. He distinguished the pathways of unsaturated and saturated fatty acids from the rate of incorporation of deuterium from labelled body water, and showed that the former cannot be intermediates in the production of the latter (1937). Schoenheimer also established that the sterol coprostanol is produced from cholesterol by gut bacteria, and that vitamin D in cows' milk is not derived from the potential dietary precursor in plants, ergosterol (1929). He published *The Dynamic State of Body Constituents* (1942). His isotopic tracer techniques, although requiring careful interpretation, became widely used for elucidating biochemical pathways (e.g. by Konrad Bloch). Schoenheimer committed suicide during World War II.

SCHÖNBEIN, Christian Friedrich
(1799–1868)

German chemist, born in Metzingen, Wurttemberg. He trained as a pharmacist, worked in two chemical factories and taught at a technical institute at Keilhau. He also taught and travelled in England and France, and was professor at Basle from 1835 until his death. He conducted research in many areas, but is mainly remembered for his discovery of ozone in 1839. He noticed that the oxygen obtained by the electrolysis of water had a peculiar smell and discovered that this was due to the presence of a second gas (ozone) produced at the anode. He found that this gas resembled chlorine and bromine in its chemical properties, but believed it to be a compound until Jean Charles Galissard de Marignac and August Arthur de la Rive showed in 1845 that the same gas can be produced by passing electric sparks through pure, dry oxygen. Schönbein then realized that ozone is an allotropic form of oxygen (the same element but in a different physical form). In 1845 he treated cotton-wool with a mixture of sulphuric acid and nitric acid washing out the excess acid. The result was nitrocellulose (guncotton), a highly inflammable fluffy white substance which was soon widely used as an explosive. It is still used as a propellant, sometimes on its own and sometimes in conjunction with nitroglycerine (invented by Sobrero and developed by Nobel). By treating cotton-wool with less nitric acid, Schonbein made collodion, which found uses in photography and medicine. He was also one of the pioneers of chromatography, which depends on the fact that the solution of any substance has a characteristic capillarity. When the edge of a paper is dipped into a solution containing a mixture of substances, the substances separate out, each rising by different amounts up the paper. This technique, which he developed from the work of Kunge, was further refined by Tswett. Schönbein died in Sauersberg, near Baden-Baden.

SCHRIEFFER, John Robert
(1931–)

American physicist, born in Oak Park Illinois. He graduated in electrical engineering and physics at MIT, and studied superconductivity for his PhD under Bardeen at Illinois University. Collaboration with Bardeen and Cooper led to the BCS (Bardeen–Cooper–Schrieffer) theory of superconductivity. The first steps towards this successful theory were made by Cooper, who showed that although the like charges of two electrons cause them to repel each other, when an electron interacts with the positive charges of a metal lattice it can deform the lattice, with the net effect that pairs of electrons experience an overall attraction. These stable couples of electrons become known as 'Cooper pairs'. Using a statistical approach. Schrieffer succeeded in generalizing the theory from a description of the properties of a single Cooper electron pair to that of a solid containing many pairs. In an intensive month of work following Schrieffer's breakthrough, Bardeen, Cooper and Schrieffer were able to show that their theory accounted for all the experimentally observed superconducting phenomena. For the development of the BCS theory all three shared the 1972 Nobel Prize for Physics. Following postdoctoral work in Europe and short-term posts in the USA, Schrieffer held professorships at the University of Pennsylvania, Cornell and the University of California. Since 1992 he has been University Professor at Florida State University. He is the author of *Theory of Superconductivity* (1964), and has also worked on dilute alloy theory, surface physics and ferromagnetism.

SCHROCK, Richard
(1942–)

American organometallic chemist, born in Indiana. He received his BA from University of California at Riverside in 1967 and his PhD from Harvard in 1971. Following a year at Cambridge University, he moved to the Central Research and Development Department of Du Pont. In 1980 he was appointed to a professorship at MIT, where he became Frederick G. Keyes Professor of Chemistry in 1989. His research centres upon the chemistry of high-oxidation-state early-transition-metal complexes. In particular he has studied those complexes which contain alkylidene, alkylidyne or dinitrogen ligands. His work is of much significance to industrially important catalytic processes, including the controlled polymerization of olefins and acetylenes. Schrock was the first recipient of the American Chemical Society Award for Organometallic Chemistry in 1985, and is a member of the US National Academy of Sciences.

SCHRÖDINGER, Erwin
(1887–1961)

Austrian physicist, born in Vienna where he was educated by a private tutor before entering Vienna University, receiving his doctorate in 1910. He remained at Vienna as a researcher until 1914, when he joined the Austrian army as an artillery officer. From 1920 he held chairs at Stuttgart (1920), Jena (1920–1921), Breslau (1921) and Zurich (1921–1927) universities. In 1926 he published a series of papers which founded the science of quantum wave mechanics, and shortly after succeeded Planck as Professor of Physics at the University of Berlin. In 1933 during Hitler's rise to power, he accepted an invitation to Oxford (1933–1936) but decided to return to Austria as professor at Ganz University (1936–1938). When Austria was invaded, he fled to Dublin where he worked at the Institute for Advanced Studies which was created for him (1938–1956). He retired in 1956 and returned to Austria as Professor Emeritus at Vienna University. Schrödinger's work was inspired by de Broglie's proposal that particles have a dual nature and in some circumstances will behave like waves: Schrödinger introduced his celebrated wave equation which describes the behaviour of such systems. When applied to the hydrogen atom it correctly predicts the observed energy levels of the electrons, without using the unacceptable assumptions of Niels Bohr's model of the atom. Schödinger showed that the wave equation was mathematically equivalent to the matrix approaches developed almost simultaneously by Heisenberg, Born and Pascual Jordan. Dirac soon developed a more complete theory of quantum mechanics from their foundations, and for this work Schrödinger and Dirac shared the 1933 Nobel Prize for Physics. Born concluded that the wavefunction solutions to Schrödinger's equation describe probability waves, so that for a particle in a given state at a particular time, it is only possible to predict the most likely state of the particle in the future. Schödinger disliked this statistical approach to quantum mechanical problems and preferred the more deterministic approach of classical physics. He wrote *What is Life* (1946) and *Science and Man* (1958).

SCHULTZE, Max Johann Sigismund
(1825–1874)

German zoologist, born in Frieburg. He studied at Greifswald and Berlin, and taught at Bonn, where he became Director of the Anatomical Institute in 1877. He studied the anatomy of a variety of animals and in particular single-cell ones. This led him to define the cell (1861), with a structure of a nucleus surrounded by protoplasm, as the basic building block of all living organisms. A histologist, he used osmic acid as a stain for nervous tissues and demonstrated the nerve endings in the basilar membrane of the ear. He also showed that the retina of birds possesses two different sensory nerve endings, the rods and cones, to which he ascribed separate functions. He proposed the duplicity theory of vision in 1866, a forerunner of modern theories.

SCHUSTER, Sir Arthur
(1851–1934)

British physicist, born in Frankfurt of Jewish parents. He studied at Heidelberg and Cambridge and became Professor of Applied Mathematics (1881) and Professor of Physics (1888) at Owens College, Manchester, resigning his chair in 1907 to make way for Rutherford. He carried out important pioneering work in spectroscopy which showed that an electric current is conducted through gases by ions, and that once 'ionized', the current could be maintained by a small potential. In terrestrial magnetism, he showed that there were two kinds of daily variations; atmospheric variations caused by electric currents in the upper atmosphere, and internal variations due to induction currents in the Earth. The Schuster-Smith magnetometer is the standard instrument for measuring the Earth's magnetic field. He led the eclipse expedition to Siam in 1875. He was elected FRS in 1879, and was the founder (and the first Secretary) of the International Research Council. He was knighted in 1920.

SCHWANN, Theodor
(1810–1882)

German physiologist, famous for the cell theory. Born in Neuss, Schwann was educated in Cologne and studied medicine, graduating in Berlin in 1834. He remained in Berlin for four years as assistant to Johannes Muller. In his Berlin years, he studied digestion, and isolated from the stomach lining the enzyme pepsin. He later showed the role of yeast cells in producing fermentation; developing the experiments of Spallanzani, he thereby cast doubt upon the idea of spontaneous generation, confirming that no micro-organisms were produced and no putrefaction ensued in a sterile broth. Some years later, Pasteur's work finally destroyed the theory of spontaneous generation. Schwann also discovered the 'Schwann cells' composing the lyelin sheath around peripheral nerve axons, and he showed an egg to be a single cell which, once fertilized, evolves into a complex organism. Schwann's most renowned work, however, was on cell theory. Following the botanist Schleiden, who had argued that all plant structures are cells, Schwann was persuaded that animal tissues are also based on cells and he became a leading advocate of cell theory. In a major book of 1839 he contended that the entire plant or animal was comprised of cells, that cells have in some measure a life of their own, but that the life of the cells is also subordinated to that of the whole organism. The cell theory became pivotal to nineteenth-century biomedicine. Virchow, who in 1855 said 'all cells arise from pre-existing cells', was to demonstrate how the study of affected cells was central to pathology and physiology. Like Schleiden, Schwann believed cells reproduced by budding from a nucleus; the concept of the formation of cells by division came only later, with Remak and Virchow. Wounded by attacks from the chemists Liebig and Wohler over his fermentation ideas, Schwann despaired of a career in Germany, in 1838 emigrating to Belgium, where he became professor at Louvain, and in 1848 at Liege. Increasingly solitary and depressed, he did little more science.

SCHWARTZ, Melvin
(1932–)

American physicist, born in New York City. He was educated at the University of Columbia where he received his PhD in 1958 and later became Professor of Physics (1963–1966, 1991–). From 1966 he held professorships at Stanford University, but in the early 1980s he left academic research to work in the computer industry until 1991, when he became Associate Director of High Energy and Nuclear Physics at Brookhaven National Laboratory. He was awarded the 1988 Nobel Prize for Physics jointly with Lederman and Steinberger for carrying out the experiment which demonstrated the existence of the muon neutrino (1962). His other interests have included the study of CP violation.

SCHWARZ, Harvey Fisher
(1905–1988)

American electrical engineer, born in Edwardsville, Illinois. He was the co-inventor (with William J. O'Brien) of the Decca radio-navigation system for ships and aircraft. He studied electrical engineering at Washington University, St Louis. Working for the General Electric Company in Schenectady, New York, he helped to develop 'Radiola 44', the first domestic radio receiver to use the newly invented screen-grid valve. As chief engineer of Brunswick Radio Corporation, he was sent to Britain in 1932 to design radios and radiograms for manufacture in the UK, and made his home there for the rest of his life although remaining a US citizen. During World War II, working for Decca, he and O'Brien developed a prototype radio-navigation system that was put into operation for the first time during the D-Day landings in the seaborne invasion of Normandy in 1944.

SCHWARSCHILD, Karl
(1873–1916)

German astronomer, born in Frankfurt; he was the first to predict the existence of 'black holes'. He became interested in astronomy as a schoolboy and had published two papers on binary orbits by the time he was 16. Educated at the universities of Strassburg and Munich, he was appointed Director of the Göttingen Observatory (1901) and the Astrophysical Observatory in Potsdam (1909). He volunteered for military service in 1914 at the beginning of World War I and returned home in 1916 after contracting a rare skin disease, from which he died. His lasting contributions are theoretical and were largely made during the last year of his life. In 1916, while serving on the Russian front, he wrote two papers on Einstein's general theory of relativity, giving the first solution to the complex partial differential equations of the theory. He also introduced the idea that when a star contracts under gravity, there will come a point at which the gravitational field is so intense that nothing, not even light, can escape. The radius to which a star of given mass must contract to reach this stage is known as the Schwarzschild radius. Stars that have contracted below this limit are now known as black holes.

SCHWEIGGER, Johann Salomo Christoph
(1779–1857)

German physicist, born in Erlangen, the son of a prominent theologian and Protestant minister. He was appointed Professor of Chemistry and Physics at the Physikotechnisches Institut in Nuremberg (1811–1816), at Erlangen (1818), and at Halle (1819), where he remained until his death. He became best known as the founder of the *Journal für Chemie und Physik* (1811–1828), and as one of the inventors of the simple galvanometer (1820), which he called his 'doubling apparatus' as the needle was deflected by a multiturn coil of insulated wire (Oersted's original magnetic needle was deflected by a single wire). Poggendorff invented a similar instrument which he called the 'condenser', analogous to Volta's electrostatic condenser for magnifying electrostatic phenomena. Improvements followed rapidly, including by Ampere, Avogadro, Oersted and Nobili, who designed the common astatic form of this instrument (1825), which was made popular in England by Wheatstone.

SCHWINGER, Julian
(1918–)

American physicist, born in New York City. He studied at Columbia University where he received his PhD in 1939. He worked at the University of California at Berkeley under Oppenheimer and during World War II at the MIT Radiation Laboratory and the Metallurgical Laboratory of the University of Chicago. He moved to Harvard in 1945 where he became professor in 1947, and since 1972 he has held chairs at the University of California at Los Angeles. Independently of Feynman and Tomonaga, he developed a theory of quantum electrodynamics describing the interaction of light and matter. Problems had arisen when Dirne's early theoretical work in this field produced infinite values in predictive calculations. Schwinger showed that by reformulating the calculations so that the infinites corresponded to measurable quantities, such as the electron mass and the electron charge, then by replacing these quantities with their known values, finite answers could be found. This mathematical process for avoiding the problem of infinities is known as renormalization and is an important tool in theoretical physics. Schwinger was awarded the 1965 Nobel Prize for Physics jointly with Feynman and Tomonaga for the development of quantum electrodynamics. He went on to study synchrotron radiation.

SCORESBY, William
(1789–1857)

English Arctic explorer and scientist, born near Whitby. As a boy he went with his father a, whaling captain, to the Greenland seas. He studied chemistry and natural philosophy at Edinburgh University, and made several voyages to the whaling grounds which he published in *An Account of the Arctic Regions* (1820), the first scientific accounts of the Arctic seas and lands. In 1822 he surveyed 400 miles of the east coast of Greenland. After a period of study at Cambridge, and after being ordained (1825), the held various charges at Exeter and Bradford but continued his scientific investigations, travelling to Australia in 1856 to study terrestrial magnetism. He became involved in the controversy about the fate of the explorer Sir John Franklin and his companions who had disappeared while searching for the North-West Passage (1845), which induced Scoresby to write *The Franklin Expedition* (1851). He also worked on improving the marine compass, in particular its magnetic needle, and the reliability of such compasses on iron ships. He was elected FRS in 1824.

SEABORG, Glen Theodore
(1912–)

American atomic scientist who synthesized plutonium, born in Ishpeming, Michigan. He studied at the University of California at Los Angeles and took his doctorate at Berkeley where he became an instructor in chemistry in 1939, professor in 1945, and chancellor from 1958 to 1961. Following the work of Soddy, his earliest work was on isotopes, discovering many previously unknown isotopes among the common elements. In 1939 he began to follow up the work of Fermi in Italy who had attempted and failed to synthesize elements heavier than uranium (the heaviest naturally occurring element, atomic number 92) by bombarding it with neutrons. Instead he had succeeded in splitting uranium into smaller atoms, with the release of a great amount of energy. In 1940 McMillan and Abelson, working with the Berkeley cyclotron, synthesized the first 'transuranic' element, neptunium (atomic number 93). The following year Seaborg synthesized the next transuranic element in the series (atomic number 94) by bombarding uranium with deuterons in the Berkeley cyclotron. Named plutonium, it has a fissile isotope plutonium-239 whose destructive power Seaborg instantly recognized. Around the same time Seaborg and his team also discovered that a rare isotope of uranium, uranium-235, is fissile. The USA therefore found itself with two possible sources of power for a nuclear weapon. Seaborg was transferred by the government to the Manhattan Project, a group of scientists gathered at the metallurgical laboratory of the Manhattan District of the Corps of Engineers at the University of Chicago to manufacture an atomic bomb. He was part of Fermi's team which achieved the first chain reaction in uranium-235 on 2 December 1942. He was also in charge of developing a technique to separate plutonium after it had been synthesized in uranium-238. It was his laboratory which, in 1945, produced enough plutonium for the first atomic bomb. Seaborg and his team continued research on further transuranic elements and in 1944 synthesized americium (atomic number 95) and cirium (96). In 1950, by bombarding these with alpha rays, they produced berkelium (97) and californium (98). They later produced einsteinium (99), fermium (100), mendelevium (101) and unnilhexium (106). The heaviest of these elements, which have a very short half-life (the half-life of unnilhexium, for example, is less than 1 second), could only be prepared in microscopic quantities. In 1951 Seaborg shared the Nobel Prize for Chemistry with MacMillan. He was Chairman of the US Atomic Energy Commission from 1961 to 1971.

SECCHI, Angelo
(1818–1878)

Italian astronomer, born in Reggio Emilia. He joined the Society of Jesus in 1833 and in 1841 was made Professor of Physics and astronomy at the Jesuit College in Loreto. In 1948 he became Professor of Astronomy at the Roman College of the society but, with all the Jesuits, he was expelled from Italy in the same year and spent a brief exile in England and the USA before returning to become director of the observatory of the Roman College in 1849. He established a new observatory on the roof of the church of Saint Ignatius with a 10-inch refractor which he used particularly for observations of the Sun. At the total solar eclipse of 1860 observed in Spain he succeeded in photographing the prominences and the corona. Using an objective prism he observed several thousand stellar spectra and divided the stars into three types – white, yellow and red – which corresponded roughly to their temperatures. He thus initiated the field of spectral classification which ensured his place as a pioneer of astronomical spectroscopy. His beautifully illustrated book on the Sun, *Le Soleil* (1875), became widely known. Apart from his astronomical researches, Secchi did much work on the phenomena of terrestrial magnetism and meteorology.

SEDGWICK, Adam
(1785–1873)

English geologist, born in Dent, Cumbria. Having graduated in mathematics from Trinity College, Cambridge (1808), he became Woodwardian Professor of Geology there in 1818. In 1831 he began his geological mapping in Wales and introduced the Cambrian system in 1835. He had carried out studies in the Lake District as early as 1822, but it was not until the Cambrian and Silurian systems had been established in Wales and the Welsh Borders that he was fully able to understand its geology. Sedgwick became embroiled in controversy with Murchison; the dispute was finally resolved with the introduction of the Ordovician system by Charles Lapworth. His best work was on *British Palaeozoic Fossils* (1854). With Murchison he studied the Lake District, the Alps and south-west England, where they identified the Devonian system.

SEEBECK, Thomas Johann
(1770–1831)

Estonian-born German physicist, born in Tallin. A member of a wealthy merchant family, he went to Germany to study medicine, qualifying in 1802, but spent his time thereafter in research in physics, Like Oersted, he was inspired by the German Romantic movement, in particular by Johann von Goethe's anti-Newtonian theory of colours, so that his first research (1806) was on the heating and chemical effects of the colours of the solar spectrum. He investigated optical polarization in stressed glass (1812), but this work had been largely anticipated by Brewster and Biot. His most significant discovery was that of thermoelectricity, which he called 'thermomagnetism' (1822) as he did not believe that an actual electric current was being generated when heat was applied to a junction of two metals, such as bismuth and copper. Apart from the impact it had on theory, the thermoelectric effect is now much used in thermocouples for temperature measurement.

SEFSTRÖM, Nils Gabriel
(1765–1829)

Swedish physician and chemist, born in Ilsbo, North Helsingland. He qualified as a physician at Uppsala in 1813. He became a lecturer in chemistry at the Royal Military Academy at Carlberg and was later appointed as professor at the Artillery School at Marieberg, while also teaching at the new School of Mines at Falun. Researching both at Falun and in Berzelius's laboratory in Stockholm, he found a hard white metal in Swedish iron and in the spoil from an ironworks. This metal had previously been described in 1801 by the Spanish mineralogist del Rio who had not recognized it as an element. Sefström called it vanadium. Only a few grams of it were found and it was Berzelius not Sefström, who succeeded in examining its compounds and who realized that it resembled chromium. It was many years before its atomic weight and other properties were accurately described by Roscoe. It is now important industrially – vanadium is alloyed to steel when very high strengths are required. Sefström died in Stockholm.

SEGRÈ, Emilio
(1905–1989)

American physicist, born in Rome. He was educated at Rome University, studying engineering then physics, and obtained his doctorate in 1928. He remained at the University of Rome working with Fermi until 1936, when he was appointed director of the physics laboratory in Palermo, but in 1938 he was dismissed from this post under Musolini's regime. He moved to the University of California at Berkeley where he remained from 1938 to 1972, apart from a period working on the Manhattan atom bomb project during World War II. In 1937 Segrè discovered the first entirely man-made element technetium. Three years later he was involved in the discoveries of astatine and plutonium (1940). He was also instrumental in devising chemical methods for dividing nuclear isomers. In 1955, using the new bevatron particle accelerator at the University of California, the research team led by Segrè discovered the antiproton, the antiparticle of the proton (with identical mass but negative electric charge) which had been predicted by Dirac. For this work he shared the 1959 Nobel Prize for Physics with Chamberlain.

SELYE, Hans Hugo Bruno
(1907–1982)

Canadian physician, born in Vienna, the son of a surgeon. He studied medicine in Prague, Paris and Rome, and was assistant in experimental pathology at the German University (1929–1931) before emigrating to the USA. After one year as a Research Fellow at Johns Hopkins University, where he worked on detoxification by the liver, he moved to McGill University in Montreal (1933–1945) and then became Director of the Institute for Experimental Medicine and Surgery at the University of Montreal in 1945. From 1932 he produced a long series of papers with Collip on the hormonal interactions involving the adrenal gland, hypothalamus and pituitary gland, and their effects on osteoblast multiplication, ion regulation and bone formation (introducing the term mineralocorticoid), the blood sugar level, gonadal stimulation by 'pregnancy urine', and lactation and menstruation in relation to development. He also observed the anti-hormone effect resulting in prolonged treatment with pituitary extract. He is best known for his 'general adaptation syndrome', defined as the 'physiological mechanism which raises resistance to damage as such' (1949), which links stress and anxiety, and their biochemical and physiological consequences, to human disorders as hypertension, nephrosclerosis and rheumatic diseases.

SEMENOV, Nikolai Nikolaevich
(1896–1986)

Russian physical chemist, born in Saratov. In 1917 he graduated from the university at Petrograd (now St Petersburg). From 1920 he worked at the Physico-Technical Institute there, becoming professor in 1928. In 1931 he became a full member of the USSR Academy of Sciences. Also in 1931, the Physico-Technical Institute became the Institute of Chemical Physics of the Academy, and Semenov was appointed its first director. He held this appointment until his death. In 1943 the institute moved to moscow, and from 1944 Semenov was also professor at Moscow State University. Semenov's earliest researches were on electrical phenomena in gases and solids, and on other topics in molecular physics. However, he is best known for his contributions to chemical kinetics, particularly in connection with chain reactions. He investigated explosion limits and many other features of combustion, flames and detonation. Much of his work was parallel to that of Hinshelwood, and they shared the 1956 Nobel Prize for Chemistry. Semenov was the author of several influential books, notably *Chemical Kinetics and Chain Reactions* (1934) and *Some Problems of Chemical Kinetics and Reactivity* (1954). He was awarded the Order of Lenin five times, and became a Foreign Member of the Royal Society in 1958.

SEMMELWEIS, Ignaz Philipp
(1818–1865)

Hungarian obstetrician, born in Buda (now Budapest). He studied at the University of Pest and in Vienna. From 1845 he worked in the first obstetrical clinic of the Vienna general hospital. He observed that the first clinic had a much higher rate of puerperal fever than the second obstetrical clinic, run by midwives. His investigations convinced him that this was caused by medical staff and students going directly from the post-mortem to the delivery rooms, spreading the putrefactive cause of the disorder. He instituted a rigorous programme of washing hands and instruments in chlorinated lime solution between autopsy work and examining patients, and the mortality rate in the first clinic was reduced to about the same level as the second. He found some support, but also much opposition to his ideas, and he left Vienna in 1850 for Pest. He published his systematic treatise on *The Aetiology, Concept and Prophylaxis of Childhood Fevers* in 1861. His last years were clouded by frustration that his ideas were not more widely accepted, and from mental instability. He died in a mental asylum, ironically from a cut in his finger which turned septic and produced a disease much like puerperal fever. His ideas were peculiar to himself, but in the later bacteriological age he came to be seen as a pioneer of antiseptic obstetrics.

SENEBIER, Jean
(1742–1809)

Swiss botanist, plant physiologist and pastor, born in Geneva. He had an early interest in natural history but his parents wished him to become a minister, in 1765, after presenting a thesis on polygamy, he was ordained pastor of the Protestant Church, Geneva. During a year in Paris he met many scientists. Charles Bonnet enabled him to begin experiments in plant physiology, and to publish a paper on the art of observing (1772) in response to a question posed in 1768 by the Netherlands Society of Sciences, Haarlem. He was pastor of Chancy near Geneva (1769–1773) and librarian for the Republic of Geneva from 1773; in 1777 he began translating the works of Spallazani. Senebier was the first to demonstrate the basic principle of photosynthesis; his most important papers were *Action de la Lumière sur la Vegetation* (1779) and *Experiences sur l'Action de la Lumière Solaire dans la Vegetation* (1788). He was also the first to precisely define experimental method, which he did in *Art d'Observer* (1775, an elaboration of his 1772 paper) and *Essai sur l'Art d'Observer et de Faire des Experiences* (1802). He taught many prominent life scientists, including Augustin Pyrame de Candolle, Jean-Antoine Colladon (1758–1830), a precursor of Mendel, and Nicolas Theodore de Saussure.

SERRE, Jean-Pierre
(1926–)

French mathematician, born in Bages. He studied at the École Normale Superieure before working at the National Centre for Scientific Research (CNRS) and the University of Nancy, and became professor at the College de France in 1956. For his early work in homotopy theory, especially on the computation of the homotopy groups of spheres, and the ideas he introduced into algebraic topology, he was awarded the Fields Medal (the mathematical equivalent of the Nobel Prize) in 1954. Later he turned to algebraic geometry, writing the definitive paper on the theory of sheaves, class field theory, group theory and number theory. He is also the author of numerous elegant books on various branches of mathematics, and was for a time a member of Bourbaki.

SERVETUS, Michael
(1511–1553)

Spanish theologian and physician. Born in Tudela, he worked largely in France and Switzerland. In *De Trinitatis Erroribus* (1531) and *Christianismi Restitutio* (1553) he denied the Trinity and the divinity of Jesus; he escaped the Inquisition but was burned by Calvin in Geneva for anti-trinitarian heresy. He lectured on geography and astronomy, and practised medicine at Charlien and Vienna (1538–1553). He appears to have prefigured William Harvey (1578–1657) in discovering the pulmonary (lesser) circulation of the blood, although his views were little known to contemporaries and carried no biomedical influence.

SEWARD, Sir Albert Charles
(1863–1941)

English palaeobotanist, born in Lancaster. He studied at Cambridge and Manchester and was a lecturer (1890–1906) and later Professor of Botany at Cambridge (1906–1936). He is best known for his works on English palaeobotany, *Wealden Flora* (1894–1895). *Jurassic Flora* (1900–1903), the four-volume *Fossil Plants* (1898–1919), and a panoramic survey, *Plant Life Through the Ages* (1931). He was knighted in 1936.

SEYFERT, Carl Keenan
(1911–)

American astronomer, born in Cleveland, Ohio, and educated at Harvard. After working at the McDonald Observatory and the Mount Wilson Observatory, he became Associate Professor of Astronomy and Physics and Director of Barnard Observatory, Vanderbilt University (1946–1951): from 1951 he was Professor of Astronomy and Director of the Arthur J. Dyer Observatory. Seyfert became famous for his work on a special group of galaxies (named after him) which have very bright bluish star-like nuclei, barely perceivable arms and spectra containing broad high-excitation emission lines; he started to study these systematically in 1943. They are now thought to be the low-luminosity cousins of quasars.

SHANNON, Claude Elwood
(1916–)

American applied mathematician and pioneer of communication theory, born in Gaylord, Michigan. He was educated at Michigan University and at MIT, where he received a PhD in mathematics. A student of Bush, he worked on the differential analyser and picked up on Bush's suggestion that he study the logical organization of its relay circuits as a thesis subject. Shannons master's thesis was one of the most influential ever written, and in 1938 he published a seminal paper (*A Symbolic Analysis of Relays and Switching Circuits*) on the application of symbolic logic to relay circuits. His work not only helped translate circuit design from an art into a science, but its central tenet – that information can be treated like any other quantity and can be manipulated by a machine – had a profound impact on the development of computing. After graduating from MIT, he worked at the Bell Telephone Laboratories (1941–1972) (where Stibitz had built a binary adder) in the esoteric discipline of information theory. He wrote *The Mathematical theory of Communications* (1949, with Warren Weaver).

SHAPLEY, Harlow
(1885–1972)

American astronomer, born in Nashville, Missouri. He worked for two years as a newspaper reporter before entering the University of Missouri where he quickly changed from journalism to astronomy. Following three years as a graduate student at Princeton (1911–1914) he was appointed to the staff of the Mount Wilson Observatory in California (1914–1921). His studies of Cepheid variable stars allowed him to use these as indicators of distance (1914). In this way he established the distances of globular star clusters and discovered that the centre of the globular cluster system is far removed from the Sun (1918). This result placed the Sun near the edge of the stellar system and not at its centre as had been the accepted view. While Shapley's new model of our galaxy was soon adopted, astronomers were not in agreement as to the structure of the universe as a whole; a memorable 'Great Debate' on the question took place between Shapley and Heber Doust Curtis in 1921. Shortly afterwards Shapley was appointed Director of Harvard College Observatory (1921) where he remained until his retirement in 1952. The prolific research output at Harvard and the number of distinguished astronomers trained there during that period are mainly due to Shapley's efforts. His own researches include the discovery of the first two dwarf galaxies, companions to our own galaxy, and investigations on the Magellanic Clouds.

SHARP, Phillip Allen
(1944–)

American molecular biologist, born in Kentucky. He was educated at Union College in Barbourville, Kentucky, where he received his BA in 1966, and at the University of Illinois where he obtained his PhD in 1969. He began his career as a Postdoctoral Fellow at Caltech (1969–1971), then became a Senior Research Investigator at Cold Spring Harbor Laboratories, New York (1972–1974), and Associate Professor at MIT (1974–1979). He was appointed Professor of Biology there in 1979, and has also served as Director of the MIT Center for Cancer Research (1985–1991) and head of the biology department since 1991. He was also a co-founder of Biogen, where he has been a director since 1978 and chairman of the scientific board since 1987. Sharp invented the technique known as S1 nuclease mapping, now used extensively to detect the size of unknown RNA molecules and to delineate the end of the RNA species. This led to his discovery in 1977 that genes are split into several sections, separated by stretches of DNA known as 'introns' which appear to carry no genetic information. The origins of this apparently redundant DNA remain a puzzle, but the discovery has prompted much research on how this phenomenon may be implicated in genetic diseases and speculation that it may provide a mechanism for rapid evolution. Sharp shared the 1993 Nobel Prize for Physiology or Medicine jointly with Richard Roberts, who had discovered split genes around the same time.

SHARPEY-SCHAFER, Sir Edward
(1850–1935)
English physiologist, born in Hornsey. He was educated at University College, London. He served as professor there (1883–1899) and later in Edinburgh (1899–1933). He gained many distinctions, becoming FRS in 1878 and president of the British Association in 1912, and receiving a knighthood in 1913. An able microscopist, his *Essentials of Histology* (1885) became a crucial textbook. One prime area of his research interests focused on neurophysiology and in particular the theory of brain localization. He experimented on nerve section and regeneration. Another field of interest lay in the emergent discipline of endocrinology. Sharpey-Schafer made extensive studies of the effects of suprarenal and of pituitary extracts. Investigating the role of the pancreas in carbohydrate metabolism, he surmised that the islet tissue must act as an organ of internal secretion of an as yet hypothetical fluid that he named 'insuline'. He championed the materialist doctrine of life and was a passionate defender of vivisection.

SHARPLESS, (Karl) Barry
(1941–)

American chemist, born in Philadelphia. After graduating from Dartmouth College (1963) he studied for a PhD at Stanford, University. He then undertook postdoctoral work (1968–1970) at Stanford and Harvard before joining the staff of MIT. He was made full professor in 1975. During 1977–1980 he held a post at Stanford, but he subsequently returned to MIT, becoming Arthur C. Cope Professor in 1987. He later moved to the Scripps Research Institute (1990). In the 1970s Sharpless worked on the metal-catalysed addition of oxygen to double bonds (epoxidation) which led to his most significant work on the asymmetric epoxidation of alkenes. More recently he has developed synthetic strategies for dihydroxylation reactions, of great value in the asymmetric synthesis of natural products. He was elected to the US National Academy of Sciences in 1985.

SHEN KUA
(1031–1095)

Chinese administrator, engineer and scientist, born in Hangchow. He made significant contributions to such diverse fields as astronomy, cartography, medicine, hydraulics and fortification. His first appointment was in 1054, and in the following few years he accomplished some notable work in land reclamation. As director of the astronomical bureau from 1072, he improved methods of computation and the design of several observational devices; in 1075 he constructed a series of relief maps of China's northern frontier area, and designed fortifications as defences against nomadic invaders. He surveyed and improved the Grand Canal over a distance of some 150 miles, using stone-filled gabions, wooden piles and long bundles of reeds to strengthen the banks and close gaps. In 1082 he was forced by intrigue to resign from his government posts, and occupied his last years in the writing of *Brush Talks from Dream Brook*, a remarkable compilation of about 600 observations which has become one of the most important sources of information on early Chinese science and technology.

SHEPARD, Francis Parker
(1897–1985)

American geological oceanographer, born in Brookline, Massachusetts. He was educated at Harvard (BA 1919) and at the University of Chicago, where he received a PhD in geology in 1922. He joined the staff of the University of Illinois and remained there until 1946, being promoted to professor in 1939. Around 1930 his interests turned towards the sea, apparently following summer cruises aboard his father's yacht off New England. There his sediment sampling failed to reveal theoretically predicted coarse to fine grain gradation, and this challenged him to study sea-bed processes. This led to a part-time appointment at Scripps Institution of Oceanography until 1948. The award of a Penrose fund of $10,000 to Shepard and colleagues in 1936 marked the beginning of Pacific marine geology and the turning point in his career. During World War II he was employed in San Diego, preparing sediment charts for the navy. In 1948 he was appointed Professor of Submarine Geology at Scripps Institution of Oceanography. Shepard put boundless energy into the study of the environmental conditions under which the ancient marine strata had been laid, making new comparisons with presently forming examples distinguishing sedimentation processes in different environments such as deltas, bays and continental shelves. He published *Submarine Geology* in 1948.

SHEPPARD, Philip Macdonald
(1921–1976)

English ecological geneticist. He was educated at Marlborough College and Worcester College, Oxford. After teaching in Ford's department at Oxford (1951–1956), he moved to Liverpool University, where he became Professor of Genetics in 1963. Sheppard extended and deepened Ford's pioneering studies but also carried out significant work on mimicry and human genetics, and on polymorphism in the land snail *Cepaea*. He showed that the banding patterns on snails, long assumed to have no effect on survival, were highly influential under certain circumstances – this discovery highlighted the dangers of assuming that traits are selectively neutral without direct evidence. He went on to demonstrate the important consequences of linked complexes of genes ('super-genes') in snails, mimetic butterflies, and human characteristics such as the Rhesus blood group system. Sheppard also played a key role in introducing genetical concepts into medicine. He was elected FRS in 1965.

SHERMAN, Henry Clapp
(1875–1955)

American biochemist, born in Ash Grove, Virginia. He trained at Maryland Agricultural College (1893–1895) and moved to Columbia University where he became successively Professor of Analytical Chemistry (1905–1907), Professor of Organic Analysis (1907–1911), Professor of Food Chemistry (1911–1924) and Mitchill Professor of Chemistry (from 1924). He devoted his research career to determining the quantitative nutritional requirements for metal ions and vitamins. In 1920 he reported the need for calcium in man, and he went on to establish the daily requirement (1931) and explore calcium's interaction with dietary phosphorus and vitamin D. He found that for rats, double the adequate amount of dietary calcium produced better development, longer reproductive capacity and increased life-span, without arterial calcification, over many generations (1941). From 1932 he studied the requirements for B vitamins in relation to the deficiency disease polyneuritis, and observed that riboflavin in excess caused tumours (1943). His major study on vitamin A identified a suitable weekly dose (1934), its storage in the body (1940) and the requirements for well-being in the rat detailed over more than 60 generations. Sherman also observed iron deficiency anaemia, and began investigating cobalt in 1946. For most of this work he developed suitable assays. Sherman's prolific publications included *Chemistry of Food and Nutrition* (1911).

SHERRINGTON, Sir Charles Scott
(1857–1952)

English physiologist, born in London. After studying at Caius College, Cambridge, and St Thomas's Hospital in London, he became a lecturer in physiology at St Thomas's and Professor-Superintendent of the Brown Animal Sanatory Institute. In 1895 he became Professor of Physiology at Liverpool, and he was later appointed Waynflete Professor of Physiology at Oxford (1913–1935). His career focused on the structure and function of the nervous system, exemplified by his analysis of the reflex arc, summarized in *The Integrative Action of the Nervous System* (1906), a book which constituted a significant landmark in modern neurophysiology. He described the reciprocal innervation of antagonistic muscles, by which the activity of one set of excited muscles is integrated with another set of inhibited muscles; he coined the word 'synapse' to describe the junction between nerve cells; he studied sense organs extensively; and he mapped the motor areas of the cerebral cortex of mammals. In addition to his neurophysiological research, Sherrington was the first person to use diphtheria antitoxin successfully in Britain; he worked during World War I in a munitions factory and chaired the Industrial Fatigue Board; and produced an influential textbook on experimental physiology, *Mammalian Physiology* (1919). He was President of the Royal Society (1920–1925), wrote poetry and medical history, and summarized much of his philosophical approach in *Man on his Nature* (1941). He shared the 1932 Nobel Prize for Physiology or Medicine with Adrian.

SHKLOVSKY, Josef Samuilovich
(1916–1985)

Soviet astronomer, born in Glukhov, the Ukraine. He worked for a time as a foreman on railway construction in Kazakhstan before becoming a student first at the Far-Eastern State University and later at the University of Moscow, where he graduated in physics in 1938. He proceeded to postgraduate research in theoretical astrophysics at the State Astronomical Institute in Moscow, and was awarded his doctorate in 1949. He was responsible for setting up in 1953 the radio astronomy division of the Astronomical Institute which led to the country's first radio telescope; he was associated with this institution for the rest of his life. He also became involved in the design of equipment for the Soviet space programme, which began with the world's first artificial satellite (Sputnik 1) in 1957. Although successful and innovative in these technical spheres, Shklovsky's outstanding contributions were in high-energy astrophysics. He was among the first to suggest the possibility of synchrotron radiation from astronomical sources, and in 1953 solved the problem of the continuous spectrum of the Crab nebula in terms of synchrotron radiation from energetic electrons in a magnetic field, the result of a supernova explosion in AD 1054. Other researches concerned radio line emission from ionized hydrogen in space, and the study of faint radio sources. In 1954 Shklovsky was appointed professor at Moscow State University, where he was an inspired teacher and a lucid expositor of astronomy. He collaborated with Sagan on the popular book *Intelligent Life in the Universe* (1966); among his other books was *Stars, their Birth, Life and Death* (1975), translated into English in 1978. He was awarded the Lenin Prize in 1959 for his contributions to space research.

SHOCKLEY, William Bradford
(1910–1989)

American physicist, born in London, the son of two American mining engineers. Brought up in California, he was educated at Caltech and MIT before starting work at the Bell Telephone Laboratories in 1936. During World War II he directed anti-submarine warfare research and he became Consultant to the Secretary for War in 1945. Returning to Bell Telephones he collaborated with Bardeen and Brattain in trying to produce semiconductor devices to replace thermionic valves. Using a germanium rectifier with metal contacts including a needle touching the crystal, they invented the point-contact transistor (1947). A month later Shockley developed the junction transistor (for transfer of current across a resistor). These devices led to the miniaturization of circuits in radio, TV and computer equipment. By the mid-1950s Shockley had founded his own company, the Shockley Semiconductor Laboratory, near Palo Alto, California, to exploit his knowledge of solid-state physics. A poor manager, he was unable to commercialize his brilliant research and many of his most gifted workers, such as Noyce, left. Shockley, Bardeen and Brattain shared the Nobel Prize for Physics in 1956. From 1963 to 1974 Shockley was Professor of Engineering at Stanford. He later dabbled in eugenics, his ideas on inherited intelligence amongst the races proving highly controversial.

SIBTHORP, John
(1758–1796)

English botanist, born in Oxford where he was educated before studying medicine at Edinburgh. He then went to Montpellier (1783), where he was elected a member of the Academy of Sciences. Shortly afterwards, he succeeded his father as Sherardian Professor of Botany at Oxford, and in 1784 unsuccessfully bid against his friend James Edward Smith for the collections of Linneaeus. He left for the Continent to plan a botanical expedition to Greece, aiming to study the plants named by Dioscorides. In Vienna, he examined the Dioscorides Codex and secured the services of the botanical artist Ferdinand Bauer. Sibthorp and Bauer reached Crete in 1786. After several months there they visited other Greek islands, Athens, western Turkey, Cyprus and Athos, returning to England in late 1787. Sibthorp (with Smith) then helped found the Linnaean Society (1788) and published his *Flora Oxoniensis* (1794). On his second expedition to Greece and Turkey (commencing 1794), he developed tuberculosis, from which he died. Apart from *Flora Oxoniensis*, Sibthorp's only work was his share in *Flora Graeca* and *Prodromus Florae Graecae*, both posthumously published by Smith.

SIDGWICK, Nevil Vincent
(1873–1951)

English chemist, born in Oxford. At Christ Church, Oxford, he was a student of Augustus Vernon Harcourt and gained a first class in chemistry in 1895. He followed this with a first in classics in 1897. After acting as demonstrator in the Christ Church laboratory for a year, he went to work in Ostwald's laboratory in Leipzig and then with Hans von Pechmann at Tübingen. He returned to be a Fellow and Tutor of Lincoln College, a position he held until retirement in 1947. In the university he became reader (from 1924) and professor (from 1935). Up to 1920 Sidgwick carried out various physico-chemical solution studies involving kinetics, ionic equilibria or phase equilibria. He was greatly stimulated, however, by the nascent electronic theory of valency, and his researches began to bear on this. The most important outcome was his book *The Electronic Theory of Valency* (1927). His later books *The Covalent Link in Chemistry* (1933, from his 1931 Baker Lectures at Cornell University) and *The Chemical Elements and Their Compounds* (2 vols, 1950) were also highly influential. His earlier book *The Organic Chemistry of Nitrogen* (1910) went through later editions under editors. Sidgwick was appointed CBE in 1935, was elected a Fellow of the Royal Society in 1922 and received its Royal Medal in 1937. He was President of the Faraday Society from 1932 to 1934 and of the Chemical Society from 1935 to 1937.

SIEGBAHN, Kai
(1918–)

Swedish physicist, born in Lurid, the son of Karl Siegbahn. Educated at Stockholm, he was Professor of Physics at the Royal Institute of Technology there until 1954, and thereafter professor at Uppsala University. In the early 1950s he started making high-precision measurements on the energies of electrons emitted from solids exposed to X-rays. Unlike previous observations which had revealed the electron energies to be distributed in bands, Siegbahn found sharp peaks in the bands at energies that were characteristic of the materials he was exposing to the X-rays. These peaks corresponded to the energies required to displace one of the inner bound electrons. Furthermore he found that the peak energies were dependent on the chemical environment of the atoms, i.e. on the type of chemical bond linking neighbouring atoms or molecules. This technique, which became known as ESCA (electron spectroscopy for chemical analysis), offered a delicate but powerful experimental method for studying the energies of electrons around bonded atoms. The method has been extended by Siegbahn's team for use with liquids and gases as well as solids, and they have also worked on the related technique of ultraviolet photoelectron spectroscopy. He shared the 1981 Nobel Prize for Physics with Bloembergen and Schawlow for his work in developing high-resolution electron spectroscopy.

SIEGBAHN, Karl Manne Georg
(1886–1978)

Swedish Physicist, born in Orebro. He was educated at the University of Lurid, and became professor there (1920) and at Uppsala (1923). From 1937 he was Professor of the Royal Academy of Sciences and Director of the Nobel Institute for Physics at Stockholm. Improving on the techniques of Barkla and Harry Moseley, he succeeded in producing X-rays of various wavelengths and penetrating power, which were labelled K, L, M, N, O, P, and Q in order of increasing wavelength. This discovery reinforced Niels Bohr's shell model of the atom, and the atomic shells were lettered in the same way. For his development of X-ray spectroscopy, Siegbahn was awarded the Nobel Prize for Physics in 1924. In the same year he showed that X-rays could be refracted by means of a prism.

SIEMENS, (Ernst) Werner von
1816–1892)

German electrical engineer, brother of William Siemens, born in Lenthe, Hanover. In 1834 he entered the Prussian Artillery, and in 1844 he took charge of the artillery workshops at Berlin. He developed the telegraphic system in Prussia, discovered the insulating property of gutta-percha, and devoted himself to the construction of telegraphic and electrical apparatus. In 1847 together with Johann Georg Halske, scientific instrument maker at the University of Berlin, he established Siemens & Halske, initially for manufacturing telegraphy equipment, but which evolved into one of the great electrical engineering firms. Besides devising numerous forms of galvanometer and other electrical instruments, he was one of the discoverers of the self-acting dynamo (1867). He determined the electrical resistance of different substances; the SI unit of electrical conductance is named after him. In 1887 he endowed the Physikalisch Technische Reichsanstalt in Berlin, of which Helmholtz was the first director. He was ennobled in 1888. One of his sons, Wilhelm (1855–1919), was a pioneer of the incandescent lamp.

SIEMENS, Sir (Charles) William, originally (Karl) Wilhelm
(1823–1883)

German-born British electrical engineer, brother of Werner von Siemens, born in Lenthe, Hanover. In 1843 he visited England to introduce a process for electro-gilding invented by Ernst Werner and himself, and in 1844 he patented his differential governor. He was naturalized in 1859. As manager in England of the firm of Siemens Brothers, he was actively engaged in the construction of telegraphs, designed the steamship *Faraday* for cable-laying, promoted electric lighting, and constructed the Portrush Electric Tramway in Ireland (1883). In 1861 he designed an open-hearth regenerative steel furnace which became the most widely used in the world. Other inventions included a water-meter, pyrometer and bathometer. Siemens was elected FRS in 1863, and was President of the Institution of Mechanical Engineers (1872), the Iron and Steel Institute (1877), and the British Association for the Advancement of Science (1882). He was also the first President of the Society of Telegraph Engineers (1872) and Chairman of the Royal Society of Arts (1882). He was knighted in 1883. He was assisted in England by another brother, Friedrich (1826–1904), who invented a regenerative smelting oven extensively used in glassmaking (1856).

SIMPSON, George Gaylord
(1902–1984)

American palaeontologist, born in Chicago. Educated at the universities of Colorado and Yale, he joined the staff of the American Museum of Natural History in New York City in 1927, and from 1959 to 1970 taught at Harvard. His early work was concerned with the Mesozoic mammals which provide important evidence for the understanding of mammalian evolution. In the 1930s he carried out palaeontological research on the Tertiary mammals from Patagonia, a unique and hitherto little-studied fauna. He is considered one of the leading twentieth-century palaeontologists and he proposed a classification of mammals which is now standard. Although mainly concerned with taxonomy, after World War II he devoted himself to demonstrating that the neo-Darwinian ideas of geneticists such as Mayr and Dobzhansky could be reconciled with the palaeontological evidence. He was particularly concerned with the circumstances which gave rise to the evolution of new species. His influential books *Tempo and Mode in Evolution* (1944) and *The Major Features of Evolution* (1953) were concerned with the fusion of palaeontology and evolutionary genetics. Some of his ideas are presented in popular form in *The Meaning of Evolution* (1949).

SIMPSON, Sir James Young
(1811–1870)

Scottish obstetrician and pioneer of anaesthesia, born in Bathgate, West Lothian, the son of a baker. With the financial support of the rest of his large family he went to Edinburgh University at the age of 14, studying arts, then medicine, and becoming Professor of Midwifery in 1840. He originated the use of ether as anaesthetic in childbirth (January 1847), and experimenting on himself and his assistants in the search for a better anaesthetic, discovered the required properties in chloroform (November 1847). He championed its use against medical and religious opposition until its employment by Queen Victoria at the birth of Prince Leopold (1853) signalled general acceptance. He founded gynaecology by his sound tests, championed hospital reform, and in 1847 became Physician to the Queen in Scotland. He was made a baronet in 1866.

SIMSON, Robert
(1687–1768)

Scottish mathematician, Professor of Mathematics at Glasgow from 1711. His life's work was dedicated to the editing and restoration of the work of the ancient Greek geometers; his edition of Euclid's *Elements* (1758) was the basis of nearly all editions for over a century.

SLIPHER, Vesto Melvin
(1875–1969)

American astronomer, born in Mulberry, Indiana. He studied at Indiana University before working for over 50 years at the Lowell Observatory, Arizona, becoming its director in 1926. He obtained the first successful photographs of Mars, and demonstrated that methane is present in the atmosphere of Neptune. The research which led to the discovery of the planet Pluto was carried out under his direction. Primarily a spectroscopist, his spectral studies revealed the presence of gaseous interstellar material, and he suggested that the nebula in the Pleaides cluster of stars is illuminated by starlight reflected off dust grains (1912). By measuring the Doppler shift in light reflected from the edges of planetary discs, he determined the periods of rotation of Uranus, Jupiter, Saturn, Venus and Mars in 1912. In his most important work, he extended this method to the Andromeda nebula, not yet perceived as an extragalactic object, and discovered that it is approaching the Earth at around 300 kilometres per second (1912). Similar studies of other nebulae showed this to be an exception, as most were found to be receding from the Earth at very high speed. His results directed Hubble to the concept of the expanding universe, in which galaxies are moving apart at relative speeds proportional to their separation.

SMAGORINSKY, Joseph
(1924–   )

American meteorologist, born in New York City. He received a master's degree from New York University (1948) and joined the Weather Bureau to work on objective analysis of weather charts and on calculating vertical motion in the atmosphere from horizontal wind and pressure fields. Charney invited him to Princeton University in 1950. It was there that the idea of using a very large computer for weather forecasting was developed. After obtaining his PhD (1953), Smagorinsky returned to the Weather Bureau to set up a research group on numerical weather prediction. He headed the General Circulation Research Laboratory, which eventually became the Geophysical Fluid Dynamics Laboratory (GFDL), for 27 years. During this time he was the driving force behind much of the enormous progress made in numerical weather prediction. From the initial, relatively simple, barotropic and primitive equation models, the science progressed so that more complex physical processes such as convection, variations in radiation due to different surfaces and ocean atmosphere interactions could be included in the programmes. This enabled the original relatively short-period forecast models to be modified for use as climate prediction models. Smagorinsky's work made GFDL a world leader in this field.

SMITH, Sir Grafton Elliot
(1871–1937)

Australian-born anatomist and anthropologist, born in Grafton, New South Wales. He graduated in medicine from the University of Sydney in 1892. In 1896 he went to England to conduct research at Cambridge, and in 1900 he was made Professor of Anatomy at the new medical school in Cairo. He was later appointed to chairs of anatomy at Manchester (1909) and University College London (1919). Smith became an authority on brain anatomy and human evolution, building his reputation on studies of cranial morphology and the Egyptian practice of mummification. His books, including *Migrations of Early Culture* (1915), *The Evolution of the Dragon* (1919) and *The Diffusion of Culture* (1933), explain similarities in culture all over the world by diffusion from pharaonic Egypt. He was knighted in 1934.

SMITH, Hamilton Othanel
(1931–)

American molecular biologist, born in New York City. He graduated from Johns Hopkins Medical School, and after teaching at the University of Michigan, returned to Johns Hopkins in 1967. He was later appointed Professor of Microbiology (1973) and Professor of Molecular Biology and Genetics (1981) there. During the late 1950s when he was drafted into the US navy, he began to read about genetics. Initially he began to work on phage (the bacterial virus), but later used the bacterium *Haemophilus influenzae*. He found that bacteria produce enzymes, called endonucleases, which can split the DNA strand of invading phage particles so that they are inactivated. Because the enzymes curtail the activity of the phage, they are also called 'restriction enzymes'. In the 1970s Smith went on to isolate endonucleases which would split a DNA strand at a specific site. More than 100 such enzymes have now been isolated. Because of their site-specificity, the use of these enzymes allows the nucleotide sequence of DNA to be established. Smith shared the 1978 Nobel Prize for Physiology or Medicine with Werner Arber and Nathans.

SMITH, Henry John Stephen
(1826–1883)

Irish mathematician, born in Dublin. He was educated at Rugby School and Balliol College Oxford, of which he was elected a Fellow. In 1860 he became Oxford's Savilian Professor of Geometry. He was the greatest British authority of his day on the theory of numbers, and wrote a fine six-part report on the subject for the British Association for the Advancement of Science which was influential in communicating new developments to British mathematicians. He was posthumously awarded the prize of the Paris Academy of Sciences for his work on the representation of integers as sums of squares (sharing the prize with the 18-year-old Hermann Minkowski). He also wrote on elliptic functions and geometry.

SMITH, Sir James Edward
(1759–1828)

English botanist, born in Norwich. Although he inherited a love of botany from his mother, he did not begin studying botanical science until the day Linnaeus died, when Smith was 18. He studied medicine at Edinburgh University and London. In 1783 aged only 24, he bought the entire natural history collection of Linnaeus from his widow, bidding against his friend Sibthorp, and brought it to London. On his return from a continental tour (1786–1787), he became a founder member and first President of the Linneaen Society of London (1788–1828). After his marriage (1796) he lived in Norwich, travelling to London annually until 1825 to lecture at the Royal Institution. He wrote 3,348 botanical articles, his most important works being *English Botany* (36 vols, 1790–1814), *Flora Britannica* (3 vols, 1800–1804), *Exotic Botany* (2 vols, 1804–1805), *Flora Graeca* (7 vols, 1806–1828), *Prodromus Florae Graecae* (2 vols, 1806–1843) and *English Flora* (4 vols, 1824–1828). He was deacon at the Octagon Chapel, Norwich, and wrote a tract, *A Defence of the Church and Universities of England* (1819), in response to Cambridge University's objection to Thomas Mertyn's invitation to him to lecture there, on the grounds of his unitarian beliefs.

SMITH, (Ernest) Lester
(1904–1992)

English biochemist, born in Teddington, Middlesex. He was educated at Chelsea Polytechnic, and in 1928 he joined the pharmaceutical firm of Glaxo where he spent his entire career, becoming Senior Research Biochemist. Initially he worked on the dietary requirements for vitamin A and its storage properties as affected by oxidation and light. During World War I he was responsible for the first commercial production of penicillin in England, using bacterial fermentation, and discovered that the biosynthesis of penicillin F (the common form), G or X could be selectively promoted by adjustment of the growth medium. He independently isolated the anti-pernicious-anaemia factor (vitamin B12) in crystalline form in the same year as Folkers in the USA (1948), and showed that it contained cobalt. He separated and characterized the various forms of the vitamin, and was the first to prepare the doubly radiolabelled vitamin, containing cobalt-60 and phosphorus-35, for metabolic tracer studies (1952). In 1958 he isolated an antibiotic complex with low host toxicity (called E129) which was effective against streptococci that had developed resistance to other known antibiotics (such as penicillin and streptomycin). He published *Intelligence Comes First* (1975), *Our Last Adventure* (1982) and *Inner Adventures* (1988), and was elected FRS in 1957.

SMITH, Michael
(1932–)

British-Canadian biochemist, born in Blackpool. Educated at the University of Manchester, in 1956 he moved to the University of British Columbia, where he is now professor and Director of the Biotechnology Laboratory. In 1978, he published his discovery of 'site-specific mutagenesis', a technique which allows scientists to alter the genetic code through mutations induced at specific locations – all previous methods of mutation, using radiation and chemicals, produced random mutations. This new method has allowed the production of a whole range of proteins with diverse functions. Smith was awarded the 1993 Nobel Prize for Chemistry jointly with Mullis.

SMITH, Theobald
(1859–1934)

American microbiologist and immunologist, born in Albany, New York, the greatest American bacteriologist of his generation. He received his medical degree from the Albany Medical College, and was subsequently associated with several US institutions, including

Harvard University (as professor, 1896–1915) and the Rockefeller Institute for Medical Research (1915–1929). He studied both animal and human diseases, and first implicated an insect vector in the spread of disease when he showed that Texas cattle fever is spread by ticks. He distinguished the forms of bacillus causing human and bovine tuberculosis, and laid the scientific foundations for a cholera vaccine. He also improved the production of smallpox vaccine and diphtheria and tetanus antitoxins, and established precise techniques for the bacteriological examination of water, milk and sewage.

SMITH, William
(1769–1839)

English civil engineer and geologist, born in Churchill, Oxfordshire, the son of a blacksmith. In 1787 he became an assistant to a surveyor, and he was later appointed engineer to the Somerset Coal Canal (1794–1799). His survey work during canal construction introduced him to a variety of rock sequences of different ages, and in 1799 he produced a coloured geological map of the country around Bath. From 1799 he was a consultant engineer and surveyor, travelling great distances in the course of his work and later settling in London (1804). Smith used fossils to aid his identification of strata and to fix their position in the succession. He produced the first geological map of England, *A Delineation of the Strata of England and Wales, with part of Scotland* (1815), and 21 coloured geological maps of the English counties (1819–1824) assisted by his nephew John Phillips (1800–1874). Smith is often regarded as the father of English geology and stratigraphy.

SMITHSON, James Louis Macie
(1765–1829)

English chemist, the benefactor of the Smithsonian Institution in Washington DC. He was born in Paris, the illegitimate son of Sir Hugh Smithson Percy, 1st Duke of Northumberland, and Elizabeth Macie. At first known as Macie, he changed his name in 1801 after the death of his mother. He was educated at Oxford, and showing an early aptitude for mineralogy, he was elected FRS in 1787. He analysed zinc ores and other minerals; zinc carbonate was later named 'smithsonite' after him. He also proposed an innovative balance to weigh very small quantities. However, he is chiefly remembered for the manner in which he disposed of his wealth, rather than for his contribution as a chemist. He left a will stipulating that his estate of £105,000 should pass to his nephew, with the proviso that if the nephew died without heirs the money should be used to found 'at Washington, under the name of the Smithsonian Institution, an Establishment for the increase and diffusion of knowledge among men'. His nephew died in 1835 leaving no children and the Smithsonian Institution was established by Act of Congress in 1846. It soon grew to be a world-famous museum and institute for scientific research.

SMYTH, Charles Piazzi
(1819–1900)

British astronomer, born in Naples, son of William Henry Smyth, a British naval officer, and godson of Giuseppe Piazzi after whom he was named. On retiring from the navy, William Smyth set up a private observatory at Bedford where he produced his highly popular *Bedford Catalogue* (1844) of double stars which earned him the Gold Medal of the Royal

Astronomical Society (1845). Charles Piazzi Smyth was educated at Bedford School and received a sound training in practical astronomy at home. At the age of only 16, he was chosen for the post of chief assistant at the Royal Observatory, Cape of Good Hope, South Africa (1827–1837). He succeeded Henderson as Astronomer Royal for Scotland and Professor of Astronomy at the University of Edinburgh, where he remained until his retirement in 1888. His most important contribution to astronomy was his expedition to the island of Tenerife (1856) to test the advantages of high-altitude sites for observational astronomy, the first ever scientifically conducted investigation into this question. His results, published by the Admiralty and recorded in his *Tenerife, an Astronomer's Experiment* (1858) proved the superiority of a mountain site for every type of observation. His ambition to erect a station on the Peak of Tenerife went unfulfilled; he did, however, make several excursions to sunny locations abroad to make spectroscopic observations of the Sun. He was a successful early photographer who began with calotypes of South African scenes in 1843, and was the inventor of a flat-fielding lens for cameras (1874). In 1864 he carried out a metrological survey of the great pyramid of Giza, but his bizarre interpretation of the pyramid as a divinely inspired monument – a belief which attracted many followers – gravely tarnished his scientific reputation.

SNELL, George Davis
(1903–)

American geneticist, born in Bradford, Massachusetts, the son of an inventor. He graduated from Dartmouth College in 1926 and conducted graduate studies at Harvard University. After teaching zoology for two years, Snell was awarded a fellowship at the University of Texas where he demonstrated, for the first time, that X-rays can induce mutations in mammals. In 1933 he became Assistant Professor at Washington University, but in 1935 he joined the Jackson laboratory in Bar Harbor, Maine, where he remained until 1973. In the late 1930s, Snell began to work on the genes responsible for rejection of tissue transplants in mice – which he called histocompatibility genes – later named the major histocompatibility complex (MHC). Research in this area took a dramatic turn when Dausset discovered that histocompatibility in humans is also controlled by an MHC. Later Benacerraf showed that the MHC controls the production of proteins on cell surfaces which allow certain white blood cells to distinguish them from foreign or abnormal cells. Snell, Dausset and Benacerraf shared the 1980 Nobel Prize for Physiology or Medicine.

SNELL, Willebrod van Rolien, (Latin Snellius)
(1580–1626)

Dutch mathematician, born in Leiden. He succeeded his father as Professor of Mathematics at Leiden (1613) and discovered the law of refraction known as Snell's law, which relates the angles of incidence and refraction of a ray of light passing between two media of different refractive index. He also extensively developed the use of triangulation in surveying.

SNOW, John
(1813–1858)

English anaesthetist and epidemiologist, born in York. He was a young general practitioner when cholera first struck Britain in 1831–1832, and his experience then convinced him that the disease was spread through contaminated water. After 1836 he practised in London, and during the cholera outbreaks of 1848 and 1854 there, he carried out some brilliant epidemiological investigations, tracing one local outbreak to a well in Broad (now Broadwick) Street, Soho, into which raw sewage seeped. His additional work implicated the Thames, into which many of London's sewers drained and from which much of London's domestic water was obtained. He showed that houses which were supplied by companies obtaining their water from downstream in the Thames had much higher incidences of cholera than those supplied by water drawn from upstream before the sewage had contaminated it. Snow was also a pioneer anaesthetist. He did fundamental experimental work on ether and chloroform, devised apparatus to administer anaesthetics, and gave chloroform to Queen Victoria in 1853, during the birth of Prince Leopold. During the last 10 years of his life, he was much in demand by surgeons as an anaesthetist.

SNYDER, Solomon Halbert
(1938–)

American psychiatrist and pharmacologist, born in Washington. He received his MD at Georgetown University (1962), worked at the Intern Kaiser Foundation Hospital, San Francisco (1962–1963), and then became a research associate at the National Institute of Medical Health in Bethesda, Maryland (1963–1965), and resident psychiatrist at Johns Hopkins Hospital in Baltimore (1965–1968). He then held a number of professorships at Johns Hopkins Medical School, where since 1980 he has been Distinguished Service Professor in Neuroscience, Psychiatry and Pharmacology. From the mid-1960s Snyder investigated the biochemistry of nervous tissue, determining the stereospecific requirements for catecholamines of synaptosomes (vesicles that contain neurotransmitter substances) from different areas of the brain (1968), and extending earlier observations that mammalian ornithine decarboxylase has an extremely rapid turnover time of 10–30 minutes (1969). This enzyme, which is involved in polyamine syntheses and possibly the regulation of RNA syntheses, links with his major interest in the effects of opiates and psychotropic drugs on the brain and the naturally occurring brain hormones, enkephalins and endorphins. In 1973 he successfully demonstrated the presence of opiate receptors in nervous tissue, opening the way for the study of endogenous algesics. He exploited his receptor binding assay for the study of the functional distribution of neurologically important compounds, such as met- and leu-enkephalin which have binding potencies similar to morphine (1976–1977). More recently he has studied transport across membranes involving cyclic adenosine monophosphate (cAMP) and its link with olfaction (1985–1986), inositol triphosphate in calcium ion transport (1988) and the calcium-regulated enzyme nitric oxide synthase which synthesizes the 'vascular endothelium derived relaxing factor', nitric oxide (1990).

SOBRERO, Ascanio
(1812–1888)

Italian chemist who invented nitroglycerine, born in Casale, Montferrato. He studied medicine at Turin, and organic chemistry at Paris and with Liebig at Giessen. From 1844 he taught at Turin, being appointed to the chair in 1847. In 1844 or 1845 he added glycerol to a mixture of concentrated nitric and sulphuric acids and created nitroglycerine, a colourless, oily, sweet-tasting liquid which is a powerful explosive. Sobrero described its explosive properties and also investigated its effects as a drug. Nitroglycerine was manufactured on a huge scale after Nobel invented dynamite. Today it is used as a fuel in rockets and missiles; it is also employed in medicine as it dilates blood vessels and thus eases cardiac pain. Sobrero was also the first person to prepare lead tetrachloride (1850). He died in Turin.

SODDY, Frederick
(1877–1965)

English radio chemist, renowned for his work on radioactivity, born in Eastbourne. He studied at the University College of Wales, Aberystwyth, and at Oxford. In 1900, after two years research at Oxford, he was appointed demonstrator in chemistry at McGill University, Montreal, where he and Rutherford studied radioactivity. They realized that the reason radioactive elements display properties at odds with their position in the periodic table is because they have decayed, in part, into other elements. As part of the decay process they emit three types of radiation (now known as alpha, beta and gamma radiation). In 1903, in London working with Ramsay, Soddy verified his prediction that radium produces helium when it decays. Lecturing at Glasgow from 1904 to 1914, he showed that uranium decays into radium, work later developed by Boltwood. He also showed that a radioactive element may have more than one atomic weight although its chemical properties are identical. Because the chemical properties are unchanged, Soddy argued that all forms of the same element belong in the same place in the periodic table and therefore he named them 'isotopes' ('the same place'). He later demonstrated that isotopes exist among elements which are not radioactive. His discovery of isotopes was of fundamental importance to all physics and chemistry, although the reason for the existence of isotopes did not become clear until Chadwick discovered the neutron in 1932. Soddy's other great discovery was that when an atom of a radioactive element emits an alpha particle, it is transformed into an isotope of an element two places down in the periodic table, but when it emits a beta particle it moves one place higher. In 1914 he was appointed to the chair at Aberdeen, where he was largely employed on chemical research connected with World War I. Moving to Oxford in 1919, where he was Dr Lee's Professor of Chemistry until 1936, Soddy reorganized the laboratory facilities and the teaching syllabus. In 1921 he was awarded the Nobel Prize for Chemistry. After his retirement he wrote on ethics, politics and economics, urging fellow scientists only to conduct research in areas which had peaceful applications. He died in Brighton.

SOKAL, Robert Reuven
(1926–   )

Austrian-born American entomologist and biometrician, born in Vienna. He studied at St John's in Shanghai and at the University of Chicago, where during 1948–1951 he worked in the department of zoology. In 1953 he moved to the University of Kansas, becoming Assistant Professor of Entomology in 1953; more recently he has worked at the State University of New York, Stony Brook. His early researches concerned the genetics of aphids and flies. He has latterly done much valuable work in the fields of biometry and biostatistics. With the English bacterial taxonomist and microbiologist Peter H. A. Sneath, he has carried out much pioneering work in the development of the principles and methods of numerical taxonomy. This has included the publication of two jointly authored standard works, *Principles of Numerical Taxonomy* (1963) and *Numerical Taxonomy: The Principles and Practice of Numerical Classification* (1973). Sokal's other main works, both jointly written with F. James Rohlf, are *Biometry* (1969) and *Introduction to Biostatistics* (1969).

SOMERVILLE, Mary Greig, née Fairfax
1780–1872)

Scottish mathematician and astronomer, born in Jedburgh. The daughter of a naval officer, she was inspired by the works of Euclid, and studied algebra and classics, despite intense disapproval from her family. From 1816 she lived in London, where she moved in intellectual and scientific circles, and corresponded with foreign scientists. In 1926 she presented a paper on *The Magnetic Properties of the Violet Rays in the Solar Spectrum* to the Royal Society. 1831 saw the publication of *The mechanism of the heavens*, her account for the general reader of Laplace's *Mécanique Celeste.* This had great success and she wrote several further expository works on physics, physical geography and microscopic science. She supported the emancipation and education of women, and Somerville College (1879) at Oxford is named after her.

SOMMERFELD, Arnold
(1868–1951)

German physicist, born in Konigsberg (now Kaliningrad, Russia). He was educated at the University of Konigsberg and appointed as professor at Clausthal (1897), Aachen (1900) and Munich (1906). With Klein he developed the theory of the gyroscope. He also researched into wave spreading in wireless telegraphy. Sommerfield generalized the quantization rules developed by Niels Bohr so that Bohr's quantum model of the atom could be applied to multi-electron atoms. He also evolved a theory of the electron in the metallic state.

SONDHEIMER, Franz
(1926–1981)

English chemist, born in Stuttgart of Jewish stock, he moved with his family to England in 1937. His chemical education began at Imperial College, London, in 1944 and he graduated with a PhD in 1948. After collaboration with Raphael, he moved to Harvard to work with Robert Woodward on steroid synthesis. In 1952 he moved to Syntex SA in Mexico City in succession to Djerassi. After four very successful years there he became head of the department of organic chemistry at the Weizmann Institute in Israel, but kept close contact with Syntex. In 1963 he moved to a Royal Society Chair at Cambridge, but he later transferred this to University College London. Although steroid synthesis, particularly those associated with oral contraceptives, had been his area of greatest success, he increasingly worked on large-ring annulene compounds. He was, above all else, a perfectionist and this became an increasing burden to him and led to many periods of deep depression. He took his own life in 1981. He received many honours and was elected FRS in 1967. He collected a fine library of early scientific books.

SORBY, Henry Clifton
(1826–1908)

English geologist and metallurgist, born in Woodbourne, Sheffield. He inherited a modest fortune which allowed him to devote his time to science. Sorby was the first to study rocks in this section under the microscope, demonstrating that this could reveal their minute structure and composition, as well as much about their mode of origin. He also adapted the technique for the study of metals by treating polished surfaces with etching materials. Using the microscope he observed the alignment of micas which causes cleavage in slates and identified fossils in chalk. In 1858 he published *On the Microscopical Structure of Crystals*, which marks a prominent landmark in petrology. He also wrote on biology, architecture and Egyptian hieroglyphics.

SORENSEN, Soren Peter Lauritz
(1868–1939)

Danish chemist, born in Havrebjerg. He was educated in Copenhagen, where he later became Director of the Carlsberg Research Laboratory. In 1901 he described how formaldehyde lowers the dissociation constant (pK) of an amino group, so that it could be titrated by standard electrometric procedures. This method, the 'formol titration', is still used today. In 1909, in *Enzyme Studies II* which described the effects of hydrogen ion concentration on enzyme activity, he introduced the term pH to represent the negative logarithm of the hydrogen ion concentration. He also used for the first time the German word '*Puffer*', which became 'buffer' in English, to describe the ability of certain solutions to resist change in pH. In 1912 he independently established the generalized relationships among acids and bases which later became known as the Henderson-Hasselbalch equation; $pH = pK + \log([base]/[acid])$, where [base] and [acid] denote concentrations. In the following decade, along with others, he prepared extensive lists of dyestuffs that change colour at predetermined hydrogen ion concentrations. In 1923, to unify the mass of data that resulted, he devised the pH scale, arguably the most important single contribution ever made to the life sciences. He made other contributions to pH measurement, the osmotic pressure of protein solutions, and on de-naturation.

SOSIGENES
(fl.c.40 BC)

Alexandrian astronomer and advisor to Julius Caesar in his reform of the calendar. At that time the error in the calendar was such that three additional months had to be interpolated to correct it. In Caesar's reform the year was to become independent of the Moon, and to consist of 365 days, an extra day being added in February every fourth year (the leap year) to make the average length of the year 365 days. The new Julian calendar began in 45 BC and remained in force until 1582 when Pope Gregory introduced the improved Gregorian system in which centuries are leap years only if divisible by 400.

SOUTHWOOD, Sir (Thomas) Richard (Edmund)
(1931–)

English zoologist and ecologist. He was educated at the University of London (1949–1955), and after teaching there he became Director of Imperial College Field Station at Silwood Park (1967–1969) and Professor of Zoology and Applied Entomology (1969–1979). He then moved to the University of Oxford, where he became Linacre Professor of Zoology in 1979 and Vice-Chancellor in 1989. His early work was concerned with the ecology and systematics of the Hemiptera. He wrote for the 'Wayside and Woodland' series *Land and Water Bugs of the British Isles* (1959, with D. Leston), a volume which married scientific rigour with a popular approach. He has carried out research on the relationships between insects and plants, particularly with regard to the factors which determine the number of species of insect that are associated with a species of plant. He is also interested in the way in which features of habitat influence the evolution of life-history strategies in insects. He was Chairman of the Royal Commission on Environmental Pollution (1981–1986) at the time when the recommendation to eliminate lead from petrol was made, Chairman of the Anglo-Scandinavian Committee for the Surface Water Acidification Programme, which is concerned with research into acid rain, and became Chairman of the National Radiological Protection Board in 1985. Elected FRS in 1977, he served as Vice-President of the Royal Society (1982–1984), and was knighted in 1984.

SPALLANZANI, Lazaro
(1729–1799)

Italian biologist and naturalist. Born in Scandiano in Modena, Spallanzani studied law at Bologna and became a priest. But his first love was natural philosophy. He rose to become Professor of Mathematics and Physics at Reggio (1757), moving on to Modena in 1763 and to the Chair of Natural History at Pavia in 1769. He was an enthusiastic traveller who enlarged the Natural History Museum at Pavia. He is remembered for his skills in experimental physiology, where he pursued wide-ranging and fruitful experimentation. Deeply interested in reproduction, he set about rebutting the long established theory of spontaneous generation. In the seventeenth century, Redi had shown that insects developed on putrefying flesh only from deposited eggs. Building on Redi's work, Spallanzani showed in 1765 that broth, boiled thoroughly and hermetically sealed, remained sterile. It was not, however, until Pasteur's time a hundred years later that the idea of spontaneous generation was finally abandoned. On the basis of animal experimentation, Spallanzani disputed the traditional opinion that digestion was a kind of cooking by stomach heat, counter-arguing that gastric juice constituted the key digestive agent. He was the first to observe blood passing from arteries to veins in a warm-blooded animal, the chick. He successfully artificially inseminated amphibians, silkworms and a spaniel bitch. A staunch preformationist in embryology, he was convinced the ovum already contained all the organs later materializing in the embryo. He also experimented on the sensory systems of bats concluding that, even if blinded, bats could still catch insects and fly well enough to avoid even small objects. He was also fascinated by electricity, especially electric fish.

SPEDDING, Frank Harold
(1902–1984)

American nuclear scientist, born in Hamilton, Ontario. He studied chemical engineering at the University of Michigan and took his doctorate at the University of California at Berkeley. He spent his working life at Iowa State University at Ames. In 1942 he was co-opted to the Manhattan Project at the University of Chicago – with Fermi, Seaborg and others – to develop the first atomic bomb. In the laboratory at Ames he found a way to separate the fissile isotope of uranium, uranium-235, from its other isotopes in sufficient quantities for the core of the first atomic pile. He and his co-workers produced 6 tons of uranium in the form of large pellets ('Spedding's eggs') which were dropped into matching holes in the graphite core. The pile went critical on 2 December 1942, in the first man-made nuclear chain reaction. After World War II, Spedding looked for cheaper ways of separating the lanthanides, the group of 'rare earth metals' which have atomic weights from 57 to 71. These elements have many industrial uses but are chemically very similar and therefore difficult to isolate. To overcome the problem, Spedding developed the ion-exchange chromatograph: a mixture of substances is passed through very fine particles of resin in a column and the different substances separate out into different bands. He used the same technique to separate the actinides, heavy metals of atomic weights of 89-103 with properties similar to the lathanides. He died in Ames.

SPEMANN, Hans
(1869–1941)

German zoologist, born in Stuttgart. Educated in Stuttgart and Heidelberg, he became professor at Rostock (1908–1914), Director of the Kaiser Wilhelm Institute of Biology in Berlin (1914–1919) and professor at Freiburg (1919–1935). He was an experimental embryologist and discovered the 'organizer function' of certain tissues during development. He studied the early stages of newt development and showed that if presumptive skin tissue (ectoderm) is transplanted into presumptive neural tissue, for example, it develops into the latter and not into skin. In other words, the fate of embryonic cells is not programmed at an early stage but rather by the tissues that they are in contact with. He wrote *Embryonic Development and Induction* (1938) and won the Nobel Prize for Physiology or Medicine in 1935.

SPENCER, Sir (Walter) Baldwin
(1860–1929)

English-born Australian anthropologist and biologist, born in Stretford. In 1884 he graduated in natural sciences from Exeter College, Oxford, and in 1886 he was elected to a fellowship at Lincoln College. Within a year he was made foundation Professor of Biology at Melbourne University, an appointment which he held for 32 years. In 1894 Spencer joined W. A. Horn's expedition to Central Australia, where he met the telegraphist Francis James Gillen. Gillen had become intimate with the local tribespeople, and Gillen and Spencer collaborated in anthropological studies resulting in a number of invaluable published works, including *Native Tribes of Central Australia* (1889) and *Northern Tribes of Central Australia* (1904). Gillen died in 1912, and in the same year Spencer was appointed Chief Protector of the Aborigines. He was knighted in 1916. He died while on a trip to the world's most southerly settlement, on Tierra del Fuego.

SPENCER, Herbert
(1820–1903)

English philosopher, born in Derby. He had a varied career as a railway engineer, teacher, journalist and subeditor at *The Economist* (1848–1853) before devoting himself entirely to writing and study. His particular interest was in evolutionary theory which he expounded in *Principles of Psychology* in 1855, four years before

Charles Darwin's *The Origin of Species*, which Spencer regarded as welcome scientific evidence for his own *a priori* speculations and a special application of them. He also applied his evolutionary theories to ethics and sociology and became an advocate of 'social Darwinism', the view that societies naturally evolve in competition for resources and that the 'survival of the fittest' is therefore morally justified. He announced in 1860 a *System of Synthetic Philosophy*, a series of volumes which were to comprehend metaphysics, ethics, biology, psychology, and sociology, and nine of these appeared between 1862 and 1893. He viewed philosophy itself as the science of the sciences, distinguished by its generality and unifying function.

SPERRY, Roger Wolcott
(1913–)

American neuroscientist, born in Hartford, Connecticut. Educated at Oberlin College he studied zoology at Chicago University (PhD 1941), then worked as a Research Fellow at Harvard and at the Yerkes Laboratory of Primate Biology (1941–1946). He taught at Chicago University (1946–1952), and was Hixon Professor of Psychobiology at Caltech from 1954 to 1984. He first made his name in the field of developmental neurobiology, his experiments helping to establish the means by which nerve cells come to be wired up in particular ways in the central nervous system. In the 1950s and 1960s he pioneered the behavioural investigation of split-brain animals, and from 1961 he was able to study human split-brain patients at the White Memorial Medical Centre in Los Angeles. From detailed observations he and his collaborators established that each hemisphere possessed specific higher functions, the left side controlling verbal activity and processes such as writing, reasoning etc; whereas the right side is more responsive to music, face and voice recognition etc. Sperry argued that each hemisphere also contained its own consciousness, and his experiments led him into studies of philosophy and consideration of the mind/brain problem. His work has stimulated new approaches and work in neurology, psychiatry and psychology, as well as the basic neurosciences. He shared the Nobel Prize for Physiology or Medicine in 1981 with Hubel and Wiesel.

SPITZER, Lyman Jr
(1914–)

American astrophysicist, born in Toledo, Ohio. He was educated at Yale and Princeton, where he was Professor of Astronomy from 1947 to 1979. His interest in energy generation in stars led to his early attempt to achieve controlled thermonuclear fusion, for which he devised a method of 'containing' ionized gas, or plasma, in a magnetic field; the principle continues to form part of experimentation in this area. In 1951, with Blitade, he suggested that the class of galaxies known as SO were formed when spiral galaxies collided. With Martin Schwarzschild, he postulated in 1956 the existence of giant molecular clouds in interstellar space long before they were observed. They suggested that these clouds were responsible for inducing random velocities in galactic disc stars, these velocities increasing with the age of the stars. The stability of these clouds when they were far from the galactic plane also indicated that hot gas exists out there. In 1958 Spitzer studied the tidal shock that occurs between cluster stars and the galactic disc; he summarized many of his ideas in *Dynamical Evolution of Globular Clusters* (1987).

SPOTTISWOODE, William
(1825–1883)

English mathematician, physicist and printer, born in London. His father was Member of Parliament for Colchester and also a printer. He was educated at Harrow and in mathematics at Balliol College, Oxford, and in 1846 he succeeded his father as head of the printing house of Eyre and Spottiswoode. Spottiswoode did original work on the polarization of light and electrical discharge in rarefied gases. The latter experiments were made with a colossal induction coil (an early form of transformer) made for him by the London instrument-maker, A. Apps. With its secondary coil of no less than 280 miles of wire, it was capable of producing discharges with a length of 42 inches. He wrote a series of original memoirs on the contact of curves and surfaces, and the first elementary mathematical treatise on determinants (1851). He was elected FRS in 1871, and served as the Royal Society's President from 1878 until his death. He was buried in Westminster Abbey.

SPRENGEL, Christian Konrad
(1750–1816)

German amateur botanist, born in Brandenburg. He became rector of Spandau, but neglected his duties to make discoveries about the part played in the pollination of plants by nectaries and insects. He was eventually removed from his clerical post for failing to provide even such basic services as sermons on Sundays. He suggested that flowers had adapted specifically to allow for insect pollination. He observed that dichotomy, the floral structure which favours cross-pollination by preventing self-pollination, is extremely frequent. His work influenced entomologists of the time, but became largely neglected until re-examined by Charles Darwin. His nephew, Kurt Sprengel (1766–1833), wrote histories of medicine (1803) and botany (1918).

SPRENGEL, Hermann Johann Philipp
(1834–1906)

British chemist and physicist, inventor of the high vacuum pump which bears his name. He was born in Schillerslage, near Hanover, and educated at the universities of Göttingen and Heidelberg. In 1859 he moved to England, later becoming a British citizen. He carried out research at Oxford and in the laboratories of several institutions in London. He mechanized the pump devised by Heinrich Geissler in 1858, making the action of the pump much swifter and more efficient. He described his invention in *On the Vacuum*. Sprengel's pump had far reaching impact. It made possible of radiation in a high vacuum (leading to the discovery of J. J. Thomson), Ramsay's and Travers's work on the rare gases, and Swan's electric light bulbs. Sprengel also developed the pyknometer, a U-shaped vessel for expansion of liquids, and researched and wrote extensively on high explosives. He was elected FRS in 1878, and died in London.

SPRUCE, Richard
(1817–1893)

English botanist, born in Ganthorpe near Malton, Yorkshire, where his father was village schoolmaster. He himself became a schoolmaster, first at Haxby, then at St Peter's Collegiate School, York. As a recreation he studied botany, collecting bryophytes and other plants from the North Yorkshire Moors; his first paper (1841) was on the bryophytes of Eskdale. In 1844, when the school at York closed, he decided to make botany his career. From 1845 to 1846 he collected in the Pyrenees, discovering many unrecorded bryophytes. In 1849 Sir William Jackson Hooker sent him to South America where, six months later, at Santarem, he met Wallace. Spruce spent the following 15 years exploring the Amazon, Orinoco, the Andes and Ecuador, until he returned to England in poor health in 1864. He brought back a collection of more than 7,000 plant specimens, maps of three previously unexplored rivers, and vocabularies of 21 Amazonian tribes. His most famous publication was *Notes of a Botanist in the Amazon*, edited by Wallace (2 vols, 1908); he also wrote *The Musci and Hepaticeae of the Pyrenees* (1850), *Palmae Amazonicae* (1869) and *Hepaticae Amazonicae et Andinae* (1884).

STAHL, Franklin William
(1929–)

American molecular biologist, born in Boston. He was educated at Harvard and the University of Rochester, and held posts at Caltech (1955–1958), the University of Missouri (1958–1959) and the University of Oregon, where he became professor in 1970. In 1958 he established with Meselson that DNA replicates by a semi-conservative mechanism.

STAHL, Georg Ernst
(1660–1734)

German chemist who developed the phlogiston theory of combustion. Born in Ansbach, he studied medicine at Jena and in 1687 was appointed court physician to the Duke of Sachsen-Weimar. From 1694 he taught at the newly founded University of Halle and in 1716 became personal physician to Frederick-William I, King of Prussia. Stahl believed that although physiological processes could largely be explained in terms of chemistry, each organism was directed by a life force or 'anima'. This doctrine, which became known as animism, brought him into conflict with his contemporaries who subscribed to the mechanistic explanations prevalent at the time. Stahl was most influential, however, in his attempts to explain combustion. He suggested that when a substance burns it loses a vital essence, which he termed 'phlogiston'. He argued that with metals the process is reversible: metals when heated lose their phlogiston and form a calx, but on further heating the phlogiston rejoins the calx to form the metal again. With organic substances the process is irreversible. Since substances do not burn without air, he argued that air was necessary to absorb the phlogiston. This theory, though erroneous, fitted the experimental facts as they were known at the time and became entrenched in chemical theory. It was only overthrown at the end of the eighteenth century by the work of Lavoisier and others. Stahl died in Berlin.

STANIER, Roger Yate
(1916–)

Canadian microbiologist, born in Victoria, British Columbia. He was educated at Shawrugan Lake School, Canada (where he was profoundly unhappy), and at the University of British Columbia, where he graduated in 1936. He then moved to the USA, obtaining his MA at the University of California at Los Angeles (1940), and his PhD at the University of Stanford (1942), where he worked with Van Niel. He was awarded a Guggenheim fellowship to work with Marjorie Stephenson at Cambridge in 1945, and became professor at Berkeley (1947–1971), then at the Pasteur Institute. Stanier researched bacterial tryptophan metabolism, and discovered the mandelate pathway and the mechanisms of action of streptomycin. He postulated that the amino acid tryptophan is metabolized to catechol via the intermediates kyrturenine and anthranilic acid. He then showed that catechol is degraded to acetyl coenzyme A via the bacterial beta-ketoadipate pathway. He also made major contributions to bacterial chromatophores (coloured pigments) – notably bacteriochlorophyll – and carotenoid pigments. Elected FRS in 1978, he has received many honours, including honorary doctorates and the Carlsberg Medal.

STANLEY, Wendell Meredith
(1904–1971)

American biochemist, born in Ridgeville, Indiana. He was educated at Earlham College and Illinois University, where he received his PhD in 1929. He was a Research Fellow in Munich (1930–1931) and in 1931 joined the Rockefeller Institute for Medical Research, Princeton, before holding a series of professorships at the University of California from 1940. He was appointed Director of the Virus Laboratory at Berkeley in 1948. After working briefly on zymosterol (yeast sterol) and potassium transport into model cells (1931–1932), he isolated the tobacco mosaic virus (1935) using the salt fractionation techniques of Northrop, and showed it to contain protein and nucleic acid (1936) – this was described by Astbury as 'the most thrilling discovery of the century'. Stanley went on to characterize the physical (shape, birefringence, molecular weight) and chemical (amino acid composition, reactive groups and ribonucleic acid composition) properties of the virus, and with Fraenkel-Conrat determined the protein amino acid sequence (1960). Stanley also isolated other plant viruses, compared virus variants using immunological techniques and independently noted that viruses can cause cancer (1949). He shared the 1946 Nobel Prize for Chemistry with Northrop and Sumner.

STANLEY, William
(1858–1916)

American electrical engineer, born in Brooklyn, New York City. He patented a carbonized filament incandescent lamp and a self-regulating dynamo which were acquired by Westinghouse's Union Switch and Signal Co (from 1886 Westinghouse Electric). After working for Hiram Percy Maxim (1840–1916), he joined Westinghouse to work on a long-range alternating current system for which one of the principal components to be developed was the transformer. He founded the Stanley Electric Manufacturing Co for the commercial exploitation of alternating current systems.

STARK, Johannes
(1874–1957)

German physicist, born in Schickenhof, the son of a farmer. Educated at Munich, he held teaching posts at Munich, Göttingen, Hanover and Gredswald, before being appointed to chairs at Aachen (1909), Gredswald (1917) and Wurzburg (1920). Because of his support for Hitler and his stand against modern theoretical physics which he viewed as a 'Jewish science', he was appointed as President of the Physikalisch Technische Reichsanstalt at Charlottenburg in 1933, and in the following year, of the Notgemeinschaft der Deutschen Wissenschaft (later renamed Deutsche Forschungsgemeinschaft). However, he did not hold these posts for long because of his quarrelsome nature and internal political struggles, and he retired in 1936. In 1947 he was sentenced by a German denazification court to four years in a labour camp. In his early years, Stark made important contributions in physics. He discovered the 'Stark effect' concerning the splitting of spectrum lines by subjecting the light source to a strong electrostatic field, and also the Doppler effect in 'canal rays' (rays of positively charged particles emitted from the anode in discharge tubes). He argued that these phenomena reinforced Einstein's theory of special relativity and Planck's quantum theory. He was awarded the Nobel Prize for Physics in 1919. In the 1920's he turned vehemently against both relativity and quantum physics.

STARLING, Ernest Henry
(1866–1927)

English physiologist, born in London. He qualified in medicine in 1889 from Guy's Hospital, where he was then appointed lecturer in physiology, and the following year began a lifelong professional and personal association with Rayliss, later his brother-in-law. He moved to University College (Jodrell Professor of Physiology 1899–1923; Foulerton Research Professor of Physiology, 1923–1927) and with Bayliss began a series of experiments on the nervous control of the viscera. In the course of this they discovered the pancreatic secretion secretin (1902), and for this and similar chemical messengers they coined the word 'hormone'. His studies of cardiovascular physiology did much to elucidate the physiology of the circulation and the mechanisms of cardiac activity, still known today as 'Starling's law of the heart', and his work on capillary function gave rise to 'Starling's equilibrium' which equated intravascular pressure and osmotic forces across the capillary membrane. During World War I he was director of research at the Royal Army Medical Corps College, and was associated with research on poison gases and chaired the Royal Society's Food Committee. He wrote many influential texts, including *Principles of Human Physiology* (1912).

STAS, Jean Servais
(1813–1891)

Belgian chemist, born in Louvain. He qualified as a doctor at Louvain and in 1837 moved to Paris. Here he investigated carbonic acid and other organic substances with Durnas. In 1840 he was appointed to the Chair of Chemistry at the Military School in Brussels, retiring due to ill-health in 1868. Stas's principal work was on atomic weights. Following Dalton's atomic theory, Prout had suggested that atomic weights are integer multiples of the atomic weight of hydrogen, and that any deviations were the result of experimental errors. For more than 20 years, Stas laboured to make accurate determinations of the atomic weight of carbon, nitrogen, chlorine, sulphur, potassium, sodium, lead and silver by chemical methods, setting up chemical reactions and weighing the reagents and their products, and proved conclusively that Prout's proposal was incorrect. The reason for this discrepancy in weights was not understood until Soddy discovered isotopes and Chadwick discovered the neutron. Stas died in Brussels.

STAUDINGER, Hermann
(1881–1965)

German chemist, born in Worms. After graduating from the local gymnasium, he studied chemistry at the University of

Halle but soon transferred to the Technical University at Darmstadt. He obtained his doctorate at Halle in 1903 for a study of malonic esters. He then became assistant to Johannes Thiele in Strassburg, where he discovered keten. On being appointed assistant professor in the Technical University at Karlsruhe he began research on the structure of rubber, and in 1910 found a new and simpler way to synthesize isoprene, the basic unit of rubber. In 1912 he succeeded Willstatter at the Federal Institute of Technology in Zurich, where he worked on the synthesis of natural products. After World War I he returned to his work on the structure of rubber and propounded the view that rubber consists of macromolecules rather than aggregates, a suggestion that caused considerable controversy. This controversy continued until 1926, when X-ray crystallographic evidence largely confirmed Staudinger's view. During the 1930s he started the study of complex biological macromolecules and in the 1940s he turned to molecular biology. He was awarded the Nobel Prize for Chemistry in 1953 for his discoveries in the field of macromolecular chemistry. A research institute was established for Staudinger at the University of Freiburg in the 1940s, and he remained there until his retirement in 1956.

STEARN, William Thomas
(1911–)

English botanist, bibliographer and horticulturist, born in Cambridge. His early career was as an apprentice antiquarian bookseller at Cambridge (1929–1932) and as librarian of the Royal Horticultural Society (1933–1941, 1946–1952); the latter post was broken by Military service during World War II in Britain, India and Myanma (Burma). From 1952 to 1976 he was a botanist at the British Museum (Natural History), London; now retired, he is still an extremely active consultant. He is the world's foremost authority of Linnaeus, with an encyclopaedic knowledge of his life and works. His immense output (some 440 works to date) includes the books, *Botanical Latin* (1966), *Gardener's Dictionary of Plant Names* (1972), *Peonies of Greece* (1984, with Peter Davis) and *Flower Artists of Kew* (1990). His botanical researches have ranged widely, including studies of Symphyton, Vinca, Epimedium and numerous monocotyledonous genera such as Ornithogalum and the difficult onion genus, Allium. Stearn has also published many papers on the history and bibliography of botany.

STEBBINS, George Ledyard
(1906–)

American botanist, born in Lawrence, New York. He studied biology at Harvard University and spent his career at the University of California, at Berkeley (before 1937–1950) and Davis (1950–1973), where he established the department of genetics. He was the first to apply modern ideas of evolution to botany, as expounded in his *Variation and Evolution in Plants* (1950). From the 1940s he used artificially induced polyploidy (the condition of having more than twice the basic number of chromosomes) to create fertile hybrids; this technique is of value both in taxonomy and in plant breeding. His other books include *Processes of Organic Evolution* (1966), *The Basis of Progressive Evolution* (1969) and *Flowering Plants: Evolution Above the Species Level* (1974). In the latter, he proposed a new classification system of flowering plants, closely modelled on that of Cronquist but with some modifications.

STEBBINS, Joel
(1878–1966)

American astronomer, born in Omaha, Nebraska. He received his first degree at the University of Nebraska and his doctorate at Lick Observatory. He was appointed as an astronomer on the staff of the University of Illinois and became Professor of Astronomy in 1913. In 1922 he moved to the University of Wisconsin and became Director of the Wasburn Observatory. Soon after the invention of the photocell, in 1911, Stebbins applied electronic photometers to astronomical sources and showed that they were much more accurate for brightness measurement than methods that relied on the eye or photography. In the 1930s he used photoelectric techniques to measure the way in which the dust and gas in the galaxy affected the transmission of starlight, causing a reddening effect. The magnitude of this interstellar reddening helped define the structure and size of the galaxy. With Albert Whitford he introduced a six-colour photometry system, extending from the ultraviolet to the infrared, which he applied with considerable success to the study of galaxies.

STEENSTRUP, Johannes Iapetus Smith
(1813–1897)

Danish zoologist, born in Vang, Norway, and educated in Aalborg and at the University of Copenhagen. From 1846 to 1873 he was Professor of Zoology at Copenhagen and Director of the Zoology Museum there. He is best known for his pioneering studies on cephalopods, but also worked on many marine invertebrates and achieved fame for his demonstration of an alternation of sexual and asexual generations in certain animals. He was a founder of scientific archaeology through his work on the animal and plant remains in Danish peat bogs, recognizing significant climatic changes in prehistoric times and explaining the origin of coastal shell mounds as Stone Age middens. He also wrote on the distribution and extinction of the great auk.

STEFAN, Josef
(1835–1893)

Austrian physicist, born near Klagenfurt. He was a brilliant experimenter who became Professor of Physics at Vienna in 1863 after being a school teacher for seven years. In 1866 he was appointed Director of the Institute for Experimental Physics founded in Vienna by Doppler in 1850. In 1879 he proposed Stefan's law (or the Stefan-Boltzmann law), that the amount of energy radiated per second from a black body is proportional to the fourth power of the absolute temperature. He used this law to make the first satisfactory estimate of the Sun's surface temperature. He also designed a diathermometer to measure heat conduction, worked on the kinetic theory of heat and on the relationship between surface tension and evaporation (1886).

STEIN, William Howard
(1911–1980)

American biochemist, born in New York City. He studied at Harvard and Columbia University, and joined the staff of the Rockefeller institute, where he became Professor of Biochemistry in 1954. Throughout his career he collaborated closely with Moore, particularly in the development of a column chromatographic method for the sequential elution, identification and quantification of amino acid n-fixtures derived from the hydrolysis of proteins or from physiological tissues. By determining the amino acid composition of bovine pancreatic ribo-nuclease (1954–1956), they complemented the structural studies of Anfinsen on this enzyme. They also automated all the steps for the analysis of the base sequence of RNA on a small sample (1958) and studied a novel protease from streptococcus (discovered by the English microbiologist Stuart D. Elliott) showing that although this enzyme possesses a similar specificity and catalytic site organization (including a so-called 'essential thiol group') to the plant protease papain, other aspects of the molecular structure are quite different. Ibis was the first example of convergent evolution – two enzymes of similar function arising by different evolutionary paths. Stein, Moore and Anfinsen shared the Nobel Prize for Chemistry in 1972.

STEINBERGER, Jack
(1921–)

German-born American physicist, born in Bad Kissingen. He moved to the USA in 1935, and was educated at the University of Chicago, where he received his PhD in 1948. He held professorships at Columbia University from 1950 to 1972, and from 1968 to 1986 was a staff member at CERN, the European centre for nuclear research in Geneva, where he was a director from 1969 to 1972. With colleagues, he proved the existence of a neutral pion by observing the coincidence of the two gamma rays from its decay at the Berkeley synchroton, and measured the spin and parity of the charged pion. Studies of muons had shown that they behaved like heavy electrons. However, Steinberger showed that, unlike the electron, the muon decay involves not one but two neutrinos. To explain this it was proposed that the muon is a heavy electron but has its own identity, the 'electron charge' being carried by the electron neutrino and the 'muon charge' being carried by the muon neutrino. In 1960 Schwartz published a paper proposing an experiment that could establish the existence of the two distinct neutrinos; it was performed at Brookhaven by Lederman, Schwartz, Steinberger and their collaborators, and in 1962 they announced that they had observed 20 muon events proving the existence of the two distinct neutrino types. Steinberger went on the study CP violation and carry out further neutrino experiments. He was spokesman for the ALEPH Collaboration, one of the four large groups of physicists that built an experiment to run at the Large Electron-Positron (LEP) collider at CERN. Since 1989 this has been precisely testing the electroweak model experimentally. Steinberger was awarded the 1988 Nobel Prize for Physics jointly with Lederman and Schwartz.

STEINER, Jakob
(1796–1863)

German-Swiss geometer, born in Utzendorf. He became professor at Berlin in 1834, and pioneered 'synthetic' geometry, particularly the properties of geometrical constructions, ranges and curves. He possessed tremendous powers of visualization, but his papers were renowned for their obscurity and he frequently withheld proofs, a practice which earned him the nickname of the 'celebrated sphinx'. He also published many ingenious proofs (not all rigorous) of the obvious but elusive result that the circle encloses the greatest area of all curves of a given length.

STEINMETZ, Charles Proteus, originally Kari August Rudolf
(1865–1923)

German-born American electrical engineer, born in Breslau (now Wroclaw,

Poland). Educated there and at the Technical High School, Berlin, he was forced to leave Germany in 1888 due to his socialist activities and he emigrated to the USA in 1889. He began work in a small electrical factory which was taken over by the General Electric Company in 1893, and he worked for that company for the rest of his life. A hunchback from birth, he was rapidly recognized as a brilliant theoretician, yet was sufficiently practical to be granted over 200 patents for his inventions. Perhaps his finest, and certainly one of his earliest achievements was to work out in complete detail, using complex numbers, the mathematical theory of alternating currents. This enabled the design of AC machines to be made more efficient, and consolidated the victory of AC over DC gained by Tesla in fierce competition with Edison. In addition to his work for General Electric, he was Professor of Electrical Engineering and later Professor of Electrophysics at Union College, Schenectady, from 1902. Among his many other discoveries were the phenomenon of magnetic hysteresis and a method of providing protection against lightning for high-power transmission lines.

STENO, Nicolaus (also known as Niels Stensen)
(1638–1686)

Danish physician, naturalist and theologian who made major advances in anatomy, geology, crystallography, palaeontology and mineralogy. Born a Protestant in Copenhagen, he converted to Catholicism and settled in Florence. He was appointed personal physician to the Grand Duke of Tuscany in 1666 and Royal Anatomist at Copenhagen in 1672. He became a priest in 1675, and gave up science on being appointed Vicar-Apostolic to North Germany and Scandinavia. He is buried in the crypt of the Medici in Florence. As a physician he discovered Steno's duct of the parotid gland, and investigated the function of the ovaries. A passionate anatomist, he showed that a pineal gland resembling the human is found in other creatures; he used this to challenge Descartes's claim that the gland was the seat of the uniquely human soul. Steno's examination of quartz crystals disclosed that, despite differences in the shapes, the angle formed by corresponding faces is invariable for a particular mineral. This constancy (Steno's law) follows from internal molecular ordering. Steno is perhaps best remembered for his contributions to geology and palaeontology. Having found fossil teeth far inland closely resembling those of a shark he had dissected, he championed the organic nature of fossils against those who believed they were 'sports of nature'. On the basis of his palaeontological views, he also contended that sedimentary strata were laid down in former seas. He sketched what are perhaps the earliest geological sections.

STEPTOE, Patrick Christopher
(1913–1988)

English gynaecologist and reproduction biologist, born in Witney. He was educated in London at King's College and St George's Hospital Medical School. Following military service, he specialized in obstetrics and gynaecology, becoming senior obstetrician and gynaecologist at the Oldham Hospitals in 1951. In 1980 he became Medical Director of the Bourn Hall Clinic in Cambridgeshire. He had long been interested in laparoscopy (a technique of viewing the abdominal cavity through a small incision in the umbilicus) and in problems of fertility. He met Edwards in 1968, and together they worked on the problem of *in vitro* fertilization of human embryos, which 10 years later resulted in the birth of a baby after *in vitro* fertilization and implantation in her mother's uterus. The ethical issues are still controversial, but the technique has become more common.

STERN, Otto
(1888–1969)

American physicist, born in Sohrau, Germany, and educated at Breslau University where he obtained his doctorate in 1912. He then held posts at Zurich, Frankfurt and Rostock before becoming Professor of Physical Chemistry at the University of Hamburg (1923–1933). With the rise of the Nazis he moved to the USA where he became Research Professor of Physics at the Carnegie Institute of Technology in Pittsburgh (1933–1945). In collaboration with Walther Gerlach, Stern carried his best-known experiment in 1920–1921. By projecting a beam of silver atoms through a non-uniform magnetic field, they showed that two distinct beams could be produced. This provided fundamental proof of the quantum theory prediction that an atom should possess a magnetic moment which can only be oriented in two fixed directions relative to an external magnetic field. For this work he was awarded the Nobel Prize for Physics in 1943. Stern also determined the magnetic moment of the proton.

STEVENS, Nettie Maria
(1861–1912)

American biologist, born in Cavendish, Vermont. She began her career as a librarian, but later entered Stanford University to study physiology. She received a PhD from Bryn Mawr College, Pennsylvania (1903), and was subsequently appointed to research posts there. The college eventually created a research professorship for her, but she died before she could take up the position. Stevens was one of the first to show that sex is determined by a particular chromosome, fertilization of an egg by a sperm carrying the X chromosome will result in female offspring and that Y-carrying sperm will produce a male embryo (a discovery made independently by Edmund Wilson). She extended this work to studies of sex determination in various plants and insects, demonstrating unusually large numbers of chromosomes in certain insects and the paired nature of chromosomes in mosquitoes and flies.

STEVIN, Simon
(1548–1620)

Flemish mathematician and engineer, born in Brugge. He held offices under Prince Maurice of Orange, wrote on fortification and book-keeping and invented a system of sluices and a carriage propelled by sails. He was responsible for introducing the use of decimals which were soon generally adopted, and which he had advocated in a somewhat cumbersome notation in his book *De Thiende* (1585). His maxim 'A wonder is not a wonder' was a rallying cry for those who advocated rational experimentation in the new sciences. He wrote influentially on statics and the law of the inclined plane is due to him.

STEWART, Balfour
(1828–1887)

Scottish physicist, born in Edinburgh, the son of a tea merchant. He studied at St Andrews and Edinburgh, and became assistant to Edward Forbes at Edinburgh and later Director of Kew Observatory (from 1859), and Professor of Physics at Owens College, Manchester (from 1870).

He carried out original work on radiant heat (or infrared radiation) and thermal radiation (1858), but unfortunately, this was overshadowed by similar investigations by Kirchhoff. Stewart was elected FRS in 1862 and awarded the Royal Society's Rumford Medal. He was one of the founders of spectrum analysis and wrote papers on terrestrial magnetism (explaining both the daily and seasonal variations) and on sunspots.

STEWART, Sir Frederick Henry
(1916–)

Scottish geologist, born in Aberdeen and educated at Aberdeen University. He was a mineralogist with ICI (1941–1943) before becoming lecturer in geology at the University of Durham (1943–1956). From 1956 to 1982 he was Regius Professor of Geology at the University of Edinburgh. Stewart's principal research interests have been in petrology and mineralogy, particularly evaporite deposits as well as fossil fish. He wrote *Marine Evaporites* (1963) and was co-editor of *The British Caledonides* (1963). He was elected FRS in 1964, and knighted in 1974.

STEWART, Ralph Randles
(1890–)

American botanist, born in West Hebron, New York. He studied botany at Columbia University and during 1911–1914 taught biology at Gordon College, Rawalpindi. He returned briefly to the USA (1914–1917) and obtained a doctorate from Columbia University for a thesis on the flora of western Tibet. From 1917 to 1960 he again taught at Gordon College, and made many plant collecting trips all over Pakistan, Kashmir and Western Tibet; his collections number some 60,000 specimens. Stewart became the foremost authority on the botany of Pakistan and the western Himalayas, publishing over 40 papers on the subject. His botany teaching inspired many students, including Syed Irtifaq Ali (b. 1930) and Eugene Nasir (b. 1908) who began the *Flora of Pakistan* (published from 1970 in over 170 parts) using Stewart's collections as a basis. During 1960–1981 he worked as a research associate at the University of Michigan Herbarium, Ann Arbor, and published two important works on the flora of Pakistan: *An Annotated Catalogue of the Vascular Plants of West Pakistan and Kashmir* (1972) which enumerated 5,783 taxa, and *History and Exploration of Plants in Pakistan and Adjoining Areas* (1982). He continues to be active and attended celebrations in Karachi to mark his hundredth birthday.

STIBITZ, George Robert
(1904–)

American mathematician and computer scientist, the son of a theology professor at a small college in Dayton, Ohio. Stibitz attended Denison University, a small liberal arts college in Granville, Ohio, and Cornell University, where he was awarded a doctorate in mathematical physics. By 1937 he was working at Bell Telephone Laboratories, where he utilized telephone relays to build a binary adder (Shannon, who later joined Bell, had also noticed the correspondence between relays, binary mathematics and symbolic logic). In 1939 Bell Laboratories supported Stibitz in the building of a more sophisticated 'complex number calculator', the Model I. Although reliable and easy to use, it was not programmable or general-purpose, and did not have a memory. Stibitz later designed program-controlled calculators for the military during World War II, but these were soon to be outmoded by the development of the electronic digital computer.

STIRLING, James, known as 'the Venetian'
(1692–1770)

Scottish mathematician, born in Garden, Stirlingshire. He studied at Glasgow and Oxford (1711–1716), but left without graduating. His first book, on Isaac Newton's classification of cubic curves, was published in Oxford in 1717. He visited Venice at about this time, returned to Scotland in 1724, and then went on to London, where he taught mathematics. From 1735 he was Superintendent of the lead mines at Leadhills, Lanarkshire, and corresponded with Maclaurin. His principal mathematical work was *Methodus differentialis* (1730), in which he made important advances in the theory of infinite series and finite differences, and gave an approximate formula for the factorial function, still in use and named after him.

STOCK, Alfred
(1876–1946)

German chemist, famous for his work on the boron hydrides (boranes). In 1912 he discovered diborane (B2H6), and over the next 20 years he pioneered the synthesis and isolation of many 'higher' boranes. In 1926 he found that the reaction of diborane with ammonia yielded the compound B3N3H6 (borazine), which is known as 'inorganic benzene' due to some of its chemical and physical properties. Stock is also famous for his development of high-vacuum apparatus for handling chemical samples – indispensable for his work on the highly air-sensitive boranes.

STOKES, Sir George Gabriel
(1819–1903)

Irish mathematician and physicist, born in Skreen, Sligo. He graduated in 1841 from Pembroke College, Cambridge, and in 1849 became Lucasian Professor of Mathematics. From 1887 to 1892 he was Conservative MP for Cambridge University. He first used spectroscopy as a means of determining the chemical compositions of the Sun and stars, published a valuable paper on diffraction (1849), identified X-rays as electromagnetic waves produced by sudden obstruction of cathode rays, and formulated Stokes' law expressing the force opposing a small sphere in its passage through a viscous fluid. He is also remembered for his derivation of Stokes' theorem, a useful result which identifies the equivalence of two particular integral operations in vector calculus. He was made a baronet in 1889.

STOMMEL, Hank (Henry Melson)
(1920–1992)

American physical oceanographer, born in Wilmington, Delaware, and educated at Yale where he graduated in Physics in 1942. As a conscientious objector during World War II he instructed navy students in geometry and celestial navigation. He subsequently moved to Woods Hole Oceanographic Institution to work in physical oceanography (1944–1960), and became Professor of Oceanography at Harvard (1960–1962) and at MIT (1963–1978). He returned to Woods Hole in 1978. Stommel's research covered the broad field of physical oceanography in both theory and experimental work. He investigated the intensification of oceanic currents due to the Coriolis force at the western end of a gyre (circuit), and mapped the thermocline (the base of the warmer surface waters), developing the concept of its action as a boundary layer. He found vertical oscillation in the thermocline as it changed depth seasonally, and was one of the first to investigate ocean circulation at great depths. His books included studies of lost islands and the Gulf Stream.

STONE, Francis Gordon Albert
(1925–   )

English chemist, born in Exeter. He received his BSc and PhD from Christ's College, University of Cambridge, and in 1963 he became the first occupant of the Chair of Inorganic Chemistry at Bristol University. He retired in 1990, when he became Robert A. Welch Distinguished Professor of Chemistry at Baylor University, Texas. Stone is an organometallic chemist whose research has centred upon synthetic and mechanistic studies of transition metal complexes; he has published more than 650 articles. One principal interest has been in the field of polynuclear metal compounds, i.e. the metal 'cluster' compounds where metal atoms occupy the vertices of a polyhedron. He developed the chemistry of metal cluster compounds which contain bridging carbene or carbyne groups and has thus been able to study the reactivity of such groups at di- and tri-metal centres. Such studies have important implications for homogeneous catalytic processes. He was editor of the major work *Comprehensive Organometallic Chemistry* which was published in 1984. Elected FRS in 1976, his achievements have been recognized by many learned societies. He was awarded the Davy Medal of the Royal Society in 1989, the Organometallic Chemistry (1972), Transition Metal Chemistry (1979) and Longstaff (1990) medals of the Royal Society of Chemistry, and the American Chemical Society Award for Inorganic Chemistry in 1985. He has held several Royal Society of Chemistry endowed lectureships.

STORK, Gilbert
(1921–   )

American chemist, born in Brussels. After having obtained his baccalaureate in France, he moved to the USA and graduated from the University of Florida in 1942. He obtained his PhD from the University of Wisconsin (1945), was appointed instructor at Harvard and later became assistant professor. He moved to Columbia University as associate professor in 1953 and was appointed to the Eugene Higgins Chair in 1967. Throughout his career he has been concerned with developing synthetic routes to many different types of compounds. He has received many honours including the National Medal of Science (1983).

STRACHEY, Christopher
(1916–1975

English computer programmer and theorist, born in Hampstead, London. Strachey read mathematics and natural sciences at King's College, Cambridge, and then worked on radar at the research laboratories of Standard Telephones & Cables Ltd (STC). After World War II, he became a school teacher at St Edmond's School, Canterbury, and then at Harrow School. While at Harrow he did programming work on Turing's ACE at the National Physical Laboratory, and on the Manchester University Mark 1 built by Kilburn and Sir Frederic Calland Williams. In 1952 he was recruited by Lord Halsbury as adviser to the National Research Development Corporation (NRDC) which placed him at the heart of the developing British computer industry. He ran a private consultancy (1959–1965) and was appointed to a personal chair at Oxford in 1971. Recognized as one of the foremost computer architects and logicians of his day, Strachey's influence was apparent in the design of several British computers such as the Ferranti Pegasus, and in many innovative concepts. He made significant contributions in the areas of time-sharing, whereby users at several consoles could enjoy the facilities of one computer, and in denotational semantics, which sought to understand the meaning of computer languages in a mathematical way.

STRASBURGER, Eduard Adolf
(1844–1912)

German botanist, born in Warsaw. He studied botany in Paris, Bonn and Jena and spent his career at Jena (1869–1880) and Bonn (1880–1912). He studied the alternation of generations in plants, the embryo sac found in gymnosperms and angiosperms, and double fertilization in angiosperms. In his book *Cell Formation and Cell Division* (1875) and its later editions he laid down the basic principle of cytology, the study of cells, for which he made Bonn the world's leading centre. His work did much to show that mitosis (normal somatic cell division) in plants is a process essentially similar to that described for animal cells by Beneden and others. He observed nuclear fusion in the ovules of gymnosperms and angiosperms and remarked on the formation of the polar nucleus in their egg-cells. He introduces the terms haploid and diploid to describe respectively the halving and doubling of chromosome numbers in plant generations. Strasburger's *Textbook of Botany for Universities*, written with other botanists (including Andreas Schimper) under his guidance, is a classic, much used and widely translated in over 30 editions from 1894 onwards. In its updated form, it is still currently in print.

STRATO, or STRATON, of Lampsacus
(d.c.269 BC)

Greek philosopher, the successor to Theophrastus as the third head of the Peripatetic School (from about 287 to 269 BC) which Aristotle founded. His writings are lost, but he seems to have worked mainly to revise Aristotle's physical doctrines. He had an original theory about the void, its distribution explaining differences in weights of objects. He also denied any role to teleological, and hence theological, explanations in nature, which led naturally to the position David Hume dubbed 'Stratonician atheism' – universe is ultimate, self-sustaining and needs no further external or divine explanation to account for it.

STROMEYER, Friedrich
(1776–1835)

German chemist, born in Göttingen, and educated there and in Paris. He taught at Göttingen from 1802, becoming Professor of Chemistry in 1810. He was one of the first teachers to insist that his students had opportunities for practical work and Leopold Gmelin and Bunsen were among his pupils. He was also the inspector of apothecaries for Hanover. Stromeyer was a noted mineralogist, and in 1817 he discovered cadmium, a very white metal element akin to tin, in a sample of zinc carbonate. A rare metal, it is important today because cadmium rods are used to absorb excess neutrons in nuclear actors. Stromeyer died in Göttingen.

STROMGREN, Bengt
(1908–1987)

Danish astronomer, born in Gotelorge, the son of Elis Stromgren, Director of the Copenhagen Observatory and a distinguished astronomer in the field of classical mathematical astronomy. Bengt studied astronomy under his father's tuition at the University of Copenhagen, and atomic and quantum theory under Niels Bohr in the nearby Institute of Theoretical Physics. The close cooperation between father and son resulted in their joint publication of an outstanding textbook on astronomy in 1933. In 1936 Stromgren joined the second Otto Struve at Yerkes Observatory in the USA in work on the physics of stellar atmospheres and the properties of interstellar gas. His most outstanding work in this field concerns 'the physics of ionized gas clouds' known as HII regions surrounding hot stars. For the period of World War II he returned to Denmark where in 1940 he succeeded his father as Director of the Copenhagen Observatory, but after the war he returned to the USA as Director of Yerkes Observatory and professor at the Princeton Institute for Advanced Studies. In 1967 he returned finally to Denmark where he was granted the palatial residence at Carlsberg formerly occupied by Bohr. In these later years his work was concerned with problems of stellar composition and its correlation with ages of stars. Stromgren was awarded many honours from universities and academies the world over, including the Gold Medal of the Royal Astronomical Society (1962), and he acted both as General Secretary and as President of the International Astronomical Union.

STRUVE, Otto
(1897–1963)

Russian-born American astronomer, born in Kharkov, the son of Ludwig Struve, Professor of Astronomy at the University of Kharkov, and grandson of Otto Wilhelm Struve. He was educated at the University of Kharkov, where he graduated in 1919 after his studies were interrupted (1916–1918) by service in World War I. In 1919 he joined the White Army in opposition to the revolution and after its defeat escaped into exile, suffering considerable privations before making contact with American astronomers and being offered a post at the Yerkes Observatory in 1921. His family, including his father, perished in the revolution. In 1932 he was appointed Director of Yerkes Observatory and in 1939 he founded the McDonald Observatory of the University of Texas, with charge of both observatories until 1947. He was head of the department of astronomy at the University of Chicago (1947–1950) and Director of the Leuschner Observatory at the University of California from 1950 until his death. He was also the first Director of the National Radio Astronomy Observatory (1959–1962), and held an appointment at the Princeton Institute for Advanced Studies. Struve was principally a stellar spectroscopist, who performed an immense volume of observational work on stars of various types, on the interstellar medium and on gaseous nebulae. He attracted as staff members or as visiting collaborators many of the world's leading astronomical spectroscopists. He was awarded the Gold Medal of the Royal Astronomical Society in 1944, the fourth member of the family in successive generations to receive this honour, and was President of the International Astronomical Union from 1952 to 1955.

STRUVE, Otto Wilhelm
(1819–1905)

Russian astronomer, born in Dorpat (now Tartu), son of Wilhelm Struve, Director of Dorpat Observatory. He entered the University of Dorpat (1835) where he studied under his father, and became assistant at the Pulkova Observatory on his father's appointment as its director in 1839. He remained on the staff of the Pulkova Observatory for the rest of his working life, first as assistant (1839–1845), than as assistant director (1845–1862) and as director in succession to his father (1862–1889). Continuing his father's researches on double stars with Pulkova's 15-inch refractor, he discovered 500 new ones. His own most important studies were his determination of the constant of precession and of the solar motion through space (1841) for which he was awarded the Gold Medal of the Royal Astronomical Society in 1850. He took part in international projects such as the transits of Venus (1874) for which he organized 31 expeditions within and beyond the Russian empire. The Struve dynasty of astronomers included Otto's sons, (Karl) Hermann (1854–1920), who also received the Gold Medal of the Royal Astronomical Society (1903) and became Director of the Konigsberg (1895–1904) and the Berlin-Babelsberg (1904–1920) observatories, and (Gustav Wilhelm) Ludwig (1858–1920), Director of Kharkov Observatory (1897–1919) and father of the second Otto Struve.

STRUVE, (Friedrich Georg) Wilhelm
(1793–1864)

German-born Russian astronomer, born in Altona near Hamburg, and educated until the age of 14 at the gymnasium where his father was rector. From 1809 to 1813 he studied at the University of Dorpat (now Tartu) in Estonia (then part of the Russian empire), and he was later appointed Professor of Mathematics and Astronomy there (1816–1839), becoming also Director of Dorpat Observatory in 1818. At Dorpat he carried out a major programme of double star observations, published in a fundamental catalogue of 3,112 double stars entitled *Micrometria Mensurae* (1837). In 1837 he also measured the parallax of the star Vega, one of the three astronomers (the others being Bessel and Henderson) who in that year first succeeded in making such an observation. In 1835 Struve was summoned by the Russian emperor Nicholas I to superintend the building and equipping of a new observatory at Pulkova near St Petersburg, completed in 1839 with Struve in charge. The lavishly endowed and magnificently equipped observatory became the astronomical capital of the world. Struve's astronomical researches at Pulkova included work on the structure of the Milky Way (1847). He also supervised a huge geodetic survey, completed in 1860, extending from the Baltic to the Caucasus along an arc of meridian through Dorpat. In 1862 he handed over the directorship of Pulkova to his son and assistant, Otto Wilhelm Struve, the first of several astronomers among his descendants. He was awarded the Gold Medal of the Royal Astronomical Society in 1827 for his early work on double stars.

STURGEON, William
(1783–1850)

English scientist, born in Whittington, North Lancashire. Initially he followed in his father's trade as a shoemaker, but then enlisted in the Royal Artillery and was stationed at Woolwich, where he studied science in his free time. He became a bootmaker at Woolwich (1820), and was appointed a lecturer at the East India Company Royal Military College of Addiscombe (1824) and of the short-lived Adelaide Gallery of Practical Sciences (1832), before becoming Superintendent of the Royal Victoria Gallery of Practical Sciences in Manchester (1840). He ended his career as an itinerant lecturer, and for his services to science was awarded an annuity by the government. He constructed the first practical electromagnet (1825), the first moving-coil galvanometer (1836) and various electromagnetic machines. His *Annals of Electricity* (1836) was the first journal of its kind in Britain.

STURM, Jacques Charles Francois
(1803–1855)

French mathematician, born in Geneva. He discovered the theorem named after him concerning the location of the roots of a polynomial equation. With his friend Liouville, he also did important work on linear differential equations. In 1862 he measured the velocity of sound in water by means of a bell submerged in Lake Geneva.

STURTEVANT, Alfred Henry
(1891–1970)

American geneticist, born in Jacksonville, Illinois. He developed an enthusiasm for heredity through devising pedigrees for his father's farm horses, and later became a student of genetics under Thomas Hunt Morgan at Columbia University. He received his BA in 1912 and a PhD in 1914. From 1928 he spent his career at Caltech, as Professor of Genetics (1928–1947) and Professor of Biology (1947–1962). As an undergraduate Morgan had suggested to him that genes which are far apart on the same chromosome are more likely to be separated by the mechanism of recombination or 'crossing-over'; crossing-over occurs when there is a break in one chromosome which then attaches, or recombines with another chromosome. Using this idea, Sturtevant drew up the first chromosome map of the fruit-fly *Drosophila* in 1911. Later, as part of Morgan's 'fly room' group, he provided the mathematical background for genetic mapping experiments on *Drosophila*. Together with Morgan, Hermann Muller and C. B. Bridges, he established the basis for the chromosomal theory of heredity in *The Mechanism of Mendelian Inheritance* (1915). He also wrote *A History of Genetics* (1965).

SUESS, Eduard
(1831–1914)

Austrian geologist, born in London, the son of a German wool merchant of Jewish extraction. His parents moved to Prague (1843) and then to Vienna (1845) where, after a spell as an assistant in the geological department of the Royal Natural History Museum (1851–1857), he rose to great eminence at the university (1857–1901), becoming assistant professor and Professor of Geology. The greater part of his life was devoted to the study of the evolution of the features of the Earth's surface, particularly the problem of mountain building, which presented itself to his mind during his many excursions into the eastern Alps. He also focused attention on the volcanic islands and associated deep-sea trenches in the Pacific. His theory that there had once been a great super-continent made up of the present southern continents was a forerunner of modern theories of continental drift. His four-volume book *Das Antlitz der Erde* (1885–1909); translated as *The Face of the Earth* (1904–1910), was his most important contribution, ranking alongside Lyell's *Principles of Geology* and Charles Darwin's *Origin of Species* in significance. A man of varied interests and enthusiasms, he was a radical politician, an economist, an educationalist and a geographer, and sat in the Austrian Lower House.

SUGDEN, Samuel
(1892–1950)

English physical chemist, born in Leeds. After he studied chemistry at the Royal College of Science, London, his career was interrupted by World War I, in which he served briefly in the Royal Army Medical Corps and then as a chemist at Woolwich Arsenal (1916–1919). In 1919 he joined Birkbeck College, London, as a lecturer and he later became Professor of Physical Chemistry there (1932–1937). He was then appointed Professor of Chemistry at University College London and remained in that post until his death. During World War II he held various scientific posts in the war effort. Sugden is mainly remembered for devising (in 1924) a function of the molecular volume and surface tension of a liquid which he called the 'parachor'. This was considered to be the molecular volume measured at a standard internal pressure. The parachor appeared to have both additive and constitutive properties in terms of molecular structure, and it was hoped that it would be an important tool in determining the structures of organic molecules. For some years the parachor seemed to be useful, as detailed in Sugden's book *The Parachor and Valency* (1930), but ultimately it did not live up to expectations. Sugden's later work involved studies of paramagnetism in inorganic and organic chemistry, measurements of dipole moments or organic molecules, and pioneering applications of radioactive isotopes in the investigation of reaction mechanisms. He was elected FRS in 1934.

SUMNER, James Batcheller
(1887–1955)

American biochemist, born in Canton, Massachusetts. Educated at Harvard, he became assistant professor (from 1914) and then Professor of Biochemistry (1929–1955) at Cornell University, and Director of the Laboratory of Enzyme Chemistry (1947–1955). In 1926 he was first to crystallize (the then ultimate criterion of purity) an enzyme (urease), and demonstrated its protein nature. He then determined its kinetic and chemical properties, showing a dependence on reactive sulfhydryl groups on the protein. He raised and purified antibodies to urease (1933–1934), and purified plant antibody-like globulins (agglutinins) from Jack bean meal (canavulin and concavulins A and B) in 1938 thereby establishing a firm basis for the serological investigation of proteins. He also purified enzymes important for carrying out oxidative processes in the body (peroxidase from fig sap, catalase and horseradish peroxidase). These enzymes contain the non-protein components haem and iron, and Sumner investigated the function of these components by the effects various modifications had on the catalytic activity of the enzymes. He introduced the name 'monoamine oxidase' (an important pharmacological enzyme), used potato phosphorylase to prepare glucose 1-phosphate (1944), and isolated bean lipoxidase and rhodanese (1945) among many other achievements. He shared the 1940 Nobel Prize for Chemistry with Northrop and Wendell Stanley.

SUOMI, Verner
(1915–   )

American meteorologist and space scientist, born in Evaleth, Minnesota, of Scandinavian parents. After graduating from a teachers' training college at Winona (1939), he went to Chicago University to work on instrument development under Rossby. He designed an automatic dewpoint recorder which was carried into the stratosphere by balloon and also worked on a sonic anemometer. In 1948 he went to Wisconsin University and received a PhD (1953) for work on boundary layer processes. A net radiometer which he invented was included in the payload for the first American satellite. His greatest achievement was to design the 'spin-scan' camera for geostationary meteorological satellites: this instrument scans about a quarter of the Earth's surface continuously from one position in space and enables meteorological features such as cloud systems and tropical storms to be monitored, and upper winds to be estimated. It was later modified to operate at infrared wavelengths. Suomi also designed a small radio altimeter for use on constant level balloons, extensively used in the Global Atmospheric Research Program (1974). In the early 1970s he played a large part in the formation of the Man-Computer Interactive Data System to deal with various types of meteorological data. He also worked on radiometers to produce vertical temperature profiles from space.

SUTCLIFFE, Reginald Cockroft
(1904–1991)

British meteorologist, born in Wrexham, Wales. He studied mathematics at Leeds University and received a DSc at Bangor, North Wales, in 1927. He joined the Meteorological Office (1928) and worked in Malta in collaboration with Bergeron (1928–1932). He later became Director of Research at the Meteorological office (1953), and in 1965 he was appointed as first Professor of Meteorology at Reading University. He was President of the World Meteorological Organisation Commission for Aerology (1957–1961) and of the Royal Meteorological Society (1955–1956). His greatest contribution to meteorology was in the theory of development. Although handicapped by lack of upper air observations, he tackled the development problem in a systematic three-dimensional way. His most famous paper was *A Contribution to the Theory of Development* (1947) in which he used pressure (instead of height) as the vertical co-ordinate in his equations. This new departure was subsequently widely adopted and the paper led to practical advances in weather forecasting. Sutcliffe was responsible for the introduction of harotropic and baroclinic numerical weather prediction models into the Meteorological Office forecasting routine and at Reading University he set up the first undergraduate course with meteorology as a principal subject. He was the recipient of the International Meteorological Organisation Prize (1963) and Symons Gold Medal winner (1955).

SUTHERLAND, Earl Wilbur Jr
(1915–1974)

American biochemist, born in Burlingame. He studied medicine at the Washington University School of Medicine in St Louis and received his medical degree in 1942. He served an internship before being called up for service as a surgeon in the army. After World War II, Sutherland became an instructor and Associate Professor in the biochemistry department of Washington University. In 1953 he became Director of the Pharmacology Department of the Western Reserve University in Cleveland. Sutherland's research concerned the conversion of glycogen (the energy store in liver and muscle) into glucose, and the stimulation of this process by the hormones glucagon and epinephrine. He showed that a molecule known as cyclic-AMP promotes the activation of phosphorylase, the enzyme responsible for the glycogen-glucose transformation, and proposed that glucagon and epinephrine act by inducing the cell to produce c-AMP. Sutherland had discovered a new principle – the second messenger theory of hormonal action. In 1963 he became Professor of Physiology at Vanderbilt University. He showed that c-AMP acts as second messenger for many mammalian hormones, and was awarded the 1971 Nobel Prize for Physiology or Medicine.

SUTHERLAND, Sir Gordon Brims Black McIvor
(1907–1980)

Scottish physicist, born in Watten, Caithness. He was educated at the University of St Andrews and at Cambridge, where he completed his PhD in 1933. He remained at Cambridge as a Fellow and Lecturer of Pembroke College (1935–1949), Assistant Director of Research in Colloid Science (1944–1947) and Reader in Spectroscopy (1947–1949). He then moved to the USA as Professor of Physics at the University of Michigan (1949–1956). Returning to the UK in 1956, he took up the directorship of the National Physical Laboratory at Teddington. In 1964 he became Master of Emmanuel College at Cambridge, where he remained until his retirement. His areas of research included infrared and Raman spectroscopy, his interest lying in the structure of molecules. During the 1930s he published widely on the absorption spectra of numerous molecules, including C2H2 and N204, and was able to explain his observations in terms of the vibrational modes of the molecular bonds. He also investigated various triatomic molecules, including O3, F2O and NO2, again successfully identifying their vibrational modes. He was elected FRS in 1949 and knighted in 1960.

SVEDBERG, Theodor
(1884–1971)

Swedish physical chemist, born in Flerang, near Valbo. He entered the University of Uppsala in 1904 to study chemistry and was associated with that university for the next 45 years. From 1912 to 1949 he was Professor of Physical Chemistry. Although beyond retiring age, he was Director of the Gustaf Werner Institute of Nuclear Chemistry from 1949 to 1967. Svedberg's early work was on colloid chemistry: he devised an improved method of making metal sols and made extensive studies of them using the ultramicroscope. He also investigated radioactivity. In the 1920s, however, his interest in colloids led him to develop the ultracentrifuge as a means of following optically the sedimentation of particles too small to be seen in the ultramicroscope. The earliest ultracentrifuge exerted a force 5,000 times that of gravity, but between 1924 and 1939 the machine was gradually improved to exert a force over 100,000 times that of gravity. The measurements of the molecular weights of proteins which Svedberg made were particularly important, and for his work on the ultracentrifuge he received the 1926 Novel Prize for Chemistry. During World War II he developed a synthetic rubber, Sweden's supplies of natural rubber being cut off by the blockade. Svedberg's work at the Werner

Institute involved the applications of a cyclotron in medicine, in radiation physics and in radiochemistry. He was made an Honorary Fellow of the Chemical Society in 1923 and elected a Foreign Member of the Royal Society in 1944.

SVERDRUP, Haraid Uirik
(1888–1957)

Norwegian oceanographer and geophysicist, born in Oslo and related to a number of prominent people, including Johan, prime minister of Norway in the 1870s, and Grieg the composer. Complying with family wishes he entered the University of Norway to study languages, but soon left to begin military service, becoming reserve officer in 1908. On his return to university he took up science and in 1911 became assistant to Wilhelm Bjerknes, following him to Leipzig to work on atmospheric circulation (1913–1918). World War I forced him to return to Norway where he was engaged by Roald Amundsen as chief scientist on the three-year *Maud* expedition to the North Pole. There he made atmospheric, oceanographic, magnetic and ethnographic observations. In 1922 the *Maud* docked in Seattle for repairs and he took the opportunity to work at the Carnegie Institute, interpreting the magnetic observations, and Arctic tidal dynamics. In 1926 he succeeded Bjerknes as Professor of Geophysics at Bergen. Later he took a research position at the Christian Michelsens Institute in Bergen (1931), and took part in a submarine expedition to the North Pole. He moved to California in 1935 to become Director of Scripps Institution of Oceanography, where his precise determinations of tides and wave heights were valuable in the Pacific during World War II. He returned to Oslo in 1948 as Director of the Norwegian Polar Institute. In 1949 he became Professor of Geophysics at the University of Oslo, and later Dean, then Vice-Chancellor. A unit of volume transport and the Sverdrup Islands in Arctic Canada are named after him.

SWALLOW, John Crossley
(1923–   )

English physical oceanographer and geophysicist, born near Huddersfield, Yorkshire. He was educated in physics at St John's College, Cambridge (BA 1945), although his studies were interrupted by military service in the East Indies during World War II. He was captivated by Buflard's lectures and later joined the Department of Geodesy and Geophysics at Cambridge, where he was awarded a PhD in 1954. He spent four years aboard HMS *Challenger* conducting seismic refraction experiments. He was then recruited by Deacon to work at the National Institute of Oceanography, where he spent the rest of his career (1954–1983), concentrating on physical oceanography. Swallow developed a method for measuring deep currents in the ocean using neutrally buoyant floats, which sink to a predetermined depth, drift freely and can be tracked acoustically from a surface ship. This led to cooperative work with Stommel, revealing the deep western boundary current in the North Atlantic (1957) and the presence of strong mesoscale eddies at all depths in mid-ocean (1960). Always a practical scientist, he took part in many oceanographic cruises in the North Atlantic and Mediterranean, observing the winter-time formation of deep water and other phenomena, and the Somali current and equatorial circulation in the Indian Ocean. He was elected FRS in 1968.

SWAMMERDAM, Jan
(1637–1680)

Dutch naturalist, born in Amsterdam, the son of an apothecary. He studied medicine at Leiden but never practised. Using a simple microscope, he made many observations of a great range of biological material, his drawings being published after his death in the great *Biblia Naturae* (1737–1738). He described in great detail the life cycles of a dozen insect types, the most famous being that of the mayfly and the most complex, the honey bee. He classified the insects on the basis of the type of metamorphosis they undergo during their life cycles, the method which is still in general use. Among his many observations was the presence of the butterfly's wing within the pupa, and he correctly surmised that on hatching the wings are expanded by blood pressure. He was first to describe valves in the lymph vessels and the ovarian follicles in mammals. He also carried out ingenious experiments which demonstrated that contracting muscles change their shape and not their volume, in contradiction with the generally accepted view at that time. He published *Historia Insectorum Generalis* in 1669. All of his considerable biological achievements were achieved between 1663 and 1673; thereafter his father withdrew financial support and he succumbed to religious fanaticism under the influence of Antoinette Bourignon.

SWAN, Sir Joseph Wilson
(1828–1914)

British chemist, inventor and industrialist, notable for his achievements in photography, synthetic textiles and electric lighting. He was born near Sunderland and after leaving school at 13 was apprenticed to a druggist. In 1846 he joined John Mawson who had a pharmaceutical business in Newcastle. In 1856 he took out a patent for improving the wet-plate collodion photographic process and the following year he invented high-speed bromide paper, having observed that heat greatly increased the sensitivity of gelatin and silver bromide emulsion. The patent was bought by George Eastman, founder of Kodak, and helped to make photography cheaper and thus widely popular. By 1848 Swan was experimenting with carbonized paper filament for electric lamps but it was not until Hermann Sprengel developed his mechanized air-pump in 1865 that it became possible to achieve the necessary vacuum in the bulb. Swan gave his first successful demonstration, with a thin carbon rod, in 1879. In 1880 he established a small factory in South Benwell, west of Newcastle. Within three years he was manufacturing 10,000 lamp bulbs a week and a number of famous institutions – for example the British Museum – were illuminated by Swan bulbs. In 1883, he amalgamated his business with Edison who had been granted a British patent, to form the Edison and Swan Electric Light Company. Searching for a better filament for his bulbs, Swan dissolved cellulose in acetic acid and extruded it through narrow jets into a coagulating fluid. Soon afterwards Chardonnet adapted this process to make rayon, and it was further developed by Cross and Bevan who, in conjunction with Courtaulds, laid the foundations of the synthetic textile industry. Swan also made significant improvements to lead-plate batteries by designing cellular lead plates which held the lead oxide more securely. He was elected FRS in 1874, knighted in 1904 and received many other honours. He died in Overhill, Warlingham, Surrey.

SWEDENBORG, Emanuel
(1688–1772)

Swedish mystic, theologian and scientist, born in Stockholm. Born to a family soon to be ennobled, Swedenborg studied at Uppsala, later travelling widely, studying engineering, and returning home in 1716 to work for the Royal Board of Mines. He wrote widely on mathematics and technical matters (longitude, docks, decimal coinage, navigation). His *Opera Philosophica et Mineralia* (1734) expressed his metaphysical interests, and huge works on physiology and anatomy followed. He also developed a religio-geological theory of creation. In 1743–1744 he underwent a religious crisis (recorded in his *Journal of Dreams*) and in consequence he resigned his scientific post, in order to be free to expound his mystical views.

SWINBURNE, Sir James, 9th Baronet
(1858–1958)

British chemist and industrialist, notable as an electrical engineer and a pioneer of the plastics industry. He was born in Inverness and educated at Clifton College, Bristol. He began work in a locomotive works in Manchester, and in the early 1880s Swan sent him to France and the USA to establish factories to manufacture electric lamps. He then managed Crompton's dynamo works and subsequently worked independently as a consultant and inventor. He was a leading authority on the design of dynamos, and developed instruments to measure alternating currents and very low pressures in highly evacuated vessels. He also wrote on thermodynamics from the point of view of engineering design. Interested in the possibilities of making thread from viscose at a very early stage, he worked with Cross and Bevan and had a share in the company set up with Courtaulds. He formed a syndicate to develop the synthetic material made by the reaction between phenol and formaldehyde, attempting to patent it in 1907, but found he had been anticipated by Baekeland, a Belgian chemist working in the USA. Instead Swinburne established a lacquer company in Birmingham. In 1926 this was bought by Bakelite Ltd and Swinburne became president of the British division of Bakelite, a post he retained until 1948. He died in Bournemouth.

SWINGS, Pol
(1906–1983)

Belgian astronomer, born in Ransart near Charleroi. He studied mathematics and physics at the University of Liege (1923–1927), finishing with a doctorate in a topic in classical astronomy. He joined the staff of Liege University in 1927 becoming professor in 1936, a post which he retained until his retirement in 1975, apart from the years of World War II spent in the USA. His primary interest was in the identification of spectra of atoms and molecules in astronomical bodies, especially comets which before the space era were amenable to analysis only by spectroscopic means. His *Atlas of Cometary Spectra* (jointly with Leo Haser) was published in 1956. A committed internationalist, he initiated the annual Liege astronomical colloquy (1949), and served as President of the International Astronomical Union (1964–1967).

SWINTON, Alan Archibald Campbell
(1863–1930)

Scottish electrical engineer and inventor, born in Edinburgh. Educated there and in France, he was interested in all things mechanical and electrical from an early age, linking two houses some distance apart by telephone at the age of 15, only two years after its invention by Alexander Graham Bell. In 1882 he began an engineering apprenticeship in the Newcastle works of William George Armstrong, for whom he devised a new method of insulating electric cables on board ship by sheathing them in lead. In 1887 he moved to London and began to practise as a consulting engineer, specializing in the installation of electric lighting. At the same time he continued to act as consultant to Armstrongs, and later became a director of Crompton and Company, and the Parsons Marine Steam Turbine Company. Having been interested in photography since childhood, he was one of the first to explore the medical applications of radiography; he published the first X-ray photograph taken in Britain in *Nature* (23 January 1896), and was soon in demand as a consultant radiographer. In a letter to *Nature* (18 June 1908) he outlined the principles of an electronic system of television, which he called 'distant electric vision', by means of cathode rays – essentially the system in use today. A member of the Institutions of Civil, Electrical and Mechanical Engineers, and of several other scientific bodies, he was elected FRS in 1915.

SYDENHAM, Thomas
(1624–1689)

English physician, 'the English Hippocrates', born in Wynford Eagle, Dorset. He served in the Parliamentarian army during the Civil War, and in 1647 went to Oxford. There he studied medicine at Wadham College, and was elected Fellow of All Souls College. In 1651 he was severely wounded at Worcester. From 1655 he practised in London. A great friend of such empiricists as Boyle and Locke, he stressed the importance of observation rather than theory in clinical medicine. Contemptuous of sterile book-learning, he urged doctors to become close observers at the bedside, where they would learn to distinguish specific diseases, and through trial and error, to find specific remedies. He was much impressed with the capacity of Jesuit's bark (the active principle of which is quinine) to cure intermittent fever (malaria), and believed that other such specific treatments might be found. He wrote *Observationes Medicae* (1667) and a treatise on gout (1683), a disease to which he himself was a martyr, distinguished the symptoms of venereal disease (1675), recognized hysteria as a distinct disorder and gave his name to the mild convulsions of children, 'Sydenham's chorea' (St Vitus's dance), and to the medicinal use of liquid opium, 'Sydenham's laudanum'. He remained in London except when the plague was at its peak (1665). He was a keen student of epidemic diseases, which he believed were caused by atmospheric properties (he called it the 'epidemic constitution') which determined which kind of acute disease would be prevalent each year. Acute diseases, he observed, account for about two-thirds of the afflictions of human beings. By the time of his death, his reputation was growing and his works, with their vivid, homespun qualities, and their astute description of diseases, were often reprinted and translated throughout the eighteenth century.

SYLVESTER, James Joseph
(1814–1897)

English mathematician, born in London. He studied at St John's College, Cambridge, became Second Wrangler in 1837 but, as a Jew, was disqualified from graduating. He became professor at University College London (1837), and the University of Virginia (1841–1845). Returning to London, he worked as an actuary, and was called to the bar in 1850. He later returned to academic life as Professor of Mathematics at Woolwich (1855–1870) and at Johns Hopkins University, Baltimore (1877–1883), where he established the first international journal of mathematics in America. Finally he became Savilian Professor at Oxford (1883–1894). With Arthur Cayley he was one of the founders of the algebraic theory of invariants which became a powerful tool in resolving physical problems. He also made important contributions to number theory. His mathematical style was flamboyant, and he wrote in haste, continually coining new technical terms, most of which have not survived.

SYLVESTER-BRADLEY, Peter
(1913–1978)

English micropalaeontologist, born in Pinhoe, Devon, and educated at Reading University. After graduating he commenced research on the major interest of his life, Jurassic ostracodes. He was initially appointed to the staff at Scale Hayne Agricultural College, where he founded a new geology department, and joined the Royal Navy during World War II. He later became a lecturer at Sheffield University. In 1959 he became F. W. Rennett Professor of Geology at the University of Leicester. Sylvester-Bradley pursued major studies of Mesozoic ostracodes and oysters, and contributed to theories of evolution and the origin of life. He was also an outstanding leader in research, teaching and administration.

SYMONS, George James
(1838–1900)

English climatologist, born in Pimlico. Symons showed an early fascination with the weather, making observations whilst at school and becoming a reporter for the Registrar-General in 1857. He was particularly interested in rainfall and established a network of voluntary observers in 1860. This became the British Rainfall Organisation and under Symons's enthusiasm the number of observers increased rapidly to reach over 3,500 by 1899. In 1863 he founded a circular which later became Symons' *Monthly Meteorological Magazine*. He was interested in instruments, and invented the brontometer for recording the sequence of phenomena in thunderstorms, as well as organizing a comparison of thermometer screens which resulted in the Stevenson screen being accepted as the world standard. He compiled a catalogue of over 60,000 meteorological books and his comprehensive collection was bequeathed to the Royal Meteorological Society; for which he had served as Secretary and President. He was a member of many scientific committees including the Royal Society Krakatoa Committee, and was elected FRS in 1878. On his death the Royal Meteorological Society opened a memorial fund and the biennial Gold Medal provided from the proceeds is the society's highest award.

SYNGE, Richard Laurence Millington
(1914–)

English biochemist, born in Chester. He trained at Cambridge, where he studied the partition of amino acids in solvent mixtures, and joined the Wool Industry Research Association in Leeds (1941–1943). There he collaborated with Archer Martin and exploited his Cambridge research in the development of partition chromatography and the counter-current liquid-liquid separation of mixtures (1941), which revolutionized analytical chemistry. They also showed that mild protein hydrolysis is required to determine the amide content of asparagine and glutamine (1941) and developed methods for the analysis of aldehydes and hydroxy-acids. Synge and Martin shared the Nobel Prize for Chemistry in 1952 for their work. In 1944 Synge demonstrated the use of powdered cellulose or potato starch packed in columns for separating amino acids, and in 1948 he moved to the Rowett Research Institute, Aberdeen, where he showed that the arbitrary division between proteins above and below 10,000 molecular weight corresponded to the division between dialysability and non-dialysability through cellophane (1949). Around this time he partially determined the structure of the peptide antibiotic gramicidin. He later identified S-methyl L-cysteine S-oxide in cabbage as the compound responsible for the sulphurous smell of boiled cabbage water (1956), and at the time of possible therapeutic interest. From 1967 until his retirement in 1976, he worked at the Food Research Institute in Norwich. He was elected FRS in 1950.

SZENT-GYÖRGYI, Albert von Nagyrapolt
(1893–1986)

Hungarian-born American biochemist, born in Budapest. He lectured at Gröningen, where he discovered hexuronic acid (vitamin C) in the adrenal cortex, and at Cambridge, working on this substance with W. H. Haworth. He became professor at Szeged (1931–1945), where he crystallized vitamin C from paprika, and in consequence, vitamin B2 (riboflavin). He was later appointed professor at Budapest (1945–1947), Director of the Institute of Muscle Research at Woods Hole, Massachusetts (1947–1975), and Scientific Director of the National Foundation for Cancer Research, Massachusetts (1975). Szent-Györgyi also discovered the reducing system oxaloacetate to malate involved in the Hans Krebs cycle (1935), and made important contributions towards understanding muscular contraction; the enzyme ATPase of myosin (1941) and inhibition by actomyosin; myosin cleavage by trypsin into heavy and light meromyosins (1952); glycerinated fibres, allowing study of a physiologically active biochemical system (1948); and muscle relaxation by adenosine triphosphate (ATP) in the absence of calcium (1953). He was awarded the Nobel Prize for Physiology or Medicine in 1937.

SZILARD, Leo
(1898–1964)

Hungarian-born American physicist, born in Budapest. He studied electrical engineering there, and physics in Berlin, working with von Lane. In 1933 he fled from Nazi Germany to England, and in 1938 emigrated to the USA, where he began work on nuclear physics at Columbia. In 1934 he had taken out a patent on nuclear fission as an energy source, and on hearing of Hahn and Meitner's fission of uranium (1938), he immediately approached Einstein in order to write together to President Roosevelt warning him of the possibility of atomic bombs. Together with Fermi, Szilard organized work on the first fission reactor, which operated in Chicago in 1942. He then went to Los Alamos to work on the Manhattan Project, leading to the nuclear fission bomb. After World War II he researched into molecular biology in experimental work on bacterial mutations and theoretical work on ageing and memory.

T

TAIT, Peter Guthrie
(1831–1901)

Scottish mathematician and golf enthusiast, born in Dalkeith. He was educated at the universities of Edinburgh and Cambridge where he graduated as Senior Wrangler in 1852. He became Professor of Mathematics at Belfast (from 1854) and Professor of Natural Philosophy at Edinburgh (1860–1901). He wrote on quaternions, thermodynamics and the kinetic theory of gases, and collaborated with Lord Kelvin on a *Treatise on natural philosophy* (1867), the standard work on the natural sciences in English for a generation. His study of vortices and smoke rings led to early work on the topology of knots. He studied the dynamics of the flight of a golf-ball and discovered the importance of 'underspin'.

TAKAMINE, Jokichi
(1834–1922)

Japanese-born American chemist, born in Takaoka. He studied chemical engineering in Tokyo and Glasgow, and in 1887 opened his own factory, the first to make superphosphate fertilizer in Japan. In 1890, having married an American, he moved to the USA and set up an industrial biochemical laboratory there. Takamine worked for some time in the laboratory of John Abel. In 1898 he published *Testing diastatic substances*, in which he described the iodine test for following the activity of saliva or other fluids in hydrolysing starch to maltose and glucose. He later published *The blood-pressure-raising principle of the suprarenal glands and its mode of preparation* (1901), the first description of adrenaline isolated in crystalline form. After 1905, when Starling first used the word hormone to describe the animal body's 'chemical messengers', it was realized that adrenaline, an intravenous injection of which produces an enormous rise in blood pressure, was the first hormone to be isolated in pure form from a natural source.

TAKHTMAN, Armen Leonovich
(1910–   )

Armenian botanist, born in Susa, Nagorno Karabakh. From 1932 to 1943 he held various posts at Tblisi, Georgia, and Yerevan Museum and Yerevan University, Armenia, where he was awarded a doctorate (1943) for a dissertation on the evolution and phylogeny of flowering plants. In 1943 he joined the palaeobotanical section of the Institute of Botany of the Academy of Sciences of the Armenian SSR, Yerevan, but later (1954–1984) he worked at the Botanical Institute of the Academy of Sciences of the USSR in Leningrad (now St Petersburg), where he was Director of the Department of Floristics, Systematics and Evolution of Higher Plants (1977–1984). He has published several versions of a new system of flowering plant classification, the most definitive being *Sistema A Magnotiophytov* (1987). This system is complementary to that of Cronquist, with whom Takhtman corresponded for many years; there are many similarities between the two systems although Takhtman has a narrower concept of the family and recognizes more of them. Other major works include *Flowering Dispersal* (1966) and *World* (1978), a unique synthesis of current thinking on world plant geography.

TAMM, Igor Yevgenyevich
(1895–1971)

Soviet physicist, born in Vladivostock, the son of an engineer. He was educated at the universities of Edinburgh and Moscow, and taught at Moscow State University (1924–1934) before moving to the Physics Institute of the Academy. Together with Frank he developed a theory to describe the 'Cherenkov effect' discovered by Cherenkov. They demonstrated that this emission of radiation is due to a particle moving through a medium faster than the speed of light in the medium, and related the type of radiation observed to the particle mass and velocity. Tamm shared the 1958 Nobel Prize for Physics with Cherenkov and Frank for this work. The effect has been utilized in particle detectors, allowing the masses and hence identities of particles to be determined.

TANSLEY, Sir Arthur George
(1871–1955)

English botanist, born in London. Educated at Cambridge, he later lectured at University College London (1893–1906), and then at Cambridge from 1906 to 1923. Sherardian Professor at Oxford (1927–1937), he founded the precursor (1904) of the Ecological Society (1914), and was founder-editor of the journal *New Phytologist* (1902). A pioneer British plant ecologist, he published *Practical Plant Ecology* (1923) and *The British Isles and their Vegetation* (1939), and contributed to anatomical and morphological botany as well as physiology. He was President of the British Ecological Society in 1913, Chairman of the Nature Conservancy from 1949 to 1953, and was knighted in 1950.

TARSKI, Alfred
(1902–1983)

Polish-born American logician and mathematician, born in Warsaw. Educated in Warsaw, he became professor there (1925–1939), then moved to the USA at the University of California at Berkeley (1942–1968). He made contributions to many branches of pure mathematics and mathematical logic, including the Banach-Tarski paradox, which seemingly allows any set to be broken up and reassembled into a set of twice the size. The paradox hinges upon the use of sets which cannot mathematically be said to have a size at all, the non-measurable sets that arose in studies of the theory of integration. He is best remembered, however, for his definition of 'truth' in formal logical languages, as presented in his monograph *Der Wahrheitsbegriff in den Formalisierten Sprachen* (1933, 'The Concept of Truth in Formalized Languages').

TARTAGLIA, (originally FONTANA, Niccolo)
(c.1500–1557)

Italian mathematician, born in Brescia. From the age of about 12, when he was injured by a French soldier during the invasion of his home town and left with a speech impediment, he was given the name Tartaglia, meaning 'stutterer'. He became a teacher of mathematics in several Italian universities, and settled in Florence in 1524. Tartaglia was one of the first to derive a general solution for cubic equations; he disclosed this result to Cardano who claimed priority in the discovery (now known as Cardano's formula), and a long controversy followed; the credit for the very first solution of a type of cubic should probably go to Scipione da Ferro, an Italian mathematician of the previous generation. Tartaglia also published an early work on the theory of projectiles, and translated Euclid's *Elements*.

TATUM, Edward Lawrie
(1909–1975)

American biochemist, born in Boulder, Colorado. He studied at the University of Wisconsin, receiving a BA in chemistry in 1931 and a PhD in biochemistry in 1934, and taught at Stanford (1937–1945, 1948–1957), Yale (1945–1948) and Rockefeller University, New York (1957–1975). Working with Beadle on the bread mould *Neurospora*, he demonstrated the role of genes in biochemical processes. They irradiated *Neurospora* spores with X-rays, and then grew them on a variety of nutritional media. When mutant spores did not grow on minimal medium but did grow when an additional nutrient was supplied, Tatum and Beadle suggested that the spore had one or more blocks in the metabolic pathway for that particular nutrient. They formulated the 'one gene, one enzyme' hypothesis, that a single gene codes for the synthesis of one protein. At Yale, Tatum collaborated with Lederberg to show that bacteria reproduce by the sexual process of conjunction. All three shared the 1958 Nobel Prize for Physiology or Medicine. In his later years, at the Rockefeller University, Tatum concentrated on the training and education of students.

TAUBE, Henry
(1915–   )

Canadian-born American inorganic chemist, born in Neudorf, Saskatchewan. He studied at Saskatchewan University and received his doctorate at the University of California at Berkeley. In 1942 he became an American citizen and subsequently taught at Cornell and Chicago. He was appointed Professor of Chemistry at Stanford in 1962. Using radioisotopes as tracers, Taube devised new methods for studying the transfer of electrons during inorganic chemical reactions in solution. He also proved the hypothesis that metal ions in solution form chemical bonds with water, and showed that when one metal ion replaces another in a solution of a metallic salt, the acid radical forms a temporary bridge by which electrons are transferred from the replacement ion to the original metallic ion. In 1969 Taube and Carol Creutz synthesized a mixed valence cation, a new type of positively charged ion consisting of two atoms of ruthenium each bonded to five molecules of ammonia and separated by a pyrazine ring. They and their colleagues used it to investigate oxidation-reduction reactions in living tissue. Taube was awarded the Nobel Prize for Chemistry in 1983.

TAUSSIG, Helen Brooke
(1898–1986)

American paediatrician, born in Cambridge, Massachusetts. She received her MD from Johns Hopkins University in 1927 and later became the first woman to become a full professor there. Her work on the pathophysiology of congenital heart disease was done partly in association with the cardiac surgeon Blalock and between them they pioneered the 'blue baby' operations which heralded the beginnings of modern cardiac surgery. The babies were blue because of a variety of congenital anomalies which meant that much blood was passing directly from the right chamber of the heart to the left without being oxygenated in the lungs. Taussig was actively involved in the diagnosis and after-care of the young patients on whom Blalock operated, and their joint efforts helped create a new speciality of paediatric cardiac surgery.

TAYLOR, Brook
(1685–1731)

English mathematician, born in Edmonton. He studied at St John's College, Cambridge, and in 1715 published his *Methodus incrementorum*, containing the theorem on power series expansions which bears his name. He also wrote on the mechanics of the vibrating string, and with real insight on the mathematics of the theory of perspective; although one of his books later inspired William Hogarth to his famous engraving of an impossible view, Taylor's books were generally found to be obscure by later mathematicians.

TAYLOR, Sir Geoffrey Ingram
(1886–1975)

English physicist and applied mathematician, born in London. His father was an artist and his mother was related to Boole and Sir George Everest. Taylor graduated from Trinity College, Cambridge in 1908. Except during the world wars, when he provided assistance to the government at Farnborough and elsewhere, he was based in Cambridge throughout his career. A temporary readership in dynamic meteorology in 1911 was followed by six months on a scientific expedition in the North Atlantic. From 1923 to 1952 he was Royal Society Yarrow Research Professor of Physics at the Cavendish Laboratory. His many original investigations on the mechanics of fluids and solids were applied to meteorology, oceanography, aerodynamics and Jupiter's Great Red Spot. A famous series of papers laid out his statistical theory of turbulence in 1935–1938. He proposed in 1934 the idea of dislocation in crystals, a form of atomic misarrangement which enables the crystal to deform at a stress less than that of a perfect crystal. Taylor had a passion for botany, small boats and foreign travel. He was knighted in 1944 and awarded the Order of Merit in 1969.

TAYLOR, Sir Hugh Stott
(1890–1974)

English physical chemist, born in St Helens, Lancashire. After graduating in chemistry from Liverpool University, he worked with Arrhenius on acid-base catalysis in Stockholm (1912–1913) and with Bodenstein on the hydrogen-chlorine reaction in Hanover (1913–1914). In 1914 he moved to the USA and thereafter Princeton University was his permanent home, although he remained a British subject. Between 1914 and 1927 he advanced from instructor to full professor, and from 1927 to 1958 he held the David B. Jones Chair of Chemistry. Like Rideal he did wartime research in London on the Haber process (1917–1919), and in their spare time they wrote *Catalysis in Theory and Practice* (1919). His pre-war and wartime background set the course of Taylor's researches thereafter. He made many studies of gas reactions involving chains of atoms and free radicals. Particularly in the 1920s, he worked extensively on the kinetics of reactions on surfaces and showed the existence of 'active centres', which are the sites on the surface of a heterogeneous catalyst at which chemical reaction actually occurs. He also identified 'activated adsorption' of a gas on a solid surface. After the isolation of deuterium by Urey, Taylor pioneered the use of the 'heavy hydrogen' to elucidate reactions mechanisms. Taylor became a Commander of the Order of Leopold I of Belgium in 1938 and was knighted in 1953. Elected FRS in 1932, he received the Longstaff Medal of the Chemical Society in 1942 and was made an Honorary Fellow in 1949. He was President of the Faraday Society in 1952–1954.

TAYLOR, Joseph Hooton Jr
(1941–)

American astronomer and physicist, born in Philadelphia. Educated at Haverford College and Harvard University, he held various posts at the University of Massachusetts before becoming Professor of Physics at Princeton in 1980. During a systematic search for pulsars, the rapidly rotating dense stars which appear on Earth to emit regular pulses of radio waves, he discovered with graduate student Hulse one interesting candidate whose pulse frequency changed periodically (1974). The characteristics of these changes revealed that this was the first discovery of an exotic 'binary pulsar', a pulsar in orbit of another dense neutron star. For this work he shared with Hulse the 1993 Nobel Prize for Physics. Taylor's subsequent observations have given strong evidence in favour of the general relativity prediction that this system of very massive compact objects will create 'gravitational waves', leaking energy from the system and causing the objects to continually move closer together.

TAYLOR, Richard Edward
(1919–)

Canadian physicist, born in Medicine Hat, Alberta, and educated a the University of Alberta in Edmonton and Stanford University. He held posts at the Linear Accelerator Laboratory at Orsay, the Lawrence Berkeley Laboratory and the Stanford Linear Accelerator Center (SLAC), California, where he became professor in 1970 and of which he was associate director from 1982 to 1986. In the 1960s, with Jerome Friedman and Henry Kendall, Taylor led a group of physicists at SLAC who investigated the structure of the nucleons (protons and neutrons) by scattering high-energy electrons from nuclear targets. These experiments provided data that established the constituents of nucleons, now known as quarks, as real dynamic entities by determining experimentally some of their properties. The three won the 1989 W. K. H. Panofsky prize and the 1990 Nobel Prize for Physics for this work.

TAZIEFF, Haroun
(1914–)

Polish-born French vulcanologist and mountaineer, born in Warsaw. He studied in Russia, France and Belgium, firstly agricultural engineering and then geology. After various short-term posts in the Belgian Congo, he became assistant professor of mining geology in Brussels in 1950. In 1967 he was made head of research at the National Centre for Scientific Research (CNRS), Paris, and subsequently director (1971–1981). He became the first French secretary of state for the prevention of natural disasters (1974–1986) and was also Mayor of Mirmande (1977–1989). Tazieff has investigated many of the world's volcanoes, both active and inactive, and from 1958 to 1974 made 26 expeditions to Nyiragongo, Zaire. He has written around 20 books on volcanoes and world tectonics including *Forecasting Volcanic Events* (1983) and *Sur L'Etna* (1984).

TEILHARD DE CHARDIN, Pierre
(1881–1955)

French Jesuit palaeontologist, theologian and philosopher, born at the castle of Sarcenat, the son of an Auvergne landowner. He was educated at a Jesuit school, lectured in pure science at the Jesuit College in Cairo, and was ordained as a priest in 1911. He was a stretcher bearer during World War I, and subsequently became Professor of Geology (1918) at the Institut Catholique in Paris. Between 1923 and 1946 he accompanied a number of palaeontological expeditions in China, where he directed the 1929 excavations at the Choukoutien Pekin Man site. He later worked in central Asia, Ethiopia, Java and Somalia. Increasingly, his anthropological researches did not conform to Jesuit orthodoxy and he was forbidden by his religious superiors to teach and publish. Nevertheless, his work in Cenozoic geology and palaeontology became known and he was awarded academic distinctions, including the Legion of Honour (1946). From 1951 he lived in the USA and worked at the Wenner-Gren Foundation for Anthropological Research in New York. Posthumously published, his philosophical speculations, based on his scientific work, trace the evolution of animate matter to two basic principles: non-finality and complexification. By the concept of 'involution' he explains why *Homo sapiens* seems to be the only species which, in spreading over the globe, has resisted intense division into further species. This leads on to transcendental speculations, which allow him original, if theologically unorthodox, proofs for the existence of God. This work, *The Phenomenon of Man* (1955), is complimentary to *Le Milieu Divin* (1957).

TEISSERENCE DE BORT, Leon Philippe
(1855–1913)

French physicist and meteorologist, born in Paris. He joined the Central Meteorological Bureau in Paris (1878) where he was in charge from 1880 to 1892. He demonstrated that weather depended greatly on the barometric pressure at certain centres of action, notably the Azores high and the Iceland low. In 1894 he helped to produce an international cloud atlas, and he founded the observatory at Paris (1889) primarily for the study of the upper air using kites and hydrogen-filled balloons carrying instruments for measuring pressure, temperature and humidity. He also obtained samples of air from up to 14 kilometres for analysis. During 1902–1903, these kites were flown day and night for nine months both in Paris and Holland. He discovered that the temperature of the atmosphere does not continue to fall with height but at a certain height becomes constant (the tropopause). He thus identified and named the stratosphere. He published a time cross-section of the isotherms above Paris to a height of 10 kilometres from 27 January to 1 March 1901, and later carried out upper air observations in kite and balloon experiments at a number of locations. From these observations he was able to show that the height of the tropopause varies with latitude, being much higher in the tropics. He was awarded the Symons Gold Medal of the Royal Meteorological Society (1908).

TELLER, Edward
(1908–)

Hungarian-American physicist, born in Budapest. He graduated in chemical engineering at Karlsruhe University, and studied theoretical physics at Munich, Göttingen and under Niels Bohr at his institute in Copenhagen. He left Germany in 1933, lectured in London and Washington (1935) and contributed profoundly to the modern explanation of solar energy, anticipating the theory behind thermonuclear explosions. In 1940, with Szilard and Wigner, he met with a US government committee to discuss the feasibility of a nuclear fission bomb. Later, in collaboration with Szilard and Fermi, he was involved in the construction of the first nuclear fission pile in Chicago before moving to the Manhattan Project at Los Alamos to develop the fission bomb. He then joined Oppenheimer's theoretical study group at the University of California at Berkeley, and later became director of the newly established nuclear laboratories at Livermore (1958–1960). From 1963 he was Professor of Physics at California University. He repudiated any moral implications of his work, stating that, but for Oppenheimer's moral qualms, the US might have had hydrogen bombs in 1947. After Russia's first atomic test (1949) he was one of the architects of President Harry S. Truman's crash programme to build and test (1952) the world's first hydrogen bomb. He wrote *Our Nuclear Future* (1958).

TEMIN, Howard Martin
(1934–)

American virologist, born in Philadelphia. As a high school student he spent summers at the Jackson Laboratory at Bar Harbor, Maine. He attended Swarthmore College, and studied with Dulbecco at Caltech, where he completed his PhD on the Rous sarcoma virus (which causes cancer in chickens) in 1959. Since 1969 he has held various professorships at the University of Wisconsin. Temin formulated the 'provirus' hypothesis, that the genetic material of an invading virus is copied into the host cell DNA. In 1970 he isolated the enzyme 'reverse transcriptase' (independently of Baltimore), which transcribes RNA into a double-stranded DNA (the provirus), enabling the new DNA to be inserted into the host cell. Viruses which contain this enzyme are 'retroviruses'; RNA forms their genetic material, and they reverse the usual process of DNA being transcribed to RNA. Reverse transcriptase is used to make copies of specific genes, clones, and is widely used for genetic engineering. In 1975, Temin shared the Nobel Prize for Physiology or Medicine with Dulbecco and Baltimore.

TEMMINCK, Coenraad Jacob
(1778–1858)

Dutch ornithologist, born in Amsterdam, the son of the treasurer of the Dutch East India Company. At the age of 17 he became an auctioneer with the company and used this contact to collect exotic birds and animals. When the company was dissolved in 1800 he turned his attentions full-time to the study of natural history and became an accomplished taxidermist. His *Catalogue Systématique du Cabinet d'Ornithologie et de la Collection de Quadrumanes* (1807) comprised a description of over a thousand specimens from his own collection. He later published *Histoire Naturelle Générales des Pigeons et des Gallinacées* (3 vols, 1813–1815), which established him as one of the leading European ornithologists. In 1820 he became the first Director of the Dutch National Museum of Natural History at Leiden which housed his own collection and grew in size and importance under his direction. His *Manuel d'Ornithologie* (1815–1840) remained for many years the standard text on European birds.

TENNANT, Charles
(1768–1838)

Scottish chemist and industrialist, born in Ochiltree, Ayrshire. He attended the parish school and was then apprenticed to a silk weaver. He studied bleaching and then set up his own bleachfields at Damley, near Paisley. At that time, traditional methods of bleaching were being replaced by chlorine, a method introduced in France by Berthollet. The chlorine was used in solution and was difficult to handle. In 1799 Tennant took out a patent for a dry bleaching powder made from chlorine and solid slaked lime, an innovation that was probably the invention of Macintosh, a fellow chemist and industrialist who was for a short time one of his partners. The powder could be conveniently transported to the expanding textile industry and the chlorine was easily regenerated when required by treating the powder with acid. Demand was such that the St Rollox works, which Tennant established in 1800, grew to be the largest chemical works in the world. By 1835 it covered more than 100 acres and produced sulphuric acid, alkali and soap as well as bleaching powder. Tennant was one of the first men to make a fortune out of the heavy chemical industry. He died in Glasgow.

TENNANT, Smithson
(1761–1815)

English chemist, born in Selby, Yorkshire. He was educated at Edinburgh University, where he was a pupil of Joseph Black, and at Cambridge where he became an early supporter of Lavoisier. He then travelled in Denmark and Sweden, meeting Scheele among other scientists. Around 1796, having bought land near Cheddar, Somerset, he embarked on agricultural research. He analysed lime from many parts of Britain and showed that some limes contain magnesium compounds and that these are injurious to plant life. In 1797 he demonstrated that diamond is a form of carbon by burning the diamond and showing that it produced the same amount of 'fixed air' (carbon dioxide) as an equal weight of charcoal. In 1800 he entered into partnership with a former fellow student, Wollaston, to produce platinum vessels, wire and electrodes for chemical research and industry, the merit of platinum being that it is resistant to heat and simple acids. In the course of their research on platina, the naturally occurring form of the metal, Tennant and Wollaston each discovered two new elements. In 1804 Tennant investigated the black residue left when platina is dissolved in aqua regia (a mixture of concentrated nitric and hydrochloric acids), separating and describing two new elements which he named iridium and osmium. Tennant was elected FRS in 1785. In 1815 he was appointed to the Chair of Chemistry at Cambridge; he was killed the following year in a riding accident near Boulogne, France.

TESLA, Nikola
(1856–1943)

Yugoslav-born American physicist and electrical engineer, born in Smiljan, Croatia. He studied at Graz, Prague and Paris, emigrating to the USA in 1884. He left the Edison Works at Menlo Park after quarrelling with Edison, worked for a short period with Westinghouse, but then concentrated on his own inventions. He was a prolific and highly innovative inventor. Among his many projects, he improved dynamos, and electric motors, invented the high-frequency Tesla coil and an air-core transformer. Tesla was firmly in favour of an alternating current electricity supply, as opposed to direct current initially favoured by Edison. By 1888 he had obtained patents on a whole polyphase AC system which he sold to Westinghouse. He again demonstrated the feasibility of AC, by lighting the 1893 Chicago World Columbian Exposition, and AC transmission was also chosen for the Niagara Falls project (1893–1895). He produced artificial lightning of a prodigiousness never since equalled, predicted wireless communication two years before Marconi, and experimented with a very low-frequency wireless communication system using the Earth as the conducting medium. Near the end of his life he became an eccentric recluse. His papers have been deposited in the Nikola Tesla Museum in Belgrade.

THALES
(c.624–545 BC)

Greek natural philosopher, astronomer and geometer, traditionally the founder of Greek and therefore European philosophy. He came from Miletus on mainland Ionia (Asia Minor), as did his intellectual successors Anaximander and Anaximenes. Thales is believed to have proposed the first natural cosmology, identifying water as the original substance and (literally) the basis of the universe. He is supposed to have visited Egypt, where he developed his astronomical techniques, and to have predicted the solar eclipse of 585 BC. Included in the traditional canon of the *Seven Wise Men*, he attracted various apocryphal anecdotes, for example as the original absent-minded professor who would fall into a well while watching the stars.

THEAETETUS
(c.414–369 BC)

Greek mathematician. He was an associate of Plato at the Academy, whose work was later used by Euclid in Books X and X111 of the *Elements*. Plato named after him the dialogue Theaetetus, which was devoted to the nature of knowledge. He is credited with being the first to prove that n is irrational whenever n is not a perfect square.

THEILER, Max
(1899–1972)

South African-born American bacteriologist, born in Pretoria. He enrolled in the premedical course at the University of Cape Town in 1916, and later studied at St Thomas's Hospital Medical School and the School of Tropical Medicine and Hygiene of the University of London. After receiving his medical degree in

1922, he went to Harvard Medical School to work on amoebic dysentery. He settled in the USA and turned his attention to yellow fever, working at Harvard until 1930 and then at the Rockefeller Institute, New York (1930–1964). He was Professor of Epidemiology and Microbiology at Yale Medical School from 1964 to 1967. In 1919 Noguchi reported that he had isolated a bacterium responsible for yellow fever. Theiler showed in 1926 that in fact yellow fever was caused by a filterable virus. While at the Rockefeller, he perfected the mouse protection test. In this test a mixture of yellow fever virus and human serum was injected into a mouse, and the survival of the mouse indicated that the serum had neutralized the virus and that the serum donor was therefore immune. This enabled an accurate survey of the worldwide distribution of yellow fever. Theiler contracted the disease in 1929, but survived and subsequently became immune. He is also remembered for his discovery of an infection in mice identical to polioencephalomyelitis, or Theiler's disease. He was awarded the 1951 Nobel Prize for Physiology or Medicine for his work in connection with yellow fever, for which he discovered the vaccine 17D in 1939. This formed the basis of vaccines now used to control the disease.

THENARD, Louis Jacques
(1777–1857)

French organic chemist and statesman, born in La Louptiere (now Louptiere-Thenard), the son of a peasant farmer. At a young age he left home for Paris in search of an education and attended lectures by Fourcroy and by Vauquelin, who gave him a home in return for his services as a bottle-washer. In 1798 he was appointed demonstrator at the École Polytechnique; he later succeeded Vauquelin in the chair at the College de France in 1804, became Dean of the Faculty of Sciences of Paris in 1821, and was Chancellor of the University of France from 1845 to 1852. He was a prominent member of many public bodies, particularly those concerned with the application of science to industry, and received many honours culminating in a peerage in 1832. He also served two terms in the Chamber of Deputies. Thenard made many important discoveries in organic chemistry. He prepared a wide range of esters (neutral products formed by the reaction between an acid and an alcohol), discovering that the reaction was analogous to the reaction between an acid and a base. He investigated cobalt and its compounds, and from alumina and copper arsenate prepared a stable brilliant blue pigment (Thenard's blue) which was used in porcelain manufacture to replace the expensive pigments made from lapis lazuli. Between 1808 and 1811 he collaborated with Gay-Lussac to study potassium, an element whose high reactivity made it very difficult to isolate. It had first been isolated by Humphry Davy; Thenard and Gay-Lussac prepared it in much larger quantities by fusing potash with iron filings in a gun barrel. While investigating its properties they discovered boron (1908). They proved that sodium hydroxide and potassium hydroxide contain oxygen and hydrogen. They also studied the photochemistry of chlorine, and the composition of organic compounds using potassium chlorate as an oxidizing agent. In 1818 Thenard announced the discovery of hydrogen peroxide, perhaps his greatest achievement. He passed oxygen over barium oxide, made the product (barium peroxide) into a paste, then precipitated the barium out with sulphuric acid. His observation that finely divided metals acted on hydrogen peroxide to produce heat and hydrogen without themselves being affected together with knowledge of Dobereiner's work on platinum, led him to the study of surface catalysis (although the term 'catalysis', coined by Berzelius, was not introduced until several years later). He was also the author of an influential textbook, *Traité Elementaire de Chimic* (4 vols, 1813–1816), which went through six editions and was much translated. He died in Paris.

THEOPHRASTUS
(c.372–c.287 BC)

Greek philosopher, born in Eresus on Lesbos, the son of a fuller. He studied at Athens under Aristotle, from whom he had inherited a library, and succeeded him as head of the Peripatetic School (Lyceum) from 322 BC. Theophrastus shared Aristotle's encyclopaedic conception of philosophy. Most of his prolific output is lost, but there are still extant important treatises on plants (representative of his interest in natural science), reconstructed fragments of his history of earlier philosophers, and the more literary, volume of *Characters*, containing 30 deft sketches of different moral types, which has been widely translated and imitated. His *Historia Plantarum* and *Plantarum Causae* mentioned around 450 species, not described in much detail, but he did allude to differences in floral structure between plants, and to seed germination. He established the fundamental difference of organization between plants and animals. In other observations, he derived insights into the essentials of plant morphology and classification, laying the foundations for the work of later botanists. Linnaeus called Theophrastus the father of botany, because so many aspects of modern botany can be traced back to his work, including morphology, anatomy, systematics, physiology, ecology, pharmacognosy, applied botany and plant pathology.

THEORELL, (Axel) Hugo Theodor
(1903–1982)

Swedish biochemist, born in Linköping. After training in medicine at the Karolinska Institute, Stockholm, he received his MD for a study of the effect of plasma lipids on the sedimentation of red blood cells (1930), and became a lecturer (1930–1932) and assistant professor at Uppsala (1932–1936), and Director of the Nobel Institute of Biochemistry at Stockholm (1937–1970). He crystallized myoglobin (oxygen storage protein of muscle) and determined its molecular weight (1932). During a period in Berlin, in 1934 he purified, the 'yellow ferment' (an electron transfer enzyme), obtained as crude extract there by Warbury, and separated the yellow coenzyme (flavine mononucleotide) from the protein. The resulting inactivation was reversed when the two components reunited. Theorell helped elucidate the nature of the yellow pigment, later identified as riboflavin (vitamin B2), and established the 1:1 linkage with protein using his newly invented electrophoresis apparatus (1935). On his return to Uppsala he purified diphtheria antitoxin (1937), and crystallized and characterized cytochrome C, another electron carrier, establishing the sulphur between haem and protein (1938–1941). He subsequently purified and studied peroxidases and dehydrogenases, and introduced fluorescence spectrometry. Theorell was awarded the 1955 Nobel Prize for Physiology or Medicine, and was elected a Foreign Member of the Royal Society in 1959.

THOM, René Frédéric
(1923–)

French mathematician, born in Montbeliard. He studied at the École Normale Superieure, and worked at Grenoble and Strasbourg, where he became professor. Since 1964 he has been at the Institut des Hautes Études Scientifiques. In 1958 he was awarded the Fields Medal (the mathematical equivalent of the Nobel Prize). His work has been in algebraic topology, where he was one of the creators of a novel and powerful theory known as cobordism theory, and on the singularity theory of differentiable manifolds, but he is best known for his book *Stabilité structurelle et morphogenese* (1972) which introduced 'catastrophe theory'. This has been applied to widely differing situations such as the development of the embryo, social interactions between human beings or animals, and physical phenomena such as breaking waves, and has attracted much publicity as well as some controversy.

THOMAS, (Edward) Donnall
(1920–)

American physician and haematologist, born in Mart, Texas, the son of a general practitioner. In 1937 Thomas entered the University of Texas at Austin, where he studied chemistry and chemical engineering. After receiving his MA in 1943, he entered Harvard Medical School, graduating MD in 1946. After an internship, training in Haematology, serving in the army, and posts at MIT and the Brigham Hospital, Boston, Thomas joined the staff of the Mary Imogene Bassett Hospital in Cooperstown. There he began work on bone marrow transplantation in dogs and in humans. This proved a difficult procedure due to problems such as graft rejection and 'graft-versus-host' disease. However, occasional successes occurred with dogs when bone marrow was transplanted between members of the same litter. In 1963 Thomas became professor at the Washington University School of Medicine, Seattle, and here he demonstrated that using new tissue-typing techniques and immunosuppressive drugs, some patients with leukaemia, aplastic anaemia and certain genetic diseases can be cured by bone marrow transplants. He joined the Fred Hutchinson Cancer Research Center in 1975. Thomas shared the 1990 Nobel Prize for Physiology or Medicine with Joseph Edward Murray.

THOMAS, Sir John Meurig
(1932–)

Welsh Physical chemist, born in Llanelli. He studied chemistry at University College, Swansea, and Queen Mary College, London. From 1958 to 1969 he advanced from assistant lecturer to reader in chemistry at the University College of North Wales, Bangor and then moved to the University College of Wales, Aberystwyth, as professor and head of department. From 1978 to 1986 he was Professor of Physical Chemistry at Cambridge. He then became Director of the Royal Institution and of its Davy-Faraday Laboratory. He vacated these posts in 1991 but is still Fullerian Professor at the Roy Institution. Since 1991 he has been Deputy Pro-Chancellor of the University of Wales. Thomas's main field of research has been surface chemistry, particularly heterogeneous catalysis. In recent years he has pioneered in the development of 'uniform' heterogeneous catalysts – solid catalysts in which the active sites are distributed uniformly throughout their bulk. He is the author of *Heterogeneous Catalysis; Theory and Applications* (1991, with W. J. Thomas) and *Michael Faraday and the Royal Institution* (1991). He was knighted in 1991. Thomas was elected FRS in 1977 and gave the Bakerian Lecture in 1990. He received the Hugo Muller Medal of the Royal Society of Chemistry in 1983 and its Faraday Medal in 1989. He gave the Baker Lectures at Cornell University in 1982–1983.

THOMAS, Sidney Gilchrist
(1850–1885)

English metallurgist, born in Canonbury in North London. Educated at Dulwich College, he intended to study medicine, but after the death of his father in 1867 he became a police-court clerk. However, he attended evening classes in chemistry at the Birkbeck Institution and studied metallurgy at the Royal School of Mines. From 1870 he decided to tackle a problem that had dogged the Bessemer steelmakers: how to remove phosphorus from iron. In May 1878 at a meeting of the Iron and Steel Institute, Thomas was able to announce that, with the help of his cousin Gilchrist and Edward Martin, he had solved the problem of 'dephosphorization' by using dolomite for the furnace lining, together with an addition of lime to produce a basic slag that allowed the removal of both phosphorus and sulphur. This method was described as the 'basic Bessemer process' in Britain, but was always known as the 'Thomas process' on the Continent. Within a few years, the same principles were applied to the Siemens open hearth furnace. Ironically, the main effect of Thomas's invention was to open up the whole range of the world's iron ores, with the loss of the UK's leadership in steelmaking. Ascetic, a pacifist and philanthropist, Thomas died prematurely from a lung complaint (probably TB) and was buried in Paris.

THOMPSON, Sir D'Arcy Wentworth
(1860–1948)

Scottish zoologist and classical scholar, born in Edinburgh and educated at Trinity College, Cambridge. He was Professor of Biology at Dundee (1884–1917) and at St Andrews (from 1917). The ideas for which he is remembered are contained in his *On Growth and Form* (1917), a book read both for its biological content and literary style. His knowledge of mathematics and physics led him to interpret the forms of organs and biological structures on the basis of the physical forces acting upon them during development. He was also able to demonstrate mathematically that the superficial dissimilarity of related animals could be accounted for by differential growth rates, and these ideas resulted in him adopting an anti-Darwinian stance. His *Glossary of Greek Birds* (1895) and *Glossary of Greek Fishes* (1945) derive from his classical interests. He was knighted in 1937.

THOMPSON, Silvanus Phillips
(1851–1916)

English physicist, born in York into a Quaker family. He sat for an external London degree (1869), and was appointed the first Professor of Physics at the newly established University of Bristol (1878) and Professor of Physics and Principal of the City and Guilds Technical College, Finsbury (1885). He was a highly proficient textbook writer on electricity, light and magnetism, wrote a witty, effective little book called *Calculus Made Easy* (1910), and was a distinguished historian of science, writing biographies of William Gilbert, Faraday, Kelvin, and Philipp Reis whom he regarded as the true inventor of the telephone. He almost beat Heinrich Hertz to the discovery of radio waves, and he made a number of technical contributions. He may well have been appointed Principal of the University of London (1901) if he had not been so outspoken about British conduct in the Boer War.

THOMSON, Elihu
(1853–1937)

English-born American inventor, born in Manchester. He emigrated to the USA and was educated in Philadelphia, where he was a chemistry teacher until he decided on a career as an inventor. He became one of the great pioneers of the electrical manufacturing industry in the USA. He cooperated in 700 patented electrical inventions, which included the three-phase AC generator and arc lighting. With Edwin James Houston, he founded the Thomson-Houston Electric Company (1883), which merged with Edison's firm in 1892 to form the General Electric Company. He declined the Presidency of MIT in 1919, but agreed to be Acting President from 1921 to 1923.

THOMSON, Sir George Paget
(1892–1975)

English physicist, son of J. J. Thomson. He was born and educated in Cambridge, where he became a Fellow of Trinity College. He served in the Royal Flying Corps during World War I, was Professor of Physics at Aberdeen (1922–1930) and Imperial College, London (1930–1952), and became Master of Corpus Christi at Cambridge (1952–1962). In 1927 Thomson and Alexander Reid were the first to notice that a beam of electrons passed through a thin metal foil in a vacuum produced circular interference fringes. This was firm evidence for de Broglie's theory that moving particles have wave-like properties. Thomson went on to analyse the diffraction characteristics of electrons, thereby opening the door to their use in the analysis of surfaces. In 1937 he shared the Nobel Prize for Physics with Davisson for the discovery, separately and by different methods, of electron diffraction by crystals. During World War II, Thomson advised the government on the possible relevance of nuclear fission to the making of a superbomb. After the war he supported the introduction of a world atomic authority for the peaceful exploitation of nuclear power. He was scientific adviser to the UN Security Council (1946–1947) and for his contributions to electrical science he was awarded the Faraday Medal by the Institution of Electrical Engineers (1960). His works included *The Atom* (1937) and *Theory and Practice of Electron Diffraction* (1939). He was elected FRS in 1930, and knighted in 1943.

THOMSON, James
(1822–1892)

Irish-Scottish physicist and engineer, elder brother of Lord Kelvin, born in Belfast. His father was a professor of mathematics there, and moved in 1832 to the same position at the University of Glasgow. James Thomson was educated at home until 1832 when he began attending classes at the University of Glasgow, where he became a matriculated student two years later at the age of 12. He graduated MA with honours in mathematics and natural philosophy in 1839, then spent several short periods as an engineering apprentice but was troubled by ill-health until in 1851 he settled as a civil engineer in Belfast. In 1857 he was appointed Professor of Civil Engineering at Queen's College, Belfast, and in 1873 moved to the same chair in succession to Rankine at the University of Glasgow. He carried out important researches in fluid dynamics, inventing or improving several types of water-wheels, pumps and turbines. Over a long period he studied the effect of pressure on the freezing point of water, and its influence on the plastic behaviour of ice and the movement of glaciers. He published many papers on that and a wide variety of other subjects including elastic fatigue, ocean under-currents and trade winds. He was elected FRS in 1877.

THOMSON, Sir J(oseph) J(ohn)
(1856–1940)

English physicist, discoverer of the electron, born in Cheetham Hill near Manchester, the son of a Scottish bookseller. He entered Owen's College, Manchester, at the age of 14 with the intention of becoming a railway engineer, but a scholarship took him to Trinity College, Cambridge, where he graduated Second Wrangler in 1880. In 1884 he was elected FRS and succeeded Lord Rayleigh (1842–1919) as Cavendish Professor of Experimental Physics, and in 1919 was himself succeeded by his brilliant student, Rutherford. Thomson's early theoretical work was concerned with the extension of Maxwell's electromagnetic theories. This led to the study of gaseous conductors or electricity and in particular the nature of cathode rays. By studying the deflections of cathode rays in a highly evacuated discharge tube when electric and magnetic fields were applied, he showed that the rays consist of rapidly moving particles, and calculated their charge-to-mass ratio. He went on to measure the particles' specific charge in an experiment similar to that carried out by Millikan, and deduced that these 'corpuscles' (electrons) must be more than 1,000 times smaller in mass than the lightest known atomic particle, the hydrogen ion. This discovery was inaugurated by his lecture to the Royal Institution in 1897, and published in the *Philosophical Magazine*. Before the outbreak of World War I, Thomson had successfully studied the nature of positive rays (1911), and this work was crowned by the discovery of isotopes, which Thomson showed could be separated from each other by deflection of positive rays in electric and magnetic fields, a technique now known as mass spectrometry. 'J J' made the Cavendish Laboratory the greatest research institution in the world. Although simplicity of apparatus was carried to 'string and sealing wax' extremes, seven of his research assistants subsequently won the Nobel Prize; Thomson himself was awarded the Nobel Prize for Physics in 1906. He was knighted in 1908, and became the first scientist to be appointed master of Trinity College (1918–1949). In 1936 he published *Recollections and Reflections*, and he was buried near Newton in the nave of Westminster Abbey.

'T HOOFT, Gerard
(1947–)

Dutch physicist. He was educated at Utrecht University, where he was later appointed professor. 'T Hooft's work has been concerned with gauge theories of particle physics which attempt to describe the various types of interaction between fundamental particles. The unified theory of the electromagnetic and weak interactions, as proposed by Steven Weinberg and Salam originally predicted a force of infinite strength. While a research student, 'T Hooft found a way of making the force both finite and calculable; this led to the universal acceptance of the electroweak theory. He has also showed that recent developments in this field predict theexistence of a heavy magnetic monopole (so far undiscovered), which Dirac also predicted by different reasoning, and has contributed to the theories of quantum gravity.

THORARINSSON, Sigurdur
(1912–1983)

Icelandic geologist and glaciologist. His scientific training was in Stockholm 1930s with field studies in Swedish Lapland (1933) and Iceland (1934), and in 1936–1938 on the Vatnjokull glacier expedition. He obtained a degree (1939) for studies on the movement and drainage of Hoffellsjokull and his doctorate (1944) for his classical work on tephrochronology, the relative dating of volcanic ash layers. In 1945 he settled in Iceland where he worked as a geologist for the National Research Council and then joined the Museum of Natural History (1947–1969). From 1969 he was Professor of Geography and Geology at the University of Iceland. Thorarinsson was the first to use tephrochronology to study the eruption history of volcanoes, the first to describe and analyse catastrophic glacier outburst (jokulhlaups), and carried out early work on glacier shrinkage and eustatic change level. He made the first determination glacier mass balance and its relation to climatic change, and was also a pioneer in nature conservation in Iceland.

THORPE, Sir (Thomas) Edward
(1845–1925)

English chemist, physicist and historian of science, born near Manchester. He was educated at Owens College, Manchester, where he spent some time as Roscoe's assistant and at the University of Heidelberg under Bunsen and also at Bonn. He was appointed to the Chair of Chemistry first at Anderson's College, Glasgow (1870), then at the Yorkshire College of Science at Leeds (1874) and the Royal College of Science, London (1885). In 1894 he became Government Chemist – the first such appointment – and he returned to the Royal College of Science from 1909 to 1912. Thorpe discovered several new compounds of chromium, sulphur and phosphorus, including phosphorus pentafluoride which demonstrated that phosphorus could have a valence of five. He also carried out determinations of atomic weights, for example of silicon, gold, titanium and radium, that were more accurate than any others of the time. He travelled to the West Indies and other places to view four eclipses of the Sun, and in collaboration with Sir Arthur Rucker made a magnetic survey of the British Isles. As Director of the Government Laboratory at Clements Inn, London, he was responsible for removing arsenic from some beers, legislating against lead in pottery glazes, eliminating paraffins from tobacco leaf, and substituting red phosphorus for yellow phosphorus in matches (yellow phosphorus having been found to be responsible for the cancer of the jaw which affected many workers in the match industry). Thorpe is also remembered for his work in the history of science, particularly his biography of Priestley. He was elected FRS in 1876, made a Companion of the Bath in 1900, and knighted in 1909. He died in Salcombe, Devon.

THORPE, William Homan
(1902–1986)

English zoologist, born in Hastings. He studied agriculture at Cambridge, where he developed an interest in agricultural entomology, and investigated the biological control of insect parasites in California (1927–1929). He was research entomologist at Farnham Royal Parasite Laboratory of the Imperial Bureau of Entomology (1929–1932) and became a lecturer in Entomology at Cambridge University in 1932. He was instrumental in setting up the Ornithological Field Station at Madingley near Cambridge which became the Sub-department of Animal Behaviour in 1960. In 1966 Cambridge's first Chair of Ethology was created for him. Thorpe had a strong interest in conservation, and served on many committees concerned with conservation matters. He was one of the founders of ethology and published the influential *Learning and Instinct in Animals* in 1956. His research was mainly concerned with bird song, and working initially with the chaffinch, he demonstrated (with Peter Marler) that song results from the integration of innate and learned components of sound patterning. In 1961 he wrote *Bird-Song: The Biology of Vocal Communication and Expression in Birds*. He was elected FRS in 1951.

THORSON, Gunnar Axel
(1906–1971)

Danish marine ecologist, born in Copenhagen. He was educated at Copenhagen University, and subsequently became a leading member of a three-year expedition to east Greenland, during which he undertook research for his PhD on the reproductive ecology of Arctic marine invertebrates. He was later appointed to the staff of the Copenhagen University Zoological Museum (1934) and founded a private laboratory on the island of Ven in the Oresund, from which stemmed the research that led to the publication of his classic and highly influential monograph *Reproduction and larval development of Danish marine bottom invertebrates* (1946). He was elected to the Royal Danish Academy of Sciences in 1955. During World War II the laboratory was moved to Helsingor, and in 1957 Thorson was promoted to full professor; he was later formally appointed Director of the Helsingor Marine Laboratory (a converted torpedo station). Thorson's 1946 monograph is still a cornerstone in marine invertebrate larval ecology, and his geographic categorization of developmental modes of benthic marine invertebrates at all latitudes and depths is now formalized in the literature as 'Thorson's rule'. A second important contribution was his geographic concept of 'parallel marine bottom communities', a hypotheses that similar biotopes at differing latitudes supported ecologically comparable assemblages characterized by the same genera (albeit different species). Despite the conceptual utility of such broad-scale hypotheses, Thorson himself acknowledged that this latter theory was not tenable on its extension to include tropical communities.

THUNBERG, Carl Per
(1743–1828)

Swedish botanical explorer, born in Jönköping. As a medical student at Uppsala he was taught botany by Linnaeus and collected plants for him; after graduation in 1770 he travelled as a ship's surgeon to South Africa, Java and Japan in the company of Francis Masson, collecting and describing 3,000 plants, 1,000 of which were new to science. In Japan, he was not allowed to collect plants personally, but he organized local collectors. From 1778 he taught botany at Uppsala, and in 1784 he succeeded Linnaeus's son as professor. His Japanese discoveries were published as *Flora Japonica* (1784), and those from South Africa as *Prodromus Plantarum Capensium* (1794–1800) and *Flora Capensis* (1807–1823, with Joseph August

Schultes). He also wrote and published monographs on Protea, Ixia, Oxalis and Gladiolus. His description of his voyage was published between 1788 and 1793; translated into English, it was published as *Travels in Europe, Africa and Asia* (1793–1795).

TILDEN, Sir William Augustus
(1842–1926)

English organic chemist, born in London. He was apprenticed to a pharmacist and studied for a year at the Royal College of Chemistry before becoming a demonstrator at the Pharmaceutical Society (1863–1872). For the following eight years he taught chemistry at Clifton College, Bristol, moving to the chair of Chemistry at Mason College and then succeeding Edward Thorpe at the Royal College of Science in 1894. Tilden showed that there is only one compound of nitric oxide and chlorine, nitrosyl chloride, and that it is a valuable reagent for investigating the terpenes, a little-understood class of hydrocarbons which occur in the essential oils of many plants. He also discovered that if the hydrocarbon isoprene was prepared from terpenes, it separated out on standing into fragments of a yellowish substance floating in a syrupy fluid. These fragments have properties identical to natural rubber. It was already known that isoprene prepared by other methods interacts with strong acids to give a tough, elastic substance resembling rubber, and Tilden did not pursue his discovery of its spontaneous conversion. In physical chemistry he worked on the solubility of salts at temperatures above 100 degrees Celsius and on the specific heats of metals. He discovered that specific heats alter with temperature, decreasing as temperatures fall (he lowered temperatures to –180 degrees Celsius) and increasing as they rise, with the extent of the shift varying inversely as the atomic weight of the element. This discovery has proved very important in many branches of industry. Tilden was elected FRS in 1880, and knighted in 1909. He died in London.

TIMOSHENKO, Stephen (Stepan Prokofyevich)
(1878–1972)

Soviet-American civil engineer, born in the Ukraine. He was educated at the two technical institutes in St Petersburg, with an intervening period working as a railway engineer, then spent some time studying under Prandd at the University of Göttingen. In 1906 he began lecturing at Kiev University. Dismissed for his pro-Jewish views in 1911, he nevertheless remained in Russia, often under conditions of great difficulty and hardship. Eventually in 1920 he fled to Yugoslavia and then to the USA in 1922. There he found that, in his view, the teaching of mechanics to engineering students was far from satisfactory, and he devoted himself to the improvement of both teaching and research in all aspects of the subject. His own thinking on the strength of materials and the theory of elasticity, as well as his technical papers, were (and remain) an important contribution towards the achievement of that end. In 1936 he joined the staff of Stanford University in California where he taught engineering mechanics and strength of materials until he was 76 years of age. In 1946 he received the James Watt Gold Medal of the (British) Institution of Mechanical Engineers, and in 1959 he was elected a member of the Soviet Academy of Sciences.

TINBERGEN, Nikolaas
(1907–1988)

Dutch ethologist, born in the Hague, brother of the Nobel prize-winning economist Jan Tinbergen. He studied zoology at Leiden, and after World War II taught at Oxford (1949–1974). With Konrad Lorenz he is considered to be the co-founder of ethology, the study of animal behaviour in relation to the environment to which it is adapted. His best-known studies were on the three-spined stickleback and the herring gull, animals which perform many stereotyped or instinctive behaviour patterns. Much of his work was centred around aspects of social behaviour. He was able to elucidate the evolutionary derivation of many of these behaviours through comparative studies. His concentration on instinctive behaviour brought him into conflict with comparative psychologists who disagreed with the separation between acquired and innate behaviours. His books included his classic *The Study of Instinct* (1951), *The Herring Gull's World* (1953), *Social Behaviour in Animals* (1953) and *The Animal in its World* (2 vols 1972–1973). In 1973 he published a controversial book *Autistic Children* (with his wife Lies) in which he proposed a behavioural causation for autism. He shared the 1973 Nobel Prize for Physiology or Medicine with Lorenz and Karl von Frisch.

TING, Samuel Chao Chung
(1936–)

American physicist, born in Ann Arbor, Michigan, where his father was a student at that time. He was raised in China, educated there and in Taiwan, and at Michigan University (1956–1962). He worked in elementary particle physics at the European nuclear research centre (CERN) in Geneva and at Columbia University. Later he led a research group as DESY, the German synchrotron project in Hamburg, and from 1967 worked at MIT. In 1974 he was head of a team at the Brookhaven National Laboratory which conducted an experiment in which protons were directed onto a beryllium target, and a product particle, with a lifetime 10,000 times longer than could be predicted by previous discoveries, was observed and named the J particle. At the same time, and independently, Burton Richter made the same discovery and named the particle. It is now named the J/ particle and for its discovery Ting and Richter shared the 1976 Nobel Prize for Physics. In the 1980s and 1990s, Ting led the L3 collaboration at CERN working at the LEP electron-positron collider where the standard model of particle interactions was precisely tested.

TISELIUS, Arne Wilhelm Kaurin
(1902–1971)

Swedish chemist, born in Stockholm. He trained and worked at the University of Uppsala, where he became assistant professor in 1930 and was later appointed Professor of Biochemistry (1938–1968). He developed an accurate method for determining diffusion constants of proteins, important for analysing ultracentrifuge sedimentation data (1934). He introduced protein analysis by moving boundary electrophoresis (1930–1937), and thus identified serum proteins as albumin and (gamma) globulins; he also showed that antibodies are globulins (1937). Tiselius's electrophoretic analysis became the best criterion of protein purity and he found multiple components in crystalline pepsin. He isolated bushy stunt and cucumber mosaic viruses (1938–1939), and invented preparative electrophoresis (1943), electrokinetic filtration (1947) and other analytical techniques. From 1944 he developed methods for the chromatographic separation and identification of amino acids, sugars and other molecules using activated charcoal, cellulose, silica, ion exchange and other media. He worked with Sanger on the chemistry of insulin (1947) and with Synge on chromatographic analysis (1950). Tiselius became Vice-President (1947–1960) and President (1960–1964) of the Nobel Foundation and was awarded the Nobel Prize for Chemistry in 1948. He was elected a Foreign Member of the Royal Society in 1957.

TITTERTON, Sir Ernest William
(1916–1990)

English atomic physicist, born in Tamworth, Staffordshire. Educated at Birmingham University, he was a research officer for the Admiralty during World War II, before becoming in 1943 a member of the British mission to the USA to participate in the Manhattan Project to develop the atomic bomb. He was a senior member of the timing team at the first atomic test in 1945, and advisor on instrumentation at the Bikini Atoll tests in 1946, before returning to Los Alamos, New Mexico, as head of the electronics division until 1947. Titterton then worked at the Atomic Energy Research Establishment at Harwell until 1950 when he became Professor of Nuclear Physics at the Australian National University, Canberra. He was involved in the British nuclear tests at Maralinga, South Australia, until 1957, and subsequently held various research and advisory appointments in the field of nuclear energy. He was knighted in 1970.

TIZARD, Sir Henry Thomas
(1885–1959)

English chemist and administrator, born in Gillingham, Kent. He studied chemistry at Magdalen College, Oxford, and worked with Nemst in Berlin (1908–1909). Returning to Oxford, he was soon appointed Fellow and Tutor of Oriel College, but within a few years his career was interrupted by World War I. After brief service in the army, he transferred to the Royal Flying Corps and spent much of the war as an experimental pilot. In 1918–1919 he was Assistant Controller of Experiments and Research for the RAF. He returned to Oxford and became reader in thermodynamics, but soon left to be Assistant Secretary of the Department of Scientific and Industrial Research, of which he became Permanent Secretary in 1927. From 1929 to 1942 he was Rector of Imperial College and from 1942 to 1946 he was President of Magdalen College, Oxford. He then became Chairman of the Defence Policy Research Committee and of the Advisory Council on Scientific Policy, from which he retired in 1952. His personal scientific work included some electrochemistry before 1914 and important work on aircraft fuels carried out with David Pye around 1920, which led ultimately to the system of octane rating, which expresses the anti-knocking characteristics of a fuel. In the 1930s and 1940s he was increasingly involved as an adviser to the British government in the scientific aspects of air defence, particularly in connection with radar. He was Chairman of the Aeronautical Research Committee from 1933 to 1943 and led a scientific mission to the USA in 1940. He received many military honours, and was elected FRS in 1926.

TOBIAS, Phillip Vallentine
(1925–)

South African anatomist and physical anthropologist, born in Durban. He studied at Witwatersrand University, receiving a BSc in 1946, and an MB and BCh in 1950. He has remained at Witwatersrand nearly all his life, becoming a lecturer in anatomy in 1951 and Professor of Anatomy and Human Biology in 1959. Since 1979 he has been Director of the Palaeoanthropological Research Unit. Tobias has worked on cytogenetics and human genetics, the human biology of the living peoples of Africa, and palaeoanthropology. He has studied and described hominid fossils in many parts of Africa including those of the Olduvai Gorge, Tanzania. With Louis Leakey and J. R. Napier he described and named the species *Homo habilis*. His work has led to about 800 publications, including three major works: *Australopithecus (Zinjanthropus) boisei* (1967), *The Brain in Hominid Evolution* (1971) and a two-volume work on *Homo habilis* (1991).

TODD, Alexander Robertus, Baron of Trumpington
(1907–)

Scottish chemist, born in Glasgow. He studied at the University of Glasgow and obtained his first doctorate at Frankfurt on the chemistry of bile acids in 1931. He was awarded an 1851 Exhibition Senior Studentship and worked with Robinson at Oxford for a second doctorate on the chemistry of natural pigments (1933). He spent two years in the medical chemistry department in Edinburgh, working on the chemistry of vitamin B, and then moved to the Lister Institute of Preventive Medicine in London. In 1938 he was offered a position at what was then a small college in Pasadena, later to be widely known as Caltech, but fortunately for British chemistry, the Chair of Organic Chemistry at Manchester became vacant at the same time and he held that post from 1938 to 1944. His final post was as professor at Cambridge where he remained until his retirement in 1971. All his researches concerned the chemistry of natural products, including vitamins B1, E and B12; the constituents of cannabis species; insect colouring matters; factors influencing obligate parasitism; and various mould products. However, the work for which he was awarded the Nobel Prize for Chemistry in 1957 concerned the structure and synthesis of nucleotides. The four chemical bases (adenine, guanine, uracil and cytosine) of DNA had been known for some time. Todd and his co-worker established the manner in which sugar molecules and phosphate groups are attached to these bases to form nucleotides, the building blocks of DNA. This work was a necessary preliminary to Crick and James Watson's proposal of the double helix as the structure of DNA, and to an understanding of the biological processes of growth and inheritance. Todd received many academic honours. He was elected to the Royal Society in 1942 and was later its president (1975–1980). He was knighted in 1954 and was awarded the Order of Merit and made a Life Peer in 1977. He was also the first Chancellor of Strathclyde University. A man of strong personality, he was known affectionately at Cambridge as Todd Almighty, later Lord Todd Almighty. As a trustee of various charities and as a member of many government committees he has played a substantial part in promoting scientific activity both in Britain and abroad.

TOLANSKY, Samuel
(1907–1973)

English physicist, born in Newcastle upon Tyne. He was educated at the university in his home town and at Imperial College, London and from 1934 held teaching posts at the University of Manchester. In 1947 he moved to become Professor of Physics and then head of department at Royal Holloway College of the University of London. His research interests included nuclear physics, in particular the areas of hyperfine structure and line-spectra, using the latter to investigate the nuclear spin of tin, bromine and iodine. He also carried out work in multiple-beam interferometry and studied the surface structure and properties of diamonds. His publications included *History and Use of Diamonds* (1962), *High Resolution Spectroscopy* (1948) and *Surface Microtopography* (1960). He was elected FRS in 1951.

TOMBAUGH, Clyde William
(1906–)

American astronomer, born in Streator, Illinois. Too poor to attend college, he built his own 9-inch telescope, and in 1929 became an assistant at the Lowell Observatory. In 1933 he won a scholarship to the University of Kansas and received an MA in 1936. Lowell had predicted the existence of an outermost planet, which he named Planet X, from his estimates of the perturbation of the orbits of Uranus and Neptune. Tombaugh joined the search team that was run by Slipher, Director of Lowell Observatory from 1926. Tombaugh devised the blink comparator which enabled him to detect if anything had moved in the sky between the taking of two celestial photographs, a few days apart. On 18 February 1930, thanks to his meticulous technique, he discovered Pluto in the constellation of Gemini. It was too faint to be the expected Planet X, and Tombaugh spent another eight years looking, but without success. In 1946 he became astronomer at the Aberdeen Ballistics Laboratories in New Mexico, and he was later appointed astronomer (1955–1959), associate professor (1961–1965) and professor (from 1965) at New Mexico University.

TOMONAGA, Sin-Itiro
(1906–1979)

Japanese physicist, born in Kyoto. He was educated at Kyoto Imperial University where he was a classmate of Yukawa. After graduating (1929) they both remained at Kyoto as unpaid assistants. Tomonaga then joined Yoshio Nishina at Riken, the Institute for Physical and Chemical Research in Tokyo (1932). Nishina had studied in Copenhagen for seven years and passed on the spirit of the Danish approach to quantum mechanics to Tomonaga. During the next five years Tomonaga published many papers on positron creation and annihilation, and one on high-energy neutrino-neutron scattering in which he emphasized the increased probability of a neutrino reaction with increasing energy. In 1937 he moved to Leipzig to work with Heisenberg, and his work here on a model of the nucleus earned him a DSc from Tokyo University (1939). He returned to Riken (1939) and during World War II did minor research on radar. During this time he produced his most important work, a relativistic quantum description of the interaction between a photon and an electron. For the resulting theory of 'quantum electrodynamics' he shared the 1965 Nobel Prize for Physics with Feynman and Schwinger, who had derived the same results independently. He increasingly became involved in scientific administration, becoming President of the Science Council of Japan (1951) and President of Tokyo University (1956).

TONEGAWA, Susumu
(1939–)

Japanese molecular biologist, born in Nagoya. Tonegawa studied chemistry at the University of Kyoto, graduating in 1963. He joined the department of biology of the University of San Diego where he received his PhD in 1968. In 1971 he accepted an appointment at the Institute for Immunology in Basle, Switzerland. Tonegawa applied the restriction enzyme and recombinant DNA techniques to the problem of the origins of antibody diversity. There were two schools of thought – one was that the genes controlling the production of all the different antibodies that an animal can make are inherited. The other was that in the formation of the antibody-manufacturing cells, the B-lymphocytes, the genes somehow undergo changes allowing them to produce a new wide range of antibodies. Tonegawa's research confirmed the second view, and provided details of the mechanism by which the genes are changed. In 1981 he returned to the USA as Professor of Biology at MIT, where he has applied the techniques of molecular biology to another aspect of the immune system – the action of the T-lymphocytes. Since 1988 he has also been Howard Hughes Medical Institute Investigator. Tonegawa was awarded the 1987 Nobel Prize for Physiology or Medicine.

TORREY, John
(1796–1873)

American botanist, born in New York City. He qualified in medicine and taught physical sciences at the West Point Military Academy, joining the US army as assistant surgeon in 1824. He was Professor of Chemistry at West Point, the US Military Academy, and at Cornell, before becoming Chief Assayer at the US Assay Office in New York from 1854 to 1873. He founded the New York Lyceum of Natural History, and became Emeritus Professor of Botany and Chemistry at Columbia College in 1856. However, throughout his life his main interest was botany, and he prepared several floras for North America and also collected over 50,000 plant species; his collection formed the basis for the herbarium of the New York Botanical Gardens. The genus *Torreya* in the yew family is named after him, as well as the Torrey Botanical Club. His publications include *A Flora of the Northern and Middle Sections of the United States* (1824), *A Flora of North America* (1838–1843, with Asa Gray) and *Flora of the State of New York* (1843).

TORRICELLI, Evangelista
(1608–1647)

Italian physicist and mathematician, probably born in Faenza. In 1627 he went to Rome, where he devoted himself to mathematical studies, and later made the acquaintance of Galileo's *Dialoghi delle nuove scienze* (1638). He was inspired by that work to develop some of its propositions in his own treatise *De Motu* (1641), which led Galileo to invite him to become his amanuensis; on Galileo's death in 1642 he was appointed mathematician to the grand-duke and professor to the Florentine Academy. He discovered that it is because of atmospheric pressure that water will not rise above 33 feet in a suction pump. To him are owed some of the fundamental principles of hydromechanics, and in a letter written in 1644 he gave the first description of a mercury barometer or 'torricellian tube'. His skill in grinding near-perfect lenses enabled him to build remarkably effective telescopes, and he even made a simple microscope by using a small glass sphere as a lens. He published a large number of mathematical papers on such topics as conic sections, the cycloid and logarithmic curves, and he determined the point still known as Torricelli's point on the plane of a triangle for which the sum of the distances from the vertices is a minimum.

TOURNEFORT, Joseph Pitton de
(1656–1708)

French botanist, born in Aix, and educated by Jesuit priests, as his family had destined him for the church. From his father's death (1677) until 1683, Tournefort studied at Montpellier. He became Professor of Botany at the Jardin du Roi, Paris (1683–1708), where he enlarged the living collections by undertaking a series of European expeditions (1685–1689), from Holland to the Iberian Peninsula. By 1689 he had become one of Europe's foremost botanists. In 1694 he published a three-volume textbook, *Elements de Botanique*, translated into Latin as *Institutiones Rei Herbariae* (1700). From 1700 to 1702 he travelled in the Levant with the artist Claude Aubriet; his account of the journey, *Relations d'un Voyage du Levant*, was published posthumously (1717). Tournefort's fundamental contribution to botany was the creation of the modern concept of the genus, defining it by diagnosis and carefully distinguishing between the describing and the naming of a genus. His generic concept was followed by Linnaeus, Adanson and Antoine Laurent de Jussieu. Most of the 725 plant genera he recognized are still accepted. Tournefort made outstanding contributions towards the establishment of objectivity in taxonomy. Although his own classification was highly artificial, his methods ultimately led to the later development of natural classification. He was also interested in mineralogy and shells, and was a noted physician who played a key role in the emancipation of botany from medicine.

TOWNES, Charles Hard
(1915–)

American physicist, born in Greenville, South Carolina. He was educated in his home town at Furman University, then at Duke University and Caltech, where he completed his PhD in 1939. During World War II he worked at the Bell Telephone Laboratories designing radar bombing systems and navigational devices. He also made the first studies of the microwave spectra of gases. In 1949 he joined Columbia University where he used short-wavelength radar to probe the electrical and magnetic interaction between the rotating motion of molecules and the 'spinning' nuclei within. In need of an intense source of microwaves to extend these investigations, Townes turned to the problem of designing such a source. This led him in 1951 to use highly energized oscillating ammonia molecules to produce microwaves. Gaseous ammonia molecules were first excited by pumping energy into them either electrically or thermally. By passing through the gas a weak beam of microwaves with the same frequency as the natural molecule oscillating frequency, Townes triggered the ammonia molecules to emit their own microwave radiation. Intense, coherent radiation was produced by the molecules with only a narrow range of frequencies, in the first operational maser (Microwave Amplification by Stimulated Emission of Radiation); the forerunner of the laser.

For his work on the maser and fundamental work in the field of quantum electronics that underpinned the maser-laser principles, Townes was joint winner of the Nobel Prize for Physics with Basov and Prokhorov in 1964. He was appointed to professorships at MIT and the University of California at Berkeley (1967–1986), and has been active in developing microwave and infrared astronomy techniques. In 1968, with his associates at Berkeley, he discovered the first polyatomic molecules in interstellar space (ammonia and water).

TOWNSEND, Sir John Sealy Edward
(1868–1957)

Irish physicist, born in Galway. The son of a civil engineering professor at Queen's College in Galway, Townsend graduated from Trinity College, Dublin, in 1890. After teaching mathematics in Ireland he went to Trinity College, Cambridge (1895), as one of J. J. Thomson's first research students at the Cavendish Laboratory. He became Wykeham Professor of Physics at Oxford in 1900. By 1897 Townsend had succeeded in making a direct determination of the elementary electrical charge, and his main area of research continued to be the kinetics of ions and electrons in gases. After 1908 he concentrated on the study of the properties of electron clouds. His investigations into the electron's mean free path were later seen to have implications for an understanding of its wave-like nature within quantum theory. Townsend was knighted in 1941.

TRAUBE, Ludwig
(1818–1876)

German pathologist, brother of Moritz Traube. Born in Ratibor, he was educated at Breslau and then at Berlin. Overcoming much petty anti-Semitism, he became professor at the Berlin Friedrich-Wilhelm Institute (1853) and at Berlin University (1872). Much influenced by Purkinje and Johannes Muller, he developed the study of experimental pathology in Germany, using animal experimentation. His most important work focused upon the pathology of fever, and the effects of various drugs and other stimuli and inhibitors upon muscular and nervous activity. He explored the effects of digitalis and other drugs in the management of heart disease and described the rhythmic variations in the tone of the vasoconstrictor centre (now known as the Traube-Hering waves). Traube won an unmatched reputation for his researches in experimental pathology, while remaining an unregenerate therapeutic nihilist.

TRAUBE, Moritz
(1826–1894)

German wine merchant and chemist, born in Ratibor, Silesia (now part of Poland). He studied chemistry in Berlin and Giessen, and was encouraged by Liebig and by his elder brother Ludwig Traube to pursue the study of fermentation. In 1849 he took over the family wine business in Ratibor, transferred it to Breslau in 1866, and ran it until 1886. Both in Ratibor and Breslau he carried out research in his private laboratory, mainly on fermentation. He showed that an 'unorganized ferment' (later called an enzyme) produced by yeast was responsible for fermentation. Traube studied and classified various enzymes. He also showed that protein was not the source of muscle energy, contrary to the views of Liebig. In 1867 he discovered that a membrane of cupric ferro-cyanide is permeable to water but not to certain solutes. Around 10 years later such a semi-permeable membrane was used by Pfeffer in his studies of osmotic pressure, which were interpreted by van't Hoff.

TRAVERS, Morris William
(1872–1961)

English chemist, renowned for his work on the rare gases. He was born in London and educated at the universities of London and Nancy, France. He was a demonstrator (1894–1898) and later assistant professor (1898–1903) at University College London, before moving to the chair at University College, Bristol. During 1906–1911 he did much to establish the Indian Institute of Science at Bangalore, of which he became director, and during World War I he was put in charge of Duroglass Ltd at Walthamstow. He was later President of the Society of Glass Technology. He worked as a consultant chemical engineer, returning to Bristol from 1927 to 1939. During World War II he was a consultant on explosives to the Ministry of Supply. At University College Travers helped Ramsay to determine the properties of argon and helium. They found helium in meteorites while heating the meteorites in search of new gases. In May 1898 Travers evaporated some liquid air and found that a spectroscopic analysis of the least volatile fraction showed fines never observed before. This was krypton. Liquefying air again in June, Ramsay and Travers collected the most volatile fraction and discovered neon. In July they discovered xenon, the least volatile of all. As a spin-off from his work on the rare gases, Travers developed an apparatus for liquefying hydrogen (although after James Dewar) and helped to set up experimental liquid air plants in several European countries. In 1920 he began work on high-temperature furnaces and fuel technology, and in 1927 he established a research group at Bristol to work on organic gases at high temperatures. He also wrote a biography of Ramsay (1956) and arranged 24 volumes of his papers. Travers was elected FRS in 1904. He died in Stroud, Gloucestershire.

TREVIRANUS, Gottfried Reinhold
(1776–1837)

German biologist and anatomist. Born in Bremen, the brother of Ludolf Christian Treviranus, he studied medicine and mathematics at the University of Göttingen, before being appointed Professor of Mathematics at Bremen. His *magnum opus, Biologie* (1802–1822), summarized all that was known at the time about the phenomena of life and proved extremely influential in introducing the new concept of 'biology' to the German public. Treviranus carried out important histological and anatomical research on vertebrates, investigating the reproductive organs of worms, molluscs and arachnids. He was famous for his anatomical study of the louse.

TRUMPLER, Robert Julius
(1886–1956)

Swiss-born American astronomer, born in Zurich and educated at the university there and at the University of Göttingen, Germany (1911). He worked with the Swiss geodetic survey before moving to the USA, where he served on the staff of the Allegheny Observatory (1915–1919) and Lick Observatory (1919–1938). From 1938 until his retirement in 1951 he was professor at the University of California at Berkeley. At Lick Observatory he studied the dimensions and brightnesses of open star clusters in the Milky Way and explained the disproportionate faintness of the more distant ones as the effect of absorption of light in interstellar space (1930). This important discovery led to a reassessment of the distance scale of our galaxy. He also demonstrated that the light of distant clusters is reddened as well as dimmed, an effect caused by small grains of dust in the spiral arms of the galaxy. Trumpler was elected to the US National Academy of Sciences in 1932.

TSIOLKOVSKY, Konstantin Eduardovich
(1857–1935)

Russian astrophysicist and pioneer of rocket propulsion, born in the village of Izheskaye in the Spassk district. Almost totally deaf from the age of nine, and largely self-educated in science, in 1881 he worked out the kinetic theory of gases, unaware that Maxwell had already done so more than a decade earlier. By 1895 his published papers had begun to mention the possibility of space flight, and three years later he pointed out that this would require the development of liquid fuel rocket engines. In 1903 he published his seminal work, *Exploration of Cosmic Space by means of Reaction Devices*, which established his reputation as the father of space flight theory. Unfortunately he lacked the resources to carry out any experimental work, and the first liquid fuel rocket was launched by Goddard in the USA in 1926. Tsiolkovsky continued to publish scientific papers, and also gave his ideas on space travel wider circulation by writing a number of works of science fiction. Towards the end of his life the Soviet government became interested in space flight and his work was belatedly given the recognition it deserved. Around 22 years after his death it was intended to launch the first Soviet satellite on the hundredth anniversary of his birth of 17 September 1857; in fact Sputnik I was 29 days late, but honoured his memory none the less.

TSWETT, or TSVETT, Mikhail Semenovich
(1872–1919)

Russian organic chemist, pioneer of chromatography. He was born in Asti, Italy, and appointed as assistant at Warsaw University in 1903. From 1908 to 1917 he taught botany and microbiology at Warsaw Technical University which moved to Nizhni Novgorod during World War I. In 1917 Tswett was appointed Professor of Botany and Director of the Botanical Gardens at Yuryev (Tartu) University, which was transferred to Voronezh in 1918. As a student he investigated chlorophyll and by 1900 had established that it contains at least two green pigments. However, traditional methods of organic analysis proved too destructive for delicate organic materials, and he began to look for a method of separating substances physically in an unchanged state. He built on earlier work by Runge and Schönbein which made use of the fact that different pigments have different capacities for being absorbed by paper. Tswett ground leaves in a mixture of ether and alcohol, shook the mixture with distilled water to remove the alcohol, and allowed the solution to filter through a tube filled with ground calcium carbonate, which after much trial and error he had found to be the best absorbent for his purpose. The pigments separated into different layers in the tube, and when the powder was extracted and cut up, they could be washed out separately for study. By this means Tswett added chlorophyll c to chlorophyll a and b, and discovered a family of yellow pigments which he called 'carotenoids'. His method, which he named 'chromatography', did not attract much interest until the 1930s. Since then it has developed into a number of highly specialized and widely used techniques which are employed when complex mixtures have to be separated or substances purified. Tswett died in Voronezh.

TULASNE, Louis René
(1815–1885) and Charles (1816–1884)

French mycologists, born in Azay-le-Rideau. Louis and his brother Charles carried out important researches on the structure and development of fungi, and wrote *Selecta Fungorum Carpologia* (1861–1865) which is notable for its many fine illustrations. Their work was the first exact study of the smut and rust fungi (Uredinales and Ustilaginales). They followed this with a long series of papers on different fungi, especially underground species. They also studied the development of the ergot fungus on rye (1853), spore formation and germination in Puccinia, Ustilago and others, and the sexual organs of *Peronospora*.

TURING, Alan Mathison
(1912–1954)

English mathematician, born in London. Educated at Sherborne, he read mathematics at King's College, Cambridge and also studied at the Institute for Advanced Study in Princeton. In 1936 Turing made an outstanding contribution to the development of computer science in his paper *On Computable Numbers* (1936), in which he outlined a theoretical 'universal' machine (or 'Turing' machine) and gave a precise mathematical characterization of the concept of computability. In World War II he was a member of the Bletchley Park code-breaking team, working on 'Colossus' (a forerunner of the modern computer), before joining the National Physical Laboratory in 1945. Here Turing was able to put his theoretical ideas on computing into practice, with his brilliant design for the Automatic Computing Engine (ACE). Frustrated with the slow progress in constructing the ACE, in 1948 he accepted a post at Manchester University, where work on the Manchester Mark I computer was in full swing under Kilburn and Sir Frederic Calland Williams. Turing made contributions to the programming of the machine, researched some complicated theories in plant morphogenesis and explored the problem of machine 'intelligence'. Prosecuted for an alleged homosexual offence in 1952, he committed suicide by swallowing cyanide two years later. His stature as a major pioneer of computer science has since grown steadily.

TURNER, James Johnson
(1935–   )

English inorganic chemist, born in Darwen, Lancashire. He was educated at King's College, Cambridge, where he received his BA (1957) and PhD (1960). He then worked for a year with Longuet-Higgins and for two years at Berkeley, California, with Pimentel. He was a lecturer in chemistry at Cambridge University until 1972, when he moved to Newcastle University as professor. Since 1979 he has been Professor of Inorganic Chemistry at Nottingham University. Turner's main discoveries have been concerned with the characterization of intermediates in organometallic photochemistry (using matrix isolation to trap such species within a solid host at very low temperatures, low-temperature solutions, and very fast infrared spectroscopy to study such species in solution at room temperature on a very rapid timescale). In this way he has been responsible for the elucidation of many reaction pathways in inorganic photochemistry. Recently he turned his attention to the use of very fast spectroscopy to monitor the excited states of co-ordination compounds, and to the study of extremely fast organometallic exchange processes by observing infrared bandwidths. He was elected FRS in 1992.

TURNER, William
(c.1510–1568)

English clergyman, physician and naturalist, known as the 'father of British botany', born in Morpeth, Northumberland. A Fellow of Pembroke Hall, Cambridge, he became a Protestant, and to escape religious persecution in England travelled extensively abroad, studying medicine and botany in Italy. He formed a close friendship with Gesner in Zurich, and became the author of the first original English works on plants, including *Libellus de re Herbaria Novus* (1538), the first book in which localities for native British plants were recorded. In his *Names of Herbes* (1548) he stated that he had intended to produce a Latin herbal, but had been persuaded to wait until he had observed the plants growing in England. His major work is *A new Herball*, published in three instalments (1551–1562). This book demonstrated Turner's independence of thought and observation, but used woodcuts prepared for the octavo edition of Fuschs's *De Ristoria Stirpium* of 1545. He was Dean of Wells (1550–1553), but left England during the reign of Mary I; he was restored to Wells in 1560. He named many plants, including goatsbeard and hawkweed. The basis he laid for 'a system of nature' was developed by Ray in the following century.

TWORT, Frederick William
(1877–1950)

English bacteriologist, born in Camberley, Surrey. He studied medicine in London, and became Professor of Bacteriology there in 1919. He was the last Superintendent of the Brown Institution of the University of London. Twort studied Johne's disease and methods of the culture of acid-fast organisms, and extracted an 'essential substance' (later shown to be of the vitamin K group) from dead tubercle bacilli. In the early part of this century he discovered a 'transmissible lytic agent', named some years later by d'Herelle as 'le bacteriophage', a virus for attacking certain bacteria. In many ways the discovery of the invasion of bacteria by viruses formed the beginnings of molecular biology.

TYNDALL, John
(1820–1893)

Irish physicist, born in Leighlin-Bridge, County Carlow. Largely self-educated, he was employed on the ordnance survey and as a railway engineer, before studying physics in England and at Marburg in Germany under Bunsen. He became professor at the Royal Institution in 1854. In 1856 he and T. H. Huxley visited the Alps and collaborated in *The Glaciers of the Alps* (1860), when he made the first ascent of the Weisshorn. In 1859 he began his researches into the action of radiant heat on gases and vapours. He later investigated the acoustic properties of the atmosphere and the behaviour of light beams passing through various substances, in the course of which he discovered in 1869 the 'Tyndall effect', the scattering of light by colloidal particles in solution, thus making the light beam visible when viewed from the side.

His suggestion that the blue colour of the sky is due to the greater scattering of the shorter wavelength blue light by the colloidal particles of dust and water vapour in the atmosphere was confirmed by the theoretical studies of Lord Rayleigh (1842–1919). A prolific writer on scientific subjects, Tyndall's presidential address to the British Association in 1874 in Belfast was denounced as materialistic. He died from accidental poisoning with chloral.

TYRRELL, George Walter
(1883–1961)

English igneous petrologist, born in Watford. He studied geology at the Royal College of Science, London, and joined the staff of the University of Glasgow (1906) as an assistant to John Gregory at the start of an active career as an outstanding teacher and researcher. His move to Scotland was rewarded with fertile ground for petrological studies which he pursued with vigour. He was a pioneer of igneous petrogenesis, and undertook important studies of rocks from Scotland, particularly Arran, as well as Rockall, Iceland, Jan Mayen, Antarctica, South America and Africa. He was the author of the widely distributed textbook *Principles of Petrology* (1926), as well as *The Geology of Arran* (1928), *Volcanoes* (1931), *The Earth and its Mysteries* (1953) and other petrological works.

TYSON, Edward
(1651–1708)

English physician, born in Bristol. Educated at Magdalen, Oxford, he graduated BA in 1670 and MA in 1673. Influenced by the naturalist Robert Plot, he undertook botanical and zoological studies while in Oxford, where he also commenced medical studies, gaining his MB in 1677. Tyson set up in practice in London, being appointed physician to Bridewell and Bethlehem hospitals, while continuing to pursue anatomical experiments and dissections. He published papers in the *Philosophical Transactions* of the Royal Society of London on pathological subjects. In 1690 he published a monograph on the anatomy of the porpoise which expatiated on his wider natural history interests. Tyson saw comparative anatomy as the discipline that would reveal the underlying structural unity of nature, whose plan was given by the 'great chain of being'. The porpoise was, he believed, transitional in morphological terms between fish and land creatures. Tyson published similar anatomical work on the lumpfish, rattlesnake and shark. In 1697, with William Cowper, he dissected an opossum, drawing attention to its peculiar reproductive system. He is remembered for his 1699 work on what he called the 'orang-outang' (in reality a chimpanzee brought back from Malaya). He viewed this primate as intermediate between man and the apes on the great chain of being. Tyson's work helped spark a fierce and continuing debate about man's relations to other primates.

U

UHLENBECK, George Eugene
(1900–1988)

Dutch-American physicist, born in Batavia (now Jakarta), Indonesia. He studied at Leiden and was awarded his PhD in 1927. From 1927 until 1960 he worked at the University of Michigan, where he was appointed Professor of Theoretical Physics in 1939. To explain the results of the Stern-Gerlach experiment which showed that when a beam of particles pass through a magnetic field some are deflected in one direction and some in the opposite direction, and the observation of close doublets of spectral lines, Uhlenbeck and his fellow student Goudsmit proposed that electrons in atoms can have intrinsic spin angular momentum as well as orbital angular momentum. Initially this was not accepted as physicists found it difficult to ascribe rotation to electrons, but later Dirac's Theory of relativistic quantum mechanics showed that spin is an intrinsic property of electrons.

ULAM, Stanislav Marcin
(1909–)

American mathematician and research scientist, born in Lwow, Poland (now Lvov, the Ukraine). Educated at the Lwow Polytechnic Institute where he received an MA in 1932 and a DSc in 1933, a common interest in set theory led to contact with von Neumann. In 1936 von Neumann invited Ulam to the USA, where he secured a post for him at the Institute for Advanced Study, and later involved him in the atomic bomb project at Los Alamos in 1944 (Ulam had been naturalized in the previous year). This required massive calculations, and Ulam and von Neumann utilized existing calculating machines and applied probabilistic (the so-called 'Monte Carlo') methods. After World War II, Ulam continued his interest in using machines to solve mathematic and scientific problems, and held several professorships. He maintained a lifelong friendship and correspondence with von Neumann, and took great interest in his later work on artificial intelligence.

ULUGH-BEG
(1394–1449)

Ruler of Turkestan from 1447. A grandson of Tamerlane who drew learned men to Samarkand, he founded an observatory there in 1420, and between 1420 and 1437 prepared new planetary tables and a new star catalogue, the latter being the first since that of Ptolemy. Positions were given with precision; this was the first time that latitude and longitude were measured to minutes of arc and not just degrees. His instruments must have been excellent and included a quadrant of 60 feet in radius. He also wrote poetry and history. After a brief reign, he was defeated and slain by a rebellious son.

UNSOLD, Albrecht Otto Johannes
(1905–)

German astrophysicist, born in Bolheim, Wurttemberg, and educated at the *real-gymnasium* in Heidenheim and at the universities of Tübingen and Munich. At Munich he obtained a doctorate in theoretical physics under Sommerfeld (1928) and continued on the academic staff until his appointment as lecturer at the University of Hamburg (1930–1932), where his colleagues included Baade and Rudolph Minkowski. In 1932 he was appointed Professor of Theoretical Physics and director of the observatory at the University of Kiel where he established a flourishing school of theoretical astrophysics and has remained, apart from short visits abroad, until now; he is currently Emeritus Professor. Unsold's principal field of research has been the physics of stellar atmospheres: he was the discoverer of the hydrogen convection zone (1931) which explains how heat energy is transported upwards to the Sun's photosphere, and his fundamental *Physik der Sternatmosphdren* ('Physics of Stellar Atmospheres', 1938) remains a classic on the subject. He is also the author of the successful college textbook *The New Cosmos*, first published in German and English in 1966. Unsold was awarded the gold medal of the Royal Astronomical Society in 1957.

UNVERDORBEN, Otto
(1806–1873)

German chemist, born in Dahme, near Potsdam. He studied for a year at the Pharmaceutical Institute at Erfurt and later entered the family manufacturing business at Dahme. All of his scientific work took place before he went into business. He experimented with the destructive distillation of organic substances, including resins, shellacs and animal oils. He discovered guaiacol by the destructive distillation of wood, and also aniline (which he termed 'crystalline') by the dry distillation of indigo. Aniline, which was later found to be a primary aromatic amine, became important to the German dye industry. It was extracted from coal tar and gave Germany a virtual monopoly of synthetic dyes until World War I. Today most of the aniline produced is used in the manufacture of urethane polymers, rubber and chemicals for agriculture. Unverdorben also investigated the fluorides, and mistakenly believed he had synthesized chromium hexafluoride. He died in Dahme.

URBAIN, Georges
(1872–1938)

French chemist who discovered many of the rare earths. He was born in Paris and educated at the École de Physique et de la Chimie in Paris, where Pierre Curie was on the staff, and at the University of Paris. He was Professor of Analytical Chemistry at the Sorbonne from 1906 to 1928, when he became Director of the Institute de Chimie de Paris. Between 1895 and 1912 he performed more than 200,000 fractional crystallizations in which he separated the elements samarium, europium, gadolinium, terbium, dysprosium and holmium. In 1907 he discovered lutetitun in ytterbium, previously thought to have been in a pure form, and in 1922 isolated hafnium at the same time as Hevesy and Coster working in Copenhagen. Urbain lent samples of the rare earths to Marie Curie who was investigating the radioactive properties of all the known elements. In parallel with his work on the rare earths, Urbain wrote on isomorphism and phosphorescence. He died in Paris.

UREY, Harold Clayton
(1893–1981)

American chemist, born in Walkerton, Indiana. He taught in rural schools (1911–1914) and then studied at Montana State University (BS, 1917) and the University of California, Berkeley (PhD in chemistry, 1923). In 1923–1924 he worked with Niels Bohr in Copenhagen. From 1924 to 1929 he was an associate in chemistry at Johns Hopkins University and he was on the chemistry faculty of Columbia University, New York, from 1929 to 1945 (as full professor from 1934). He worked at the chemistry department and Institute for Nuclear Studies of the University of Chicago from 1945 to 1958, and thereafter continued to be scientifically active in retirement for many years at the University of California, La Jolla. Urey's earliest researches were on atomic and molecular spectra and structure, but he is chiefly remembered for his discovery in 1932 of heavy hydrogen (deuterium), jointly with Ferdinand Brickwedde and George Murphy. Subsequently he made many studies of the separation of isotopes and isotopic exchange reactions. During World War II he was prominent in the attempts to separate uranium-235 for the atomic bomb as part of the Manhattan Project. He later advocated an international ban on nuclear weapons. After 1945 his research interests moved to geochemistry and cosmochemistry, and he wrote *The Planets* (1952) and *Some Cosmochemical Problems* (1961). Urey was awarded the Nobel Prize for Chemistry in 1934. He received the Davy Medal of the Royal Society in 1940 and became a Foreign Member in 1947. He was made an Honorary Fellow of the Chemical Society in 1945, and also received the Priestley Medal of the American Chemical Society.

USSING, Hans Henrikson
(1911–   )

Danish biophysicist who graduated in physiology from the University of Copenhagen in 1934. He joined an expedition to study plankton, but came across a technical problem which he discussed with his former professor Krogh. Krogh used the opportunity to divert him instead into a study of permeability problems using 'heavy water', water containing the deuterium isotope of hydrogen. After examining permeability parameters in frog skin, Ussing started to analyse protein structure and synthesis using deuterium as a tracer, but was disrupted by the German invasion of Denmark. During World War II he became a lecturer in biochemistry, and after liberation returned to the zoophysiology department to study active and passive transport mechanisms across biological membranes. He developed a comprehensive theoretical framework for working with radioactive tracers, techniques he pioneered as isotopes were then coming into widespread use in biology. His experiments particularly utilized a frog skin preparation mounted in a specially constructed apparatus, known as an Ussing chamber, so that each surface of the skin was bathed in a separate fluid. The chemical manipulation and analysis of the medium on each side provided information about transport processes across the skin. In 1951 Ussing proposed a cyclical carrier mechanism of permeability, linking active sodium and potassium transport across the membrane, and during the following decade active transport of sodium ions was demonstrated across many membrane preparations. The fundamental properties of biological tissues that Ussing discovered have been important in several areas of medicine, such as absorption, diffusion and secretion in tissues like the gut and kidneys. Apart from a year as a Rockefeller Fellow at the University of California (1948) he remained in Copenhagen, as Research Professor and head of the isotope division of the department of zoophysiology (1951–1960) and head of the Institute of Biological Chemistry (1960–1980) where he continues as Emeritus Research Professor.

V

VAN ALLEN, James Alfred
(1914–)

American physicist, born in Mount Pleasant, Iowa. He graduated from Iowa Wesleyan College in 1935, then spent time at numerous institutions before becoming head of the physics department at the State University of Iowa. During World War II he developed the radio proximity fuse, a device fitted to explosive projectiles that made use of radio waves to detect the proximity of targets, detonating the explosives once the target came within a certain distance. In anti-aircraft fire this meant that a direct hit was no longer necessary to detonate a shell. This work gained Van Allen expertise in the miniaturization of electronics, which was of great use to him in experiments carried out after the war into the properties of the Earth's upper atmosphere. Using V2 rockets acquired from Germany at the end of the war, Van Allen measured the cosmic-ray intensity at altitudes of around a hundred miles, using radio telemetry to convey the experimental data to Earth. In the 1950s he began work on 'rockoons', rockets launched from balloons. He was involved in the launching of the USA's first satellite, Explorer 1 (1958), and had ensured that his cosmic-ray detector was amongst the payload. This detector and those in the subsequent Explorer satellites launched in the same year revealed the startling result that above a certain altitude there was much more high-energy radiation than anyone had expected. The satellite observations showed that the Earth's magnetic field traps Earth-bound high-speed charged particles in two doughnut-shaped zones which have become known as the Van Allen belts. Van Allen, who has produced a large number of scientific papers and received numerous scientific awards, has been a member of several US governmental committees concerned with space exploration.

VAN ANDEL, Tjeerd Hendrik
(1923–)

Dutch-born American marine geologist and geological archaeologist, born in Rotterdam. He was raised in Indonesia and educated at the University of Gröningen. His career began with an academic appointment at the State Agricultural University at Wageningen (1948–1950) followed by a spell in industry as a sedimentologist with Shell Oil Company (1950–1956). From 1957 to 1964 he was associated with the Scripps Institute of Oceanography, University of California, before becoming Professor of Geology at the School of Oceanography of Oregon University (1968–1976). He was appointed Wayne Loel Professor of

Geology at Stanford University in 1976. Van Andel has devoted his professional life to the investigation of the undersea world and has been involved in tectonic ocean mapping, deep sea drilling, mineral resource assessment and palaeo-oceanography. He has published extensively on recent sediments of the continents and oceans, the origin and nature of the continental shelf, the geology and geophysics of the mid-ocean ridges, palaeoclimatology, mineral resource assessment and geo-archaeology. He was a member of the first scientific expedition to view and map the Mid-Atlantic Ridge from a deep-sea submersible (1974).

VAN DE GRAAFF, Robert Jernison (1901–1967)

American physicist, born in Tuscaloosa, Alabama. An engineering graduate of the University of Alabama (1923), he continued his studies at the Sorbonne (1924) where Maric Curie's lectures inspired him to study physics. During his PhD research at Oxford, he conceived the design of an improved type of electrostatic generator, in which electric charge could be built up on a hollow metal sphere. At Princeton in 1929 he constructed the first working model of this generator (later to be known as the Van de Graaff generator) in which the charge was carried to the sphere by means of an insulated fabric belt; in this way, potentials of over a million volts could be achieved. At MIT, Van de Graaff adapted his generator for use as a particle accelerator using the high voltages to precisely control the acceleration of charged nuclear particles and electrons to high velocities. This Van de Graaff accelerator became a major research tool of atomic and nuclear physicists. The generator was also employed to produce high-energy X-rays, useful in the treatment of cancer. During World War II he was the Director of the MIT High Voltage Radiographic Project which developed X-ray sources for the examination of the interior structure of heavy ordnance. He was a co-founder of the High Voltage Engineering corporation which manufactured particle accelerators, and resigned from MIT in 1960 to devote his time to the corporation.

VAN DE HULST, Hendrik Christoffell (1918–)

Dutch astronomer, born in Utrecht, where he was educated and obtained his PhD in 1946. After several years at universities in the USA, he became Professor of Astronomy at Leiden and Director of Leiden Observatory in 1970. In 1944 he suggested that interstellar hydrogen might be detectable at radio wavelengths, due to the 21.1 centimetre (1420.4 megahertz) radiation emitted when the orbiting electron of a hydrogen atom flips between its two possible spin states. Due to World War II it was not until 1951 that such emissions were first detected, by Purcell and Harold Ewen. The technique has since proved invaluable in detecting neutral hydrogen in both our own and other galaxies. Van de Hulst was also an expert on the scattering and absorption of dust in the interstellar medium, and was the first to suggest that the diameter of a typical interstellar dust grain is around the same as the wavelength of visible fight. With C. Allen, he showed that the solar F (Fraunhofer) corona is due mainly to forward scattering by dust (1946), this dust being the innermost component of the zodiacal cloud.

VAN DER MEER, Simon
(1925–)

Dutch physicist and engineer, born in the Hague and educated at the Technical University, Delft. He worked at the Philips research laboratories in Eindhoven (1952–1955) before becoming senior engineer for CERN, the European nuclear research centre in Geneva. He developed a method known as 'stochastic cooling' to produce a higher intensity beam of anti-protons in accelerators than had been produced before. This technology made possible the experiments which led to the discovery of the field particles W and Z, which transfer the weak nuclear interaction. Van der Meer shared the 1984 Nobel Prize for Physics with Rubbia for their separate contributions to this discovery.

VAN DER WAALS, Johannes Diderik
(1837–1923)

Dutch physicist, born in Leiden. The son of a carpenter, at 25 he entered the University of Leiden graduating in 1865. After teaching physics at Deventer and the Hague, he studied again at Leiden: his doctoral dissertation *On the Continuity of the Liquid and Gaseous States* (1873) was quickly seen to be of major significance in the study of fluids and academic recognition followed. Van der Waals convincingly accounted for many phenomena concerning vapours and liquids observed by Thomas Andrews and others, notably the 'critical temperature'. Simple kinetic theories of gases assumed non-interacting molecules with no volume; by postulating the existence of intermolecular forces and a finite molecular volume, van der Waals derived a new equation of state (the van der Waals equation) which agreed much more closely with experimental data, particularly under extreme conditions. This work led to a Nobel Prize for Physics (1910), and was a guide for the future liquefaction of permanent gases such as hydrogen and helium. The weak attractions between molecules (van der Waals forces) were named in his honour. As Professor of Physics at Amsterdam University (1877–1907) he gained distinction as a teacher and as an advocate of the chemical thermodynamics of Gibbs.

VANE, Sir John Robert
(1927–)

English pharmacologist, born in Tardebigg, Worcestershire. He studied at Birmingham and Oxford before taking up pharmacology appointments at Yale (1953–1955) and the Institute of Basic Medical Sciences, Royal College of Surgeons, London (1955–1973), where he was appointed professor in 1966. He subsequently moved to the Wellcome Research Laboratories, Kent (1973–1985). Working on adrenergic receptors of the nervous system and the role of the lung in drug uptake and metabolism, he devised a bioassay for the detection (1967) and (in conjunction with a blood platelet aggregometer) characterization of labile and bioactive arachidonic acid (essential fatty acid) metabolites. He investigated the interconversion of prostaglandins (associated with the swelling, reddening and pain of tissue damage) by isomerases (1970), and reported the inhibition of prostaglandin synthesis by aspirin (1971). In 1976, studying thromboxane biosyntheses (prostaglandin metabolites causing platelet aggregation and vasoconstriction), he discovered prostacyclin (PGI2), the short-lived antagonist of platelet aggregation (with a half-life of approximately 20 seconds), which is synthesized for prostaglandin endoperoxidase in the blood vessel wall. The ratio of PGI2 to PG endoperoxidase appears crucial for the control of thrombus formation in coronary disease and (with renin) for the control of blood pressure in the kidney; in the latter connection Vane also worked on angiotensin and bradykinin. For this work he shared the 1982 Nobel Prize for Physiology or Medicine with Bergstrom and Samuelsson. Van was elected FRS in 1974, served as the Royal Society's Vice-President (1985–1987) and was awarded its Royal Medal (1989). He was knighted in 1984.

VAN NIEL, Cornelis Bernardus Kees
(1898–   )

Dutch microbiologist, born in Haarlem. He studied chemistry at the Technological University in Delft, and later worked with Kluyver on iron and sulphur bacteria, and propionic acid bacteria. In 1929 he accepted a position at the Hopkins Marine Station of Stanford University. He made major contributions to the study of photosynthesis in bacteria, particularly in the *Thiorhodaceae* and *Athiorhodaceae*. Van Niel showed that the green and purple sulphur bacteria do not use water as the exclusive hydrogen donor (as in plants), but use hydrogen sulphide and other reduced sulphur compounds instead; this explains both their dependence on these reduced compounds and their inability to produce oxygen. He was able to delineate the light and dark reactions in *Athiorhodaccae*, where oxidation proceeds by way of reactions which do not require the presence of light.

VAN'T HOFF, Jacobus Henricus
(1852–1911)

Dutch physical chemist, born in Rotterdam. He was educated at the polytechnic in Delft, the University of Leiden, Bonn (under Kekule), Paris (under Wurtz) and Utrecht, where he took his doctorate in 1874. After a brief period teaching physics and chemistry at the veterinary school in Utrecht, he became Professor of Chemistry, Mineralogy and Geology at the University of Amsterdam (1878). In 1896 he moved to Berlin as a member of the Prussian Academy of Sciences and as professor at the university there. Van't Hoff is rightly regarded as one of the founders of physical chemistry, but his first work was in organic chemistry. He postulated that the four bonds of carbon are directed towards the corners of a tetrahedron (suggested independently by Le Bel around the same time). This idea, which provided a basis for explaining the optical activity of certain organic compounds, was published in 1874, but was not recognized as of fundamental importance until later. From 1877 he began to devote himself to physical chemistry and in his *Études de Dynamique Chimiquc* (1884), he developed the principles of chemical kinetics and applied thermodynamics to chemical equilibria. The equation for the effect of temperature on equilibria is commonly called the van't Hoff isochore. His important work on osmotic pressure was published in 1886, and this work was further developed in the next decade in connection with the theory of electrolytic dissociation of Arrhenius. With Ostwalk he founded the journal *Zeitschrift für Physikalische Chemie* in 1887. He later studied the phase relationships of the Stassfurt salt deposits. He was awarded the first Nobel Prize for Chemistry in 1901. Van't Hoff was elected a Foreign Member of the Royal Society in 1897, having been awarded its Davy Medal in 1893. He became an Honorary Fellow of the Chemical Society in 1888.

VAN TIEGHEM, Philippe
(1839–1914)

French botanist and biologist, born in Bailleul. Under Pasteur he produced a dissertation on ammoniacal fermentation. His second Dissertation was on the plant family *Araceae*. He became professor at the École Normale Superieure in 1864, where he began work on pure cultures of isolated bacterial strains. Well known for his studies of myxomycetes and bacteria, he produced a new classification of plants based on their gross anatomy. He established the relationship of the blue algae to the bacteria and showed that coal originated by a fermentation process. He also elucidated the principles of plant symmetry, and produced two important textbooks in which his ideas on plant anatomy and plant physiology were expounded. The *Traité de Botanique* was published in 1884, and the *Elements de Botanique* in two volumes between 1886 and 1888.

VAN VLECK, John Hasbrouck
(1899–1980)

American physicist, born in Middletown, Connecticut, the son of a mathematics professor. He largely founded the modern theory of magnetism, by applying Dirac's theory of quantum mechanics to the magnetic properties of atoms. Unusually for this field, he was trained in the USA; after study at Wisconsin and Harvard, he took up posts at Minnesota, Wisconsin and finally Harvard (1934–1969). In the late 1920s and early 1930s his research was in dielectric and magnetic susceptibilities culminating in his classic text, *The Theory of Electric and Magnetic Susceptibilities* (1932). He also elucidated chemical bonding in crystals and studied the crystal fields and ligand fields, electric fields experienced by the electrons of an ion or atom due to the presence of neighbouring ions or atoms. These fields influence the energy levels permitted in the system and therefore have far reaching effects on the optical, magnetic and electrical properties of the material. During World War II van Vleck contributed to the exploitation of radar, showing that atmospheric water and oxygen molecules would cause troublesome absorption at certain wavelengths. In 1977 his pioneering research was recognized with the joint award of the Nobel Prize for Physics, with Philip Mott.

VARMUS, Harold Elliot
(1939–)

American molecular biologist, born in Oceanside, New York. He was educated at Amherst College, Virginia, where he received a BA in 1961, Harvard University where he received his MA in 1962, and Columbia University, New York, where he graduated MD in 1966. He began his career as a surgeon in the US Public Health Service (1968–1970) then moved to the University of California Medical Centre, San Francisco, as a lecturer (1970–1972), assistant professor (1972–1974) and associate professor (1974–1979). Since 1979 he has been Professor of Microbiology and Immunology there, simultaneously holding the posts of Professor of Biochemistry and Biophysics (from 1982) and American Cancer Society Professor of Molecular Virology (from 1984). In 1989 he was awarded the Nobel Prize for Physiology or Medicine (jointly with Bishop) for his contribution to the discovery of oncogenes. Oncogenes are normal cellular genes which direct various aspects of cellular growth and differentiation. If their production is altered in some way, for example by mutation or viral activation, the faulty protein gives rise to cancer in the cell. The discovery of these genes has been of vital importance in understanding the mechanisms of cancer.

VASKA, Lauri
(1928–)

American chemist, born in Rakvere, Estonia. He attended the universities of Hamburg and Göttingen in Germany before moving to the USA where he received his PhD at the University of Texas. In 1964 he was appointed as professor at Clarkson College of Technology, Potsdam, New York. Vaska is famous for his work on oxygen complexes of transition metals, including his synthesis of the so-called Vaska's complex, (PPh3)2C1(CO)(O2). This complex takes up O2 in a reversible manner and is thus used as a model for the uptake and transport of O2 by haemoglobin in the blood.

VAVILOV, Nikolai Ivanovich
(1887–1943)

Russian botanist and plant geneticist, born in Moscow, brother of the physicist Sergei Vavilov (1891–1951). He trained in Moscow and at the John Innes Agricultural Institute, Merton, Surrey. During World War I, he was asked to go to Iran, where he discovered many crop plant varieties and took seed samples back to Russia – the initial nucleus of his internationally renowned World Collection. He undertook further journeys to the Pamir Mountains and to Afghanistan, and then to the Mediterranean, Abyssinia, the Far East and the Americas. He assembled the world's largest collection of seeds, numbering some 200,000 specimens, over 40,000 of them being varieties of wheat. In 1923 he established a network of 115 experimental stations across the USSR to sow the collection over the widest possible range. He published extensively on the centres of origin of crop plants, formulating the principle of diversity which postulates that, geographically, the centre of greatest diversity represents the origin of a cultivated plant. His commanding international reputation was challenged by the politico-scientific 'theories' of Lysenko, who denounced him at a genetics conference (1937) and gradually usurped his position. Arrested in 1940, he died of starvation in a Siberian labour camp.

VENING MEINESZ, Felix Andries
(1887–1966)

Dutch geophysicist, born in the Hague. In 1910 he graduated in civil engineering from Delft Technical University. His first appointment was with the Netherlands State Committee to undertake a gravity survey of the Netherlands where, in order to overcome the problems of making measurements on vibrating peaty subsoil, he designed his classical pendulum apparatus with two pendulums swinging in the same plane; for the theory of this work he was awarded a doctorate in 1915. This led him to investigate means of measuring gravity at sea. Working on a naval submarine to avoid wave turbulence during a cruise to Indonesia in 1923, he achieved marine gravity determination to an accuracy of 1 mgal, comparable to land measurements. In other long submarine voyages which followed, he aimed to determine the Earth's shape, but also discovered that Airy isostasy prevails over the oceans, and that a belt of negative isostatic anomalies exists parallel to trenches, where he calculated an elastic crustal thickness of 35 kilometres. His speculative thoughts on mantle convection, downbuckling and crustal shear failure showed great insights into Earth processes, which with his book *The Earth's Crust and Mantle* (1964) were preludes to modern theories of plate tectonics, although he did not believe that continental drift had occurred during recent geological times. He was Professor of Geophysics at the Rijksuniversiteit Utrecht (1937–1966) where the geophysical laboratory is named after him.

VENN, John
(1834–1923)

English logician, born in Drypool, Hull. A Fellow of Caius College, Cambridge (1857), he developed Boole's symbolic logic, and in his *Logic of Chance* (1866) the frequency theory of probability. He is best known for 'Venn diagrams', pictorially representing the relations between sets, although similar diagrams had already been used by Leibniz and Leonhard Euler.

VENTURI, Giovanni Battista
(1746–1822)

Italian physicist, born in Bibiano, near Reggio. He was ordained priest at the age of 23 and in 1773 became Professor of Geometry and Philosophy at the University of Modena. Later he was appointed professor at Pavia, although from 1796 he lived mainly in Paris. In addition to work on sound and colours, he published studies of the geological material contained in the notebooks of Leonardo da Vinci (1797). Venturi is remembered for his work on hydraulics (published 1797), particularly for the effect named after him (the decrease in the pressure of a fluid in a pipe where the diameter has been reduced by a gradual taper) which he first investigated in 1791. The Venturi flow-meter, based on this phenomenon, was invented by the American engineer Clemens Herschel (1842–1930).

VERNADSKY, Vladimir Ivanovich
(1863–1945)

Russian mineralogist, geochemist and biogeochemist, born in St Petersburg. He studied at the University of St Petersburg, where Mendeleyev's brilliant lectures awakened a strong desire for knowledge and its application. From 1886 he was curator of the university's mineral collection, and in 1888 he travelled abroad to work in Munich and with Le Chatelier in Paris. He returned to Russia in 1890 to take up a research post at Moscow University, and became Professor of Mineralogy there in 1898. In 1914 he was appointed director of the geological and mineralogical museum of the Academy of Sciences in St Petersburg; he also spent a period in the Ukraine (1917–1921), where he founded the Ukrainian Academy of Sciences. Vernadsky conducted important research on rock-forming silicates and aluminosilicates and their structures, and on the relationship between crystal form and physiochemical structure. Later in life he concentrated on the structure of the chemical composition of plants and animals. He introduced the term 'biosphere' in 1926, and as a result of his biogeochemical studies concluded that all of the main gases in the Earth's atmosphere were generated by living organisms. He was one of the first to recognize the potential of nuclear power as an important source of energy and argued for responsibility on the part of scientists.

VERNON HARCOURT, Augustus George
(1834–1919)

English chemist, born in London. While at Balliol College, Oxford, he was first a student of Benjamin Collins Brodie and then his assistant. From 1859 to 1902 he was Dr Lee's Reader in Chemistry at Christ Church, Oxford, where he converted a former anatomy museum into a chemical laboratory. He had many students who were later distinguished, including Dixon, David Chapman and Sidgwick. Vernon Harcourt was elected FRS in 1868 and was President of the Chemical Society from 1895 to 1897. In various capacities he also served the British Association for the Advancement of Science (he was a nephew of William Vernon Harcourt, who was essentially the founder of the association). His lifelong research interest was the study of rates of chemical change, pursued in collaboration with the mathematician William Esson. Around 1864–1866 they discovered the law of mass action, at about the same time as the Norwegians Guldberg and Waage. Vernon Harcourt also did much work related to the manufacture of coal-gas, and his pentane lamp provided the British standard of luminosity for many years. He also devised an inhaler for the controlled administration of chloroform as an anaesthetic.

VERNON HARCOURT, William Venables
(1789–1871)

English chemist and clergyman, one of the founders of the British Association for the Advancement of Science, born in Sudbury, Derbyshire. He was educated at home and then spent five years in the Royal Navy before going to Christ Church, Oxford, in 1807. At Bishopsthorpe, which was his first parish, he set up his own chemical laboratory, taking advice from Wollaston and Humphry Davy. Later he moved to other ecclesiastical posts. He worked on the effect of heat on inorganic compounds, and on methods of making achromatic lenses by combining glasses with different dispersions; in 1824 he was elected FRS. His principal importance, however, lay in the encouragement he gave to others. He became the first President of the Yorkshire Philosophical Society and played a major part in the establishment of the British Association for the Advancement of Science. At the time, the Royal Society was becoming more restrictive and professional in its membership: the idea behind the British Association was that it should be open to everyone interested in science. Today any member of the general public may still join simply by contacting the association's headquarters in London and paying the small annual subscription. The association held its first meeting in 1831 and Vernon Harcourt was elected its president in 1839. In 1861 he succeeded to the family estates at Nuneham, Oxfordshire, where he died 10 years later.

VESALIUS, Andreas
(1514–1564)

Belgian anatomist, one of the first dissectors of human cadavers. Born in Brussels, a pharmacist's son, Vasalius studied in Paris, Louvain and Padua, where he took his degree in 1537. He was appointed Professor of Surgery at Padua University. An ardent champion of dissection, his lectures were somewhat novel, in that he performed dissections himself, instead of following the normal custom of leaving this to an assistant while the professor read from a textbook. He made use of drawings in his lectures. In 1538 he published his six anatomical tables, still largely Galenist, and in 1541 he edited Galen's works. Comprehensive anatomizing enabled him to point out many errors in the traditional medical teachings derived from Galen. For instance, Vesalius insisted he could find no passage for blood through the ventricles of the heart, as Galen had assumed. His greatest work, the *De Humani Corporis Fabrica* ('On the Structure of the Human Body', 1543), was enriched by magnificent illustrations. With its excellent descriptions and drawings of bones and the nervous system, the book set a completely new level of clarity and accuracy in anatomy. Many structures are described and drawn in it for the first time, e.g. the thalamus. Perhaps upset by criticism, Vesalius left Padua to become Court Physician to the emperor Charles V and his son Philip II of Spain. He died on the way back from a pilgrimage to Jerusalem.

VIETA, Franciscus, or VIÉTE, François
(1540–1603)

French mathematician, born in Fontenay-le-Comte. He became a privy councillor to Henri IV of France and decoded an important Spanish cypher. His *In Artem Analyticam Isagoge* (1591) is probably the earliest work on symbolic algebra, and he devised methods for solving algebraic equations up to the fourth degree. He also wrote on trigonometry and geometry, and obtained the value of π as an infinite product. He believed that his algebra was essentially the method used by the Greeks, but not transmitted in their published works. His influence on Descartes is a matter of dispute, and may have been slight as Descartes himself claimed, because Vieta's work was mostly published posthumously and was not widely available; however, Vieta undoubtedly was a major influence on Fermat, whose earliest work was written in Vieta's notation.

VILLEMIN, Jean-Antoine
(1827–1892)

French physician and experimentalist, born in Prey, Vosges. He studied medicine in Strasbourg and Paris, where he received his MD in 1853. A modest man, he continued in medical practice in Paris, but in addition operated a private laboratory where he worked assiduously in his spare time. Among his most fundamental observations was the discovery, in the 1860s, that material taken from the lung of a person with tuberculosis would, when inoculated into an animal, produce tuberculosis in the animal. This work pointed towards a specific infective agent, which Koch discovered in 1882.

VINE, Frederick John
(1939–1988)

English geophysicist, educated at St John's College, Cambridge. He became associate professor at the department of geological and geophysical sciences at Princeton University (1967–1970), then reader (1970–1974) and Professor of Environmental Science (1974–1988) at the University of East Anglia. He undertook important work with Matthews in the interpretation of marine magnetic anomalies and their use in confirmation of the sea-floor spreading hypothesis, as well as palaeomagnetism, plate tectonics and energy resources.

VIRCHOW, Rudolf
(1821–1902)

German pathologist and politician, and founder of cellular pathology. Born in Schivelbein, Pomerania, Virchow graduated in medicine in Berlin, and then secured a junior post in Berlin's great hospital, the Charité. He rose to become Professor of Pathological Anatomy at

Wurzburg (1849–1856). He quickly proved himself a skilful pathologist. In 1845 he recognized leukaemia, and proceeded to study animal parasites, inflammation, thrombosis and embolism. Deeply politically committed, his involvement on the liberal side in the revolutions of 1848 led to his dismissal from his Wilmburg post, although he was subsequently reinstated. In 1856 he returned to Berlin as Professor of Pathological Anatomy. In the 1850s Virchow enthusiastically adopted Schwann and Schleiden's cell theory, applying it to pathological investigation. He argued that disease originated in cells, or at least was the response of cells to abnormal circumstances. His suggestions led to much fertile work, aided by refinements in microscopes, new dyes for selective staining and the development by him of the microtome for making thin sections. With Virchow modern pathology begins. However, he did not see eye-to-eye with his French contemporary Pasteur on the germ theory of disease. Virchow saw disease as a matter of the continuous cellular change, rather than as a result of attack by invasive agents. His *Cellularpathologié* (1858) established that tumours and all other morbid structures contained cells derived from previous cells. Virchow was also sceptical about Charles Darwin's evolutionary theory, treating it as an hypothesis only. He remained lastingly politically active. Sitting as a liberal member of the Reichstag (1880–1893), he opposed the Chancellor so forcefully that Bismarck challenged him to a duel in 1865 (it was not fought). In practical politics, his efforts in public health in Berlin led to improved water and sewage purification. An enthusiastic archaeologist, he worked on the 1879 dig to discover the site of Troy.

VIRTANEN, Artturi Ilmari
(1895–1973)

Finnish biochemist, born in Helsinki. He was educated at the University of Helsinki and after holding a number of industrial posts associated with the butter and cheese industry, he became Professor of Biochemistry, first at Finland Institute of Technology (1931–1939) and then at the University of Helsinki (1939–1948). He studied the bacterial metabolism of sugars to form succinate and lactate, and observed that proteases are released into the growth medium (1931). This related to his earlier finding that root nodules of leguminous plants release nitrogous substances (1927). He discovered aspartase (1932), that legume bacteria can convert aspartate to beta-alanine (1937) and the transamination of aspartate with pyruvate to give alanine (1940). He implicated the formation of hydroxylamine at an early stage in nitrogen fixation by legume root nodules, and established its conversion to aspartate via an oxime intermediate (1938). For these discoveries he was awarded the Nobel Prize for Chemistry (1945). Virtanen also worked on the nutritional requirements of plants, and the plant biosynthesis of carotene and vitamin A. He also isolated and characterized haemoglobin and other pigments from legume nodules, and showed that silage can be preserved by dilute hydrochloric acid. He published *Cattle Fodder and Human Nutrition* in 1938.

VOGEL, Hermann Carl
(1841–1907)

German astronomer, born in Leipzig and educated at the university in his native city. He was appointed assistant at the Leipzig Observatory, where Johann Karl Friedrich Zollner stimulated his interest in astrophysics and in 1870 recommended him to a position at a newly founded private observatory at Bothkamp near Kiel. Using a 28-centimetre refractor, Vogel made some pioneering studies of the spectra of the major planets which established his reputation as an astrophysicist. In 1874 he was called in by the Prussian government as adviser in the planning of the new astrophysical observatory in Potsdam, and in 1882 was appointed its director. Introducing photographic methods into stellar spectroscopy, Vogel was the first to achieve sufficient accuracy to measure radial velocities of stars. In 1889 he discovered the binary nature of Algol from such measurements, thereby opening up the important field of spectroscopic double stars. He had somewhat less satisfaction with the 'great Potsdam double refractor' with an aperture of 80 centimetres and a focal length of 12 metres installed in 1899 whose performance did not match that of the large mirror telescopes being built in California. Nevertheless, the Potsdam astrophysical observatory became under his guidance one of the world's leading centres of astrophysics. Vogel continued in office until his death, although in his later years he became a recluse whose chief interest was to play the magnificent organ installed in his official residence at the observatory.

VOGEL, Hermann Wilhelm
(1834–1898)

German chemist, known for his contribution to colour photography and colour printing. He was born in Doberlug, Lower Lusataia, and was Professor of Photochemistry, Spectrum Analysis and Photography at the Royal Technical College, Berlin, from 1879 until his death. In 1874 he discovered that certain organic dyes can make silver bromide dry plates sensitive to light in the wavelengths that they themselves absorb. He then developed the orthochromatic photographic plate which was sensitive to red and yellow light. He also took out a patent for three-colour printing which specified that each colour should be printed at a slightly different position, thus avoiding the blurred result when the dots from the different colours of screen were printed on top of each other. He died in Berlin.

VOGT, Peter
(1932–)

German-born American microbiologist. He was educated at the University of Tübingen and moved to the USA as an assistant professor of pathology (1962–1966) at the University of Colorado and associate professor until 1967, when he became associate professor (1967–1969) and then Professor of Microbiology (1969–1971) at the University of Washington. He was Hastings Distinguished Professor of Microbiology at the University of Southern California from 1978 to 1980, and has been Chairman of the Microbiology Department there since 1980. A major field of interest for Vogt has been oncogene transduction by retroviruses. Oncogenes are normal cellular genes which can be activated in a variety of ways to become carcinogenic. One way this can occur is via infection of a cell by a retrovirus. Retroviruses are RNA viruses which take over a cell's machinery to produce a DNA copy which then becomes incorporated into cellular DNA. When this happens the retrovirus occasionally picks up a cellular oncogene or a corrupted copy of this gene that may drastically affect subsequently infected host cells. Vogt showed that there are basically two ways by which oncogenes can be activated by retroviral transduction. The gene sequence may be altered so that it codes for protein with abnormal function or it can be brought under control of powerful viral enhancers or promoters to overproduce a normal gene product. In either case the involvement of retroviruses is clearly of vital importance in certain carcinogenic processes.

VOLTA, Alessandro Giuseppe Anastasio, Count
(1745–1827)

Italian physicist, inventor of the electric battery, born in Como. He was appointed Professor of Physics at Como (1775) and at Pavia (1778). In 1795 he became Rector of Pavia University, but he was dismissed in 1799 for political reasons; later reinstated by the French, he retired in 1815. He was summoned to show his discoveries to Napoleon, and received medals and titles at home and abroad. His main contributions were in electrostatics, current electricity and gas chemistry. He invented the electrophorus (1775), the precursor of the induction machine of which the Wimshurst of the early 1880s became the best-known example, the condenser (1778), the candle flame collector of atmospheric electricity (1787), the calibrated straw electrometer (1780s), and the electrochemical battery, or 'voltaic pile' (1800), which was the first source of continuous or current electricity. It was inspired by a controversy he had with Galvani concerning the nature of animal electricity. He also invented an electric spark eudiometer (1776), and his famous 'inflammable air' (hydrogen) electric pistol (1777). Lavoisier followed Volta's suggestion that the mixtures of air and hydrogen should be sparked over mercury (and not water), and identified the resultant to be water (1782). His name is given to the SI unit of electrical potential difference, the volt.

VOLTERRA, Vito
(1860–1940)

Italian mathematician, born in Ancona. He was professor at Pisa, Turin and Rome. In 1931 he was dismissed from his chair at Rome for refusing to sign an oath of allegiance to the Fascist government, and he spent most of the rest of his life abroad. He worked on integral equations, where he introduced the idea of studying spaces of functions that proved exceptionally fertile. He also worked on mathematical physics and the mathematics of population change in biology, where he put forward the Lotka-Volterra equations, a pair of differential equations that describe a simple predator-prey population model. Through his breadth of interest and energy, he became a leading representative of Italian mathematics abroad.

VON BRAUN, Wernher
(1912–1977)

German-born American rocket pioneer, born in Wirsitz. He studied engineering at Berlin and Zurich and founded in 1930 a society for space travel which carried out experiments at a rocket-launching site near Berlin. Since rockets were outside the terms of the Versailles Treaty, the German army authorities became interested and by 1936, with Hitler's backing, von Braun was director of a rocket research station at Peenemunde, where he perfected the infamous V-2 rockets first launched against Britain in September 1944. A total of 4,300 were fired, more than a thousand of which landed on London. At the end of World War II he surrendered, with his entire development team, to the Americans. He became a naturalized American in 1955 and a director of the US army's Ballistic Missile Agency at Huntsville, Alabama, and was chiefly responsible for the manufacture and successful launching of the first American artificial earth satellite, Explorer 1, in 1959. From 1960 to 1970 he was Director of the Marshal Space Flight Center, where he developed the Saturn rocket for the Apollo 8 Moon landing (1969). His books include *Conquest of the Moon* (1953) and *Space Frontier* (1967).

VON KLITZING, Klaus
(1943–)

German physicist, born in Schroda. He was educated at the Technical University in Munich and at Wurzburg University, where he received his doctorate in 1972. He was then appointed professor at Munich in 1980, and in 1985 became Director of the Max Planck Institute, Stuttgart. In 1977 he presented a paper on two-dimensional electronic behaviour in which the quantum Hall effect, which occurs in semiconductor devices at low temperatures, was clearly implied, but few realized its significance and von Klitzing only appreciated what had occurred in 1980. This caused a major revision of the theory of electric conduction in strong magnetic fields and also provided a highly accurate laboratory standard of electrical resistance. For this work he was awarded the 1985 Nobel Prize for Physics.

VON LAUE, Max Theodor Felix
(1879–1960)

German physicist, born near Koblenz. In 1905 he was offered an assistantship, by Planck; they became lifelong friends, and an especially appreciated honour must have been the award of the Max Planck Medal from the German Physical Society (1932). After university posts in Zurich, Frankfurt (1914) and Berlin (1919) he was appointed advisor to the Physikalisch Technische Reichsanstalt, and Deputy Director of the Kaiser Wilhelm Institute for Physics. He lost his influence during the Nazi era because of his opposition to the regime, but was reinstated after World War II, and aged 71 was appointed Director of the former Kaiser Wilhelm Institute for Chemistry and Electrochemistry in Berlin-Dalhlem (1951). He applied the concept of entropy to optics, and demonstrated that Fizeau's formula for the velocity of fight in flowing water followed from Einstein's theory of special relativity (of which he was an early adherent). In 1912 he discovered that X-rays are diffracted by the three-dimensional array of atoms in crystals; for this work he was awarded the 1914 Nobel Prize for Physics. He died as the result of a car accident.

VON MISES, Richard
(1883–1953)

American mathematician and philosopher, born in Lemberg, Austria-Hungary (now Lvov, the Ukraine). He was professor at Dresden (1919), Berlin (1920–1933), Istanbul, and from 1939 at Harvard, where he became Gordon McKay Professor of Aerodynamics and Applied Mathematics in 1944. An authority in aerodynamics and hydrodynamics, he set out in *Wahrscheinlichkeit, Statistik und Wahrheit* ('Probability, Statistics and Truth' 1928) a frequency theory of probability which has had wide influence, even though not generally accepted.

VONNEGUT, Bernard
(1914–   )

American physicist, born in Indianapolis. Educated at MIT, he worked under Schaefer at the General Electric Company (1945–1952) then at the A. D. Little Company until 1967, when he became Professor of Atmospheric Science at New York State University. In 1947 he improved a method for artificially inducing rainfall, by using silver iodide as a cloud-seeding agent.

VON NEUMANN, John (Johann)
(1903–1957)

Hungarian-born American mathematician, born in Budapest. Educated in Berlin and Budapest, he taught at Berlin (1927–1929), Hamburg (1929–1930) and Princeton (1930–1933) before becoming a member of the newly founded Institute for Advanced Study at Princeton (1933). In 1933 he became a consultant to the Manhattan Project at Los Alamos for the construction of the first atomic bomb, and in 1954 he joined the US Atomic Energy Commission. His best-known mathematical work was on the theory of linear operators, but he also had a new axiomatization of set theory, later used by Godel. In addition he formulated a precise mathematical description of the recently developed quantum theory (1932), and worked on Lie groups. His work during World War II led him to study the art of numerical computation and to design some of the earliest computers, and his theoretical description of a programmable computer governed computer architecture until quite recently. In *The Theory of Games and Economic Behaviour* (1944), written with Oskar Morgenstern, he created a theory applicable both to games of chance and to games of pure skill, such as chess. These ideas have since become important in mathematical economics and operational research. He went on to invent the idea of self-replicating machines (arguably later shown to be exemplified through the actions of DNA in cells) and cellular automata.

VORONOFF, Serge
(1866–1951)

Russian physiologist, born in Voronezh. Educated in Paris, he became Director of Experimental Surgery at the Collège de France. Later working in Switzerland, he built on Metchnikoff's work on longevism and developed a theory connecting gland secretions with senility. Pioneering endocrinological surgery, Voronoff specialized in grafting animal glands (especially monkey glands) into the ageing human body, with a view to restoring potency and ensuring long life. These experiments won him considerable notoriety. He also grafted the thyroid from monkeys into mentally backward and defective children in an attempt to restore them to normalcy.

W

WADATI, Kiyoo
(1902–)

Japanese seismologist. After graduating from Tokyo Imperial University in 1925 and receiving a DSc in 1932, he entered the Central Meteorological Observatory (now the Japan Meteorological Agency). His most important contributions have been to advances in the detection of deep earthquakes, particularly those lying on an inclined plane dipping deep within the Earth beneath Japan. Similar deep, inclined, seismic zones were being located by Benioff around the same time. These are now known as Wadati-Benioff zones and show the motion of downgoing oceanic crust as it subducts into the mantle at an island arc. Also interested in Antarctic research, Wadati carried out fieldwork at the Showa base between 1973 and 1974.

WADDINGTON, Conrad Hal
(1905–1975)

English embryologist and geneticist, born in Evesham. He graduated from the University of Cambridge in 1926 with a degree in geology, and after a brief interlude as a palaeontologist, turned to embryology. From 1947 to 1970 he was Professor of Animal Genetics at Edinburgh University. He studied the effects of chemical messengers in inducing embryonic cells to form particular tissues during development, and was especially concerned with the ways in which both genes and environmental influences control the development of embryos in a stepwise manner, the process of epigenesis. He held the unfashionable view that environmentally induced effects could be incorporated in a heritable manner. Waddington introduced the term 'canalization' for the process of developmental stability where a particular phenotype is expressed in spite of the presence of genes with a different potential. His *Organisers and Genes* (1940) is concerned with the relationship between Mendelian genetics and experimental embryology; he also wrote a standard textbook of embryology, *Principles of Embryology* (1956). He was interested in the popularization of science and in forging links between science and the arts, particularly the visual arts. His more popular books included *The Ethical Animal* (1960) and *Biology for the Modern World* (1962).

WAERDEN, Bartel Leendert van der
(1903–)

Dutch mathematician, born in Amsterdam, where he obtained his doctorate in 1926. He was professor at Gröningen (1929–1931), Leipzig (1931–1945), Johns Hopkins (1947–1948), Amsterdam (1948–1951) and Zurich

(1951–1962). Waerden worked in algebra, algebraic geometry and mathematical physics, and published books on the history of science and mathematics in the ancient world such as *Science Awakening* (1954). His classic textbook *Moderne Algebra* (1931) was influential in publicizing the new algebra developed by Hilbert, Ernst Steinitz, Artin, Noether and others, and his book on the application of group theory to quantum mechanics (1932) showed its relevance to physics. He also wrote a series of 20 papers on algebraic geometry devoted to showing the power and rigour of the new algebraic methods in resolving old questions about curves and surfaces that had been bedevilled by overgenerous intuition.

WAGER, Lawrence Rickard
(1904–1965)

English geologist, petrologist and explorer, born in Batley, Yorkshire. He was educated at Pembroke College, Cambridge, where he became interested in petrology under the influence of Harker, and was appointed as a lecturer at the University of Reading (1929–1943). He was an excellent rock-climber and mountaineer, and in 1933 climbed Everest to 28,000 feet without oxygen, some 20 years before John Hunt's successful summit attempt. He took part in a series of major scientific expeditions including several to East Greenland (1930–1953), where he undertook his most significant geological work, mapping a major area and carrying out his classical study of the petrology and geochemistry of the Skaergaard layered igneous complex. He served with distinction in the photographic reconnaissance section of the Royal Air Force during World War II, and subsequently resumed his geological career as Professor of Geology at Durham (1944–1950) and then at Oxford from 1950. He was elected FRS in 1946. Wager's important research on various aspects of igneous and metamorphic petrology included notable studies of crystal nucleation and the origin of rhythmic layering, in the Rhum Tertiary igneous complex of Scotland, and the nature of the Marscoite hybrid rock suite on Skye. In 1955 he was instrumental in the establishment of the Oxford radiometric age determination laboratory.

WAGNER-JAUREGG, or WAGNER VON JAUREGG, Julius
(1857–1940)

Austrian neurologist and psychiatrist, born in Wels and educated in Vienna. He became professor at Graz (1889) and Vienna (1893). Although his chairs were in psychiatry, he remained more interested in general medical aspects of psychiatric disorders, such as the relationship between cretinism and goitre. He won the 1927 Nobel Prize for Physiology or Medicine for his discovery in 1917 of a treatment for general paralysis (a late stage of syphilis) by infection with malaria. This was based on an older observation that patients with a variety of serious mental disorders occasionally improved after they had suffered from a bout of febrile illness. He devised a series of experiments on patients suffering from dementia and other forms of psychiatric disease, noting that the best results were obtained in those with general paralysis, several of whom showed sufficient amelioration that they could be discharged from the asylum. This 'fever therapy' was hardly ideal and was abandoned when antibiotics and other better treatments became available.

WAKSMAN, Selman Abraham
(1888–1973)

American biochemist, born in Priluka, the Ukraine. After receiving private tuition in Russia he moved to the USA in 1910 and became a US citizen in 1915, graduating in the same year at Rutgers University, where he spent most of his research life, becoming Professor of Microbiology in 1930. From 1915 he worked on the microbial breakdown of organic substances in the soil and the nature of humus. From this work emerged a new classification of microbes (1922) and methods for their scientific cultivation (1932). He began a metabolic characterization of actinimycete fungi, and observed that they produce antibacterial substances (1937). From 1939 he searched for antibiotics of medical importance and discovered the rather toxic anti-cancer drug actinomycin (1941), the first anti-tuberculosis drug streptomycin (1944), another anti-streptococcal drug neomycin (1949), the anti-trichomonad streptocin and several other anti-bacterial agents. For these worldly benefits he was awarded the Nobel Prize for Physiology or Medicine in 1952. He also worked extensively on marine bacteria and the enzyme alginase, and wrote *Enzymes* (1926), *Principles of Soil Microbiology* (1938), the autobiographical *My Life with the Microbes* (1954), *The Actinomycetes* (3 vols, 1959–1962), *The Conquest of Tuberculosis* (1964) and other works.

WALD, George
(1906–   )

American biochemist, born in New York City. He studied zoology at New York University and at Columbia University, and worked under Warburg in Berlin. Subsequently he worked at Harvard (1932–1977), where he became Professor of Biology in 1948. His early work was divided between the possible role of carotenoids in photosyntheses (soon abandoned) and the way in which visual purple, the retinal pigment of the eye, responds to stimulation by light. He established in 1935 that visual purple is converted by light to a yellow compound which slowly changes to a colourless compound (vitamin A). His subsequent work was directed mainly towards elucidating the details of this process, and in 1956 he made the key discovery that only one of six geometric isomers of vitamin A (11-cis-retinal) combines with the protein opsin to form visual purple (rhodopsin). He found that light transformed 11-cis-retinal to all-trans-retinal (retinenel) which caused the opsin to change shape and release the all-trans-retinal, which was reduced to vitamin A, then slowly oxidized back to rhodopsin. He discovered a similar system using vitamin A2 in fish, the visual mechanism proving common to the eyes of all known animals. Wald also established the nutritional relationship between vitamin A, night blindness and vitamin-deficient retinopathy. For these discoveries he shared the 1967 Nobel Prize for Physiology or Medicine with Granit and Hartline.

WALDEN, Paul
(1863–1957)

Latvian chemist, born in Rosenbeck. He studied at Riga Polytechnical School where he collaborated with Ostwald on ionic conductivity in solution. He became professor at the school in 1894, and two years later discovered the Walden inversion for which he is best known. From a study of changes in optical activity he found that substitution on an enantiomer brings about inversion of configuration, i.e. the R form is converted into the S form. A complete explanation of this came only with the mechanistic studies of Christopher Ingold. He also produced many electrochemical studies. He became director of St Petersburg Academy of Sciences chemical laboratory, and in 1919 moved to Germany and became professor at Rostock. He continued lecturing at Tübingen until the age of 90.

WALDEVER-HARTZ, Wilhelm
(1839–1921)

German histologist and anatomist, born in Hehlen and educated in Göttingen, Greifswald and Berlin. Inspired to a career in medical science by Henle's lectures, Waldeyer-Hartz became professor at several universities, including Breslau, Strassburg and Berlin. He established his reputation with his histological studies of cancers, which he classified according to their embryological cells of origin. Thus, carcinomas come from epithelial cells, whereas sarcomas originate from the connective (mesodermal) tissues. He was asked to give a diagnosis for Emperor Frederick III's tumour of the vocal cords. His other work included studies of the histology of the spinal cord, comparative neuroanatomy of the ape's brains and the embryological development of the tonsils, and a synopsis of surgical anatomy. He coined the words 'neuron' and 'chromosome'.

WALKER, Sir James
(1863–1935)

Scottish physical chemist, born in Dundee. He began his career in the flax and jute industry, but forsook this to study science at Edinburgh University, from which he graduated in 1885. From 1887 to 1889 he worked with Baeyer at Munich and then with Ostwold at Leipzig. He returned to Edinburgh as an assistant, but went on to University College London, to work with Ramsay in 1892. From 1894 to 1908 he was Professor of Chemistry at University College, Dundee, and he was then professor at Edinburgh until his retirement in 1928. Walker's researches were mainly on the physical chemistry of aqueous solutions, and his papers covered such topics as ionization constants, kinetics and osmotic pressure. In 1895 he carried out a pioneering study of the kinetics and mechanism of the conversion of ammonium cyanate into urea, a topic to which he subsequently reverted many times. His *Introduction to Physical Chemistry* (1899) was influential in shaping chemical education. During World War I he organized the manufacture of TNT in Edinburgh. Walker was elected a Fellow of the Royal Society in 1900 and received its Davy Medal in 1923. In 1921–1923 he was President of the Chemical Society. He was knighted in 1921.

WALLACE, Alfred Russel
(1823–1913)

Welsh naturalist, born near Usk, Gwent. He worked as a surveyor and as a teacher in Leicester, but his passion was natural history, and he was inspired by Charles Darwin's *Origin of Species* and Robert Chambers's evolutionary treatise *Vestiges of Creation*. Together with Bates, he travelled and collected in the Amazon basin (1848–1852). During the return voyage, the ship caught fire and sank, and his vast collection of living and preserved specimens was lost. Undaunted, he planned a new expedition, this time to the Malay Archipelago (1854–1862). He was an indefatigable collector, describing thousands of new tropical species, and was the first European to observe orang-utans in the wild. He also made pioneering contributions to the ethnology and linguistics of the native peoples in the regions he visited. His *Geographical Distribution of Animals* (1876) became the founding text of zoogeography, and he is remembered for 'Wallace's line', the division in the Malay Archipelago between the Asian and Australian floras and faunas. While in Malaysia, he conceived the idea of the 'survival of the fittest' as the key to evolution, applying Malthus's ideas on checks to population growth to the natural variation he had observed as a naturalist. In 1858 he wrote up his discovery in an article entitled *On the tendency of varieties to depart indefinitely from the original type*, and sent it to Darwin, who was sent into turmoil at the duplication of his own unpublished ideas. This precipitated the joint Darwin-Wallace paper at the Linnaean Society in 1858, and the hurried publication of Darwin's *Origin of Species* the following year. On his return to England, Wallace generously allowed Darwin the credit, and even entitled his own book on evolution *Darwinism* (1889). He wrote several other highly influential books, including *Contributions to the Theory of Natural Selection* (1870) and *Island Life* (1882), as well as inspired accounts of his voyages, *Travels on the Amazon* (1869) and *The Malay Archipelago* (1872). He was an outspoken advocate of socialism, pacifism, women's rights and other causes, and encapsulated his views in *Social Environment and Moral Progress* (1913). He was also active in psychic and spiritualist circles, ultimately causing many scientific colleagues to distance themselves from him. Although he remained a staunch believer in natural selection, he did not believe that this process had created the higher faculties of the human brain, which he regarded as miraculously endowed.

WALLACH, Otto
(1847–1931)

German chemist, born in Konigsberg, Prussia (now Kaliningrad, Russia). After study at the Potsdam Gymnasium, he entered the University of Göttingen in 1867. He worked there for his doctorate with Hans Hubner on the positional isomerism of toluene compounds. In 1870 he moved to Bonn as assistant to Kekule and remained there for 19 years. From 1989 until his retirement in 1915 he worked as Director of the Chemical Institute in Göttingen. Throughout his life he studied the composition of essential oils obtained from plants, a topic which Kekule suggested to him. From these oils he isolated many compounds belonging to a class he called terpenes, showing that they consisted of five carbon atom fragments known as isoprene units. The compounds differ in the way the units are arranged. One of his greatest achievements was the elucidation of the structure of a-terpineol (1895). Even in his own lifetime the value of his work was greatly appreciated, and it initiated much work which continues to this day. He was awarded the Nobel Prize for Chemistry in 1910. As well as a distinguished chemist he was a serious art collector.

WALLER, Augustus Voiney
(1816–1870)

English physiologist, born near Faversham. Walker studied medicine in Paris, receiving his MD in 1840. He set up in general practice in Kensington. Elected FRS in 1851, he devote himself to full-time research for five years, working in Bonn, and later under Flourens at the Jardin des Plantes in Paris, before being appointed Professor of Physiology at Queen's College, Birmingham. Waller was a fine microscopist, remembered for his patient anatomical investigations of the nervous system. He paid special attention to the processes of nerve degeneration and regeneration, discovering the Wallerian degeneration of nerve fibres. He also conducted innovative investigations of the autonomic nervous system, particularly analysing the dilation of the iris under light stimuli. His experiments on the vasoconstrictor properties of nerves from the ciliospinal region extended the researches of Bernard and Brown-Sequard.

WALLICH, Nathaniel
(1786–1854)

Danish botanist, born in Copenhagen, where he studied under Martin Vahl and graduated in medicine. In 1807 he became surgeon to the Danish colony at Serampore, India. This became British in 1813 and Wallich joined the British medical staff. In 1815 he was appointed Superintendent of the Calcutta Botanical Garden, and so began a very active, distinguished botanical career. In 1820, with William Carey (1761–1834), he began to publish Roxburgh's *Flora Indica*, with much additional matter written by himself. He collected many plants on expeditions to Nepal (1820) and western India (1825), and was one of the earliest botanists to collect in Myanma (Burma) (1826–1827). Returning home on the grounds of ill-health in 1828, he brought back some 8,000 specimens; 9,148 species are represented in his *A Numerical List of Dried Specimens of Plants in the East India Company's Museum* (1828). Between 1830 and 1832 he published his most important work, *Plantae Asiaticae Rariores* (3 vols). He went back to India (1832–1847) and explored Assam, paying particular attention to wild tea plants, before returning to England.

WALLIS, Sir Barnes Neville
(1887–1979)

English aeronautical engineer and inventor, born in Derbyshire. After winning a scholarship to Christ's Hospital, London, and training as a marine engineer at Cowes, he joined the Vickers Company in 1911 and two years later was transferred to the design office of Vickers Aviation. After a short period of military service during World War I, he rejoined the company as chief designer in their airship department at Barrow-in-Furness, for whom he later designed the airship R100 which made its maiden flight in 1929 and successfully crossed the Atlantic. From 1923 he also acted as chief designer of structures at Vickers Aviation, Weybridge, where he designed the Wellesley and Wellington bombers with their revolutionary geodetic fuselage structure, the bombs which destroyed the German warship Tirpitz and V-rocket sites, and the 'bouncing bombs' which destroyed the Mohne and Eder dams in Germany during World War II (1943). From 1945 to 1971 he was chief of aeronautical research and development for the British Aircraft Corporation at Weybridge. In the early 1950s he was responsible for the design of the first variable-geometry (swing-wing) aircraft, the experimental Swallow; the same design principle was later incorporated in the US Air Force's General Dynamics F-11 multi-purpose fighter which first flew in December 1964, and in the Panavia Tornado which is in service with many of the world's air forces. He was elected FRS in 1945, and knighted in 1968.

WALLIS, John
(1616–1703)

English mathematician, born in Ashford, Kent. He graduated at Cambridge, and took holy orders, but in 1649 became Savilian Professor of Geometry at Oxford. During the Civil War he sided with the parliament, was secretary in 1644 to the Westminster Assembly, but favoured the Restoration. Besides the *Arithmetica Infinitorum* (1656), in which he offered a remarkable method for finding areas under curves in terms of infinite sums (soon replaced by the more rigorous calculus), he wrote on the binomial theorem and gave an infinite product for pi. He also wrote on proportion, mechanics, the quadrature of the circle (in opposition to Thomas Hobbes), grammar, logic, theology, and the teaching of the deaf and dumb. In addition Wallis was an expert on deciphering, and edited the work of some of the Greek mathematicians. He wrote a tendentious and xenophobic history of algebra, and is also remembered as one of the founders of the Royal Society.

WALTON, Ernest Thomas Sinton
(1903–)

Irish physicist, born in Dungarvan, Waterford. He studied at Trinity College, Dublin, graduating in 1926. In 1927 he went to the Cavendish Laboratory, Cambridge, where he studied under Rutherford and was awarded his PhD in 1934. He later became Professor of Natural and Experimental Philosophy at Trinity College, Dublin (1947–1974). With Cockcroft, he produced the first artificial disintegration of a nucleus by bombarding a lithium nucleus with protons accelerated across a potential of 710 kilovolts (1932). This was the first successful use of a particle accelerator, and by studying the energies of the two alpha particles produced they were able to verify Einstein's theory of mass-energy equivalence. The use of particle accelerators was crucial for the understanding of the substructure of nuclei and later the nucleons themselves. Cockcroft and Walton were awarded the 1951 Nobel Prize for Physics in recognition of this work. In 1952 he was appointed chairman of the School of Cosmic Physics and the Dublin Institute for Advanced Studies.

WANG, An
(1920–1990)

Chinese-born American physicist and computer company executive, born in Shanghai. He graduated in science from Jiao Tong University in Shanghai (1940), and in 1945 emigrated to the USA, where he studied applied physics at Harvard. Known as 'the Doctor', Wang had both technological genius and entrepreneurial ability. He played a major role in inventing magnetic core memories for computers, and his patents provided enough income for him to break into the commercial world. In 1951 he founded Wang Laboratories in Boston, Massachusetts, which expanded rapidly through the success of another invention attributed to Wang – the electronic calculator. Wang became a leading manufacturer of minicomputers, competing alongside the likes of Olsen's DEC. His company was also renowned for its word-processing software. However in the 1980s personal computers and workstations ate into Wang's business; in 1983 he handed control of the business to others, including his son Frederick, who was appointed Wang's president in 1987. It was a disastrous move: after Frederick was ousted in 1989, An Wang returned briefly after cancer surgery to attempt to turn the company round. However, in 1992 (two years after his death) the company which had enjoyed revenues of $3 billion and employed 31,000 in its heyday filed for bankruptcy.

WARBURG, Otto Heinrich
(1883–1970)

German biochemist born in Freiburg, Baden. Educated at Berlin and Heidelburg, he won the *Pour le Merite* (the German equivalent of the Victoria Cross) during World War I. He worked at the Kaiser Wilhelm (later Max Planck) Institute in Berlin from 1913, becoming director there in 1953. He was the first to discover the important role of iron, in association with oxidase enzymes, in nearly all cells. He called the iron-containing enzyme that catalysed direct electron transfer to oxygen *Atmungsferment* and showed that its 'reduction' was inhibited by cyanide, and that it was probably a haem protein. In 1926 he demonstrated that oxygen uptake by yeast is inhibited by carbon monoxide and exploited J. B. S. Haldane's finding that carboxyhaemoglobin is dissociated by light to determine from its absorption characteristics that this is also a haem protein, subsequently equated by Keilin with cytochrome oxidase. He also discovered that the oxidation of glucose 6-phosphate by a yeast preparation ('old yellow enzyme') requires the cofactor riboflavin (1932), and that the 'new yellow enzyme', discovered by Hans Krebs, contains a different electron carrier (flavinadeninedinucleotide). With E. Negelin he showed that the green alga *Chlorella* produces one molecule of oxygen from the absorption of four quanta of red light (1923) – an efficiency of 65 percent, disputed by others but confirmed by Warburg in 1950. The gas manometer developed by Warburg in 1926, from Barcroft and Haldane's blood-gas manometer for measuring metabolic reactions by the amount of oxygen or carbon dioxide taken up or released, was crucial to the discoveries of Krebs and others, and continued in use until replaced by the oxygen electrode around 1970. Warburg also engaged in cancer research. He was awarded the 1931 Nobel Prize for Physiology or Medicine, but as a Jew was prevented from accepting it by Hitler.

WARD, Nathaniel Bagshaw
(1791–1868)

English physician and botanist, born in London. He was involved with the administration of the Chelsea Physic Garden, where he worked on methods of plant cultivation. He published *On the Growth of Plants in Closely Glazed Cases* (1842), and invented the 'Wardian case' which enabled five plants to be transported successfully on long voyages, and also their cultivation in Victorian drawing rooms. Among the many plants which owe their establishment to the Wardian case, the most significant is the tea plant, which Fortune successfully brought from China to India. Ward published *Aspects of Nature* in 1864.

WARMING, (Johannes) Eugenius Bulow
(1841–1924)

Danish botanist, born in the North Frisian island of Manö. He studied botany in Munich under Nageli and others. Professor at Stockholm (1882–1885) and Copenhagen (1885–1911), he wrote important works on systematic botany (1879) and plant ecology (*Plantesamfund,* 1895), being regarded as a founder of the latter, although the term had been coined earlier by Haeckel. His work demonstrated that groups of species could form a well-defined unity, such as a meadow ecosystem. He produced two excellent textbooks, *Haandbog i den Systemadyke Botanik* (1879) and *Den Almindelige Botanik* (1880). Between 1863 and 1866 he travelled in Brazil, making the most detailed and thorough study of a tropical area then produced.

WASSERMAN, August Paul von
(1866–1925)

German bacteriologist, born in Bamberg. He studied medicine at Erlangen, Vienna, Munich and Strasbourg, where he graduated in 1888 and worked on bacteriology and chemotherapy at the Robert Koch Institute in Berlin from 1890. In 1906 he discovered and gave his name to a blood-serum test for syphilis. An infected person will produce syphilis antibodies in the blood; in the Wasserman test these will react with known antigens to form a chemical complex. The test is still widely used in diagnosis.

WATERSTON, John James
(1811–1883)

Scottish natural philosopher and engineer, born in Edinburgh. He attended the University of Edinburgh, and following an engineering apprenticeship he practised as a surveyor in London (1832), later moving to a position in the hydrographer's department of the Admiralty. As naval instructor to the cadets of the East India Company in Bombay he taught navigation and gunnery from 1839, only returning to Scotland permanently in 1857. He wrote on astronomy, solar radiation, chemistry, the physiology of the central nervous system, sound and a novel kinetic theory of gases and liquids. Waterston's famous speculative memoir on gases linked heat with molecular motion and included a calculation of the ratio of specific heats at constant temperature and constant volume. Submitted to the Royal Society in 1845, it was dismissed by the referees as 'nothing but nonsense'. In 1892 it was reproduced in complete form by Lord Rayleigh (1842–1919). Many of the key ideas had by then been published by Clausius and Maxwell.

WATSON, David Meredith Seares
(1886–1973)

English zoologist, educated in chemistry and geology at the University of Manchester. After early research in palaeobotany, his interests drifted to palaeontology, to which he devoted his research, despite having no formal zoological training. Before World War I he travelled widely in South Africa and Australia collecting fossil vertebrates. From the reptilian remains in this material he traced the evolution of the mammalian skeleton from that of primitive reptiles. He later became a leading authority on early amphibians. From 1921 he was Jodrell Professor of Zoology and Comparative Anatomy at University College London, where he remained until his retirement in 1951. He was elected FRS in 1922.

WATSON, James Dewey
(1928–)

American biologist, born in Chicago. He received a BSc in zoology from the University of Chicago, and as a postgraduate at Indiana University he studied under Hermann Muller and Luria. He spent the period 1951–1953 at the Cavendish Laboratory in Cambridge, UK, and from 1955 taught at Harvard, where he became Professor of Biology in 1961. Since 1968 he has been Director of Cold Spring Harbor Laboratory in New York, and he has also served as Director of the National Center for Human Genome Research (1989–1992). While in Cambridge in 1951, Watson worked with Crick on the structure of DNA, the biological molecule contained in cells which carries the genetic information. They published their model of a two-stranded helical molecule in 1953, showing that each strand consists of a series of the nucleotide bases (adenine, thymidine, guanine and cytosine) wound around a common centre. The strands were shown to be linked together with hydrogen bonds, adenine on one strand pairing with thymidine on the other, with similar pairing between guanine and cytosine. For this work, Watson was awarded the 1962 Nobel Prize for Physiology or Medicine jointly with Crick and Wilkins. Since the mid-1980s, he has been an active supporter of the Human Genome Initiative, which aims to locate all genes in the human body and determine their DNA sequences. He wrote a personal account of the discovery of DNA structure in *The Double Helix* (1968), and the textbooks *The Molecular Biology of the Gene* (1965) and *Recombinant DNA* (1984).

WATSON, Sir William
(1715–1787)

English scientist, born in London, the son of a cornchandler. He became an apothecary and was one of the earliest experimenters on electricity. He discovered (among others) the importance of insulating conductors to increase charge, improved the Leyden jar (the first form of capacitor or condenser), was an early supporter of Benjamin Franklin's electrical theories (1748), and was the first to investigate the passage of electricity through a rarefied gas. In botany, he did much to introduce the Linnaean system to Britain. He was elected FRS (1741), and awarded a licentiate of the Royal College of Physicians, and was appointed physician to the Foundling Hospital (1761). He was chiefly interested in epidemic children's diseases, about which he wrote a pamphlet comparing methods of inoculation against smallpox. He was knighted in 1786.

WATT, James
(1736–1819)

Scottish engineer and inventor, born in Greenock. He went to Glasgow in 1754 to learn the trade of a mathematical instrument maker, and there, after a year in London, he set up in business. The university made him its mathematical instrument maker from 1757 to 1763. He was employed on surveys for the Forth and Clyde canal (1767), the Caledonian and other canals, and was engaged in the improvement of canals and in the deepening of the Forth, Clyde and other rivers. As early as 1759 his interest had been aroused in the possibilities of developing the use of steam as a motive force. In 1763–1764 a model of Thomas Newcomen's engine was sent to his workshop for repair. He easily put it into order, and seeing the defects in the working of the machine, hit upon the expedient of the separate condenser. This was probably the greatest single improvement ever made to the reciprocating steam engine, enabling its efficiency to be increased to about three times that of the old atmospheric engines. After an abortive enterprise funded by John Roebuck, he entered into a partnership with Matthew Boulton of Soho, near Birmingham (1774), when (under a patent of 1769) the manufacture of the new engine was commenced at the Soho Engineering Works. Watt's soon superseded Newcomen's machine as a pumping engine, and between 1781 and 1785 he obtained patents for the sun-and-planet motion, the expansion principle, the doubleacting engine, the parallel motion, a smokeless furnace, and the governor. He described a steam locomotive in one of his patents (1784), but discouraged William Murdock from further experiments with steam locomotion. In 1785 he was elected FRS. The SI unit of power is named after him, and horse-power, the original unit of power, was first experimentally determined and used by him in 1783. He retired in 1800, and died at Heathfield Hall, his seat near Birmingham. His son, James (1769–1848), a marine engineer, fitted the engine to the first English steamer to leave port (1817), the *Caledonia*.

WEATHERALL, Sir David John
(1933–)

English molecular geneticist, educated at the University of Liverpool and Oxford University, where he received his MD in 1974. After holding various medical posts in the UK and the Far East, he became a researcher at the Johns Hopkins Medical School, Baltimore (1960–1962, 1963–1965). Subsequently he moved to a lectureship at the University of Liverpool, where he was later appointed Professor of Haematology (1971–1974). He later became Nuffield Professor of Clinical Medicine at the University of Oxford (1974–1992), where he has been Regius Professor of Medicine since 1992. Weatherall has worked for many years on the genetics and clinical aspects of the thalassaemias, a group of inherited anaemias in which regulation of the globin genes is perturbed in some way. The clinical outcome is greatly influenced by the detailed knowledge of the causative lesion, by early detection of the problem and by the ability to predict the outcome of a pregnancy. He was elected FRS in 1977, knighted in 1987, and received the Royal Medal of the Royal Society in 1989.

WEBER, Ernst Heinrich
(1795–1878)

German physiologist, born in Wittenberg, brother of Wilhelm Weber. He studied at the University of Wittenberg and later at Leipzig. In 1818 he was appointed to the Chair of Human Anatomy; the Chair of Physiology was added in 1840. Weber undertook extensive comparative embryological and palaeontological studies, especially on the middle ear of mammals. He also demonstrated that the digestive juices are the specific products of glands, thereby opining up major new fields of physiological and chemical research. A key focus of his interest lay in the study of sensory functions, especially skin sensitivity. Novel investigations probed what was later to be called the sensory 'threshold'. He devised a method of determining and quantifying the sensitivity of the skin, enunciated in 1834, and gave his name to the Weber-Fechner law of the increase of stimuli.

WEBER, Wilhelm Eduard
(1804–1891)

German scientist, born in Wittenberg. His father was Professor of Theology at the University of Wittenberg. He shared with his two brothers an interest in science. With his elder brother Ernst Heinrich Weber he published a treatise on wave motion, and with his younger brother, Eduard (1806–1871), who became Professor of Anatomy at Leipzig, he studied the mechanism of walking (1836). The brothers were brought up in a scientific environment; one of the family's acquaintances in Wittenberg was the acoustician Chladni, and at Halle University Weber studied under Schweigger. He was appointed as a lecturer in physics at Halle (1828), and from 1831 to 1837 he was Professor of Physics at Göttingen. He left because of political troubles, was appointed to a chair at Leipzig in 1843, but returned to his old position in Göttingen in 1849, and also became the director of the astronomical observatory. His initial research was on the acoustical theory of reed organ pipes, but Gauss inspired Weber to join him in the study of geomagnetism. Their laboratories in Halle were among the first establishments to be connected by electric telegraph (1833), and they established a network of magnetic observatories to collate the measurements. In 1845 Weber developed an electrodynamometer for the absolute measurement of an electric current, in 1849 the mirror galvanometer (the technique of using a light beam as a 'weightless pointer' had already been used by Poggendorff (1826), and was to be further developed by Kelvin), and he proposed a system of electrical units analogous to those proposed by Gauss for magnetism. His most important contribution was to determine with Kohlrausch the ratio of the electrodynamic and electrostatic units of charge; later used by Maxwell to support his electromagnetic theory. His name is given to the SI unit of magnetic flux.

WEDDERBURN, Joseph Henry Maclagan
(1882–1948)

Scottish-born American mathematician, born in Forfar. He graduated in mathematics at Edinburgh in 1903, visited Leipzig, Berlin and Chicago, and returned to Edinburgh as a lecturer (1905–1909). In 1909 he moved to Princeton, New Jersey, but he returned to fight in the British army during World War I. After the war he settled at Princeton University until his retirement in 1945. His work on algebra included two fundamental theorems known by his name, one on the classification of semi-simple algebras, and the other on finite division rings.

WEEKS, Willy (Wilford Frank)
(1929–)

American glaciologist and geophysicist, born in Champaign, Illinois. He graduated from the University of Illinois with a BA in geology (1951), and received a PhD in geochemistry (1956) from the University of Chicago. Having already received a commission in the US Air Force, he was called to active service in 1955 at the Cambridge Research Center, Boston. While there an opportunity arose for him to study sea ice along the Labrador coast; this marked the start of a long-term interest in this topic. Upon receiving his discharge in 1957, he accepted an assistant professorship of Earth sciences at Washington University, St Louis (1957–1962), and he later transferred to the Cold Regions Research and Engineering Laboratory in Hanover, New Hampshire (1962–1986). Since 1986 he has been professor at the Geophysical Institute of the University of Alaska, Fairbanks, and chief scientist of the Alaska Synthetic Aperture Radar Facility. He has taken part in numerous field studies near both poles and made extensive studies of a wide variety of aspects of ice and snow.

WEGENER, Alfred Lothar
(1880–1930)

German meteorologist and geophysicist, born in Berlin and educated a the universities of Heidelberg, Innsbruck and Berlin. He first worked as an astronomer, then joined his brother Kurt at the Prussian Aeronautical Observatory in

Tegel where they undertook a 52½ hour balloon flight to test the accuracy of a clinometer. In 1906 he joined a Danish expedition to north-east Greenland and learned the technique of polar travel while making meteorological observations. On his return he became a lecturer in astronomy and meteorology at the University of Marburg. His second expedition to Greenland (1912) was almost wrecked by calving of the ice. During army service in World War I Wegener was wounded and was never fit for active service again. After the war he joined the German Marine Observatory in Hamburg and was also Professor of Meteorology in Graz, Austria (1924). *Die Entstehung der Kontinente und Ozeane* ('The Origin of Continents and Oceans') for which he is famous was first published in 1915, based upon his reservations that the continents may once have been joined into one supercontinent (Pangaea), which later broke up, the fragments drifting apart to form the continents as they are today. Wegener provided historical, geological, geomorphological, climatic and palaeontological evidence, but at that time no logical mechanism was known by which continents could drift and the hypothesis remained controversial until the 1960s, when the structure of oceans became understood. He died in Greenland during his fourth expedition there.

WEIDENREICH, Franz
(1873–1948)

German anatomist and anthropologist, born in Edenkoben. He studied medicine at Munich, Kiel, Berlin and Strassburg. He taught anatomy at Strassburg (1903–1918) and Heidelberg (1919–1924), and was Professor of Anthropology at Frankfurt from 1928 to 1933. In 1934 he left Nazi Germany and worked for seven years in China (1935–1941) at the Peking Union Medical College, collaborating with Teilhard de Chardin on fossil remains of Peking Man. From 1941 to 1948 he worked at the American Museum of Natural History in New York City. His early work was concerned with blood, bone, teeth and connective tissue. Later studies of hominid fossil remains led him to espouse an orthogenetic view of human evolution which he summarized in *Apes Giants and Man* (1946).

WEIERSTRASS, Karl Theodor Wilhelm
(1815–1897)

German mathematician, born in Ostenfelde. Educated at the universities of Bonn and Munster, he became professor at Berlin in 1856. He published relatively little but became famous for his lectures, in which he gave a systematic account of analysis with previously unknown rigour basing complex function theory on power series in contrast to the approach of Cauchy and Riemann. He made important advances in the theory of elliptic and Abelian functions, constructed the first accepted example of a continuous but nowhere-differentiable function, and showed that every continuous function could be uniformly approximated by polynomials. His name became a byword among mathematicians for rigour, and many of his most profound ideas grew out of his attempts to present a completely systematic, self-contained account of contemporary mathematics. As with many such attempts, later students found Weierstrass's paths reasonably easy to follow but difficult to extend, and the ideas of Cauchy and Riemann, once excluded from Berlin, had to be reintroduced.

WEIL, André
(1906–)

French mathematician, brother of the philosopher and religious writer Simon Weil, born in Paris. He studied at the University of Paris, and spent two years in India and some time in Strasbourg (1933–1940), the USA (1941–1942, 1947–1958) and Brazil (1945–1947), before settling at Princeton in 1958. One of the most brilliant mathematicians of the century, he has worked in number theory, algebraic geometry and topological group theory. He was one of the founders of the Bourbaki group, and has written on the history of mathematics. A study of Gauss's work helped lead him to conjectures concerning arithmetic on algebraic varieties (in simple terms the solution of polynomial equations in integers) that attracted the attention of the best of Bourbaki and others. The last and most profound of these, a generalized Riemann hypothesis, was solved by Pierre Deligne in work that earned him the Fields Medal (the mathematical equivalent of the Nobel Prize), while the original Riemann hypothesis still remains unresolved. Weil did much to extend the theory of algebraic geometry to varieties of any dimension, and to define them over fields of arbitrary characteristics.

WEINBERG, Robert Allan
(1942–)

American biochemist, born in Pittsburgh, Pennsylvania. He received his PhD from MIT and then held a series of fellowships in the USA and Europe. From 1973 to 1982 he was first assistant professor and then associate professor in the Department of Biology and Center for Cancer Research at MIT, where he is currently Professor of Biochemistry. Since 1984 he has also been a member of the Whitehead Institute for Biomedical Research in Cambridge, Massachusetts. Weinberg studies the causes of cancer, both as a result of acquisition of cancer-susceptible genes (oncogenes) and the loss of tumour suppresser genes. He discovered the tumour suppresser gene Rb1, whose loss is associated with a rare childhood cancer in which tumours develop in the retina. The Rb1 gene has now been cloned and shown to encode a protein normally expressed in the retina and possessing similar features to those found in many DNA binding proteins; it also binds to the proteins encoded by oncogenes of some DNA viruses. Although the retina cancer is rare, there is increasing evidence that loss or inactivation of tumour suppresser genes also plays a part in many common cancers occurring in adult life.

WEINBERG, Steven
(1933–)

American physicist, born in New York City. He was educated at Cornell before spending a year at the Niels Bohr Institute in Copenhagen, then obtained his doctorate at Princeton University. He then held appointments at Columbia University, the University of California at Berkeley, MIT and Harvard University, before becoming Professor of Physics at the University of Texas in 1986. In 1967 he produced a theory that unified the electromagnetic and weak forces, an achievement analogous to that of Maxwell and Einstein in classical electromagnetism. This incorporated the prediction of a new weak interaction due to 'neutral currents', whereby a chargeless particle, now known as the Z, is exchanged giving rise to a force between particles. This was duly observed in 1973, giving strong support to the theory, also independently developed by Salam and now known as the Weinberg-Salam or electroweak theory. Glashow extended the theory to include a new concept known as 'charm', and all three shared the 1979 Nobel Prize for Physics. The combined theory is known as 'the standard model' and has recently been precisely tested by experiments at the LEP electron-positron collider at the European nuclear research centre, CERN, in Geneva. Weinberg is also noted for his ability to explain complex physical concepts in non-technical terms, a gift demonstrated in his prize-winning popular book, *The First Three Minutes* (1977), which explains how the physics of the first three minutes of the Universe have shaped what we observe today.

WEISMANN, August Friedrich Leopold
(1834–1914)

German biologist, born in Frankfurt. He studied medicine at Göttingen from 1852 to 1856. In 1867 he became Professor of Zoology at the medical school of the University of Freiburg and subsequently at a new Institute of Zoology there. He investigated the development of the two-winged flies, the Diptera, describing the neuro-humoral organ which bears his name, the Weismann ring. He was a strong supporter of Charles Darwin's theory of evolution through natural selection. In an attempt to disprove the idea of acquired characters proposed by Lamarck, he amputated the tails of mice during five successive generations and found, nor surprisingly, that there was no reduction in the propensity to grow tails. His early work on the development of the Hydrozoa led him to develop his germ-plasm theory. He appreciated that the information required for the development and final form of an organism must be contained within the germ cells, the egg and sperm, and be transmitted unchanged from generation to generation. He realized that this would account for the phenomenon of sex. He also noted that some form of reduction division, which we now know to occur during meiosis, must occur if the genetic material were not to double on each generation. His theories were developed in a series of essays, translated as *Essays upon Heredity and Kindred Biological Problems* (1889–1892). His *Vortrage über Descendenztheorie* (1902) was an important contribution to evolutionary theory.

WEISS, Robin (Robert Anthony)
(1940–)

English molecular biologist. He was educated at University College London, and became a lecturer in embryology there from 1963 to 1970. After holding short-term research posts in the USA, he joined the staff of the Imperial Cancer Research Fund Laboratories in London (1972–1980). From 1980 to 1989 he was Director of the Institute of Cancer Research, and since 1990 he has been head of the institute's Chester Beatty Laboratories. Weiss has made important contributions to studies of the role of retroviruses in the causation of cancer and of the HIV virus, particularly the mechanisms by which the virus enters the mammalian cell.

WEISSMAN, Charles
(1931–)

Swiss molecular biologist, born in Budapest. He was educated at the Kantonales Gymnasium in Zurich and at Zurich University, where he became assistant to Karrer (1960–1961). In 1963 he moved to the USA, where he became Assistant Professor of Biochemistry (1964–1967) at the New York University School of Medicine. In 1967 he returned to

Switzerland as Professor Extraordinarius in Molecular Biology at the University of Zurich (1967–1970), where he is currently Director of the Institute of Molecular Biology. He was also President of the Roche Research Foundation (1971–1977) and has been a member of the scientific board of Biogen since 1978. Weissman worked extensively on the mechanisms by which genes are transcribed, in particular identifying the DNA sequences which are recognized by RNA polymerase and other transcription factors to give correct high-level expression of genes. More recently, he has concentrated on the structure and function of the insulin genes, and was responsible for cloning them – this allowed the therapeutic use of highly purified gene copies in the treatment of diabetes.

WEIZSACKER, Baron Carl Friedrich von
(1912–)

German physicist, born in Kiel and educated at the universities of Berlin, Göttingen and Leipzig, where he was Heisenberg's assistant, before moving on to Berlin, where he became assistant to Meitner. He was appointed associate professor at the University of Strasbourg before becoming Professor of Philosophy at Hamburg. Independently of Bethe, he proposed that the source of energy in stars is chain nuclear fusion reactions (1938) and described the 'carbon cycle' sequence of reactions involved in a development of Laplace's nebular hypothesis, he also suggested a possible scenario for the formation of the planets.

WELCH, William Henry Harvard
(1850–1934)

American pathologist and bacteriologist, born in Norfolk, Connecticut. Descended from a line of physicians, Welch studied medicine at the College of Physicians and Surgeons in New York, and received his MD in 1875. After a period of research in Germany with Ludwig and Cohnhelm, he became Professor of Pathology and Anatomy at Bellevue Hospital Medical College in New York in 1879. Later he was appointed Professor of Pathology at Johns Hopkins University (1884), and following his retirement he became founding-director of the university's School of Hygiene and Public Health. Welch did much to establish a thriving world-renowned medical research centre at Johns Hopkins, introducing many new techniques from Europe. Around 1891 he demonstrated the pathological effects induced by diphtheria toxin; his greatest personal discovery came in 1892 with his identification of the causative agent of gas gangrene, later named *Clostridium welchii* in his honour.

WELLER, Thomas Huckle
(1915–)

American virologist, born in Ann Arbor, Michigan, the son of a pathology professor at Michigan University. Weller studied medical zoology at Michigan and graduated in 1936. He entered Harvard University Medical School where he conducted research under Enders, who was working on methods for the cultivation of animal cells. After graduating in 1940, Weller was appointed to the staff of the Children's Hospital in Boston, and in 1942 joined the US Army Medical Corps. During World War II he conducted research in the tropics on the blood nematode Schistosoma, a subject he continued to investigate in his years at Harvard. After the war Enders invited Weller and Robbins to join him at the newly created Infectious Diseases Research Laboratory at the Boston

Children's Hospital. Weller and his colleagues developed new techniques for cultivating the poliomyelitis virus which made it possible for other workers to develop the polio vaccine. For this achievement Weller, Enders and Robbins shared the 1954 Nobel Prize for Physiology or Medicine. Weller isolated the aetiological virus of chickenpox and of shingles, demonstrating that the same virus caused both diseases. He also isolated the causative agent of German measles, and discovered a new viral aetiology of congenital damage, a virus he named 'cytomegalovirus'. In 1954 he was named Strong Professor and head of the Department of Tropical Public Health, a post which he held until his retirement in 1981.

WERNER, Abraham Gottlob
(1749–1817)

German geologist, born in Wehrau, Silesia (now in Poland). A teacher at Freiburg in Saxony from 1775, he was one of the first to frame a classification of rocks and gave his name to the Wernerian (or Neptunian) theory of deposition. The controversy between the Neptunists and Plutonists became one of the great geological debates of the late eighteenth century. In essence Werner advocated that crystalline igneous rocks were formed by direct precipitation from seawater, as part of his overall system of strata from the crystalline 'primitive rocks' succeeded by the 'transition rocks', resting on highly inclined strata, the well-stratified and flat-lying 'floetz rocks' and finally the poorly stratified alluvial series. The Plutonists, led by Hutton, were able to demonstrate the intrusive nature of such rocks. Werner was one of the great geological teachers of his time and many scholars, including Johann von Goethe and Buch, travelled to Freiburg to study under him.

WERNER, Alfred
(1866–1919)

Swiss inorganic chemist, the founder of co-ordination chemistry. He was born in Mulhouse, France, and studied at the Polytechnical School in Zurich, returning to France (1891–1892) to work with Berthelot at the Collège de France. He was appointed assistant professor at Zurich in 1893 and full professor from 1895 to 1915. Werner was the first person to demonstrate that isomerism applies to inorganic as well as to organic chemistry (isomerism being the phenomenon in which compounds with the same chemical formula differ in the arrangement of their atoms within the molecules). In Werner's day, theories of valence in inorganic chemistry were unsatisfactory, being dominated by theories of ionic dissociation and the idea that valence was constant. Werner suggested that metals have a primary electrovalence in the centre of the molecule, surrounded by a fixed number of secondary valences which bind neutral molecules such as ammonia, water or organic amines. The secondary, or co-ordinate, bonds formed one of several possible geometric shapes – for example tetrahedra or octahedra – leading to the possibility of isomerism in inorganic compounds. Searching for evidence for his theory between 1893 and 1896, Werner and Arturo Miolati electrolysed salts and measured their conductivity. Over the next 20 years Werner worked towards further proof of his theory and was able, finally, to demonstrate different spatial arrangements in cobalt salts in 1914. His views, at first regarded with hostility, gradually won acceptance and were confirmed later by X-ray diffraction. They revolutionized inorganic chemistry and opened up many new areas of research. Werner was awarded the Nobel Prize for Chemistry in 1913. He died in Zurich.

WERNICKE, Carl
(1848–1905)

German neurologist and psychiatrist, born in Tamowitz, Upper Silesia (now in Poland). He qualified in medicine at the University of Breslau, and trained under the distinguished neuropathologist Theodor Meynert, who also influenced Sigmund Freud. Wernicke moved to Karl Westphal's psychiatric and neurological clinic in 1876, two years after publishing his major work, *Der Aphasische Symtomencomplex* ('The Aphasic Syndrome'). The form of aphasia, loss of speech, which he described was marked by a severe defect in the understanding of speech, and it became known as sensory aphasia. This was in contrast to the motor aphasia proposed by Broca, which involved loss or defect in the expression of speech. Sensory aphasia includes a wide range of symptoms, including disorders in word usage and word choice, which in severe cases degenerates to incomprehensible gibberish. From post-mortem studies of brains of his patients he showed that his type of aphasia was typically localized in the left temporal lobe, now known as Wernicke's area, although he also noted bilateral lesions. He tried to correlate his anatomical and functional findings to produce a theory of language and its disorders, incorporating Broca's work in an attempt to describe language generation and dysfunction within the brain. Using his results and those of Broca, Hitzig, Gustav Theodor Fritsch and others, he suggested that fundamental neural properties were discretely localized, and that more complex functions arose out of interactions and connections between different areas. He established a clinic in Berlin specializing in diseases of the nervous system, where he worked until 1885 when he returned to Breslau (Associate Professor of Neurology and Psychiatry 1885–1890; Professor 1890–1904). In 1904 he moved to Halle as professor, but he died as a result of an accident the following year. As well as papers on aphasia, Wernicke wrote on a wide variety of psychiatric and neurological issues and described a form of encephalopathy resulting from dietary thiamine deficiency (common in alcoholics), which bears his name.

WESTINGHOUSE, George
(1846–1914)

American engineer, born in Central Bridge, New York, the son of a manufacturer of farm machinery. At the age of 15 he ran away from school to fight for the North in the American Civil War, and later served for a short time in the US navy but decided in 1865 to return to his father's workshop. In October of the same year he took out the first of his more than 400 patents, for a railway steam locomotive. He invented many other devices connected with this, most important was the air-brake system he patented in 1869, which became known as the Westinghouse air brake. This allowed the brakes on all the coaches or wagons throughout the length of a train to be applied simultaneously under the engine driver's control and greatly increased the speed at which trains could safely travel. In the same year he founded the Westinghouse Air Brake Company, and subsequently devised a number of improvements which made the air brake even more effective. His works was situated in Pittsburgh, and he took an active part in the distribution and utilization of the natural gas deposits then beginning to be exploited in the area. He was a pioneer in the use of alternating current for distributing electric power, and founded the Westinghouse Electric

Company in 1886, attracting Tesla to work for him after leaving the employment of Edison. In 1895 he successfully harnessed the power of the Niagara Falls to generate sufficient electricity for the town of Buffalo, 22 miles away.

WEYL, Hermann
(1885–1955)

German mathematician, born in Elmshorn. He studied at Göttingen under Hilbert, and became professor at Zurich (1913), and Göttingen (1930). Refusing to stay in Nazi Germany, he moved to Princeton in 1933. Among important contributions to the theory of Riemann surfaces, he gave the first rigorous account of the surfaces while extending the original insights of Riemann. Inspired by a brief period with Einstein in Zurich, he wrote on the mathematical foundations of relativity and quantum mechanics, and subsequently on the representation theory of Lie groups. Modern attempts to create a gauge theory of particle interactions may be traced back to Weyl's concepts of measurement, and he wrote profoundly on how mathematics and quantum mechanics interrelate. He also wrote on the philosophy of mathematics, on the spectral theory of integral operators (which have played an essential role in quantum theory), and on algebraic number theory. His book *Symmetry* (1952) is an elegant and largely non-technical account for the general reader of the relation between group theory and symmetry in pattern and design.

WHEATSTONE, Sir Charles
(1802–1875)

English physicist, born in Gloucester. He first became known as a result of his work in acoustics. He invented the concertina (1829), and published important papers on the figures of Chladni (the patterns produced by a fine powder spread over the surface of a vibrating plate), the resonance of columns of air, the transmission of musical sound through rigid linear conductors, and on the experimental proof of Daniel Bernoulli's theory of the vibration of air in musical instruments. In 1834 he was appointed Professor of Experimental Physics at King's College, London, a position he held for the rest of his life, in spite of his unwillingness to lecture. In 1837 he and Sir William Cooke took out a patent for an electric telegraph, and in conjunction with the new London and Birmingham Railway Company installed a demonstration telegraph line about a mile long. With the needs of the rapidly expanding railways providing the impetus and the finance, by 1852 more than 4,000 miles of telegraph line were in operation throughout Britain. Wheatstone built the first printing telegraph in 1841, and in 1845 devised a single-needle instrument. In 1838, in a paper to the Royal Society of which he had become a Fellow in 1836, he explained the principle of the stereoscope (later improved by Brewster). He invented a sound magnifier for which he introduced the term 'microphone'. Wheatstone's bridge, a device for the comparison of electrical resistances, was brought to notice (though not invented) by him. He was knighted in 1868.

WHEELER, David John
(1927–   )

English computer scientist and programmer. He was educated at Trinity College, Cambridge, and as a Research Fellow at Cambridge between 1951 and 1957, he joined Wilkes's EDSAC team in establishing the world's first computing service. Wheeler's expertise on the programming side, alongside the work of Stanley Gill, was of paramount importance, especially since a keynote of the Cambridge approach was user-convenience. The pioneering work of Wilkes, Wheeler and Gill was later published in an influential book, *The Preparation of Programs for an Electronic Digital Computer* (1951). The economy and elegance of the EDSAC programming, largely the work of Wheeler, was much in advance of any US or British group and the book was very influential. Wheeler worked in the USA at the University of Illinois between 1951 and 1953, before returning to Cambridge University, where he was appointed Professor of Computer Science in 1978. He was elected FRS in 1981.

WHEELER, John Archibald
(1911–)

American theoretical physicist, born in Jacksonville, Florida. He was educated at Johns Hopkins University, where he received his PhD in 1933, and spent the following two years in Copenhagen. He was appointed professor at Princeton (1947) and later at the University of Texas (1976). Wheeler worked with Niels Bohr on the theory of nuclear fission, and on the hydrogen bomb project. He also contributed to the search for a unified field theory, and studied with Feynman the concept of action at a distance.

WHIPPLE, Fred Lawrence
(1906–)

American astronomer, born in Red Oak, Iowa. He studied at California University, and became Professor of Astronomy at Harvard in 1945. An expert on the solar system (his *Earth, Moon and Planets* published in 1941 is a standard work), he is known especially for his work on comets. In 1950 he put forward his icy conglomerate model of the cometary nucleus in which the fount of cometary activity is a single 'dirty snowball', a few kilometres across. The spin of this nucleus, coupled with the delay in the transmission of heat through the dusty crust, leads to a jet effect which can change the orbit of the comet. He was also the first to define the term micrometeorite, realizing that below a certain size, a dust particle incident on Earth's atmosphere would be such an efficient heat radiator that it would be retarded by atmospheric friction without becoming molten. He used the rate of decay of meteors as an indicator of the temperature profile of the atmosphere. Whipple (with Fletcher Watson) was responsible for Harvard's two station meteor programme. This showed that the vast majority of meteor orbits were similar to those of comets and asteroids and that hyperbolic (i.e. interstellar) meteoroids were not an obvious component of the meteoroid flux. Whipple was the prime-mover behind the production and use of the Baker Super-Schmidt meteor cameras. He was also a pioneer in the use of these cameras to observe the decay of satellite orbits, and from this obtained measurements of atmospheric density. His meteoroid bumper shield was used to dissipate the energy of impacting dust particles on the Giotto mission to Halley's comet, which confirmed his 'dirty snowball' model. In 1967 he wrote a keynote paper on the origin and evolution of dust in the solar system.

WHIPPLE, George Hoyt
(1878–1976)

American pathologist, born in Ashland, New Hampshire, the son of a general practitioner. Whipple graduated with a

BA from Yale University in 1900, and received a medical degree from Johns Hopkins University in 1905. He joined the staff at Johns Hopkins as an assistant in pathology and began experiments on liver damage in dogs which developed into studies of the relationships between the liver, bile and haemoglobin. In 1914 he was appointed Director of the Hooper Foundation for Medical Research at the University of California. With his colleagues, he developed methods of making dogs anaemic. They showed that feeding liver to such anaemic animals was followed by a pronounced increase in haemoglobin regeneration. In 1921 Whipple became Professor of Pathology and Dean of Medicine at University of Rochester in New York. Until 1925 he was preoccupied with administrative duties, but his co-worker Frieda Robbins supervised further research on haemoglobin during this period. In 1926 a liver extract was produced in cooperation with Eli Lilly. Whipple laid the groundwork for Minot and Murphy's successful treatment of pernicious anaemia with liver in 1926 – until then this had been a fatal disease. He shared the 1934 Nobel Prize for Physiology or Medicine with Minot and Murphy, and remained as Dean of Medicine at Rochester until 1952.

WHITEHEAD, Alfred North
(1861–1947)

English mathematician and Idealist philosopher, born in London. He was educated at Sherborne and Trinity College, Cambridge, where he was Senior Lecturer in Mathematics until 1911. He became Professor of Applied Mathematics at Imperial College, London (1914–1924), and Professor of Philosophy at Harvard (1924–1937). Extending the Boolean symbolic logic in a highly original *Treatise on Universal Algebra* (1898), he contributed a remarkable memoir to the Royal Society, *Mathematical Concepts of the Material World* (1905). Profoundly influenced by Peano, he collaborated with his former pupil at Trinity, Bertrand Russell, in the *Principia Mathematica* (1910–1913), the greatest single contribution to logic since Aristotle. In his Edinburgh Gifford Lectures, 'Process and Reality' (1929), he attempted a metaphysics comprising psychological as well as physical experience, with events as the ultimate components of reality. Other more popular works included *Adventures of Ideas* (1933) and *Modes of Thought* (1938). He was awarded the first James Scott Prize of the Royal Society of Edinburgh (1922).

WHITNEY, Josiah Dwight
(1819–1896)

American geologist, born in Northampton Massachusetts. He graduated at Yale, and in 1840 joined the New Hampshire Survey. He worked in Michigan from 1847 to 1849, and in the Lake Superior region with James Hall (1811–1898). Following his studies of mining problems in Illinois, he published *Mineral Wealth of the United Stated* (1854). He was appointed professor at Iowa University in 1855, State Geologist of California in 1860 and professor at Harvard in 1865. Whitney produced important work on the *Auriferous Gravels of the Sierra Nevada* (1879–1880) in which he recognized that the gold deposits were not marine deposits as had been supposed, but were the products of erosion and deposition of pre-existing gold bearing mineral veins. He also wrote on the *Climate Changes of Later Geological Time* (1880–1882). Mount Whitney in southern California is named in his honour.

WHITTAKER, Sir Edmund Taylor
(1873–1956)

British mathematical physicist, born in Birkdale, Lancashire. He was educated at the University of Cambridge, where he held a teaching post from 1896, and in 1906 was appointed Professor of Astronomy at Dublin University and Astronomer Royal for Ireland. He later became Professor of Mathematics at Edinburgh (1912–1946). In his earlier studies of differential equations, Whittaker formulated a general solution to Laplace's equation in three dimensions. He gave great stimulus to the mathematical development of relativity theory, and introduced an important general integral representation of harmonic functions. His publications covered a diverse range of topics, including quantum mechanics, electromagnetism, the history of science and philosophy. Elected FRS in 1905, he was knighted in 1945.

WHITTINGTON, Harry Blackmore
(1916–)

English palaeontologist. After studying at the University of Birmingham, his active research career commenced with a fellowship at Yale Peabody Museum (1938), where he worked on North American and European trilobites. He became a lecturer at Judson College, Rangoon (1940), until forced to flee from the advancing Japanese army to Szechuan. He then became lecturer and later professor at Ginling Women's College (1943). From 1945 he held posts at the University of Birmingham (1945–1950) and Harvard (1950–1966) before returning to Britain as Woodwardian Professor of Geology at Cambridge (1966–1983). He wrote *The Burgess Shale* (1985) and *Trilobites* (1992).

WHITTLE, Sir Frank
(1907–)

English aeronautical engineer and inventor, born in Coventry. Son of a designer and craftsman, his interests in engineering and invention were fostered during the times he spent as a boy in his father's workshop. He developed an early interest in aeronautics, and conceived the idea of trying to develop a replacement for the conventional internal combustion aero engine. He joined the RAF as an apprentice (1923), and studied at the RAF College, Cranwell, and at Cambridge University (1934–1937). By the time he was 21 he could see that the most promising new form of propulsion for aircraft would probably consist of a high-speed jet of hot gases. He began research into jet propulsion before 1930, while still a student, and after a long fight against official inertia his engine was first flown successfully in a Gloster aircraft, code-named E.28/39, in May 1941. The actual aircraft has been preserved in the Science Museum, London, although it was in Germany in 1939 that the world's first flights of both turbo-jet and rocket-powered aircraft took place. Whittle was elected FRS in 1947, and knighted in 1948 on his retirement from the RAF. He has since acted as consultant and technical adviser to a number of British firms, and in 1977 was elected a member of the faculty of the US Naval Academy at Annapolis, Maryland.

WIELAND, Heinrich Otto
(1877–1957)

German chemist, born in Pforzheim. He studied chemistry at the universities of Munich, Berlin and Stuttgart and received a PhD from the University of Munich (1901), where he subsequently became lecturer, and in 1909, associate professor. Four years later he was appointed professor at Munich Technical University. During World War I, while on leave of absence, he worked on chemical weapons at the Kaiser Wilhelm Institute. After the war he returned to Munich until 1921, when he moved to the University of Freiburg for three years. He returned to the University of Munich in 1924 as chairman, a position he held until his retirement in 1950. He made many contributions to the development of organic chemistry. His initial studies involved nitrogen compounds and he also made extensive investigations of oxidation reactions. However, he is most famous for his studies of the bile acids (substances stored in the bladder which aid the digestion of lipids). It was for this work that he received the Nobel Prize for Chemistry in 1927. He subsequently studied the chemistry of a number of naturally occurring substances, including butterfly pigments.

WIEN, Wilhelm Carl Werner Otto Fritz Franz
(1864–1928)

German physicist, born in Gaffken in East Prussia, the son of a farmer. He was a slow starter academically and was taken out of school to learn agriculture, but he resumed his studies and went to the universities of Göttingen, Berlin (where he studied under Helmholtz) and Heidelberg. He became Helmholtz's assistant at the Physikalisch Technische Reichanstalt in Charlottenberg (1890), was appointed to professorships at Aachen (1896), Giessen (1899), Wurzburg (1899) and finally Munich (1920). He was the successor of Röntgen at both Wurzburg and Munich. His chief contribution was on black body radiation. Advancing the work of Boltzmann (1884), he showed that the wavelength at which maximum energy is radiated is inversely proportional to the absolute temperature of the body (1893). His attempt to formulate an equation that would fit the observed distribution of all possible frequencies (both short-wavelength/high-frequency and long-wavelength/low-frequency radiation) was unsuccessful (1896), but cleared the way for Planck to resolve this with the quantum theory (1900). In 1911 Wien was awarded the Nobel Prize for Physics for his work on the radiation of energy from black bodies. His researches also covered hydrodynamics, and X-rays and cathode Rays. He showed that cathode rays consisted of negatively charged particles moving at a very high velocity (1897–1898), and that 'canal rays' were positively charged particles which were deflected by electric and magnetic fields (1905). He edited the *Annalen der Physik* (1906–1928).

WIENER, Norbert
(1894–1964)

American mathematician, born in Columbia, Missouri. A youthful prodigy, he studied zoology at Harvard and philosophy at Cornell; in Europe he studied with Bertrand Russell at Cambridge and at Göttingen. He was later appointed Professor of Mathematics at MIT (1932–1960). At MIT he worked on stochastic processes and harmonic analysis, inventing the concepts later called the Wiener integral and Wiener measure for application to physical problems such as Brownian motion. During World War II he studied mathematical communication theory applied to predictors and guided missiles. His study of the significance of feedback in the handling of information by electronic devices led him to compare this with analogous mental processes in animals in *Cybernetics, or control and communication in the animal and the machine* (1948) and other works. His frankly egocentric autobiography *I am a mathematician – the later life of a prodigy* was published in 1956. For all his gifts, he seems to have had deep doubts about his own mathematical ability.

WIESEL, Torsten Nils
(1924–)

Swedish neurophysiologist, born in Uppsala. He studied medicine at the Karolinska Institute in Stockholm, and then went to the USA for postdoctoral work with Kuffler, initially at Johns Hopkins Medical School (Fellow in Ophthalmology, 1955–1958; Assistant Professor of Ophthalmic Physiology, 1958–1959), and then at Harvard Medical School (Assistant Professor of Neurophysiology, 1959–1960; Associate Professor, 1960–1967; Professor of Physiology, 1967–1968; Professor of Neurobiology, 1968–1974; Robert Winthrop Professor of Neurobiology, 1974–1983). Working with Kuffler he met Hubel, and together they studied the way in which the brain processes visual information, following on from the work of Hartline, Granit and Kuffler himself. They demonstrated that there is an hierarchical processing pathway, of increasingly sophisticated analysis of visual information by nerve cells from the retina to the cerebral cortex. Their detailed results were not only scientifically important, they also had almost immediate clinical relevance for the treatment of children with visual problems. In 1981 Wiesel and Hubel shared the Nobel Prize for Physiology or Medicine with Sperry.

WIGGLESWORTH, Sir Vincent Brian
(1899–)

English biologist, born in Kirkham, Lancashire. He was educated at the University of Cambridge and St Thomas's Hospital, and became Reader in Entomology at London (1936–1944) and Cambridge (1945–1952), where he was subsequently appointed Quick Professor of Biology (1952–1966). From 1943 to 1967 he was also Director of the Agricultural Research Council Unit of Insect Physiology. In a series of remarkable studies of insect metamorphosis, Wigglesworth demonstrated the production and secretory cell sources of hormones which selectively activate different genetic components of insects during various stages of their life cycles. He also succeeded in artificially inducing such changes experimentally by manipulating the associated hormone levels. This work and his many other wide-ranging studies of insects led to a much greater understanding of their physiology and interactions with the environment. Wigglesworth was elected FRS in 1939, and knighted in 1964.

WIGNER, Eugene Paul
(1902–)

Hungarian-born American theoretical physicist, born in Budapest. He was educated at Berlin Technische Hoshschule where he was awarded a degree in chemical engineering (1924) and a doctorate in engineering (1925). He moved to the USA in 1930, and apart from two years at the University of Wisconsin (1936–1938) he worked at Princeton University throughout his academic career, becoming Thomas D. Jones Professor of Theoretical Physics in 1938. Wigner made a number of important contributions to nuclear physics and quantum theory. In 1927 he introduced the idea that the quantum property known as parity is conserved in nuclear interactions. He believed this to be true for all nuclear reactions, but Lee and

Yang later demonstrated parity non-conservation in the weak interaction. In the 1930s, Wigner demonstrated that the strong nuclear force which binds protons and neutrons in nuclei has very short range, and is independent of any electric charge. He is especially known for the Breit-Wigner formula which describes cross-sections (probabilities) for resonant nuclear reactions, and the Wigner theorem concerning the conservation of the angular momentum of electron spin. His name is also given to the most important class of mirror nuclides (Wigner nuclides). Wigner's calculations were used by Fermi in building the first nuclear reactor in Chicago (1942) and he was instrumental in convincing the US government of the need for a nuclear bomb. He received the Fermi award in 1958, the Atom for Peace award in 1959, and the Nobel Prize for Physics in 1963 for his work in furthering quantum mechanics and nuclear physics.

WILCKE, Johan Carl
(1732–1796)

Swedish physicist, born in Wismar, Germany. Wilcke moved to Sweden with his parents in 1739. He entered Uppsala University to study theology in 1750 but concentrated instead on mathematics and physics. Whilst in Berlin he investigated electrical phenomena with his friend Aepinus, preparing a doctoral dissertation which he defended at Rostock in 1757. From 1759 he lectured on experimental physics at the Royal Swedish Academy in Stockholm. A painstaking evaluation of existing data culminated in his comprehensive map of the Earth's magnetic inclination (1768). Wilcke is also known for scientific instrument design, and above all, for experimental work into the nature of heat. In 1772 he measured the heat required to melt snow at its freezing point (the latent heat of fusion); and in 1781 he drew up a list of specific heats for different substances, giving precise details of the experimental methods he had used to determine them. Wilcke began these experiments on specific heats independently of Joseph Black.

WILKES, Maurice Vincent
(1913–)

English computer scientist, born in Dudley. He was educated at the University of Cambridge, where he was a Mathematical Tripos Wrangler at St John's College. He conducted research in physics at the Cavendish Laboratory, and after the outbreak of World War II worked in radar (as did many of the computer pioneers, such as Kilburn and Sir Frederic Calland Williams). Wilkes directed the Mathematical (later Computer) Laboratory at Cambridge (1946–1980), where he was best known for his pioneering work with the EDSAC (Electronic Delay Storage Automatic Calculator). This stored-program computer was unabashedly based on the US designs of Eckert and Mauchly and the theoretical work of von Neumann, which Wilkes had become familiar with on a US visit in 1946. Wilkes's intention – besides building the computer – was to provide a useful and reliable computing service. Helped by a team which included David Wheeler, this service, the first in the world, was available by early 1950 – ahead of the Americans. Besides important software advances, Wilkes's work also included fundamental work on processor controls (microprogramming). The EDSAC influenced the design of the Lyons LEO through the laboratory's close links with the catering firm of J. Lyons. Developments at the laboratory continued into the late 1950s, when a new computer, EDSAC II, was designed and built. After 1980 Wilkes became a computer engineer for Olsen's Digital Equipment Corporation until 1986. He was elected FRS in 1956, and published his memoirs in 1985.

WILKINS, Maurice Hugh Frederick
(1916–)

British physicist, born in Pongaroa, New Zealand. Educated at St John's College, Cambridge, he worked on uranium isotope separation at the University of California in 1944. Opposed to the use of the atomic bomb, he turned away from nuclear physics after World War II. After reading Schrödinger's book *What is Life* he developed an interest in biology and the application of physical techniques to biological research. In 1946 he joined the Medical Research Council's Biophysics Research Unit at King's College, London, where he applied the techniques of X-ray crystallography developed by William and Lawrence Bragg to biological molecules. Crick and James Watson deduced their double helix method of DNA from Wilkins and Rosalind Franklin's X-ray data of DNA fibres. Crick, Watson and Wilkins were awarded the 1962 Nobel Prize for Physiology or Medicine for their work on DNA. Wilkins went on to become director of the Medical Research Council's Biophysics Research Unit at King's College, London (1970–1972).

WILKINSON, Sir Geoffrey
(1921–)

English chemist, renowned for his work on organometallic compounds. He was born in Springside near Manchester and studied at Imperial College, London. During World War II he worked on the Canadian branch of the atomic bomb project with the National Research Council of Canada. He then moved to the Lawrence Radiation Laboratory of the University of California at Berkeley and was appointed assistant professor at Harvard in 1951. From 1955 to 1988 he was Professor of Inorganic Chemistry at Imperial College. Wilkinson's early research was on the chemistry of the transition elements. While at Harvard he studied ferrocene, a new type of compound synthesized by Thomas Kealy and Peter Ludwig Pauson, which consists of an iron atom attached to two five-sided rings of carbon and hydrogen but whose exact spatial structure and bonding had not then been determined. Using nuclear magnetic resonance spectroscopy, Wilkinson and Robert Woodward showed that the iron atom is sandwiched between the two rings and bonded to each of the five carbon atoms in both rings, a type of structure entirely new to chemistry which explained the great stability of the molecule. Wilkinson and his colleagues went on to synthesize other organometallic sandwich molecules, often using the transition elements. Since then many thousands of similar compounds have been synthesized in other laboratories. they have led to new lines of research in organic, inorganic and theoretical chemistry, and to the development of new catalysts used in the production of plastics and low-lead fuels. They are also employed in pharmaceuticals, for example L-dopa, used to treat Parkinson's disease. Wilkinson was elected FRS in 1965 and was knighted in 1976. In 1973 he shared the Nobel Prize for Chemistry with Ernst Fischer who had worked independently in Germany on organometallic sandwich compounds.

WILKINSON, James Hardy
(1919–1986)

English mathematician and computer scientist. He was educated at Sir Joseph Williamson's Mathematical School, Rochester, and at Trinity College, Cambridge. During World War II he worked at the Mathematical Laboratory, Cambridge, and at the Armament Research Department, Fort Halstead. In 1946 he joined the Mathematical Division of the National Physical Laboratory (NPL) and became involved with Turing's construction of the ACE. His chief effort was directed to producing programs for the several versions of the ACE designed by Turing and resulted in some of the earliest floating-point programs. He was also involved in the logical and electronic design of the Test Assembly, a preliminary version of the pilot ACE initiated by Huskey. Before he left the NPL in 1980, Wilkinson also published work on rounding errors in algebraic processes. In 1977 he became Professor of Computer Science at Stanford University.

WILLIAMS, Sir Alwyn
(1921–)

Welsh geologist and palaeontologist, born in south Wales and educated at the University of Wales in Aberystwyth. He was awarded a Harkness fellowship to work at the US National Museum of Washington DC (1948–1950), and subsequently became a lecturer (1950–1954), Professor of Geology (1954–1974) and Pro-Vice-Chancellor (1967–1974) at Queen's University, Belfast. In 1974 he was appointed Lapworth Professor of Geology at the University of Birmingham, and from 1976 until his retirement in 1988 he was Principal and Vice-Chancellor of the University of Glasgow. Williams's principal research interests are in Scottish Lower Palacozoic statigraphy and palaeontology and he has made important contributions to the studies of brachiopods. He was elected FRS in 1967 and knighted in 1983.

WILLIAMS, Sir Frederic Calland
(1911–1977)

English electrical engineer, born in Romily, Cheshire. After attending Stockport Grammar School, he studied engineering at Manchester University, where he received his DSc in 1939, and Oxford University. During World War II Williams worked for the Telecommunications Research Establishment at Malvern, where he was recognized as a world authority on radar. While there he began experimental work on cathode-ray tube storage, which he put to brilliant effect when he accepted the Chair in Electrotechnics at Manchester University in 1946. With the help of Kilburn, he utilized cathode-ray tubes to build the world's first electronic random access memory for a computer. The aptly named Williams tube became the basis for a prototype machine, built by Williams and Kilburn, which ran the world's first stored-program on 21 June 1948. This event, recorded Williams, was the breakthrough and sparks flew in all directions. This prototype was developed by Williams and Kilburn into the famous Manchester University Mark I, a commercial version of which Ferranti produced to Williams's specification. Williams tubes came to be widely used in early commercial computers during the 1950s in Britain and the USA. Soon, however, Williams – once described as a typical example of the British string-and-sealing-wax inventor – lost interest in computers and handed the work to Kilburn. He transferred his attention to other electrical engineering projects, such as induction machinery. He was elected FRS in 1950, and knighted in 1976.

WILLIAMS, Robert Joseph Paton
(1926–)

English inorganic chemist, born in Wallasey, Merseyside. He attended Merton College, Oxford, and published his first paper (on the Irving-Williams series describing the stabilities of certain transition-metal complexes) before obtaining his BA degree in 1948. His DPhil was completed in 1950. He then went for one year to Uppsala, Sweden, where he contributed to the development of the use of gradient elution analysis in chromatography. His lifelong interest in bioinorganic chemistry – the function of inorganic compounds in biology – already begun in his DPhil thesis became his major research topic on his return to Oxford in 1951. He published his first paper on the subject in 1953 when the field was only starting to become a recognized area of research. In 1956 he became a lecturer in chemistry at Oxford University and carried out work on the stability, redox equilibria and spectra of metal complexes, and electron transfer in mixed oxidation state compounds. He also investigated the physical, chemical and biochemical properties of vitamin B12. During a year at Harvard Medical School he worked on copper- and zinc-containing enzymes. He has proposed the way in which protons can drive biological energy capture as polymerized phosphate, and has developed nuclear magnetic resonance (NMR) spectroscopy as a means to study proteins in solution. He was elected FRS in 1972, and from 1974 to 1991 served as Napier Royal Society Research Professor.

WILLIAMSON, Alexander William
(1824–1904)

English chemist, born in Wandsworth. He studied chemistry at Heidelberg and Giessen (1844–1846), and then moved to Paris where he had a private laboratory. He was appointed Professor of Analytical Chemistry (1849), and later Professor of General Chemistry, at University College London. He was an inspiring teacher, but in 1855 his enthusiasm began to decline and his interest turned from chemistry to sources of hot air: steam engines and university committees. He resigned in 1887. He is most famous for his work on the synthesis of ethers. He supported the radical theory of organic chemistry and saw ether as of the water type. He was elected FRS in 1855.

WILLIAMSON, William Crawford
(1816–1895)

English botanist, surgeon, zoologist and palaeontologist, born in Scarborough, the son of the Curator of Scarborough Museum. He trained in medicine and became Curator of the Museum of the Manchester Natural History Society (1835–1838). He was later appointed the first Professor of Natural History and Geology (later of botany) at Owens College, Manchester (1851–1892). Williamson was the first to investigate thoroughly the plant remains (coal balls) in coal. At the time, however, the full significance of his work in fossil botany was not appreciated, and after 41 years of teaching at Owens College (later Manchester University) he was refused a pension. He is regarded as the founder of modern palaeobotany. He made several contributions to *Fossil Flora of Great Britain* (1831–1837).

WILLIS, John Christopher
(1868–1958)

English botanist, born in Birkenhead. He was educated at University College, Liverpool, and Cambridge University. In 1892 he became a lecturer there and began investigating the origin of gynodioecism in the *Labiatae*; this research led into more general studies on floral biology. In 1894 he became senior assistant to Bower at Glasgow University and began compiling his *Dictionary of the Flowering Plants and Ferns* (1896). This was originally an assemblage of taxonomic facts based on his own summaries made as a student, but in its later editions, by Airy Shaw, it became primarily a list of all known vascular plant generic names. Willis succeeded Henry Trimen (1843–1896) as director of the botanic gardens at Peradeniya, Ceylon (1896–1900). While there, he monographed the family *Podostemaceae*; this work profoundly altered his views on natural selection, which he discredited as a basis for the origin for the family's endemic species. This led to the development of his 'age and area' hypothesis in which he claimed that, other things being equal, the area occupied by a taxon is directly proportional to its age. His theories were published as *Age and Area* (1922), *The Course of Evolution* (1940) and *The Birth and Spread of Plants* (1949).

WILLIS, Thomas
(1621–1673)

English physician, one of the founders of the Royal Society (1662). Born in Great Bedwyn, Wiltshire, Willis studied classics and then medicine at Oxford, graduating BA at Christ Church, Oxford, in 1636 and MB in 1646. He briefly served in the Royalist army in the Civil War. He was one of the small group of natural philosophers including Boyle who met in Oxford in 1648–1649 and who were to become founder members of the Royal Society of London. He practised medicine in Oxford, becoming Sedleian Professor of Natural Philosophy there (1660–1675); however, his fame and wealth derived from a fashionable medical practice in the metropolis. His main work was on the anatomy of the brain and on diseases of the nervous system. His *Cerebrianatome* (1664) was the principal study of brain anatomy of its time. In this he offered new delineations of the cranial nerves and described cerebral circulation, discovering the ring of vessels now called the 'circle of Willis'. He overcame some of the difficulties of studying soft brain tissues by injecting the vessels with wax. Willis produced faithful descriptions of such fevers as typhus, typhoid and puerperal fever. He also pioneered the clinical and pathological analysis of diabetes, demonstrating that the excessive urine manufactured by the diabetic possesses a sweet taste. He was also the first to recognise that spasm of the bronchial muscles was the essential characteristic of asthma.

WILLSTÄTTER, Richard
(1872–1942)

German chemist, born in Karlsruhe. After graduation from the Realgymnasium in Nuremberg, he moved to the University of Munich and produced a doctoral thesis on the chemistry of cocaine. From Munich he was appointed to a chair at the Eidgendssische Technische Hoshschule in Zurich. He also worked (1912–1916) at the Kaiser Wilhelm Institute in Dahlem, Berlin, before returning to become professor in Munich. As a protest against antisemitism he resigned this post in 1924 and later moved to Locarno, Switzerland, where he remained until his death. Although a man of wide scientific interests, most of his chemical studies concerned the structure and synthesis of natura! products. Following his doctoral work on cocaine, a series of studies during 1894–1898 led him to revise many formulae of the tropine alkaloids, a topic to which he occasionally returned. A by-product of this work was the synthesis of the cyclic but completely non-aromatic compound 1,3,5,7-cyclo-octate traene. His most significant work concerned the structure of the chlorophylls. He showed that they were magnesium complexes of porphyrins, and laid the basis for the complete elucidation of their structures by Hans Fischer. He also studied the anthocyanin pigments of flowers. His final researches concerned enzymes, which he erroneously thought were small molecular weight compounds absorbed onto colloids. He was awarded the Nobel Prize for Chemistry in 1915 for his work on the chemistry of natural products.

WILSON, Charles Thomson Rees
(1869–1959)

Scottish pioneer of atomic and nuclear physics born in Glencorse near Edinburgh. Educated at Manchester and at Cambridge, where later he became Professor of Natural Philosophy (1925–1934), he was noted for his study of atmospheric electricity, one by-product of which was the successful protection from lightning of Britain's wartime barrage balloons. His greatest achievement was to devise the cloud chamber. While working in Scotland, he observed that shadows cast by the sun on the mountain mist were surrounded by a coloured halo. In trying to reproduce this in the laboratory he discovered that water droplets would form around ions in air that was saturated with water vapour. He used this effect to develop the cloud chamber, in which water droplets form around the track of ionization left by radiation passing through a chamber of saturated water vapour. The movement and interaction of ionizing radiations can thus be followed and photographed. This principle was also used by Glaser to develop the bubble chamber. Wilson shared the 1927 Nobel Prize for Physics with Compton, and in 1937 received the Copley Medal of the Royal Society.

WILSON, Edmund Beecher
(1856–1939)

American zoologist and embryologist, born in Geneva, Illinois, and considered to be one of the founders of modern genetics. He studied at Yale and Johns Hopkins universities, and after several teaching posts became Da Costa Professor of Zoology at Columbia University in New York. His early research was concerned with cell lineage and the formation of tissues from precursor cells. He demonstrated that in molluscs, isolated cleavage cells eventually developed into the tissues they would have become in the intact embryo. He emphasized the significance of cells as the building blocks of life and argued against the vitalistic ideas prevalent at the end of the nineteenth century. His major contribution was to show the importance of the chromosomes, particularly the sex chromosomes, in heredity and cell structure. His *The Cell in Development and Inheritance* (1896, 1925) was instrumental in the synthesis of cytology and Mendelian genetics.

WILSON, Edward Osborne
(1929–)

American biologist, born in Birmingham, Alabama. He studied there and at Harvard, where since 1956 he has been Baird Professor of Science and Curator of Entomology at the Museum of Comparative Zoology. He has been a major figure in the development of sociobiology, the investigation into the biological basis of social behaviour. His early researches into the social behaviour, communication and evolution of ants resulted in the publication of *The Insect Societies* (1971), an overview of insect social behaviour in which Wilson outlines his belief that the same evolutionary forces have shaped the behaviours of insects and other animals including human beings. His book *Sociobiology: The New Synthesis* (1975) was acclaimed for its detailed compilation and analysis of social behaviour in a wide range of animals. However, the last chapter, 'From Sociobiology to Sociology', proved controversial. In it he emphasized the importance of the genetic component in controlling a range of human behaviours, including aggression, homosexuality, altruism and differences between the sexes. His ideas probably redressed the balance that previously existed in favour of environmental determinants of human behaviour and they stimulated considerable research efforts. His books include *On Human Nature* (1978), for which he received the Pulitzer Prize, *The Ants* (1990, with B. Holldobler) and *The Diversity of Life* (1992).

WILSON, Kenneth Geddes
(1936–)

American theoretical physicist, born in Waltham, Massachusetts, son of Harvard Professor of Chemistry E. Bright Wilson. He was educated at Harvard University and Caltech, where he received his PhD in 1961. He worked at Harvard and at the European nuclear research centre, CERN, in Geneva (1962–1963). In 1963 he moved to Cornell University where he became professor in 1971. He is now professor at Columbus, Ohio. Wilson applied ingenious mathematical methods to the understanding of the magnetic properties of atoms, and later used similar methods in the study of phase changes between liquids and gases, and in alloys. Analysis of how a system changes from one phase to another, such as the onset of ferromagnetism in a magnet cooling below its Curie point, had until then been virtually impossible due to the length-scale of the interactions varying over several orders of magnitude. Leo Kadanoff had suggested that the effective spin of a block of atoms should be found and then a renormalization, or scaling, transformation made to calculate the effective spin of a larger block composed of known smaller blocks. Wilson developed this method towards a general theory, allowing the properties of a system of large numbers of interacting atoms to be predicted from observations of individual atoms, and his technique is used in many differing fields of study. For this work he was awarded the Nobel Prize for Physics in 1982. More recently, he has applied his technique to the strong nuclear force that binds quarks in the nucleus.

WILSON, Robert Woodrow
(1936–)

American physicist, born in Houston, Texas. He was educated at Rice University and Caltech. He then joined Bell Laboratories in New Jersey and became head of the radiophysics research department in 1976. There he collaborated with

Penzias in using a large radio telescope designed for communication with satellites; they detected in 1964 a radio noise background coming from all directions with an energy distribution corresponding to that of a black body at a temperature of 3.5 kelvin. Dicke and Peebles suggested that this radiation is the residual radiation from the Big Bang at the universe's creation, which has cooled to 3.5 kelvin by the expansion of the universe. Such a cosmic background radiation had been predicted to exist by Gamow, Alpher, Bethe and Robert Herman in 1948. Wilson and Penzias (jointly with Kapitza) shared the 1978 Nobel Prize for Physics for their work, which can reasonably be claimed to be one of the most important contributions to cosmology in this century. In 1970 he continued his collaboration with Penzias and they discovered (with K. B. Jefferts) the 2.6 millimetre wavelength radiation from interstellar carbon monoxide.

WILSON, (John) Tuzo
(1908–1993)

Canadian geophysicist, born in Ottawa. He was educated at the University of Toronto, where by changing courses to physics and geology he became the first Canadian geophysics graduate (1930). On a scholarship to Cambridge (1931–1932), he studied under Harold Jeffreys then went on to Princeton, where he received a PhD in geology in 1936. His thesis fieldwork included making the first solo ascent of Mount Hague (1935), and his first appointment was with the Geological Survey of Canada. World War II was spent in the Canadian army (1939–1948), where he reached the rank of Colonel. In 1946 he was appointed professor at the University of Toronto. When palaeomagnetic and heatflow evidence supported Harry Hess's 1960 theory of sea-floor spreading, Wilson was not only converted, but became a central figure in its promotion. Expert at the overview, his ideas about permanent hotspots in the Earth's mantle (1963), his elucidation of oceanic transform faults (1965) and his hypothesis on mountain building (1966) were major steps towards plate tectonic theory. He became principal of the new Erindale campus at Toronto (1967–1974) then Director-General of the Ontario Science Centre. He co-authored *Physics and Geology* (1959), one of the first geophysical textbooks. The Wilson Range in Antarctica is named after him, and the cyclical opening and closing of oceans is referred to as the Wilson cycle. He was awarded an OBE in 1946 and elected FRS in 1968. Mount Tuzo in British Columbia is named after his mother Henrietta Tuzo, who was the first to climb it (1906).

WINDAUS, Adolf
(1876–1959)

German chemist, born in Berlin. He commenced medical studies at the University of Berlin in 1895, but both there and subsequently at the University of Freiburg, he became increasingly interested in chemistry. He abandoned medicine and wrote a doctoral thesis on the cardiac poisons of digitalis in 1899. Following a year of military service he returned to Freiburg where he became lecturer, and later professor. In 1913 he was appointed Professor of Medical Chemistry at the University of Innsbruck, but two years later he moved to the University of Göttingen, and remained there for the rest of his professional life. His most important research was on the structure of cholesterol, a substance widely distributed in the body and associated with a number of cardiovascular disorders. At this point his work overlapped with that on bile acids by Wieland. In the 1920s he turned to a study of vitamin D, which is structurally related to cholesterol. For his work on cholesterol and vitamins he was awarded the Nobel Prize for Chemistry in 1928. During the 1930s he continued to study the structure of natural products, including vitamin B, and colchicine (a drug used in cancer chemotherapy). In 1938 he ceased his research and he retired in 1944.

WINKLER, Clemens Alexander
(1838–1904)

German chemist, born in Freiberg. His father managed a plant, for extracting cobalt, and after studying at the School of Mines in Freiberg, Winkler held various industrial posts. He was appointed Professor of Analytical and Technical Chemistry at the School of Mines in 1873, retiring in 1902. Winkler looked for a way to make sulphuric acid from the sulphurous gases produced by smelting cobalt ore. Finding methods of gas analysis inadequate, he designed a gas burette, later known by his name. He prepared sulphuric acid for the dye industry, making use of the well-known properties of platinum as a catalyst, but instead of using the metal in a finely divided state, he increased its surface area (and hence its efficiency) further by presenting it in the form of platinized asbestos. In 1885, while analysing argyrodite (a newly discovered silver sulphide from the Freiberg mines), he discovered that silver and sulphur accounted for only 93 percent of the compound and suspected the existence of a new element. The following year he isolated germanium, a silvery grey metalloid, which had all the properties which Mendeleyev had predicted for the missing element at number 32 in the periodic table. The discovery demonstrated the predictive powers of the periodic table and helped to make it more generally accepted. Winkler was a respected teacher and wrote many useful books on analytical chemistry. He died in Dresden.

WINSTEIN, Saul
(1912–1969)

American chemist, born in Montreal, Canada. He moved to the USA in 1923, graduated BA and MA from the University of California at Los Angeles (UCLA) and obtained his PhD from Caltech in 1938. After two postdoctoral years and a year as instructor at Illinois Institute of Technology, he returned to UCLA where he was professor from 1947 until his sudden death in 1969 at the height of his career. Following in the footsteps of Sir Christopher Ingold he made substantial contributions to our understanding of organic solvolytic reactions including neighbouring group participation, solvent ionizing power, nucleophilicity, salt effects and ion-pair phenomena. He pioneered the concept of homoaromaticity but is most famous for his work on non-classical carbonium ions, over which he sustained a long controversy with Herbert Brown. Although the controversy was never fully resolved, Winstein's concept of delocalized positive charge is clearly of considerable importance. He received many honours, including the US National Medal of Science (1970).

WISLICENUS, Johannes Adolf
(1835–1902)

German chemist, born in Klein-Eichstedt, near Querfurt. While he was a student at the University of Halle, an attempt was made to arrest his father for publishing a liberal book of Biblical criticism, and the family fled (1853) to the USA. He worked in a laboratory at Harvard but returned to Germany to complete his studies in 1856. Subsequently he became professor at the Zurich Oberen Industrieschule (1870). In 1872 he moved to Wurzburg and in 1885 to Leipzig. Some of his early work was on human physiology, and between 1863 and 1873 he studied lactic acid and related compounds. He enthusiastically applied the ideas of van't Hoff and Le Bel on the tetrahedral bonding to carbon to explain geometrical isomerism. He also developed a number of important synthetic procedures.

WITHERING, William
(1741–1799)

English physician, born in Wellington in Shropshire and educated at the University of Edinburgh. He practised medicine at the County Infirmary in Stafford, and later in Birmingham, where he became Chief Physician in the General Hospital. In 1776 he published a British Flora, *Botanical Arrangement of all the Vegetables Naturally Growing in Great Britain*, which was arranged according to Linnaeus's system, and which also contained an introduction to botany. He became acquainted with Priestley, and began studying chemistry and mineralogy as well as botany. He wrote *An Account of the Foxglove* (1785), introducing digitalis, extracted from the plant, as a drug for cardiac disease. He was the first to see the connection between dropsy and heart disease. Following his involvement in riots after the French Revolution, he fled from his job and home in Birmingham, and concentrated on bringing out a third volume of the *Botanical Arrangement* in 1792, dealing mainly with fungi and other cryptogamic plants.

WITTIG, Georg
(1897–1987)

German chemist, born in Berlin. He entered the University of Tübingen in 1916 but soon left to serve in World War I. In 1920 he recommenced the study of chemistry for a degree at the University of Marburg where he later joined the staff. He was appointed associate professor at the technical University of Brunswick in 1932 and five years later moved to the University of Freiburg. In 1944 he was made professor at the University of Tübingen, and in 1956 he moved to Heidelberg. His early work on the solution chemistry of radicals and carbanions established him as a chemist of great skill, but he is most famous for a serendipitous discovery that some ylides (organometallic compounds containing both positive and negative charges) react smoothly with aldehydes and ketones with the creation of an olefinic double bond. This procedure has been of enormous value in the laboratory synthesis of numerous important compounds, including vitamin A, vitamin D, steroids, and prostaglandin precursors. For this work he shared the 1979 Nobel Prize for Chemistry with Herbert Brown. He continued publishing until the age of 90.

WÖHLER, Friedrich
(1800-1882)

German chemist, born in Eschersheim. He was educated at the universities of Marburg and Heidelberg, and qualified as a doctor of medicine in 1823, specializing in gynaecology, but he never practised. He taught chemistry at industrial schools in Berlin (1825–1831) and Kassel (1831–1836), and became Professor of Chemistry at Göttingen in 1836, remaining there until his death. However, one of the formative experiences of

Wöhler's life was the year he spent with Berzelius in Stockholm, and this led to a lifelong friendship. His friendship with Liebig was equally important to him, and from their common interest in cyanates came Wöhler's most famous discovery. In 1828 he attempted to prepare ammonium cyanate from silver cyanate and ammonium chloride, but instead obtained urea. This was an example of the production of a natural product from non-organic materials, a result which did not accord with the theory of vitalism, which claimed that natural products could be made only in living things. Wöhler was excited by his discovery but subsequent claims that the theory of vitalism was overturned by this one experiment are exaggerations. Wöhler's urea was not the first natural product to be made in the laboratory and it was soon followed by many more, leading to a gradual erosion of the theory of vitalism. Equally important as the synthesis of urea was Wöhler's preparation of aluminium in 1827. The Danish scientist Oersted claimed to have extracted the metal from alumina in 1825, but it is doubtful whether the metal he obtained was pure aluminium. Wöhler used a different procedure and the product (still extant) is essentially pure metal. For this work he was honoured by Napoleon III but there are still arguments, mainly nationalistic, about priority. Wöhler collaborated with Liebig on a number of occasions and together they made a substantial contribution to the foundation of modern organic chemistry. From 1840 onwards he undertook numerous administrative and government duties and his research output diminished. He was an inspiring teacher and maintained a keen interest in chemistry well into his old age. He was responsible for the translation of the influential annual reports prepared by Berzelius, occasionally moderating the latter's strident language.

WOLF, Emil
(1922–)

American optical physicist, born in Prague. He moved to the UK in 1940, and was educated at the University of Bristol. His first research was into the characteristics of high performance optical systems, partly using methods due to Zernike; subsequently he worked on optical coherence in Edinburgh, Manchester and Rochester (USA). He was appointed to a full professorship at the University of Rochester in 1961, and became Professor of Physics there and Professor of Optics at the nearby Institute of Optics in 1978. His work on coherence theory paved the way for the development of quantum optics by Glauber in 1963. A reformulation of coherence theory by Wolf in 1982 led to the remarkable prediction that the spectrum of a radiation field may change as it propagates – even through the vacuum of interstellar space. This prediction of 'spectral non-invariance' contradicted unstated presuppositions which were strongly held to in every branch of physics, but it was almost immediately confirmed experimentally in optics, and shortly thereafter in acoustics. Wolf was co-author with Born of *Principles of Optics* (1959), and editor of the series *Advances in Optics* (30 vols, 1961–1992). In 1977 he was Frederick Ives medallist of the Optical Society of America, and he served as the society's president from 1978 to 1981.

WOLF, Maximilian Franz Joseph Cornelius
(1863–1932)

German astronomer, born in Heidelberg, where while still a schoolboy he began his well-known observations of asteroids. After completing his mathematical studies at the University of Heidelberg (1888) followed by two years at Stockholm University, he established a private observatory in which he continued his searches for asteroids using photographic methods. On being appointed Professor of Astronomy at the University of Heidelberg (1893) he established a well equipped observatory on the Konigstuhl which became famous for the high quality of its photographs of star clusters, bright and dark nebulae, and star clouds of the Milky Way. The 'Wolf diagram', a useful and widely used method of discovering interstellar dust by star counting, owes its name to him.

WOLLASTON, William Hyde
(1766–1828)

English chemist, born in East Dereham, Norfolk, into a family of scientists and physicians. He graduated in medicine at Cambridge in 1793 and was awarded a fellowship at Caius College. He practised in London until 1800, when he entered into a partnership with Smithson Tennant to produce platinum. Platinum is resistant to heat and simple acids, and therefore had uses in some chemical apparatus and in boilers used during the manufacture of sulphuric acid. At that time, nobody – except for Chabaneau on a very small scale – had succeeded in producing it in a malleable form. By 1805 Wollaston, who was always the more active partner, had evolved a successful technique: the platinum sponge obtained by heating the ore slowly was gently powdered, sieved, washed, allowed to settle in water, pressed, dried, heated and hammered. He built up a lucrative business making platinum boilers, wire and other apparatus, keeping the process secret until just before his death when he described it in a lecture to the Royal Society. He also conducted experiments in many areas including physiology, pathology, botany and pharmacology. In the course of studying crystallography he made improvements to the goniometer, an instrument designed by Flatly for measuring the angles between crystal faces. He wrote extensively on atomic theory, sometimes opposing and sometimes modifying the new and epoch-making theories of Dalton. Wollaston added to Dalton's advances by being one of the first scientists to realize that the arrangement of atoms in a molecule must be three-dimensional, and he also came close to formulating the 'law of definite proportions' which is credited to Proust. He was elected FRS in 1793. He died in London, leaving some of his considerable wealth to the Royal Society and the Geological Society of London to promote scientific research. The mineral wollastonite, one of the three forms of calcium silicate, was named in his honour.

WOOD, Robert Williams
(1868–1955)

American physicist, born in Concord, Massachusetts. Educated at Harvard, in Chicago and in Berlin, he was Professor of Experimental Physics at Johns Hopkins University from 1901 to 1938. He carried out research on optics, atomic and molecular radiation and sound waves. One of his most notable researches was his study of resonance radiation, which led to the spectroscopic technique of optical pumping developed by Kastler. Wood also pioneered the production of phase gratings and zone-plate lenses, anticipating the much later science of 'diffractive optics'. He wrote *Physical Optics* (1905), some fiction, and illustrated nonsense verse in *How to Tell the Birds from the Flowers* (1907).

WOODS, Donald Devereux
(1912–1964)

British microbiologist. He was educated in Ipswich and at Trinity Hall, Cambridge, where he received his PhD in 1937. After a period as reader in microbiology at the University of Oxford, he became Iveagh Professor at Oxford in 1955. His work resulted in significant advances in chemotherapy, particularly the introduction of the concept of competitive inhibition, and the elucidation of the mode of action of the sulphanilamide drugs. He was elected FRS in 1952.

WOODWARD, Sir Arthur Smith
(1864–1944)

English geologist, born in Macclesfield and educated at the University of Manchester. He was appointed as assistant (1881–1892), assistant keeper (1892–1901) and keeper of geology (1901–1924) at the Natural History section of the British Museum. Woodward did notable work on fossil fish developing from his initial brief of cataloguing the museum's fossil fish collection. From 1885 to 1923 he made some 30 expeditions abroad, in his vacation time and largely at his own expense, visiting museums, meeting palaeontologists and making collections throughout Europe, the USA, the Middle East and South America. One of Britain's most prolific ever geologists, the four volumes of his *Catalogue of Fossil Fishes in the British Museum (Natural History)* (1889, 1891, 1895, 1901) were numbered amongst some 650 publications. In spite of this, Woodward is chiefly remembered for his part in the controversy over the Piltdown Man. The amateur geologist Charles Dawson gave him the cranial fragments found at Piltdown from 1908 to 1912 for identification. Together with parts of a jawbone unearthed later, these were accepted by many anthropologists as the 'missing link' in Charles Darwin's theory of evolution, and as such one of the greatest discoveries of the age. Following scientific tests, the skull was denounced as a fake in 1953; Woodward's firm conviction that the remains were human had been the main reason for the success of the hoax. In 1898 he published *Outlines of Vertebrate Palaeontology*. He was knighted in 1924.

WOODWARD, Robert Burns
(1917–1979)

American chemist, born in Boston. Having been interested in chemistry since the age of eight when he was given a chemistry set, his formal chemical education at MIT resulted in a PhD by the age of 20. He then moved to Harvard (1937) and by 1944 had become associate professor, working on the chemistry of the antimalarial drug quinine and the new wonder drug penicillin. For the next 20 years he executed the syntheses of a dazzling array of biological compounds, including strychnine, cholesterol, lanosterol, lysergic acid, reserpine, chlorophyll, colchicine and cephalosporin C. His feel for the art and architecture of constructing complex molecules was astounding. At the same time he became famous for his lectures, delivered without a single note or a visual aid apart from blackboard and coloured chalk, sometimes lasting more than three hours with only a short pause at half-time while the speaker drank a refreshing martini. The audience rarely took notes but photographed the blackboard at the end. He was awarded the Nobel Prize for Chemistry in 1965 for the totality of his work in the art of synthetic chemistry, structure determination and theoretical analysis. But his finest work was still to come. Woodward and the Swiss chemist Eschenmoser set out to synthesize vitamin B12 in a collaborative venture. The work took over 10 years and was completed in 1976. In the course of this synthesis Woodward conceived the idea that molecular orbitals could affect the products obtained in cyclization reactions and he invited a young Harvard theoretician Hoffmann to collaborate. This led eventually to the Woodward-Hoffmann rules for the conservation of orbital symmetry. Unfortunately Woodward had died from a heart attack before the award of a Nobel Prize for Chemistry for this work. He received almost every honour and award possible for an organic chemist and the pharmaceutical company Ciba-Geigy established the Woodward Research Institute for him in Basle. Many think of Woodward as the greatest synthetic organic chemist of all time. It is reported that at the 1978 meeting of the Association of Harvard Chemists he was carried into the lecture hall in a fashion no less regal than that of a thirteenth-century caliph.

WOOLLEY, Sir Richard van der Riet
(1906–1986)

British astronomer, born in Weymouth. In his early years he lived in South Africa and studied at the University of Cape Town before returning to England and entering Cambridge University (1925–1928). On graduating in mathematics as Wrangler he became a research student under Eddington and was awarded the Isaac Newton studentship (1931). He was appointed chief assistant at the Royal Observatory, Greenwich (1933–1937), returning to Cambridge as John Couch Adams Astronomer (1937–1939). In 1939 just before the outbreak of World War II he was appointed Government Astronomer and Director of the Commonwealth Observatory, later known as the Mount Stromlo Observatory, in Canberra, Australia. During the years of World War II his energies were diverted to optical design of military instruments, but at the same time he was able to establish the observatory as an important institution for observations of the southern skies. His researches at Greenwich and Canberra were mainly concerned with solar and stellar atmospheres; he published two important books on that subject, *Eclipses of the Sun and Moon* (1937, with Sir Frank Dyson) and *The Outer Layers of a Star* (1953, with Douglas Walter Noble Stibbs). In 1956 Woolley, as 11th Astronomer Royal, succeeded Jones with charge of the Royal Greenwich Observatory at Herstmonceux in Sussex, where he supervised the completion and erection there of the 98-inch (2.5-metre) Isaac Newton telescope (now re-sited on the Canary island of La Palma). Woolley was involved in two further major telescopic projects; the Anglo-Australian 3.8-metre telescope and the United Kingdom 1.2-metre Schmidt telescope, both sited at the Sidings Springs Observatory in New South Wales, Australia. On retiring from the Royal Greenwich Observatory in 1971, Woolley who had received a knighthood on becoming Astronomer Royal, accepted the appointment of director of the newly established South African Astronomical Observatory (1972–1976).

WRIGHT, Sir Almroth Edward
(1861–1947)

English bacteriologist, born in Yorkshire. Educated in Dublin, Leipzig, Stassburg, Marburg and Sydney, he was subsequently appointed to an army medical school, where he developed a vaccine against typhoid fever. He later became Professor of Experimental Pathology at

St Mary's Hospital, London (1902). Wright was known specially for his work on the parasitic diseases, and for his research on the protective power of blood against bacteria. He was an important influence on the work of his student Sir Alexander Fleming, and took considerable pride in the publicity it afforded for St Mary's Hospital Medical School. He was knighted in 1906.

WRIGHT, Orville
(1871–1948) and
Wilbur
(1867–1912)

American pioneers of aviation. Orville was born in Dayton, Ohio, and his elder brother Wilbur was born near Millville, Indiana. They were the sons of a bishop of the United Brethren Church, who encouraged them to be independent and use their talents to the full, and they both developed a keen interest in mechanical devices at an early age. To earn a living they founded in 1892 the Wright Cycle Company, but all the time their real interests lay in emulating and surpassing the aerial exploits of the German gliding pioneer Otto Lilienthal, killed when one of his gliders crashed in 1896. Convinced that it was vital to put as much effort into learning how to control aircraft as would be needed to build them, they embarked in 1899 on a careful programme of model glider flights, followed in 1900 by short piloted flights in a glider of 5 metres wing span. They then built a small wind tunnel in which they tested more than 200 models, enabling them to compile the first accurate tables of lift and drag for different wing configurations. Their third piloted glider was thoroughly tested at Kitty Hawk, North Carolina, in September 1902, and it was so successful that they decide to attempt powered flight the following summer. After searching in vain for a suitable engine and propeller they set about making their own, and on 17 December 1903 Wilbur and Orville each made two flights, the last and longest of 59 seconds duration covering 255 metres. For the next three years the Wrights were the only men in the world to achieve a succession of controlled, powered and sustained flights; in October 1905 Wilbur flew non-stop for a distance of almost 40 kilometres. Encouraged by this, they abandoned their cycle business, and having patented their flying machine and its controls, formed an aircraft production company (1909). In 1915, Wilbur having died of typhoid fever three years earlier, Orville sold his interests in the business in order to devote himself to aeronautical research. The world acknowledged their success with numerous honours and awards.

WRIGHT, Sewall
(1889–1988)

American geneticist, born in Melrose, Massachusetts. He was educated at Lombard College, the University of Illinois and Harvard. He began his career with the US Department of Agriculture, and following studies of breeding methods to improve livestock quality, he developed a mathematical description of evolution. He showed that within small isolated populations, certain genetic features may be lost randomly if the few individuals possessing the genes happen not to pass the genes on to the next generation. This 'Sewall Wright effect' allows evolution to occur without the involvement of natural selection.

WRIGHT, Thomas
(1711–1786)

English natural philosopher, born in Byer's Green near Durham. He was trained as a scientific instrument-maker, and also gained a reputation as a teacher of mathematics. He is known for his cosmological speculations in his *Original Theory of the Universe* (1750) in which he put forward a model of the Milky Way as an indefinitely long slab filled with stars, with the Sun inside it. The model explained the high density of stars in the great circle of the Milky Way and the low density in directions at right angles to it. In later models he described the universe in terms of spheres of stars, and in terms of rings like those of Saturn which he predicted were composed of many small satellites. His work which was of a metaphysical nature was known to Kant who referred to it in his writings. Though Wright preceded Herschel in point of time, the latter's disc theory of the stellar system, based on observational data, was quite independently developed.

WRÓBLEWSKI, Zygmunt Florenty von
(1845–1888)

Polish physicist, born in Grodno (now in Byelorussia). The son of a lawyer, he entered Kiev University in 1862 but as an activist in the Polish insurrection of 1863 he was exiled to Siberia for six years. After being amnestied he settled in Berlin, studying in the laboratory of Helmholtz. He obtained a PhD in Munich (1874) and then, whilst assistant at the University of Strassburg (1875–1876), studied gaseous diffusion. Financial assistance from the Krakow Academy of Sciences enabled Wróblewski to study in Paris under Sainte-Claire Deville and to visit London, Oxford and Cambridge. He became Professor of Physics at Krakow in 1882, although he spent the first year in Paris with Deville working on aqueous solutions of carbon dioxide under high pressure. This was the starting point for his famous series of investigations on the liquefaction of gases. With the chemistry professor at Krakow, he liquefied air on a large scale for the first time (1883). A year later he obtained a mist of liquid hydrogen. Wróblewski died in 1888 from severe burns after a kerosene lamp set light to his clothes.

WU, Chien-Shiung
(1912–   )

Chinese-born American physicist, born in Shanghai. She studied at the National Centre University in China, and from 1936 in the USA, at the University of California in Berkeley. From 1946 she was on the staff of Columbia University, New York, where she was appointed professor in 1957. In 1956 Wu and her colleagues carried out an experiment to test Tsung-Dao Lee and Yang's hypothesis that parity is not conserved in weak decays; this would result in the decay not being mirror symmetric, i.e. having a preferred direction. The experiment was performed by cooling cobalt-60 nuclei in a magnetic field so that their spins were aligned. They observed that the electrons resulting from the beta decay of the nuclei were emitted preferentially in one direction, proving that the principle of parity conservation was indeed violated in weak interactions. This was later explained by the V-A theory of weak interactions proposed by Feynman and Gell-Mann.

WYCKOFF, Ralph Walter Graystone
(1897–   )

American biophysicist, born in Geneva, New York. He studied at Cornell

University, gaining a PhD in chemistry. He worked as a physical chemist in the geophysics laboratory of the Carnegie Institute (1919–1927) before moving to the subdivision of biophysics at the Rockefeller Institute (1927–1938). After a spell in commercial laboratories he moved to the University of Michigan and then to the National Institutes of Health until 1959. His post-war career included a period on attachment to the American Embassy in London (1952–1954) and in 1959 he was appointed Professor of Physics and Microbiology at the University of Arizona. While at the Rockefeller, Wyckoff developed new ultracentrifugation techniques to purify viruses, such as that causing equine encephalomyelitis, which facilitated the production of pure virus preparations from which effective vaccines against the disease could be developed. In 1944, while working in epidemiology, he collaborated with his Michigan colleague, the astronomer Robley Williams, in developing the metal shadowing method for providing three-dimensional imaging of viruses in the electron microscope, which has since been widely used. His achievements include several studies of molecular structure, such as an examination of crystals, other macromolecules and several viruses. He investigated the effects of radiation on cell structure and continued to refine investigative techniques such as centrifugation methods and electron microscopy procedures. He also developed improved routines for the purification of viruses.

WYNNE-EDWARDS, Vero Copner
(1906–)

English zoologist, born in Leeds. He was educated at the University of Oxford and the Marine Biology Laboratory in Plymouth (1927–1929), and taught at Bristol (1929–1930) and at McGill University, Montreal (1930–1946). He travelled widely in Canada carrying out research in avian biology on Baffin Island, in the Yukon Territory and at the Mackenzie river. He returned to Britain in 1946 and became Regius Professor of Natural History at Aberdeen. In 1962 he published *Animal Dispersion in Relation to Social Behaviour*, in which he argued that animal dispersal had evolved in order to regulate population density. He suggested that individuals of a species altruistically reduce their birth rates so as to benefit the species as a whole. To support his theory he produced numerous examples from throughout the animal kingdom. The mechanism he proposed to explain this was group selection; groups which exhibited such altruistic behaviour would be at an advantage in competition with populations whose members behaved selfishly and bred without regard to population pressure. Altruism he believed was mediated via territorial behaviour and dominance hierarchies. These ideas aroused considerable controversy and adverse criticism but were nevertheless instrumental in stimulating advances in sociobiology and behavioural ecology. It is generally accepted that while group selection may occur under exceptional circumstances, it has not been a major factor in evolution and most of Wynne-Edwards's examples have subsequently been explained adequately on the basis of selection acting at the level of the individual. He wrote *Evolution through Group Selection* in 1986. He served on the Red Deer Commission as Vice-Chairman (1959-1968) and the Royal Commission on Environmental Pollution (1970–1974), and was elected FRS in 1970.

X

XENOCRATES
(c.395–314 BC)

Greek philosopher and scientist, born in Chaleedon on the Bosphorus. He was a pupil of Plato and in 339 BC became head of the Academy which Plato had founded. He is recorded as travelling with Aristotle after Plato's death in 348 BC to do research under the patronage of Hermeias, tyrant of Atarneus in north-west Asia Minor, and as joining some Athenian embassies of foreign diplomatic missions. He wrote prolifically on natural science, astronomy and philosophy, but only fragments of this output survive. He generally systematized and continued the Platonic tradition but seems to have had a particular devotion to threefold categories, perhaps reflecting a Pythagorean influence: philosophy is subdivided into logic, ethics and physics; reality is divided into the objects of sensation, belief and knowledge; he distinguished gods, men and demons; he also probably originated the classical distinction between mind, body and soul.

Y

YALOW, Rosalyn, née Sussman
(1921–   )

American biophysicist, born in New York City. Yalow was the first woman to graduate in physics from Hunter College, New York (1941). She obtained a PhD from the College of Engineering of the University of Illinois in 1945. She taught physics at Hunter College until 1950, and in 1947 became consultant to the Radio-isotope Unit at the Bronx Veterans Administration (VA) Hospital. From 1950, Yalow collaborated with Solomon Berson, and during the course of research on diabetes they developed 'radioimmunoassay' (RIA). This is an ultrasensitive method of measuring concentrations of substances in the body which relies upon 'labelling' molecules with radioactive isotopes. Yalow and Berson found that in adult diabetics, the rate of clearance of injected labelled insulin from the blood is surprisingly low, and suggested that anti-bodies inactivating the insulin are formed. In 1977, for her work on RIA, Yalow shared the Nobel Prize for Physiology or Medicine with Gufflemin and Schally. Besides diabetes, Yalow has used RIA in work on dwarfism, leukaemia, peptic ulcers and neurotrans-mitters in the brain. Since 1969 she has been Chief of the Radioimmunoassay Reference Laboratory and since 1973 Director of the Solomon A. Berson Research Laboratory of the VA Medical Centre.

YANG, Chen Ning
(1922–   )

Chinese-born American physicist, born in Hofei, the son of a professor of mathematics. He was educated in Kuming, China, before gaining a scholarship to Chicago in 1945 to work under Teller. He became professor at the Institute for Advanced Studies, Princeton (1955–1965), and from 1965 was Director of the Institute for Theoretical Physics at New York State University, Stony Brook. In 1956 with Tsung-Dao Lee, who had been his fellow student at Chicago, he made a thorough analysis of all the known data in particle physics and concluded that the quantum property known as parity was unlikely to be conserved in weak interactions, and they suggested a simple experiment which would prove it. In the same year (1956), a similar experiment by a group of physicists from Columbia University and the American National Bureau of Standards, headed by Wu, confirmed that the 'law' of parity is indeed violated in the case of weak interactions. For this prediction, Lee and Yang were awarded the 1957 Nobel Prize for Physics, and the Einstein Commemorative Award from Yeshiva University in the same year. With Robert Mills,

Yang also developed a non-Abelian gauge theory which proved to be an important development in the theories of fundamental interactions for elementary particles and fields.

YANOFSKY, Charles
(1925–)

American geneticist, born in New York City. He studied at New York's City College and at Yale, and since 1961 has been Professor of Biology at Stanford, working on gene mutations, particularly on the trytophan operon of *Escherichia coli*. He has shown that the sequence of bases in the genetic material DNA acts by determining the order of the amino acids which make up proteins, including the enzymes that control biochemical processes. In 1967 he showed that the amino acid sequence of the protein for synthesizing tryptophan is collinear with the genetic map of the gene. He went on to describe how the production of the amino acid tryptophan is controlled by a process called attenuation (1977). Attenuation results in the synthesis of a smaller modified tryptophan molecule, and occurs when the levels of tryptophan are high.

YERSIN, Alexandre Emile John
(1863–1943)

Swiss-born French bacteriologist, born in Rougemont and educated at Lausanne, Marburg and Paris. He carried out research at the Pasteur Institute in Paris, working along with Emile Roux on diphtheria antitoxin. In Hong Kong in 1894 he discovered the plague bacillus at the same time as Kitasato. He developed a serum against it, and founded two Pasteur institutes in China. He also introduced the rubber tree into Indo-China.

YODER, Hatten Schuyler Jr
(1921–)

American experimental petrologist, born in Cleveland, Ohio. He was educated at the University of Chicago and MIT. Thereafter he spent his career as Petrologist (1948–1971) and Director (1971–1986) at the Geophysical Laboratory of the Carnegie Institution of Washington, where he has been Director Emeritus since 1986. Yoder undertook important experimental petrological studies of phase equilibria in mineral systems, properties of minerals at high pressures and temperatures, hydrothermal mineral synthesis and experimental heat transfer in silicates.

YONGE, Charles Maurice
(1899–1986)

English zoologist and marine biologist, born in Wakefield. Educated at Oxford, where he started reading history, and Edinburgh, where he graduated in zoology, he carried out research in marine laboratories at Millport, Plymouth and Naples. At Plymouth he studied the feeding behaviour and nutritional physiology of the oyster, a long-standing interest culminating in publication of the book *Oysters* in 1960. He was leader of the Great Barrier Reef Expedition in 1928 during which 20 scientists spent 13 months of the reef studying all aspects of the ecology and physiology of the marine organisms. Yonge researched coral physiology, in particular the role of the symbiotic zooanthellae. He edited the six volumes on the work of the expedition published by the British Museum (Natural History). He was Professor of Zoology at Bristol (1933–1944) then Glasgow (1944–1970), and on retiral from the latter he continued his research at Edinburgh. His major interest was in the bivalve molluscs. Among the aspects of their biology he investigated were the role of the crystalline style in digestion, filter feeding, structure and function of the mantle cavity and ability to bore into rock. He wrote several popular books including *The Sea Shore* (1949) and *Collins Pocket Guide to the Sea Shore* (1958, with J. H. Barrett). He was elected FRS in 1946. His public services included membership of the Fisheries Advisory Committee (1936–1956) and the Colonial Fisheries Advisory Committee (1949–1960, Chairman from 1955).

YOUNG, John Zachary
(1907–)

English zoologist, born in Bristol and educated at Magdalen College Oxford, where he was strongly influenced by his tutor Julian Huxley. He graduated in 1928 and went to the Stazione Zoologica, Naples, as Oxford's biological scholar. There, whilst working on the innovation of the gastrointestinal tract of fish, he became interested in the nervous system of cephalopods, the class of animals that included octopuses, squids and their relatives. He returned to Oxford as university demonstrator in zoology (1933–1945) and was then appointed to the Chair of Anatomy at University College London (1945–1974), the first non-medically qualified Professor of Anatomy in Britain. Young's research has centred on the anatomy and physiology of the nervous system. In 1933 while visiting the Woods Hole Marine Laboratory in Massachusetts, he reported that the squid had an unusually large diameter nerve fibre about one hundred times greater than mammalian nerves. This giant nerve fibre subsequently became the basic research material of later neurophysiologists such as Alan Hodgkin and Andrew Huxley. During World War II he studied regeneration in damaged nerves, and subsequently returned to investigations of the central nervous system, and mechanisms of learning and memory, especially in the octopus. He has proposed neural models to account for memory processes, summarized in *The Memory System of the Brain* (1966) and *Programs of the Brain* (1978). In addition, he has written several influential textbooks including *The Life of Vertebrates* (1950), *The Life of Mammals* (1957) and *Introduction to the Study of Man* (1971). He was elected FRS in 1945.

YOUNG, Thomas
(1773–1829)

English physicist, physician and Egyptologist, born in Milverton, Somerset, the son of a mercer and banker. A precocious student, he was interested in classical languages, higher mathematics, natural history and natural philosophy (physics), and being of independent means, he could follow his varied interests as he wished. He studied medicine at London, Edinburgh, Göttingen and Cambridge, and became a physician in London in 1800, but devoted himself to scientific research. In 1801 he was appointed Professor of Natural Philosophy at the Royal Institution, but three years later was forced to resign as his lectures were found to be too technical for a popular audience, and he resumed his medical practice. He was appointed physician to St George's Hospital (1811), and held this position until his death. He held several public offices related to science, spent several periods as Consultant to the Admiralty, and was Secretary to the Royal Commission on Weights and Measures (1816–1821), Secretary of the Board of Longitude (1818–1828) and Superintendent of the *Nautical Almanac* (1818–1829). He supplemented his income as an anonymous author of a wide variety of scientific articles. He was elected FRS (1794), and was involved in the Royal Society's affairs as Foreign Secretary and Member of Council. He became best known in the nineteenth century for his undulatory (wave) theory of light as expounded in his *Outlines of Experiments and Enquiries respecting Sound and Light* (1800) and in *A Course of Lectures on Natural Philosophy and the Mechanical Arts* (1807). He combined the wave theory of Huygens and Isaac Newton's theory of colours to explain the interference phenomenon of colours produced by ruled gratings, thin plates, and the colours of the supernumerary bows of the rainbow, but his theory was unacceptable during his lifetime as it was regarded 'anti-Newtonian'. He also did valuable work in insurance, haemodynamics and Egyptology, and made a fundamental contribution to the deciphering of the inscriptions on the Rosetta Stone.

YUKAWA, Hideki, originally Hideki Ogawi
(1907–1981)

Japanese physicist, born in Tokyo. He was educated at Kyoto Imperial University and after graduating in 1929, remained at Kyoto where he was appointed lecturer. He married Sumi Yukawa in 1932 and assumed her family name. In 1933 he became a lecturer at Osaka Imperial University and received his doctorate there in 1938. The following year he returned to Kyoto as Professor of Theoretical Physics (1939–1950) and he later became Director of the Kyoto Research Institute for Fundamental Physics (1953–1970). In 1935 he published his first paper in which he proposed his theory of nuclear forces, suggesting a strong short-range attractive interaction between nucleons (neutrons and protons) that would overcome the electrical repulsion between protons. This interaction was propagated by the exchange of massive particles between the nucleons, the mass of the exchange particle determining the range of the force. The existence of these intermediate particles was confirmed by Cecil Powell's discovery in 1947 of the pi-meson or pion. Yukawa also predicted that a nucleus may absorb one of the innermost electrons in the atom (1936); such 'K-capture' was soon observed. During World War II he played a minor role in military research whilst continuing his scientific work. For his theory of strong nuclear forces, and his work on quantum theory and nuclear physics, he was awarded the Nobel Prize for Physics in 1949, the first Japanese to be so honoured.

Z

ZEEMAN, Pieter
(1865–1943)

Dutch physicist, born in Zonnemaire, Zeeland. He was educated under Lorentz at the University of Leiden, where he received his doctorate in 1893. He became a lecturer at Leiden in 1897, and in 1900 was appointed as Professor of Physics at Amsterdam University. In 1896 he studied the effects of a magnetic field on sodium and lithium fight sources, and observed that the spectral emission lines were broadened. He demonstrated that this was due to the splitting of spectrum lines into two or three components and the phenomenon became known as the Zeeman effect. This was consistent with Lorentz's classical theory of light produced by vibrating electrons. More generally, atoms display several closely spaced lines; this was later explained with the development of quantum theory. The magnetic field differentiates a single atomic energy level into several components of slightly different energy associated with different quantized orientations of the total electron magnetic moment with respect to the field. Zeeman also investigated the absorption and motion of electricity in fluids, magnetic fields on the solar surface, the Doppler effect and the effect of nuclear magnetic moments on spectral lines. In 1902 he shared with Lorentz the Nobel Prize for Physics for the discovery and explanation of the Zeeman effect, and he was awarded the Rumford Medal of the Royal Society in 1922.

ZEL'DOVICH, Yakov Borisovich
(1914–   )

Soviet astrophysicist, born in Minsk. He graduated from the University of Leningrad in 1931 and moved to the Soviet Academy of Sciences, becoming a full Academician in 1958. In the late 1930s he concentrated on nuclear physics and specifically (with Y. B. Khariton) the chain reaction in uranium fission. During the 1940s he investigated the mechanisms responsible for the oxidation of nitrogen during an explosion, and the problems of flame propagation and gas dynamics. In the 1950s he turned to cosmology, and studied the production of the initial hydrogen-to-helium ratio and the degree of isotropy in the early stages of the universe. He also predicted that it would be possible to find black holes associated with X-ray emitting binary stars. In 1972, with Rashid Sunyaev, he discovered how the total energy carried by the microwave background radiation could be increased as it passed through the medium between the galactic clusters. This effect has important cosmological applications and can be used to independently estimate the Hubble constant, which measures the rate at which the expansion of the universe varies with distance.

ZENO OF ELEA
(c.430–c.490 BC)

Greek philosopher and mathematician. Little is known of his life: he was a native of Elea, a Greek colony in southern Italy, where he lived all or most of his life. He was a disciple of Parmenides of Elea, and in defence of his monistic philosophy against the Pythagoreans he devised his famous paradoxes which purported to show the impossibility of motion and of spatial division, by showing that space and time could be neither continuous or discrete. The paradoxes are: 'Achilles and the Tortoise', 'The Flying Arrow', 'The Stadium' and 'The Moving Rows'. Aristotle attempted a refutation but they were revived as raising serious philosophical issues by Bertrand Russell.

ZERMELO, Ernst Friedrich Ferdinand
(1871–1953)

German mathematician, born in Berlin. He studied mathematics, physics and philosophy at Berlin, Halle and Freiburg, and was professor at Göttingen (1905–1910) and Zurich (1910–1916). From 1926 to 1935 he was an honorary professor at Freiburg im Breisgau. Although he worked in physics and the calculus of variations among other subjects, he is now best remembered for his work in set theory. Following Georg Cantor's pioneering work, Zermelo gave the first axiomatic description of set theory in 1908; although later modified to avoid the paradoxes discovered by Bertrand Russell and others, it remains one of the standard methods of axiomatizing the theory. He also first revealed the importance of the axiom of choice, when he proved in 1904 that any set could be well-ordered, a key result in many mathematical applications of set theory.

ZERNIKE, Frits
(1888–1966)

Dutch physicist, born in Amsterdam. He was educated at Amsterdam University and became Professor of Physics at Gröningen (1910–1958). Zernike developed (from 1935) the phase-contrast technique for the microscopic examination of transparent – frequently biological – objects. For this work he was awarded the Nobel Prize for Physics in 1953. Related to this was his invention of the 'coherent background' technique for revealing the presence of phase variations in interference and diffraction patterns, and for studying very weak fringes in such patterns. He made important pioneering contributions to the understanding of optical coherence showing how it could be measured from observable features of the light field. He also discovered a modified subset of the Jacobi polynomials, which is particularly appropriate for the study of optical system performance and led to a new formulation of the wave theory of lens aberrations; these expressions are now known as Zernike polynomials.

ZIEGLER, Karl
(1898–1973)

German chemist, born in Helsa. The son of a Lutheran minister, he studied chemistry at the University of Marburg where he obtained his PhD in 1920. He was briefly a lecturer in Frankfurt and in 1927 took up the post of professor at the University of Heidelberg. He moved to the Chemical Institute Halle-Saale in 1936 and became Director of the Kaiser Wilhelm Institute for Coal Research at

Mulheim-in-Ruhr in 1943. He remained there until his retirement in 1969. His initial research showed that organo-metallic compounds effect polymerization by the generation of radicals. After World War II he concentrated on studies of organo-aluminium compounds, and found that zirconium acetyl acetonate added to triethyluminium catalysed the polymerization of ethene to give poly-ethylene of very high molecular weight. Equally important was his discovery that trialkylaluminiums and titanium chloride polymerized ethene at ambient tempera-ture and pressure. Collaboration with Natta led to the development of a family of catalysts and gave birth to a new era in the plastics industry. The royalties from these industrial processes partly fund an endowment at the Kaiser Wilhelm Institute. Ziegler shared the Nobel Prize for Chemistry with Natta in 1963.

ZINDER, Norton David
(1928–)

American geneticist, born in New York City. He studied at Columbia University and with Lederberg at Wisconsin and became professor of Genetics in 1964. In 1951, studies using mutants of the bacterium Salmonella led him to describe the process of bacterial transduction. Transduction refers to the transmission of a bacterial gene from one bacterium to another by means of a viral phage particle which carries the gene. It led to an explanation for the spread of drug resistance in bacteria, and offered a mechanism of inserting specific genes into a host cell bacterium. Recently, Zinder has been Chairman of the Program Advisory Committee for the Human Genome Project in the USA.

ZINSSER, Hans
(1878–1940)

American bacteriologist and immunolo-gist, born in New York City, the son of a prosperous German immigrant. He was educated at Columbia University and its College of Physicians and Surgeons (MD 1903). His predilection for science led him into bacteriology and immunology, which he taught at Columbia and Stanford universities before going to Harvard in 1923. Zinsser worked on many scientific problems, including allergy, the measurement of virus size and the cause of rheumatic fever. Above all, however, he clarified the rickettsial disease typhus, differentiating epidemic and endemic forms (the endemic form is still known as Brill-Zinsser's disease), researches which he brilliantly described in his popular book *Rats, Lice and History* (1935). His *Textbook of Bacter-iology* (1910) and *Infection and Resistance* (1914) became classics. A highly cultured man, he wrote poetry and essays, and left an evocative autobiography, *As I Remember Him* (1940), written in the third person while he was dying from leukaemia.

ZITTEL, Karl Alfred von
(1839–1904)

German palaeontologist, born in Bahlingen, Baden, the youngest son of a clergyman. He was educated at Heidel-berg and Paris, and commenced geolo-gical research in Dalmatia during a spell as a voluntary assistant with the Geolo-gical Survey of Austria. In 1862 he became an assistant at the Mineralogical Museum in Vienna where he undertook some teaching. The following year he was appointed professor at Kahlsruhe. From 1880 he held professorships at Munich, and in 1890 he became Keeper of the

State Geological Collections. During his career he also served as President of the Bavarian Academy. A distinguished authority on his subjects and their history, he was a pioneer of evolutionary palaeontology and was widely recognized as the leading teacher of palaeontology in the nineteenth century. His five-volume *Handbuch der Palaeontologie* (1876–1893) was arguably his greatest service to science, and it remains as one of the most comprehensive and trustworthy palaeontological reference books. His important textbook *Grundziige der Palaeontologie* first appeared in 1895, with an English translation *Textbook of Palaeontology* being produced in 1900. It was later revised by Arthur Smith Woodward (1925).

ZOHARY, Michael
(1898–1983)

Israeli botanist and phytogeographer, born in Galicia (now part of Poland). He emigrated to Palestine in 1920 and worked as a road builder. In 1922 he entered Jerusalem Seminary and became interested in botany. He later joined the botany department at the Institute of Agriculture, Tel Aviv (1925); with Naomi Feinbrun and Alexander Eig, he formed the first team of botanists to study the flora and vegetation of Palestine. From 1929, Zohary's team was based at the Hebrew University of Jerusalem. He obtained a doctorate from the University of Prague (1936), based on studies of seed dispersal ecology in Palestinian plants in which he introduced the theory of antiteleochory, whereby seed germination of desert plants is ensured by dispersal near the parent plant. He became interested in evolutionary trends in seed dispersal (especially in Cruciferae, Compositae and Luguminosae), studied the taxonomy of many genera, and co-authored a world monograph of *Trifolium* (1984). His fundamental geobotanical researches culminated in the publication of the monumental *Geobotanical Foundations of the Middle East* (2 vols, 1973). Zohary's other main publications were *Flora Palaestina* (4 vols, 1966–1986, with Feinbrun), *Plant Life of Palestine* (1962) and *Plants of the Bible* (1982).

ZONDEK, Bernhard
(1891–1967)

Israeli gynaecologist and endocrinologist, born in Wronke, Germany. He trained in medicine at the University of Berlin, where he became a member of staff in the Department of Obstetrics and Gynaecology (lecturer, 1923–1926; associate professor, 1926–1929; director, 1929–1933). He left Nazi Germany in 1933 to become Professor of Obstetrics and Gynaecology at the Hebrew University, Jerusalem (1934–1961). His research interests, influenced by his elder brother Hermann Zondek, an endocrinologist, were predominantly in reproductive endocrinology. In collaboration with Selmar Aschheim he developed the first reliable pregnancy test in 1928. They later discovered that the anterior pituitary gland produced hormones called gonadotrophins, which in turn stimulated other endocrine glands, such as the ovary, to release their hormones. This work provided important evidence of control mechanisms in reproduction, which has had widespread significance in the development of medical and social attitudes towards questions on fertility, infertility, contraception and abortion.

ZOTTERMAN, Yvnge
(1898–1982)

Swedish neurophysiologist, born in Stockholm and educated in medicine there at the Karolinska Institute. After clinical studies in the University of Uppsala he did national service in the Swedish Royal Navy, and for many years subsequently spent part of each year with the navy working on physiological problems of deep-sea diving. In 1919 he travelled to Cambridge for a few months to study physiology, and returning the following year he began working with Adrian, on recording and analysing the nerve impulses from sensory nerve endings. He returned to Sweden in 1927, continuing his research on the sensory function of the skin, particularly thermal and pain sensation, and also did pioneering work on the physiology of taste. During World War II he went to the USA on a short mission to gain support for Finland and began work on industrial physiology and on human physiology, studying the Swedish lumberjack. After the war he became Head of the Department of Physiology at the Veterinary College of Stockholm, and after formal retirement he initiated a multi-disciplinary working group in 1973 studying the problems of the elderly. Much of his life and work is summarized in his two-volume autobiography *Touch, Tickle and Pain* (1969).

ZSIGMONDY, Richard Adolf
(1865–1929)

Austrian chemist, famous for his contribution to colloidal chemistry. He was born in Vienna and attended the university there, moving to Munich to take his doctorate. He worked briefly in Berlin on the chemistry of glass, studying the colloidal inclusions which give glass its colour and opacity. After teaching in Graz for five years he was employed by the Schott Glass Manufacturing company in Jena (1897–1900), where he invented Jena milk glass. From 1900 to 1907 he worked independently in his own laboratory and as a result of his discoveries was appointed Professor of Inorganic Chemistry at Göttingen (1900–1929). Zsigmondy made colloidal solutions, particularly gold sols, his life's study. He invented the ultramicroscope, in which intense light projected from the side shows, against a dark background, the light scattered by particles. This made it possible to see individual particles (in the same way that dust is visible in a sunbeam). This microscope, developed in conjunction with Zeiss Company of Jena, led to great advances in colloidal chemistry, then in its infancy, although it has now been replaced by the electron microscope and ultracentrifuge. With his microscope Zsigmondy discovered that changes of colour occur in colloidal solutions when particles coagulate, and that protective agents such as gum arabic and gelatin preserve colour by preventing coagulation. He also showed that the valuable paint known as purple of Cassius, whose composition had long puzzled chemists, contains a mixture of gold and stannic acid particles in colloidal solution. Zsigmondy's later work was on silica and soap gels which he investigated by another new technique, ultrafiltration, in which the substances to be separated are drawn through a membrane by a decrease in pressure. In 1925 he became the first person to be awarded a Nobel Prize for colloidal chemistry. He died in Göttingen.

ZUCKERMAN, Solly, Baron
(1904–1993)

British zoologist, born in Cape Town, South Africa. He carried out research with chacma baboons near Cape Town before moving to England, where he taught at Oxford University from 1932. During World War II he investigated the biological effects of bomb blasts. He became Professor of Anatomy at Birmingham (1946–1968) and Secretary of the Zoological Society of London in 1955. He was chief scientific adviser to the British government from 1964 to 1971, and wrote official reports on aspects of farming, natural resources, medicine and nuclear policy. The results of his research on chacma baboons and captive hamadryas baboons at London Zoo were published in two influential books, *The Social Life of Monkeys and Apes* (1932) and *Functional Affinities of Man, Monkeys and Apes* (1933). He was the first primatologist to consider that such studies could provide insights into the origins and behaviour of humans. He argued that as female primates were prepared to mate over a prolonged period, sex was the original social bond, a view reinforced by his observations of male hamadryas fighting to the death over females at London Zoo. Subsequent research has shown that the majority of primates other than humans are seasonal breeders and fatal encounters are rare in the wild. Zuckerman published his autobiography, *From Apes to Warlords*, in 1978.

ZUSE, Konrad
(1910–)

German computer pioneer, born in Berlin. He was educated at the Berlin Institute of Technology before joining the Henschel Aircraft Company in 1935. In the following year, tiring of the drudgery involved in solving complex linear equations, he began building a calculating machine in his spare time, a task which occupied him until 1945. With the help of a friend, Helmut Schreyer, he built a number of prototypes, the earliest of which were constructed in the living room of his parents' home. Zuse's most historic machine was the Z3. Consisting of a tape reader, an operator's console and two cabinets packed with 2,600 relays, it had a small memory (capable of storing only 64 22-bit numbers), but was fast enough to multiply two rows of digits in only 3–5 seconds. It was the first operational general-purpose program-controlled calculator. Its inventor built up his own firm, Zuse KG, until it was bought out by another firm in the 1960s. Zuse became Honorary Professor at Göttingen University in 1966.

ZWEIG, George
(1937–)

Russian-born American physicist, born in Moscow. He was educated at the University of Michigan and Caltech, where he received his PhD in 1963. He then worked at the European nuclear research centre, CERN, in Geneva (1963–1964), before returning to Caltech, where he became professor in 1967. Independently of Gell-Mann, he developed the theory of quarks as the fundamental building blocks of hadrons, the particles which experience strong nuclear forces. They suggested that three types exist, giving rise to the three different observed properties associated with new particles that were being discovered at that time, although there are now believed to be six types of quark.

ZWICKY, Fritz
(1898–1974)

American-Swiss astronomer and physicist, born in Varna, Bulgaria. He was educated at the Federal Institute of Technology at Zurich, graduating with a PhD in physics in 1922. In 1925 he went on a fellowship to Caltech, where he remained all his life, becoming successively Professor of Theoretical Physics

(1927–1942) and Professor of Astrophysics until his retirement in 1968. Zwicky's fruitful and wide ranging researches included work on cosmic rays and rocket design as well as many branches of astronomy. He was one of the first to recognize the power of the recently invented Schmidt telescope as a means of exploring the universe on a large scale, and from 1936 onwards used the 18-inch Schmidt telescope on Mount Palomar to produce, among other researches, his catalogue of clusters of galaxies. He was the author of a highly original book on that subject, *Morphological Cosmology* (1957), and the discoverer of compact galaxies (1963), objects of exceptionally high surface brightness which are hence intrinsically very luminous. He was awarded the Gold Medal of the Royal Astronomical Society for his cosmological researches in 1973.

ZWORYKIN, Vladimir Kosma
(1889–1982)

American physicist, born in Mourom, Russia, the son of a river-boat merchant. After graduating in electrical engineering at the Petrograd Institute of Technology (1912), he studied under Langevin in Paris, served as a radio officer with the Russian army during World War I, but settled in the USA because of the Russian Revolution (1919). He joined the Westinghouse Electric and Manufacturing Co in Pittsburgh (1920), and took a doctorate at the University of Pittsburgh (1926). From 1929 he pursued a career with the Radio Corporation of America, rising to Director of Electronic Research (1946) and Vice-President and Technical Consultant (1947–1954). He is chiefly remembered for applying the cathode-ray tube to television, a development which he patented in 1928. Paul Gottfried Nipkow patented the first practical television system in Germany in 1884, followed up in England by Baird in 1926, but in the final analysis such electromechanical systems were too slow to be successful. What was required was electronic scanning based on the cathode-ray tube invented by Ferdinand Braun in 1897. Before Zworykin, this had already been proposed by the Russian physicist Boris Rosing in 1907 and independently by the English physicist Alan Campbell-Swinton in 1908, but the technology was not available to make it a reality. By 1938 Zworykin had developed the first practical television camera which he called the 'iconoscope', but its application was delayed by World War II. He made a number of other important contributions to electronic optics with his scientific team at RCA, including the electron microscope (1939), the sniperscope (1941) and secondary-emission multipliers in scintillation coulters for measuring radioactivity. In his retirement he worked primarily on the medical application of electronics.